Geometry
FOR
DUMMIES®
2ND EDITION

by Mark Ryan

Wiley Publishing, Inc.

Geometry For Dummies®, 2nd Edition

Published by
Wiley Publishing, Inc.
111 River St.
Hoboken, NJ 07030-5774
www.wiley.com

Copyright © 2008 by Wiley Publishing, Inc., Indianapolis, Indiana

Published by Wiley Publishing, Inc., Indianapolis, Indiana

Published simultaneously in Canada

No part of this publication may be reproduced, stored in a retrieval system, or transmitted in any form or by any means, electronic, mechanical, photocopying, recording, scanning, or otherwise, except as permitted under Sections 107 or 108 of the 1976 United States Copyright Act, without either the prior written permission of the Publisher, or authorization through payment of the appropriate per-copy fee to the Copyright Clearance Center, 222 Rosewood Drive, Danvers, MA 01923, 978-750-8400, fax 978-646-8600. Requests to the Publisher for permission should be addressed to the Permissions Department, John Wiley & Sons, Inc., 111 River Street, Hoboken, NJ 07030, (201) 748-6011, fax (201) 748-6008, or online at http://www.wiley.com/go/permissions.

Trademarks: Wiley, the Wiley Publishing logo, For Dummies, the Dummies Man logo, A Reference for the Rest of Us!, The Dummies Way, Dummies Daily, The Fun and Easy Way, Dummies.com and related trade dress are trademarks or registered trademarks of John Wiley & Sons, Inc. and/or its affiliates in the United States and other countries, and may not be used without written permission. All other trademarks are the property of their respective owners. Wiley Publishing, Inc., is not associated with any product or vendor mentioned in this book.

LIMIT OF LIABILITY/DISCLAIMER OF WARRANTY: THE PUBLISHER AND THE AUTHOR MAKE NO REPRESENTATIONS OR WARRANTIES WITH RESPECT TO THE ACCURACY OR COMPLETENESS OF THE CONTENTS OF THIS WORK AND SPECIFICALLY DISCLAIM ALL WARRANTIES, INCLUDING WITHOUT LIMITATION WARRANTIES OF FITNESS FOR A PARTICULAR PURPOSE. NO WARRANTY MAY BE CREATED OR EXTENDED BY SALES OR PROMOTIONAL MATERIALS. THE ADVICE AND STRATEGIES CONTAINED HEREIN MAY NOT BE SUITABLE FOR EVERY SITUATION. THIS WORK IS SOLD WITH THE UNDERSTANDING THAT THE PUBLISHER IS NOT ENGAGED IN RENDERING LEGAL, ACCOUNTING, OR OTHER PROFESSIONAL SERVICES. IF PROFESSIONAL ASSISTANCE IS REQUIRED, THE SERVICES OF A COMPETENT PROFESSIONAL PERSON SHOULD BE SOUGHT. NEITHER THE PUBLISHER NOR THE AUTHOR SHALL BE LIABLE FOR DAMAGES ARISING HEREFROM. THE FACT THAT AN ORGANIZATION OR WEBSITE IS REFERRED TO IN THIS WORK AS A CITATION AND/OR A POTENTIAL SOURCE OF FURTHER INFORMATION DOES NOT MEAN THAT THE AUTHOR OR THE PUBLISHER ENDORSES THE INFORMATION THE ORGANIZATION OR WEBSITE MAY PROVIDE OR RECOMMENDATIONS IT MAY MAKE. FURTHER, READERS SHOULD BE AWARE THAT INTERNET WEBSITES LISTED IN THIS WORK MAY HAVE CHANGED OR DISAPPEARED BETWEEN WHEN THIS WORK WAS WRITTEN AND WHEN IT IS READ.

For general information on our other products and services, please contact our Customer Care Department within the U.S. at 877-762-2974, outside the U.S. at 317-572-3993, or fax 317-572-4002.

For technical support, please visit www.wiley.com/techsupport.

Wiley also publishes its books in a variety of electronic formats. Some content that appears in print may not be available in electronic books.

Library of Congress Control Number: 2007940107

ISBN: 978-0-470-08946-0

Manufactured in the United States of America

10 9 8 7 6

About the Author

A graduate of Brown University and the University of Wisconsin Law School, **Mark Ryan** has been teaching math since 1989. He runs The Math Center (www.themathcenter.com) in Winnetka, Illinois, where he teaches high school math courses, including an introduction to geometry and a workshop for parents based on a program he developed, *The 10 Habits of Highly Successful Math Students.* In high school, he twice scored a perfect 800 on the math portion of the SAT, and he not only knows mathematics, he has a gift for explaining it in plain English. He practiced law for four years before deciding he should do something he enjoys and use his natural talent for mathematics. Ryan is a member of the Authors Guild and the National Council of Teachers of Mathematics.

Geometry For Dummies, 2nd Edition, is Ryan's fifth book. *Everyday Math for Everyday Life* (Grand Central Publishing) was published in 2002; *Calculus For Dummies* (Wiley), in 2003; *Calculus Workbook For Dummies* (Wiley), in 2005; and *Geometry Workbook For Dummies* (Wiley), in 2006. His math books have sold over a quarter of a million copies.

Also a tournament backgammon player and a skier and tennis player, Ryan lives in Chicago.

Author's Acknowledgments

Putting *Geometry For Dummies,* 2nd Edition, together entailed a great deal of work, and I couldn't have done it alone. I'm grateful to my intelligent, professional, and computer-savvy assistants. Celina Troutman helped with some editing and proofreading of the highly technical manuscript. She has a great command of language. This is the second book Veronica Berns has helped me with. She assisted with editing and proofreading both the book's prose and mathematics. She's a very good writer, and she knows mathematics and how to explain it clearly to the novice. Veronica also helped with the technical production of the hundreds of mathematical symbols in the book.

I'm very grateful to my business consultant, Josh Lowitz. His intelligent and thoughtful contract negotiations on my behalf and his advice on my writing career and on all other aspects of my business have made him invaluable.

The book is a testament to the high standards of everyone at Wiley Publishing. Joyce Pepple, Acquisitions Director, handled the contract negotiations with intelligence, honesty, and fairness. Acquisitions Editor Lindsay Lefevere kept the project on track with intelligence and humor. She very skillfully handled a number of challenging issues that arose. It was such a pleasure to work with her. Technical Editor Alexsis Venter did an excellent and thorough job spotting and correcting the errors that appeared in the book's first draft — some of which would be very difficult or impossible to find without an expert's knowledge of geometry. Copy Editor Danielle Voirol also did a great job correcting mathematical errors; she made many suggestions on how to improve the exposition, and she contributed to the book's quips and humor. The layout and graphics teams did a fantastic job with the book's thousands of complex equations and mathematical figures. Finally, the book would not be what it is without the contributions of Senior Project Editor Alissa Schwipps and her assisting editors Jennifer Connolly and Traci Cumbay. Their skillful editing greatly improved the book's writing and organization. Alissa made some big-picture suggestions about the book's overall design that unquestionably made it a better book. Alissa and her team really tore into my first draft; responding to their hundreds of suggested edits markedly improved the book.

I wrote the entire book at Cafe Ambrosia in Evanston, Illinois. I want to thank Owner Mike Renollet, General Manager Matt Steponik, and their friendly staff for creating a great atmosphere for writing — good coffee, too!

Finally, a very special thanks to my main assistant, Alex Miller. (Actually, I'm not sure *assistant* is accurate; she often seemed more like a colleague or partner.) Virtually every page of the book bears her input and is better for it. She helped me with every aspect of the book's production: writing, typing, editing, proofreading, more writing, more editing . . . and still more editing. She also assisted with the creation of a number of geometry problems and proofs for the book. She has a very high aptitude for mathematics and a great eye and ear for the subtleties and nuances of effective writing. Everything she did was done with great skill, sound judgment, and a nice touch of humor.

Publisher's Acknowledgments

We're proud of this book; please send us your comments through our Dummies online registration form located at www.dummies.com/register/.

Some of the people who helped bring this book to market include the following:

Acquisitions, Editorial, and Media Development

Senior Project Editor: Alissa Schwipps

Acquisitions Editor: Lindsay Lefevere

Copy Editor: Danielle Voirol

Editorial Program Coordinator: Erin Calligan Mooney

Technical Editor: Alexsis Venter

Senior Editorial Manager: Jennifer Ehrlich

Editorial Assistants: Joe Niesen, Leeann Harney, David Lutton

Cover Photos: © George Diebold/Getty Images

Cartoons: Rich Tennant (www.the5thwave.com)

Composition Services

Project Coordinator: Erin Smith

Layout and Graphics: Carrie A. Cesavice, Joyce Haughey

Anniversary Logo Design: Richard Pacifico

Proofreaders: Melissa D. Buddendeck, John Greenough, Stephanie D. Jumper, Jessica Kramer

Indexer: Potomac Indexing, LLC

Special Help

Jennifer Connolly, Traci Cumbay

Publishing and Editorial for Consumer Dummies

Diane Graves Steele, Vice President and Publisher, Consumer Dummies

Joyce Pepple, Acquisitions Director, Consumer Dummies

Kristin A. Cocks, Product Development Director, Consumer Dummies

Michael Spring, Vice President and Publisher, Travel

Kelly Regan, Editorial Director, Travel

Publishing for Technology Dummies

Andy Cummings, Vice President and Publisher, Dummies Technology/General User

Composition Services

Gerry Fahey, Vice President of Production Services

Debbie Stailey, Director of Composition Services

Contents at a Glance

Introduction 1

Part I: Getting Started with Geometry Basics 9
Chapter 1: Introducing Geometry 11
Chapter 2: Building Your Geometric Foundation 21
Chapter 3: Sizing Up Segments and Analyzing Angles 35

Part II: Introducing Proofs 47
Chapter 4: Prelude to Proofs 49
Chapter 5: Your Starter Kit of Easy Theorems and Short Proofs 59
Chapter 6: The Ultimate Guide to Tackling a Longer Proof 79

Part III: Triangles: Polygons of the Three-Sided Variety 89
Chapter 7: Grasping Triangle Fundamentals 91
Chapter 8: Regarding Right Triangles 107
Chapter 9: Completing Congruent Triangle Proofs 123

Part IV: Polygons of the Four-or-More Sided Variety ... 149
Chapter 10: The Seven Wonders of the Quadrilateral World 151
Chapter 11: Proving That You've Got a Particular Quadrilateral 173
Chapter 12: Polygon Formulas: Area, Angles, and Diagonals 187
Chapter 13: Similarity: Same Shape, Different Size 203

Part V: Working with Not-So-Vicious Circles 225
Chapter 14: Coming Around to Circle Basics 227
Chapter 15: Circle Formulas and Theorems 243

Part VI: Going Deep with 3-D Geometry 263
Chapter 16: 3-D Space: Proofs in a Higher Plane of Existence 265
Chapter 17: Getting a Grip on Solid Geometry 273

Part VII: Placement, Points, and Pictures: Alternative Geometry Topics 287
Chapter 18: Coordinate Geometry 289
Chapter 19: Changing the Scene with Geometric Transformations 307
Chapter 20: Locating Loci and Constructing Constructions 325

Part VIII: The Part of Tens 343
Chapter 21: Ten Things to Use as Reasons in Geometry Proofs 345
Chapter 22: Ten Cool Geometry Problems 351

Part IX: Appendixes 359
Appendix A: Formulas and Other Important Stuff You Should Know 361
Appendix B: Glossary 367

Index 375

Table of Contents

Introduction 1

About This Book 1
Conventions Used in This Book 2
What You're Not to Read 2
Foolish Assumptions 3
How This Book is Organized 3
    Part I: Getting Started with Geometry Basics 4
    Part II: Introducing Proofs 4
    Part III: Triangles: Polygons of the Three-Sided Variety ... 4
    Part IV: Polygons of the Four-or-More-Sided Variety 5
    Part V: Working with Not-So-Vicious Circles 5
    Part VI: Going Deep with 3-D Geometry 5
    Part VII: Placement, Points, and Pictures:
        Alternative Geometry Topics 6
    Part VIII: The Part of Tens 6
    Part IX: Appendixes 6
Icons Used in This Book 6
Where to Go from Here 7

Part I: Getting Started with Geometry Basics 9

Chapter 1: Introducing Geometry 11

Studying the Geometry of Shapes 12
    One-dimensional shapes 12
    Two-dimensional shapes 12
    Three-dimensional shapes 13
Getting Acquainted with Geometry Proofs 14
    Easing into proofs with an everyday example 15
    Turning everyday logic into a proof 15
    Sampling a simple geometrical proof 16
When Am I Ever Going to Use This? 18
    When you'll use your knowledge of shapes 18
    When you'll use your knowledge of proofs 19
Why You Won't Have Any Trouble with Geometry 19

Chapter 2: Building Your Geometric Foundation 21
Getting Down with Definitions 21
A Few Points on Points 25
Lines, Segments, and Rays Pointing Every Which Way 26
    Singling out horizontal and vertical lines 26
    Doubling up with pairs of lines 27
Investigating the Plane Facts 29
Everybody's Got an Angle 30
    Goldilocks and the three angles: Small, large, and just "right" 30
    Angle pairs: Often joined at the hip 31

Chapter 3: Sizing Up Segments and Analyzing Angles 35
Measuring Segments and Angles 35
    Measuring segments 35
    Measuring angles 36
Adding and Subtracting Segments and Angles 39
Cutting in Two or Three: Bisection and Trisection 40
    Bisecting and trisecting segments 40
    Bisecting and trisecting angles 42
Proving (Not Jumping to) Conclusions about Figures 43

Part II: Introducing Proofs 47

Chapter 4: Prelude to Proofs 49
Getting the Lay of the Land: The Components of a Formal Geometry Proof 49
Reasoning with If-Then Logic 51
    If-then chains of logic 52
    You've got your reasons: Definitions, theorems, and postulates 53
    Bubble logic for two-column proofs 55
Horsing Around with a Two-Column Proof 56

Chapter 5: Your Starter Kit of Easy Theorems and Short Proofs 59
Doing Right and Going Straight: Complementary and Supplementary Angles 59
Addition and Subtraction: Eight No-Big-Deal Theorems 63
    Addition theorems 63
    Subtraction theorems 67
Like Multiples and Like Divisions? Then These Theorems Are for You! 70
The X-Files: Congruent Vertical Angles Are Out There 73
Pulling the Switch with the Transitive and Substitution Properties 75

Chapter 6: The Ultimate Guide to Tackling a Longer Proof 79
Making a Game Plan 80
Using All the Givens 81
Making Sure You Use If-Then Logic 81
Chipping Away at the Problem 83
Jumping Ahead and Working Backward 85
Filling In the Gaps 86
Writing Out the Finished Proof 88

Part III: Triangles: Polygons of the Three-Sided Variety 89

Chapter 7: Grasping Triangle Fundamentals 91
Taking In a Triangle's Sides 91
Scalene triangles: Akilter, awry and askew 92
Isosceles triangles: Nice pair o' legs 93
Equilateral triangles: All parts are created equal 93
Introducing the Triangle Inequality Principle 94
Getting to Know Triangles by Their Angles 96
Sizing Up Triangle Area 96
Scaling altitudes 96
Determining a triangle's area 98
Locating the "Centers" of a Triangle 102
Balancing on the centroid 102
Finding three more "centers" of a triangle 104

Chapter 8: Regarding Right Triangles 107
Applying the Pythagorean Theorem 107
Perusing Pythagorean Triple Triangles 113
The Fab Four Pythagorean triple triangles 113
Families of Pythagorean triple triangles 115
Getting to Know Two Special Right Triangles 118
The 45°- 45°- 90° triangle — half a square 118
The 30°- 60°- 90° triangle — half of an equilateral triangle 120

Chapter 9: Completing Congruent Triangle Proofs 123
Introducing Three Ways to Prove Triangles Congruent 123
SSS: Using the side-side-side method 124
SAS: Taking the side-angle-side approach 126
ASA: Taking the angle-side-angle tack 128
CPCTC: Taking Congruent Triangle Proofs a Step Further 131
Defining CPCTC 131
Tackling a CPCTC proof 132
Eying the Isosceles Triangle Theorems 135

Trying Out Two More Ways to Prove Triangles Congruent 137
 AAS: Using the angle-angle-side theorem 137
 HLR: The right approach for right triangles 140
Going the Distance with the Two Equidistance Theorems 141
 Determining a perpendicular bisector 142
 Using a perpendicular bisector 143
Making a Game Plan for a Longer Proof 145
Running a Reverse with Indirect Proofs 147

Part IV: Polygons of the Four-or-More Sided Variety 149

Chapter 10: The Seven Wonders of the Quadrilateral World 151

Getting Started with Parallel-Line Properties 151
 Crossing the line with transversals: Definitions and theorems ...152
 Applying the transversal theorems 153
 Working with more than one transversal 155
Meeting the Seven Members of the Quadrilateral Family 157
 Looking at quadrilateral relationships 158
 Working with auxiliary lines 159
Giving Props to Quads: The Properties of Quadrilaterals 162
 Properties of the parallelogram 162
 Properties of the three special parallelograms 166
 Properties of the kite 169
 Properties of the trapezoid and the isosceles trapezoid 171

Chapter 11: Proving That You've Got a Particular Quadrilateral ...173

Putting Properties and Proof Methods Together 173
Proving That a Quadrilateral Is a Parallelogram 175
 Surefire ways of ID-ing a parallelogram 176
 Trying some parallelogram proofs 177
Proving That a Quadrilateral Is a Rectangle, Rhombus, or Square 180
 Revving up for rectangle proofs 181
 Waxing rhapsodic about rhombus proofs 182
 Squaring off with square proofs 184
Proving That a Quadrilateral Is a Kite 185

Chapter 12: Polygon Formulas: Area, Angles, and Diagonals 187

Calculating the Area of Quadrilaterals 187
 Setting forth the quadrilateral area formulas 188
 Getting behind the scenes of the formulas 188
 Trying a few area problems 190
Finding the Area of Regular Polygons 195
 Presenting polygon area formulas 195
 Tackling more area problems 196

Using Polygon Angle and Diagonal Formulas 199
    Interior and exterior design: Exploring polygon angles 199
    Handling the ins and outs of a polygon angle problem 200
    Criss-crossing with diagonals 201

Chapter 13: Similarity: Same Shape, Different Size 203

Getting Started with Similar Figures 204
    Defining and naming similar polygons 204
    How similar figures line up 205
    Solving a similarity problem 207
Proving Triangles Similar 209
    Tackling an AA proof 210
    Using SSS~ to prove triangles similar 211
    Working through an SAS~ proof 212
CASTC and CSSTP, the Cousins of CPCTC 213
    Working through a CASTC proof 214
    Taking on a CSSTP proof 215
Splitting Right Triangles with the Altitude-on-Hypotenuse
  Theorem 216
Getting Proportional with Three More Theorems 219
    The Side-Splitter Theorem: It'll make you split your sides 219
    Crossroads: The Side-Splitter Theorem extended 221
    The Angle-Bisector Theorem 223

Part V: Working with Not-So-Vicious Circles 225

Chapter 14: Coming Around to Circle Basics 227

The Straight Talk on Circles: Radii and Chords 228
    Defining radii, chords, and diameters 228
    Introducing five circle theorems 229
    Working through a proof 229
    Using extra radii to solve a problem 231
Pieces of the Pie: Arcs and Central Angles 232
    Three circle definitions for your mathematical pleasure 232
    Six scintillating circle theorems 233
    Trying your hand at some proofs 235
Going Off on a Tangent about Tangents 237
    Introducing the tangent line 237
    The common-tangent problem 238
    Taking a walk on the wild side with a walk-around problem 241

Chapter 15: Circle Formulas and Theorems 243
    Chewing on the Pizza Slice Formulas 243
        Determining arc length 244
        Finding sector and segment area 247
        Pulling it all together in a problem 248
    Digesting the Angle-Arc Theorems and Formulas 249
        Angles on a circle 250
        Angles inside a circle 252
        Angles outside a circle 254
        Keeping your angle-arc formulas straight 256
    Powering Up with the Power Theorems 257
        Striking a chord with the Chord-Chord Power Theorem 257
        Touching on the Tangent-Secant Power Theorem 259
        Seeking out the Secant-Secant Power Theorem 260
        Condensing the power theorems into a single idea 262

Part VI: Going Deep with 3-D Geometry 263

Chapter 16: 3-D Space: Proofs in a Higher Plane of Existence 265
    Lines Perpendicular to Planes 265
    Parallel, Perpendicular, and Intersecting Lines and Planes 269
        The four ways to determine a plane 269
        Line and plane interactions 270

Chapter 17: Getting a Grip on Solid Geometry 273
    Flat-Top Figures: They're on the Level 273
    Getting to the Point of Pointy-Top Figures 279
    Rounding Things Out with Spheres 285

Part VII: Placement, Points, and Pictures: Alternative Geometry Topics 287

Chapter 18: Coordinate Geometry 289
    Getting Coordinated with the Coordinate Plane 289
    The Slope, Distance, and Midpoint Formulas 291
        The slope dope 291
        Going the distance with the distance formula 294
        Meeting each other halfway with the midpoint formula 295
        The whole enchilada: Putting the formulas
            together in a problem 295
    Proving Properties Analytically 298
        Step 1: Drawing a general figure 298
        Step 2: Solving the problem algebraically 300

Deciphering Equations for Lines and Circles 302
    Line equations 302
    The standard circle equation 303

Chapter 19: Changing the Scene with Geometric Transformations 307

Some Reflections on Reflections 308
    Getting oriented with orientation 309
    Finding a reflecting line 310
Not Getting Lost in Translations 312
    A translation equals two reflections 312
    Finding the elements of a translation 314
Turning the Tables with Rotations 317
    A rotation equals two reflections 317
    Finding the center of rotation and the equations of two reflecting lines 318
Third Time's the Charm: Stepping Out with Glide Reflections 321
    A glide reflection equals three reflections 322
    Finding the main reflecting line 322

Chapter 20: Locating Loci and Constructing Constructions 325

Meeting the Conditions with Loci 326
    The four-step process for locus problems 326
    Two-dimensional locus problems 326
    Three-dimensional locus problems 332
Drawing with the Bare Essentials: Constructions 334
    Three copying methods 334
    Bisecting angles and segments 337
    Two perpendicular line constructions 339
    Constructing parallel lines and using them to divide segments 341

Part VIII: The Part of Tens 343

Chapter 21: Ten Things to Use as Reasons in Geometry Proofs 345

The Reflexive Property 345
Vertical Angles Are Congruent 346
The Parallel-Line Theorems 346
Two Points Determine a Line 347
All Radii of a Circle Are Congruent 347
If Sides, Then Angles 348
If Angles, Then Sides 348
The Triangle Congruence Postulates and Theorems 349
CPCTC 349
The Triangle Similarity Postulates and Theorems 350

Chapter 22: Ten Cool Geometry Problems 351
Eureka! Archimedes's Bathtub Revelation 351
Determining Pi 352
The Golden Ratio 353
The Circumference of the Earth 354
The Great Pyramid of Khufu 354
Distance to the Horizon 355
Projectile Motion 355
Golden Gate Bridge 356
The Geodesic Dome 357
A Soccer Ball 357

Part IX: Appendixes 359

Appendix A: Formulas and Other Important Stuff You Should Know 361
Triangle Stuff 361
Polygon Stuff 362
Circle Stuff 363
3-D Geometry Stuff 365
Coordinate Geometry Stuff 365

Appendix B: Glossary 367

Index 375

Introduction

Geometry is a subject full of mathematical richness and beauty. The ancient Greeks were into it big time, and it's been a mainstay in secondary education for centuries. Today, no education is complete without at least some familiarity with the fundamental principles of geometry.

But geometry is also a subject that bewilders many students because it's so unlike the math that they've done before. Geometry requires you to use deductive logic in formal proofs. This process involves a special type of verbal and mathematical reasoning that's new to many students. Seeing where to go next in a proof — or even where to start — can be challenging. The subject also involves working with two- and three-dimensional shapes — knowing their properties, finding their areas and volumes, and moving them around on the coordinate plane. This spatial reasoning element of geometry is another thing that makes it different and challenging.

Geometry For Dummies, 2nd Edition, can be a big help to you if you've hit the geometry wall. Or if you're a first-time student of geometry, it can prevent you from hitting the wall in the first place. When the world of geometry opens up to you and things start to click, you may come to really appreciate this topic, which has fascinated people for millennia — and which continues to draw people to careers in art, engineering, architecture, city planning, photography, and computer animation, among others. Oh boy, I bet you can hardly wait to get started!

About This Book

Geometry For Dummies, 2nd Edition, covers all the principles and formulas you need to analyze two- and three-dimensional shapes, and it gives you the skills and strategies you need to write geometry proofs. These strategies can make all the difference in the world when it comes to constructing the somewhat peculiar type of logical argument required for proofs. The non-proof parts of the book contain helpful formulas and tips that you can use anytime you need to shape up your knowledge of shapes.

My approach throughout is to explain geometry in plain English with a minimum of technical jargon. Plain English suffices for geometry because its principles, for the most part, are accessible with your common sense. I see no reason to obscure geometry concepts behind a lot of fancy-pants mathematical mumbo-jumbo. I prefer a street-smart approach.

This book, like all *For Dummies* books, is a reference, not a tutorial. The basic idea is that the chapters stand on their own as much as possible. So you don't have to read this book cover to cover — although, of course, you might want to.

Conventions Used in This Book

Geometry For Dummies, 2nd Edition, follows certain conventions that keep the text consistent and oh-so-easy to follow:

- Variables and names of points are in *italics*.
- Important math terms are often in *italics* and are defined when necessary. Italics are also sometimes used for emphasis.
- Important terms may be **bolded** when they appear as keywords within a bulleted list. I also use bold for the instructions in many-step processes.
- As in most geometry books, figures are not necessarily drawn to scale — though most of them are.
- I give you game plans for many of the geometry proofs in the book. A *game plan* is not part of the formal solution to a proof; it's just my way of showing you how to think through a proof. When I don't give you a game plan, you may want to try to come up with one of your own.

What You're Not to Read

Focusing on the *why* in addition to the *how-to* can be a great aid to a solid understanding of geometry — or any math topic. With that in mind, I've put a lot of effort into discussing the underlying logic of many of the ideas in this book. I strongly recommend that you read these discussions, but if you want to cut to the chase, you can get by with reading only the example problems, the step-by-step solutions, and the definitions, theorems, tips, and warnings next to the icons.

I find the gray sidebars interesting and entertaining — big surprise, I wrote them! But you can skip them without missing any essential geometry. And no, you won't be tested on that stuff.

Foolish Assumptions

I may be going out on a limb, but as I wrote this book, here's what I assumed about you:

- ✔ You're a high school student (or perhaps a junior high student) currently taking a standard high school–level geometry course.
- ✔ You're a parent of a geometry student, and you'd like to be able to explain the fundamentals of geometry so you can help your child understand his or her homework and prepare for quizzes and tests.
- ✔ You're anyone who wants anything from a quick peek at geometry to an in-depth study of the subject. You want to refresh your recollection of the geometry you studied years ago or want to explore geometry for the first time.
- ✔ You remember some basic algebra — you know, all those rules for dealing with *x*'s and *y*'s. The good news is that you need very little algebra for doing geometry — but you do need some. In the problems that do involve algebra, I try to lay out all the solutions step by step, which should provide you with some review of simple algebra. If your algebra knowledge has gone completely cold, however, you may need to do a little catching up — but I wouldn't sweat it.
- ✔ You're willing to do a little work. (Work? Egad!) As unpopular as the notion may be, understanding geometry does require some effort from time to time. I've tried to make this material as accessible as possible, but it is math after all. You can't learn geometry by listening to a book-on-tape while lying on the beach. (But if you are at the beach, you can hone your geometry skills by estimating how far away the horizon is — see Chapter 22 for details.)

How This Book is Organized

As you might expect, the topics in *Geometry For Dummies,* 2nd Edition, are presented in the same order as the topics in a standard high school geometry course. So you may want to go through the book in order, using it as a supplement to your course textbook. Many of the explanations, methods, tips, and strategies in this book will make problems that you find difficult or confusing in class seem much easier.

This book is divided into parts, chapters, and sections. Here's a quick glance at what the parts of this book are all about.

Part I: Getting Started with Geometry Basics

Part I gives you a lot of introductory ideas. Here you get to know geometry as a subject: what it is, why it's important, and why you should care. Then, in Chapter 2, you study some basic definitions and properties of geometric objects, such as lines, line segments, and angles.

After you learn the basics about line segments and angles, you're ready to actually do something with them. In Chapter 3, you measure segments and angles, add them, subtract them, and multiply and divide them.

Part II: Introducing Proofs

Part II gives you your first taste of geometry proofs, one of the main topics — or perhaps *the* main topic — of this book. I introduce you to theorems and postulates, the building blocks of proofs, and you see how a proof's chain of logic all fits together. You also get many fantastic proof tips and strategies that, if I do say so myself, make proofs much easier.

Chapter 6, a special bonus chapter in this part, takes you through a geometry proof step-by-step and carefully explains the logic involved in each statement and each reason. Geometry proofs aren't easy, so if you don't have much experience with them (or if you'd like a review of the proof thought process), make sure to check out this chapter.

Part III: Triangles: Polygons of the Three-Sided Variety

In Part III, you begin by reading up on the parts of a triangle: its area, altitudes, angle bisectors, medians, perpendicular bisectors, and so on. Then you find out all sorts of things about right triangles, which are probably the most important type of triangle you see in this book. Does the Pythagorean Theorem ring a bell?

Next, you start doing proofs that involve triangles. Most geometry proofs contain triangles in one way or another. Triangle proofs can be a bit daunting; sometimes solving them is sort of like a finding a route through a long maze. But if you approach them with some patience and stick-to-itiveness, you'll find that they're not so bad after all.

Part IV: Polygons of the Four-or-More-Sided Variety

In this part, you see many interesting things about quadrilaterals (four-sided figures) until visions of parallelograms start dancing around in your head. (Actually, if you start to see dancing parallelograms, I'd advise you to put down this book and take a quick break from geometry.) You study the properties of quadrilaterals such as rhombuses, rectangles, squares, kites, trapezoids, and parallelograms, and you figure out how to prove that a quadrilateral qualifies as one or more of these shapes. Then you move on to computing the area, the angles, and the number of diagonals in polygons like pentagons (5 sides), hexagons (6 sides) and even hexadecagons (16 sides!).

In the last chapter of Part IV, you look at *similar* figures — figures that have the exact same shape but different sizes.

Part V: Working with Not-So-Vicious Circles

After Part IV, it'll probably feel good to get away from polygons and study a shape with no sides at all: the circle. Circles are very simple shapes, but they have many fascinating properties. The number π (3.14159265...), one of the most significant numbers in mathematics, is derived from the relationship of the circumference of a circle to its radius. But you'll know all about that if you read the chapter on circle formulas.

In this part, you also look at proofs that involve circles and their central angles, tangents, chords, inscribed angles, and the like. Don't you just want to open the book up to this section right now? Seriously, though, these concepts really are quite interesting.

Part VI: Going Deep with 3-D Geometry

In this part, you first look at 2-D figures that are "standing up" in the third dimension. You check out parallel planes and discover how a line can have a foot. After that, you look at real, 3-D shapes such as cones, prisms, pyramids, and cylinders. I like to call these the *flat-top* and *pointy-top* shapes — read Part VI to see why.

Part VII: Placement, Points, and Pictures: Alternative Geometry Topics

In Part VII, you study some geometry topics that are a little bit off the beaten path. This chapter is not for geometry weaklings. But if you have a good handle on the basics, then you're surely ready to tackle coordinate geometry, transformations, loci, and constructions. Coordinate geometry and transformations show you how to work with figures in the *x-y* coordinate system. Then, when you check out loci, you discover how geometric shapes can be described as collections of points. Finally, the material about constructions shows you how to create geometric objects using just a compass and a straightedge.

Part VIII: The Part of Tens

In Chapter 21, you see the top ten ideas to use for justifying each logical step in a proof. This list of ideas is a good thing to refer to if you get stuck when working on a proof. Go down the list and check whether any of the ideas help. Finally, the last chapter in the book gives you a break from the formal study of geometry and lets you read up on ten famous geometry problems and wonders of the geometric world, both past and present.

Part IX: Appendixes

In Part IX, you find a list of geometry formulas and a glossary of definitions of geometry terms. Check this section when you need to remember something from a previous chapter or when you're trying to answer a question or solve a problem that involves an unfamiliar concept.

Icons Used in This Book

The following icons can help you quickly spot important information.

Next to this icon are definitions of geometry terms, explanations of geometry principles, and a few other things you should remember as you work through the book.

This icon highlights shortcuts, memory devices, strategies, and so on.

Ignore these icons, and you may end up doing lots of extra work or getting the wrong answer or both. Read carefully when you see the bomb with the burning fuse!

This icon identifies the theorems and postulates — little mathematical truths — that you use to form the logical arguments in geometry proofs.

Where to Go from Here

If you're a geometry beginner, you should probably start with Chapter 1 and work your way through the book in order, but if you already know a fair amount of the subject, feel free to skip around. For instance, if you need to know about quadrilaterals, check out Chapter 10. Or if you already have a good handle on geometry proof basics, you may want to dive into the more advanced proofs in Chapter 9.

You can also go to the excellent companion to this book, *Geometry Workbook For Dummies,* to do some practice problems.

And from there, naturally, you can go

- To the head of the class
- To Go to collect $200
- To chill out
- To explore strange new worlds, to seek out new life and new civilizations, to boldly go where no man (or woman) has gone before

If you're still reading this, what are you waiting for? Go take your first steps into the wonderful world of geometry!

Part I
Getting Started with Geometry Basics

In this part . . .

You ease into geometry in Chapter 1 by picking up a bit of its long and rich history — Pythagoras, Euclid, and Archimedes are a few of the superstars. Then you go over what geometry is all about, where it's used, and why you should bother to study it. You also get a brief introduction to geometry proofs, the important topic which over half of this book is devoted to. Then, in Chapters 2 and 3, you pick up some definitions of basic things (such as lines, segments, rays, angles, and planes) and some simple concepts (including congruence, perpendicularity, bisection, trisection, and how to add and subtract segments and angles). You use these building blocks throughout the book.

Chapter 1

Introducing Geometry

In This Chapter

▶ Surveying the geometric landscape: Shapes and proofs

▶ Finding out "What is the point of geometry, anyway?"

▶ Getting psyched to kick some serious geometry butt

Studying geometry is sort of a Dr. Jekyll-and-Mr. Hyde thing. You have the ordinary, everyday geometry of shapes (the Dr. Jekyll part) and the strange world of geometry proofs (the Mr. Hyde part).

Every day, you see various shapes all around you (triangles, rectangles, boxes, circles, balls, and so on), and you're probably already familiar with some of their properties: area, perimeter, and volume, for example. In this book, you discover much more about these basic properties and then explore more-advanced geometric ideas about shapes.

Geometry proofs are an entirely different sort of animal. They involve shapes, but instead of doing something straightforward like calculating the area of a shape, you have to come up with an airtight mathematical argument that proves something about a shape. This process requires not only mathematical skills but verbal skills and logical deduction skills as well, and for this reason, proofs trip up many, many students. If you're one of these people and have already started singing the geometry-proof blues, you might even describe proofs — like Mr. Hyde — as monstrous. But I'm confident that, with the help of this book, you'll have no trouble taming them.

This chapter is your gateway into the sensational, spectacular, and super-duper (but sometimes somewhat stupefying) subject of this book: geometry. If you're tempted to ask, "Why should I care about geometry?" this chapter will give you the answer.

Studying the Geometry of Shapes

Have you ever reflected on the fact that you're literally surrounded by shapes? Look around. The rays of the sun are — what else? — rays. The book in your hands has a shape, every table and chair has a shape, every wall has an area, and every container has a shape and a volume; most picture frames are rectangles, CDs and DVDs are circles, soup cans are cylinders, and so on and so on. Can you think of any solid thing that doesn't have a shape? This section gives you a brief introduction to these one-, two-, and three-dimensional shapes that are all-pervading, omnipresent, and ubiquitous — not to mention all around you.

One-dimensional shapes

There aren't many shapes you can make if you're limited to one dimension. You've got your lines, your segments, and your rays. That's about it. But it doesn't follow that having only one dimension makes these things unimportant — not by any stretch. Without these one-dimensional objects, there'd be no two-dimensional shapes; and without 2-D shapes, you can't have 3-D shapes. Think about it: 2-D squares are made up of four 1-D segments, and 3-D cubes are made up of six 2-D squares. And it'd be very difficult to do much mathematics without the simple 1-D number line or without the more sophisticated 2-D coordinate system, which needs 1-D lines for its x- and y-axes. (I cover lines, segments, and rays in Chapter 2; Chapter 18 discusses the coordinate plane.)

Two-dimensional shapes

As you probably know, two-dimensional shapes are flat things like triangles, circles, squares, rectangles, and pentagons. The two most common characteristics you study about 2-D shapes are their area and perimeter. These geometric concepts come up in countless situations in the real world. You use 2-D geometry, for example, when figuring the acreage of a plot of land, the number of square feet in a home, the size and shape of cloth needed when making curtains or clothing, the length of a running track, the dimensions of a picture frame, and so on. The formulas for calculating the area and perimeter of 2-D shapes are covered in Parts III through V.

I devote many chapters in this book to triangles and *quadrilaterals* (shapes with four sides); I give less space to shapes that have more sides, like pentagons and hexagons. Shapes of any number of straight sides, called *polygons*, have more-advanced features such as diagonals, apothems, and exterior angles, which you explore in Part IV.

Historical highlights in the study of shapes

The study of geometry has impacted architecture, engineering, astronomy, physics, medicine, and warfare, among other fields, in countless ways for well over 5,000 years. I doubt anyone will ever be able to put a date on the discovery of the simple formula for the area of a rectangle (Area = length · width), but it likely predates writing and goes back to some of the earliest farmers. Some of the first known writings from Mesopotamia (in about 3500 B.C.) deal with the area of fields and property. And I'd bet that even pre-Mesopotamian farmers knew that if one farmer planted an area three times as long and twice as wide as another farmer, then the bigger plot would be 3 · 2, or 6 times as large as the smaller one.

The architects of the pyramids at Giza (built around 2500 B.C.) knew how to construct right angles using a 3-4-5 triangle (one of the right triangles I discuss in Chapter 8). Right angles are necessary for the corners of the pyramid's square base, among other things. And of course, you've probably heard of Pythagoras (circa 570–500 B.C.) and the famous right-triangle theorem named after him (see Chapter 8). Archimedes (287–212 B.C.) used geometry to invent the pulley. He developed a system of compound pulleys that could lift an entire warship filled with men (for more of Archimedes's accomplishments, see Chapter 22). The Chinese knew how to calculate the area and volume of many different geometric shapes and how to construct a right triangle by 100 B.C.

In more recent times, Galileo Galilei (1564–1642) discovered the equation for the motion of a projectile (see Chapter 22) and designed and built the best telescope of his day. Johannes Kepler (1571–1630) measured the area of sections of the elliptical orbits of the planets as they orbit the sun. René Descartes (1596–1650) is credited with inventing coordinate geometry, the basis for most mathematical graphing (see Chapter 18). Isaac Newton (1642–1727) used geometrical methods in his *Principia Mathematica,* the famous book in which he set out the principle of universal gravitation.

Closer to home, Ben Franklin (1706–1790) used geometry to study meteorology and ocean currents. George Washington (1732–1799) used trigonometry (the advanced study of triangles) while working as a surveyor before he became a soldier. Last but certainly not least, Albert Einstein discovered one of the most bizarre geometry rules of all: that gravity warps the universe. One consequence of this is that if you were to draw a giant triangle around the sun, the sum of its angles would actually be a little larger than 180°. This contradicts the 180° rule for triangles (see Chapter 7), which works until you get to an astronomical scale. The list of highlights goes on and on.

You may be familiar with some shapes that have curved sides, such as circles, ellipses, and parabolas. The circle is the only curved 2-D shape covered in this book. In Part V, you investigate all sorts of interesting circle properties involving diameters, radii, chords, tangent lines, and so on.

Three-dimensional shapes

I cover three-dimensional shapes in Part VI. You work with prisms (a box is one example), cylinders, pyramids, cones, and spheres. The two major characteristics of these 3-D shapes, which you study in Chapter 17, are their *surface area* and *volume*.

Three-dimensional concepts like volume and surface area come up frequently in the real world; examples include the volume of water in a fish tank or backyard pool. The amount of wrapping paper you need to wrap a gift box depends on its surface area. And if you wanted to calculate the surface area and volume of the Great Pyramid in Egypt — you've been dying to do this, right? — you couldn't do it without 3-D geometry.

Here are a couple of ideas about how the three dimensions are interrelated. Two-dimensional shapes are enclosed by their sides, which are 1-D segments; 3-D shapes are enclosed by their faces, which are 2-D polygons. And here's a nifty real-world example of the relationship between 2-D area and 3-D volume: A gallon of paint (a 3-D volume quantity) can cover a certain number of square feet of area on a wall (a 2-D area quantity). (Well, okay, I have to admit it — I'm playing a bit fast and loose with my dimensions here. The paint on the wall is actually a 3-D shape. There's the length and width of the wall, and the third dimension is the thickness of the layer of paint. If you multiply these three dimensions together, you get the volume of the paint.)

Getting Acquainted with Geometry Proofs

Geometry proofs are an oddity in the mathematical landscape, and just about the only place you find geometry proofs is in a geometry course. If you're in a course right now and you're wondering what's the point of studying something you'll never use again, I get back to that in a minute in the section "When Am I Ever Going to Use This?" For now, I just want to give you a very brief description of what a geometry proof is.

A *geometry proof* — like any mathematical proof — is an argument that begins with known facts, proceeds from there through a series of logical deductions, and ends with the thing you're trying to prove.

Mathematicians have been writing proofs — in geometry and all other areas of math — for over 2,000 years. (See the sidebar about Euclid and the history of geometry proofs.) The main job of a present-day mathematician is proving things by writing formal proofs. This is how the field of mathematics progresses: As more and more ideas are proved, the body of mathematical knowledge grows. Proofs have always played, and still play, a significant role in mathematics. And that's one of the reasons you're studying them. Part II delves into all the details on proofs; in the sections that follow, I get you started in the right direction.

Easing into proofs with an everyday example

You probably never realized it, but sometimes when you think through a situation in your day-to-day life, you use the same type of deductive logic that's used in geometry proofs. Although the topics are different, the basic nature of the argument is the same.

Here's an example of real-life logic. Say you're at a party at Sandra's place. You have a crush on Sandra, but she's been dating Johnny for a few months. You look around at the partygoers and notice Johnny talking with Judy, and a little later you see them step outside for a few minutes. When they come back inside, Judy's wearing Johnny's ring. You weren't born yesterday, so you put two and two together and realize that Sandra's relationship with Johnny is in trouble and, in fact, may end any minute. You glance over in Sandra's direction and see her leaving the room with tears in her eyes. When she comes back, you figure it might not be a bad idea to go over and talk with her.

(By the way, this story about a party gone bad is based on Lesley Gore's No. 1 hit from the '60s, "It's My Party." The sequel song, also a hit, "Judy's Turn to Cry," relates how Sandra got back at Judy. Check out the lyrics online.)

Now, granted, this party scenario might not seem like it involves a deductive argument. Deductive arguments tend to contain many steps or a chain of logic like, "If A, then B; and if B, then C; if C, then D; and so on." The party fiasco might not seem like this at all because you'd probably see it as a single incident. You see Judy come inside wearing Johnny's ring, you glance at Sandra and see that she's upset, and the whole scenario is clear to you in an instant. It's all obvious — no logical deduction seems necessary.

Turning everyday logic into a proof

Imagine that you had to explain your entire thought process about the party situation to someone with absolutely no knowledge of how people usually behave. For instance, imagine that you had to explain your thinking to a hypothetical Martian who knows nothing about our Earth ways. In this case, you *would* need to walk him through your reasoning step by step.

Here's how your argument might go. Note that each statement comes with the reasoning in parentheses:

1. Sandra and Johnny are going out (this is a given fact).
2. Johnny and Judy go outside for a few minutes (also given).

3. When Judy returns, she has a new ring on her finger (a third given).

4. Therefore, she's wearing Johnny's ring (*much* more probable than, say, that she found a ring on the ground outside).

5. Therefore, she's going out with Johnny (because when a boy gives a girl his ring, it means they're going out).

6. Therefore, Sandra and Johnny will break up soon (because a girl will not continue to go out with a guy who's just given another girl his ring).

7. Therefore, Sandra will soon be available (because that's what happens after someone breaks up).

8. Therefore, I should go over and talk with her (duh).

This eight-step argument shows you that there really is a chain of logical deductions going on beneath the surface, even though in real life your reasoning and conclusions about Sandra would come to you in an instant. And the argument gives you a little taste for the type of step-by-step reasoning you use in geometry proofs. You see your first geometry proof in the next section.

Sampling a simple geometrical proof

Geometry proofs are like the party argument in the preceding section, only with a lot less drama. They follow the same type of series of intermediate conclusions that lead to the final conclusion: Beginning with some given facts, say A and B, you go on to say *therefore,* C; then *therefore,* D; then *therefore,* E; and so on till you get to your final conclusion. Here's a very simple example using the line segments in Figure 1-1.

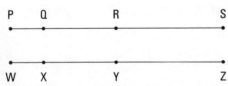

Figure 1-1: $\overline{PS}$ and $\overline{WZ}$, each made up of three pieces.

For this proof, you're told that segment $\overline{PS}$ is *congruent to* (the same length as) segment $\overline{WZ}$, that $\overline{PQ}$ is congruent to $\overline{WX}$, and that $\overline{QR}$ is congruent to $\overline{XY}$. (By the way, instead of saying *is congruent to* all the time, you can just use the symbol ≅ to mean the same thing.) You have to prove that $\overline{RS} \cong \overline{YZ}$.

Hate proofs? Blame Euclid.

Euclid (circa 385–275 B.C.) is usually credited with getting the ball rolling on geometry proofs. (If you're having trouble with proofs, now you know who to blame!) His approach was to begin with a few undefined terms such as *point* and *line* and then to build from there, carefully defining other terms like *segment* and *angle*. He also realized that he'd need to begin with some unproved principles (called *postulates*) that he'd just have to assume were true.

He started with ten postulates, such as "a straight line segment can be drawn by connecting any two points" and "two things that each equal a third thing are equal to one another." After setting down the undefined terms, the definitions, and the postulates, his real work began. Using these three categories of things, he proved his first *theorem* (a proven geometric principle), which was the side-angle-side method of proving triangles congruent (see Chapter 9). And then he proved another theorem and another and so on.

Once a theorem had been proved, it could then be used (along with the undefined terms, definitions, and postulates) to prove other theorems. If you're working on proofs in a standard high school geometry course, you're walking in the footsteps of Euclid, one of the giants in the history of mathematics — lucky you!

Now, you may be thinking, "That's obvious — if $\overline{PS}$ is the same length as $\overline{WZ}$ and both segments contain these equal short pieces and the equal medium pieces, then the longer third pieces have to be equal as well." And of course, you'd be right. But that's not how the proof game is played. You have to spell out every little step in your thinking so your argument doesn't have any gaps. Here's the whole chain of logical deductions:

1. $\overline{PS} \cong \overline{WZ}$ (this is given).
2. $\overline{PQ} \cong \overline{WX}$ and
   $\overline{QR} \cong \overline{XY}$ (these facts are also given).
3. Therefore, $\overline{PR} \cong \overline{WY}$ (because if you add equal things to equal things, you get equal totals).
4. Therefore, $\overline{RS} \cong \overline{YZ}$ (because if you start with equal segments, the whole segments $\overline{PS}$ and $\overline{WZ}$, and take away equal parts of them, $\overline{PR}$ and $\overline{WY}$, the parts that are left must be equal).

In formal proofs, you write your statements (like $\overline{PR} \cong \overline{WY}$ from Step 3) in one column, and your justifications for those statements in another column. Chapter 4 shows you the setup.

When Am I Ever Going to Use This?

You'll likely have plenty of opportunities to use your knowledge about the geometry of shapes. And what about geometry proofs? Not so much. Read on for details.

When you'll use your knowledge of shapes

Shapes are everywhere, so every educated person should have a working knowledge of shapes and their properties. The geometry of shapes comes up often in daily life, particularly with measurements.

In day-to-day life, if you have to buy carpeting or fertilizer or grass seed for your lawn, you should know something about area. You might want to understand the measurements in recipes or on food labels, or you may want to help a child with an art or science project that involves geometry. You certainly need to understand something about geometry to build some shelves or a backyard deck. And after finishing your work, you might be hungry — a grasp of how area works can come in handy when you're ordering pizza: a 20-inch pizza is four, not two, times as big as a 10-incher, and a 14-inch pizza is twice as big as a 10-incher. (Check out Chapter 15 to see why this is.)

Careers that use geometry

Here's a quick alphabetical tour of careers that use geometry. Artists use geometry to measure canvases, make frames, and design sculptures. Builders use it in just about everything they do; ditto for carpenters. For dentists, the shape of teeth, cavities, and fillings is one big geometry problem. Diamond cutters use geometry every time they cut a stone. Dairy farmers use geometry when calculating the volume of milk output in gallons.

Eyeglass manufacturers use geometry in countless ways whenever they use the science of optics. Fighter pilots (or quarterbacks or anyone else who has to aim something at a moving target) have to understand angles, distance, trajectory, and so on. Grass-seed sellers have to know how much seed customers need to use per square yard or per acre. Helicopter pilots use geometry (actually, their computerized instruments do the work for them) for all calculations that affect taking off and landing, turning, wind speed, lift, drag, acceleration, and the like. Instrument makers have to use geometry when they make trumpets, pianos, violins — you name it. And the list goes on and on . . .

When you'll use your knowledge of proofs

Will you ever use your knowledge of geometry proofs? In this section, I give you two answers to this question: a politically correct one and a politically incorrect one. Take your pick.

First, the politically correct answer (which is also *actually* correct). Granted, it's extremely unlikely that you'll ever have occasion to do a single geometry proof outside of a high school math course (college math majors are about the only exception). However, doing geometry proofs teaches you important lessons that you can apply to non-mathematical arguments. Among other things, proofs teach you the following:

- Not to assume things are true just because they seem true at first glance
- To very carefully explain each step in an argument even if you think it should be obvious to everyone
- To search for holes in your arguments
- Not to jump to conclusions

And in general, proofs teach you to be disciplined and rigorous in your thinking and in how you communicate your thoughts.

If you don't buy that PC stuff, I'm sure you'll get this politically incorrect answer: Okay, so you're never going to use geometry proofs, but you do want to get a decent grade in geometry, right? So you might as well pay attention in class (what else is there to do, anyway?), do your homework, and use the hints, tips, and strategies I give you in this book. They'll make your life much easier. Promise.

Why You Won't Have Any Trouble with Geometry

Geometry, especially proofs, can be difficult. Mathwise, it's foreign territory with some very rocky terrain. But it's far from impossible, and you can do a bunch of things to make your geometry experience as easy as pi (get it?):

- **Powering through proofs.** If you get stuck on a proof, check out the helpful tips and warnings that I give you throughout each chapter. You may also want to look at Chapter 21 to make sure you keep the ten most important ideas for proofs fresh in your mind. Finally, you can go to Chapter 6 to see how to reason your way through a long, complicated proof.

- **Figuring out formulas.** If you can't figure out a problem that uses a geometry formula, you can look in Appendix A at the back of the book to make sure that you have the formula right.
- **Coming to terms with the lingo.** If you need help with any geometry terms, the glossary in Appendix B can come in handy.
- **Sticking it out.** My main piece of advice to you is never to give up on a problem. The greater number of sticky problems that you finally beat, the more experience you gain to help you beat the next one. After you take in all my expert advice — no brag, just fact — you should have all the tools you need to face down whatever your geometry teacher or math-crazy friends can throw at you.

Chapter 2

Building Your Geometric Foundation

In This Chapter
▶ Examining the basic components of complex shapes
▶ Understanding types of points, lines, rays, segments, angles, and planes
▶ Pairing up with angle pairs

*I*n this chapter, you go over the groundwork that gets you geared up for some grueling and gut-wrenching geometry. (That's some carefully crafted consonance for you. And no, the rest of the geometry in this book isn't really grueling or gut-wrenching — I just needed some *g* words.) These building blocks also work for merely-moderately-challenging geometry and do-it-in-your-sleep geometry.

All kidding aside, this chapter should be pretty easy, but don't skip it — unless you're already a geometry genius — because many of the ideas you see here are crucial to understanding the rest of this book.

Getting Down with Definitions

The study of geometry begins with the definitions of the five simplest geometric objects: point, line, segment, ray, and angle. And I throw in two extra definitions for you (plane and 3-D space) for no extra charge. Collectively, these terms take you from no dimensions up to the third dimension.

> ### Defining the undefinable
>
> Definitions typically use simpler terms to explain the meaning of more-complex ones. Consider, for example, the definition of the median of a triangle: "a segment from a vertex of a triangle to the midpoint of the opposite side." You use the basic terms *segment, vertex, triangle, midpoint,* and *side* to define the new term, *median.* If you don't know the meaning of, say, *midpoint,* you can look up its definition and find its meaning explained in terms of *point, segment,* and *congruent.* And then you can look up one of those terms if you have to, and so on.
>
> But with the word *point* (and *line*), this strategy just doesn't work. Try to define *point* without using the word *point* or a synonym of *point* in the definition. Any luck? I didn't think so. You can't do it. And using *point* or a synonym of *point* in its own definition is circular and therefore not valid — you can't use a term in its own definition because to be able to understand the definition, you'd have to already know the meaning of the word you're trying to figure out! That's why some words, though they appear in your dictionary, are technically undefined in the world of math.

Here are the definitions of *segment, ray, angle, plane,* and *3-D space* and the "undefinitions" of *point* and *line* (these two terms are technically undefinable — see the nearby sidebar for details):

- **Point:** A point is like a dot except that it actually has no size at all; or you can say that it's infinitely small (except that even saying *infinitely small* makes a point sound larger than it really is). Essentially, a point is zero-dimensional, with no height, length, or width, but you draw it as a dot, anyway. You name a point with a single uppercase letter, as with points *A, D,* and *T* in Figure 2-1.

- **Line:** A line is like a thin, straight wire (although really it's infinitely thin — or better yet, it has no width at all). Lines have length, so they're one-dimensional. Remember that a line goes on forever in both directions, which is why you use the little double-headed arrow as in $\overleftrightarrow{AB}$ (read as *line AB*).

  Check out Figure 2-1 again. Lines are usually named using any two points on the line, with the letters in any order. So $\overleftrightarrow{MQ}$ is the same line as $\overleftrightarrow{QM}$, $\overleftrightarrow{MN}$, $\overleftrightarrow{NM}$, $\overleftrightarrow{QN}$, and $\overleftrightarrow{NQ}$. Occasionally, lines are named with a single, italicized, lowercase letter, such as lines *f* and *g*.

- **Line segment (or just segment):** A segment is a section of a line that has two endpoints. See Figure 2-1 yet again. If a segment goes from *P* to *R*, you call it *segment PR* and write it as $\overline{PR}$. You can also switch the order of the letters and call it $\overline{RP}$. Segments can also appear within lines, as in $\overline{MN}$.

  Note: A pair of letters without a bar over it means the length of a segment. For example, *PR* means the length of $\overline{PR}$.

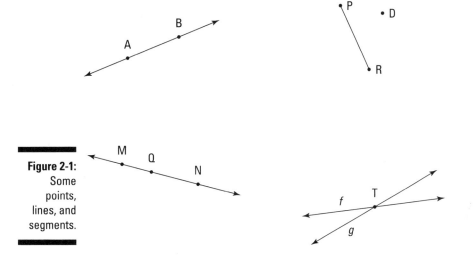

Figure 2-1: Some points, lines, and segments.

- **Ray:** A ray is a section of a line (kind of like half a line) that has one endpoint and goes on forever in the other direction. If its endpoint is point *K* and it goes through point *S* and then past it forever, you call the "half line" *ray KS* and write $\overrightarrow{KS}$. See Figure 2-2.

  The first letter always indicates the ray's endpoint. For instance, $\overrightarrow{AB}$ can also be called $\overrightarrow{AC}$ because either way, you start at *A* and go forever past *B* and *C*. $\overrightarrow{BC}$, however, is a different ray.

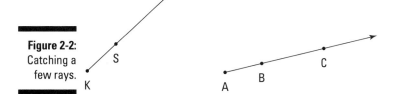

Figure 2-2: Catching a few rays.

- **Angle:** Two rays with the same endpoint form an angle. Each ray is a *side* of the angle, and the common endpoint is the angle's *vertex*. You can name an angle using its vertex alone or three points (first, a point on one ray, then the vertex, and then a point on the other ray).

  Check out Figure 2-3. Rays $\overrightarrow{PQ}$ and $\overrightarrow{PR}$ form the sides of an angle, with point *P* as the vertex. You can call the angle ∠*P*, ∠*RPQ*, or ∠*QPR*. Angles can also be named with numbers, such as the angle on the right in the figure, which you can call ∠4. The number is just another way of naming the angle, and it has nothing to do with the size of the angle.

  The angle on the right also illustrates the *interior* and *exterior* of an angle.

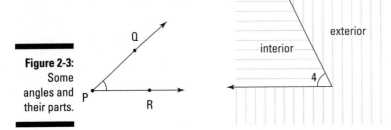

Figure 2-3: Some angles and their parts.

- **Plane:** A plane is like a perfectly flat sheet of paper except that it has no thickness whatsoever and it goes on forever in all directions. You might say it's infinitely thin and has an infinite length and an infinite width. Because it has length and width but no height, it's two-dimensional. Planes are named with a single, italicized, lowercase letter or sometimes with the name of a figure (a rectangle, for example) that lies in the plane. Figure 2-4 shows plane *m*, which goes out forever in four directions.

- **3-D (three-dimensional) space:** 3-D space is everywhere — all of space in every direction. First, picture an infinitely big map that goes forever to the north, south, east, and west. That's a two-dimensional plane. Then, to get 3-D space from this map, add the third dimension by going up and down forever.

  There's no good way to draw 3-D space (Figure 2-4 shows my best try, but it's not going to win any awards). Unlike a box, 3-D space has no shape and no borders.

  Because 3-D space takes up *all* the space in the universe, it's sort of the opposite of a point, which takes up no space at all. But on the other hand, 3-D space is like a point in that both are difficult to define because both are completely without features.

Figure 2-4: A two-dimensional plane and three-dimensional space.

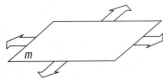
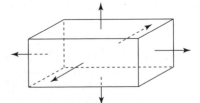

Chapter 2: Building Your Geometric Foundation

REMEMBER

Here's something a bit peculiar about the way objects are depicted in geometry diagrams: Even if lines, segments, rays, and so on, don't appear in a diagram, they're still sort of there — as long as you'd know where to draw them. For example, Figure 2-1 contains a segment, $\overline{PD}$, that goes from *P* to *D* and has endpoints at *P* and *D* — even though you don't see it. (I know that may seem a bit weird, but this idea is just one of the rules of the game. Don't sweat it.)

A Few Points on Points

There isn't much that can be said about points. They have no features, and each one is the same as every other. Various *groups* of points, however, do merit an explanation:

- **Collinear points:** See the word *line* in *collinear*? Collinear points are points that lie on a line. Any two points are always collinear because you can always connect them with a straight line. Three or more points can be collinear, but they don't have to be. See Figure 2-5.

- **Non-collinear points:** These points, like points *X, Y,* and *Z* in Figure 2-5, don't all lie on the same line.

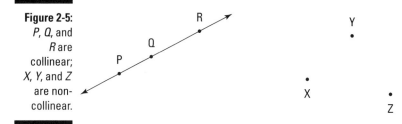

Figure 2-5: *P, Q,* and *R* are collinear; *X, Y,* and *Z* are non-collinear.

- **Coplanar points:** A group of points that lie in the same plane are coplanar. Any two or three points are always coplanar. Four or more points might or might not be coplanar.

  Look at Figure 2-6, which shows coplanar points *A, B, C,* and *D.* In the box on the right, there are many sets of coplanar points. Points *P, Q, X,* and *W,* for example, are coplanar; the plane that contains them is the left side of the box. Note that points *Q, X, S,* and *Z* are also coplanar even though the plane that contains them isn't shown; it slices the box in half diagonally.

✔ **Non-coplanar points:** A group of points that don't all lie in the same plane are non-coplanar.

 See Figure 2-6. Points *P, Q, X,* and *Y* are non-coplanar. The top of the box contains *Q, X,* and *Y,* and the left side contains *P, Q,* and *X,* but no flat surface contains all four points.

Figure 2-6: Coplanar and non-coplanar points.

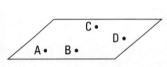
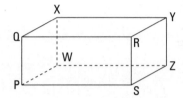

Lines, Segments, and Rays Pointing Every Which Way

In this section, I describe different types of lines (or segments or rays) or pairs of lines (or segments or rays) based on the direction they're pointing or how they relate to each other. People usually use the terms in the next two sections to describe lines, but you can use them for segments and rays as well.

Singling out horizontal and vertical lines

Giving the definitions of *horizontal* and *vertical* may seem a bit pointless. You probably already know what the terms mean, and the best way to describe them is to just show you a figure. But, hey, this is a math book, and math books are supposed to define terms. Who am I to question this age-old tradition? So here are the definitions (also check out Figure 2-7):

 ✔ **Horizontal lines, segments, or rays:** Horizontal lines, segments, and rays go straight across, left and right, not up or down at all — you know, like the horizon.

 ✔ **Vertical lines, segments, or rays:** Lines or parts of a line that go straight up and down are vertical. (Rocket science this is not.)

Figure 2-7: Some horizontal and vertical lines, segments, and rays.

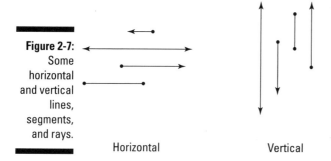

Horizontal   Vertical

Doubling up with pairs of lines

In this section, I give you five terms that describe pairs of lines. The first four are about coplanar lines — you use these a lot. The fifth term describes non-coplanar lines. This term comes up only in 3-D problems, so you probably won't have a chance to use it much.

Coplanar lines

I define *coplanar points* in a previous section — as points in the same plane — so I absolutely refuse to define *coplanar lines*. Well, okay, I suppose I don't want to be turned in to the math-book-writers' disciplinary committee, so here it is: Coplanar lines are lines in the same plane. Here are some ways coplanar lines may interact:

- **Parallel lines, segments, or rays:** Lines that run in the same direction and never cross (like two railroad tracks) are called parallel. Segments and rays are parallel if the lines that contain them are parallel. If $\overleftrightarrow{AB}$ is parallel to $\overleftrightarrow{CD}$, you write $\overleftrightarrow{AB} \parallel \overleftrightarrow{CD}$. See Figure 2-8.

Figure 2-8: Four pairs of parallel lines, segments, and rays — and never the twain shall meet.

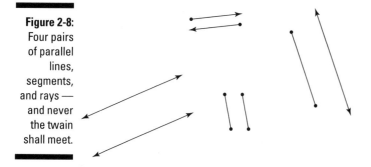

✔ **Intersecting lines, segments, or rays:** Lines, rays, or segments that cross or touch are intersecting. The point where they cross or touch is called the *point of intersection*.

- **Perpendicular lines, segments, or rays:** Lines, segments, or rays that intersect at right angles (90° angles) are perpendicular. If $\overline{PQ}$ is perpendicular to $\overline{RS}$, you write $\overline{PQ} \perp \overline{RS}$. See Figure 2-9. The little boxes in the corners of the angles indicate right angles. (You use the definition of perpendicular in proofs. See Chapter 4.)

- **Oblique lines, segments, or rays:** Lines or segments or rays that intersect at any angle other than 90° are called oblique. See Figure 2-9, which shows oblique lines and rays on the right.

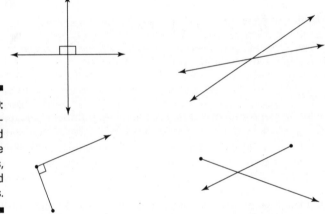

Figure 2-9: Perpendicular and oblique lines, rays, and segments.

Because lines extend forever, a pair of coplanar lines must be either parallel or intersecting. (However, this is not true for coplanar segments and rays. Segments and rays can be nonparallel and at the same time non-intersecting, because their endpoints allow them to stop short of crossing.)

Non-coplanar lines

In the preceding section, you can check out lines that lie in the same plane. Here, I discuss lines that aren't in the same plane.

Skew lines, segments, or rays: Lines that don't lie in the same plane are called skew lines — *skew* simply means *non-coplanar*. Or you can say that skew lines are lines that are neither parallel nor intersecting. See Figure 2-10. (You probably won't ever hear anyone refer to skew segments or rays, but there's no reason they can't be skew. They're skew if they're non-coplanar.)

Figure 2-10: Skew lines are non-coplanar.

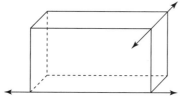

Here's a good way to get a handle on skew lines. Take two pencils or pens, one in each hand. Hold them a few inches apart, both of them pointing away from you. Now, keep one where it is and point the other one up at the ceiling. That's it. You're holding skew lines.

Investigating the Plane Facts

Look! Up in the sky! It's a bird! It's a plane! It's Superman! Wait . . . no, it's just a plane. In this short section, you discover a couple of things about planes — the geometric kind, that is, not the flying kind. Unfortunately, the geometric kind of plane is much less interesting because there's really only one thing to be said about how two planes interact: Either they cross each other, or they don't. I wish I could make this more interesting or exciting, but that's all there is to it.

Here are two no-brainer terms for a pair of planes (see Figure 2-11):

- **Parallel planes:** Parallel planes are planes that never cross. The ceiling of a room (assuming it's flat) and the floor are parallel planes (though true planes extend forever).

- **Intersecting planes:** Hold onto your hat — intersecting planes are planes that cross, or intersect. When planes intersect, the place where they cross forms a line. The floor and a wall of a room are intersecting planes, and where the floor meets the wall is the line of intersection of the two planes.

Figure 2-11: Parallel and intersecting planes.

Parallel planes

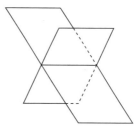
Intersecting planes

Everybody's Got an Angle

Angles are one of the basic building blocks of triangles and other polygons (segments are the other). You can see angles on virtually every page of any geometry book, so you gotta get up to speed about them — no ifs, ands, or buts. In the first part of this section, I give you five terms that describe single angles. In the following part, I discuss four types of pairs of angles.

Goldilocks and the three angles: Small, large, and just "right"

Are these geometry puns fantastic or what? Tell me: Where else can you get so much fascinating math *plus* such incredibly good humor? As a matter of fact, I liked this pun so much that I decided to use it even though it's not exactly accurate. In this section, I give you *five* types of angles, not just three. But the first three are the main ones; the last two are a bit peculiar.

Check out the following five angle definitions and see Figure 2-12 for the visual:

- **Acute angle:** An acute angle is less than 90°. Think "a-*cute* little angle." Acute angles are kind of like an alligator's mouth not opened very wide.
- **Right angle:** A right angle is a 90° angle. Right angles should be familiar to you from the corners of picture frames, tabletops, boxes, and books, the intersections of most roads, and all kinds of other things that show up in everyday life. The sides of a right angle are *perpendicular* (see the earlier "Coplanar lines" section).
- **Obtuse angle:** An obtuse angle has a measure greater than 90°. These angles are more like pool chairs or beach chairs — they open pretty far, and they look like you could lean back in them. (More comfortable than an alligator's mouth, right?)
- **Straight angle:** A straight angle has a measure of 180°; it looks just like a line with a point on it (seems kinda weird for an angle if you ask me).
- **Reflex angle:** A reflex angle has a measure of more than 180°. Basically, a reflex angle is just the other side of an ordinary angle. For example, consider one of the angles in a triangle. Picture the large angle on the *outside* of the triangle that wraps around the corner — that's what a reflex angle is.

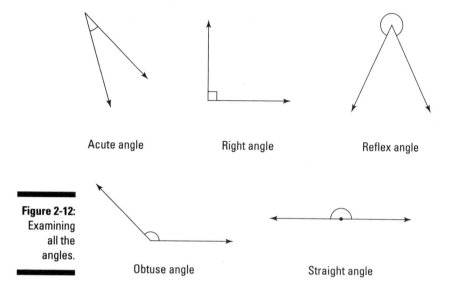

Figure 2-12: Examining all the angles.

Angle pairs: Often joined at the hip

Unlike the solo angles in the preceding section, the angles here have to be in a relationship with another angle for these definitions to mean anything. Yeah, they're a little needy. Adjacent angles and vertical angles always share a common vertex, so they're literally joined at the hip. Complementary and supplementary angles can share a vertex, but they don't have to. Here are the definitions:

- **Adjacent angles:** Adjacent angles are neighboring angles that have the same vertex and that share a side; also, neither angle can be inside the other. I realize that's quite a mouthful. This very simple idea is kind of a pain to define, so just check out Figure 2-13 — a picture's worth a thousand words.

  In the figure, ∠BAC and ∠CAD are adjacent, as are ∠1 and ∠2. However, neither ∠1 nor ∠2 is adjacent to ∠XYZ because they're both inside ∠XYZ. None of the unnamed angles to the right are adjacent because they either don't share a vertex or don't share a side.

  Warning: If you have adjacent angles, you can't name any of the angles with a single letter. For example, you can't call ∠1 or ∠2 (or ∠XYZ, for that matter) ∠Y because no one would know which one you mean. Instead, you have to refer to the angle in question with a number or with three letters.

Part I: Getting Started with Geometry Basics

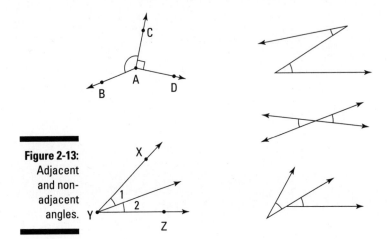

Figure 2-13: Adjacent and non-adjacent angles.

- **Complementary angles:** Two angles that add up to 90° or a right angle are complementary. They can be adjacent angles but don't have to be. In Figure 2-14, adjacent angles ∠1 and ∠2 are complementary because they make a right angle; ∠P and ∠Q are complementary because they add up to 90°. (The definition of *complementary* is sometimes used in proofs. See Chapter 4.)

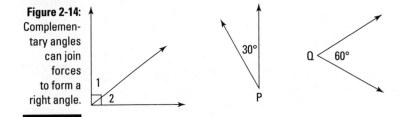

Figure 2-14: Complementary angles can join forces to form a right angle.

- **Supplementary angles:** Two angles that add up to 180° or a straight angle are supplementary. They may or may not be adjacent angles. In Figure 2-15, ∠1 and ∠2, or the two right angles, are supplementary because they form a straight angle. Such angle pairs are called a *linear pair*. Angles A and Z are supplementary because they add up to 180°. (The definition of *supplementary* is sometimes used in proofs. See Chapter 4.)

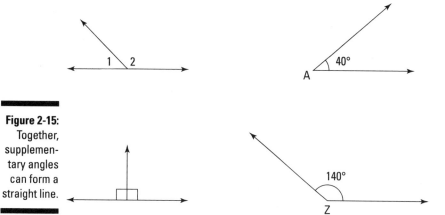

Figure 2-15: Together, supplementary angles can form a straight line.

- **Vertical angles:** When intersecting lines form an X, the angles on the opposite sides of the X are called vertical angles. See Figure 2-16, which shows vertical angles ∠1 and ∠3 and vertical angles ∠2 and ∠4. Two vertical angles are always the same size as each other. By the way, as you can see in the figure, the *vertical* in *vertical angles* has nothing to do with the up-and-down meaning of vertical.

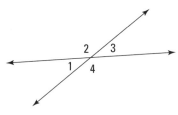

Figure 2-16: Vertical angles share a vertex and lie on opposite sides of the X.

Chapter 3
Sizing Up Segments and Analyzing Angles

In This Chapter
▶ Measuring segments and angles
▶ Doing addition and subtraction with segments and angles
▶ Cutting segments and angles in two or three congruent pieces
▶ Making the correct assumptions

This chapter contains some pretty simple stuff about the sizes of segments and angles, how to measure them, how to add and subtract them, and how to bisect and trisect them. But despite the simple nature of these ideas, this is important groundwork material, so skip it at your own risk! Because all polygons are made up of segments and angles, these two fundamental objects are the key to a great number of geometry problems and proofs.

Measuring Segments and Angles

Measuring segments and angles — especially segments — is a piece o' cake. For a segment, you measure its length; and for an angle, you measure how far open it is (kind of like measuring how far a door is open). Whenever you look at a diagram in a geometry book, paying attention to the sizes of the segments and angles that make up a shape can help you understand some of the shape's important properties.

Measuring segments

To tell you the truth, I thought measuring segments was too simple of a concept to put in this book, but my editor thought I should include it for the sake of completeness, so here it is. The *measure* or size of a segment is simply its

Part I: Getting Started with Geometry Basics

length. What else could it be? After all, length is the only feature a segment has. You've got your short, your medium, and your long segments. (No, these are *not* technical math terms.) Get ready for another shock: If you're told that one segment has a length of 10 and another has a length of 20, then the 20-unit segment is twice as long as the 10-unit segment. Fascinating stuff, right? (I call these 10-*unit* and 20-*unit* segments because you often don't see real units like feet, inches, or miles in a geometry problem.)

Congruent segments are segments with the same length. If $\overline{MN}$ is congruent to $\overline{PQ}$, you write $\overline{MN} \cong \overline{PQ}$. You know that two segments are congruent when you know that they both have the same numerical length or when you don't know their lengths but you figure out (or are simply told) that they're congruent. In a figure, giving each matching segment the same number of tick marks indicates congruence. In Figure 3-1, for instance, the fact that both $\overline{WX}$ and $\overline{YZ}$ have three tick marks tells you that they're the same size.

Congruent segments (and congruent angles, which I get to in the next section) are essential ingredients in the proofs you see in the rest of the book. For instance, when you figure out that a side (a segment) of one triangle is congruent to a side of another triangle, you can use that fact to help you prove that the triangles are congruent to each other (see Chapter 9 for details).

When you write the name of a segment without the bar over it, it means the *length of the segment,* so AG indicates the length of $\overline{AG}$. If, for example, $\overline{XY}$ has a length of 5, you'd write $XY = 5$. And if $\overline{AB}$ is congruent to $\overline{CD}$, their lengths would be equal, and you'd write $AB = CD$. Note that in the two preceding equations, an equal sign is used, not a congruent symbol.

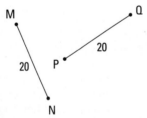
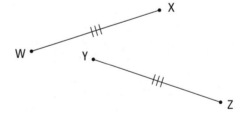

Figure 3-1: Two pairs of congruent segments.

Measuring angles

Measuring angles is pretty simple, but it can be a bit trickier than measuring segments because the size of an angle isn't based on something as simple as length, which you see all the time in your everyday life. The size of an angle is based, instead, on how wide the angle is open. In this section, I introduce you to some points and mental pictures that help you understand how angle measurement works.

Chapter 3: Sizing Up Segments and Analyzing Angles 37

Degree: The basic unit of measure for angles is the degree. One degree is $\frac{1}{360}$ of a circle, or $\frac{1}{360}$ of one complete rotation.

A good way to start thinking about the size and degree-measure of angles is by picturing an entire pizza — that's 360° of pizza. Cut a pizza into 360 slices, and the angle each slice makes is 1° (I don't recommend slices this small if you're hungry). For other angle measures, see the following list and Figure 3-2:

- If you cut a pizza into four big slices, each slice makes a 90° angle (360° ÷ 4 = 90°).
- If you cut a pizza into four big slices and then cut each of those slices in half, you get eight pieces, each of which makes a 45° angle (360° ÷ 8 = 45°).
- If you cut the original pizza into 12 slices, each slice makes a 30° angle (360° ÷ 12 = 30°).

So $\frac{1}{12}$ of a pizza is 30°, $\frac{1}{8}$ is 45°, $\frac{1}{4}$ is 90°, and so on. The bigger the fraction of the pizza, the bigger the angle.

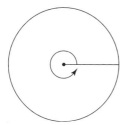

There are 360° in a circle (or a pizza).

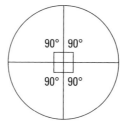

If you cut a pizza into four big slices, each slice makes a 90° angle (360° ÷ 4 = 90°).

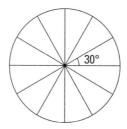

Figure 3-2: Larger angles represent bigger fractions of the pizza.

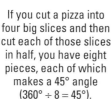

If you cut a pizza into four big slices and then cut each of those slices in half, you have eight pieces, each of which makes a 45° angle (360° ÷ 8 = 45°).

If you cut the original pizza into 12 slices, each slice makes a 30° angle (360° ÷ 12 = 30°).

The fraction of the pizza or circle is the only thing that matters when it comes to angle size. The length along the crust and the area of the pizza slice tell you nothing about the size of an angle. In other words, $\frac{1}{6}$ of a 10-inch pizza represents the same angle as $\frac{1}{6}$ of a 16-inch pizza, and $\frac{1}{8}$ of a small pizza has a larger angle (45°) than $\frac{1}{12}$ of a big pizza (30°) — even if the 30° slice is the one you'd want if you were hungry. See Figure 3-3.

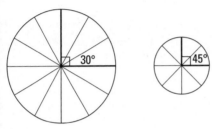

Figure 3-3: A big pizza with little angles and a little pizza with big angles.

Another way of looking at angle size is to think about opening a door or a pair of scissors or, say, an alligator's mouth. The wider the mouth is open, the bigger the angle. As Figure 3-4 shows, a baby alligator with its mouth opened wide makes a bigger angle than an adult alligator with its mouth opened less wide, even if there's a bigger gap at the front of the adult alligator's mouth.

Big angle   Little angle

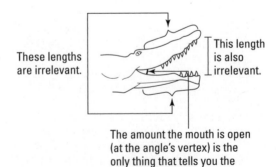

Figure 3-4: The adult alligator is bigger, but the baby's mouth makes a bigger angle.

These lengths are irrelevant.

This length is also irrelevant.

The amount the mouth is open (at the angle's vertex) is the only thing that tells you the size of an angle.

An angle's sides are both rays (see Chapter 2), and all rays are infinitely long, regardless of how long they look in a figure. The "lengths" of an angle's sides in a diagram aren't really lengths at all, and they tell you nothing about the angle's size. Even when a diagram shows an angle with two segments for sides, the sides are still technically infinitely long rays.

Congruent angles are angles with the same degree measure. In other words, congruent angles have the same amount of opening at their vertices. If you were to stack two congruent angles on top of each other with their vertices together, the two sides of one angle would align perfectly with the two sides of the other angle.

You know that two angles are congruent when you know that they both have the same numerical measure (say, they both have a measure of 70°) or when you don't know their measures but you figure out (or are simply told) that they're congruent. In figures, angles with the same number of tick marks are congruent to each other. See Figure 3-5.

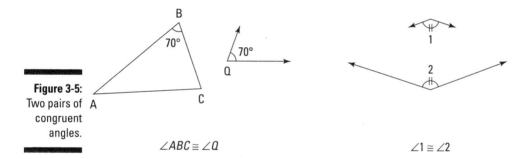

Figure 3-5: Two pairs of congruent angles.

Adding and Subtracting Segments and Angles

The title of this section pretty much says it all: You're about to see how to add and subtract segments and angles. This topic — like measuring segments and angles — isn't exactly rocket science. But adding and subtracting segments and angles is important because this geometric arithmetic comes up in proofs and other geometry problems. Here's how it works:

- **Adding and subtracting segments:** To add or subtract segments, simply add or subtract their lengths. For example, if you put a 4" stick end-to-end with an 8" stick, you get a total length of 12". That's how segments add. Subtracting segments is like cutting off 3" from a 10" stick. You end up with a 7" stick. Brain surgery this is not.

Part I: Getting Started with Geometry Basics

> ✓ **Adding and subtracting angles:** To add or subtract angles, you just add or subtract the angles' degrees. You can think of adding angles as putting two or more pizza slices next to each other with their pointy ends together. Subtracting angles is like starting with some pizza and taking a piece away. Figure 3-6 shows how this works.

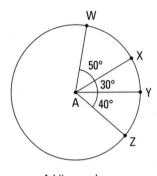
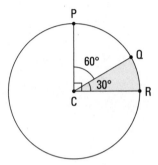

Figure 3-6: Adding and subtracting angles.

Adding angles
$\angle WAX + \angle XAY + \angle YAZ = \angle WAZ$
$50° + 30° + 40° = 120°$

Subtracting angles
$\angle PCR - \angle QCR = \angle PCQ$
$90° - 30° = 60°$

Cutting in Two or Three: Bisection and Trisection

For all you fans of bicycles and tricycles and bifocals and trifocals — not to mention the biathalon and the triathalon, bifurcation and trifurcation, and bipartition and tripartition — you're really going to love this section on bisection and trisection: cutting something into two or three equal parts.

The main point here is that after you do the bisecting or trisecting, you end up with *congruent* parts of the segment or angle you cut up. In Chapter 5, you see how this comes in handy in geometry proofs.

Bisecting and trisecting segments

Segment *bisection,* the related term *midpoint,* and segment *trisection* are pretty simple ideas. (Their definitions, which follow, are used frequently in proofs. See Chapter 4.)

> ✓ **Segment bisection:** A point, segment, ray, or line that divides a segment into two congruent segments *bisects* the segment.

✓ **Midpoint:** The point where a segment is bisected is called the *midpoint* of the segment; the midpoint cuts the segment into two congruent parts.

✓ **Segment trisection:** Two things (points, segments, rays, lines, or any combination of these) that divide a segment into three congruent segments *trisect* the segment. The points of trisection are called — check this out — the *trisection points* of the segment.

I doubt you'll have any trouble remembering the meanings of *bisect* and *trisect*, but here's a mnemonic just in case: A *bi*cycle has two wheels, and to *bi*sect means to cut something into two congruent parts; a *tri*cycle has three wheels, and to *tri*sect means to cut something into three congruent parts.

Students often make the mistake of thinking that *divide* means to *bisect*, or cut exactly in half. I suppose this error is understandable because when you do ordinary division with numbers, you are, in a sense, dividing the larger number into equal parts ($24 \div 2 = 12$ because $12 + 12 = 24$). But in geometry, to *divide* something just means to cut it into parts of any size, equal or unequal. *Bisect* and *trisect*, of course, *do* mean to cut into exactly equal parts.

Here's a problem you can try using the triangle in Figure 3-7. Given that rays $\overrightarrow{AJ}$ and $\overrightarrow{AZ}$ trisect $\overline{BC}$, determine the length of $\overline{BC}$.

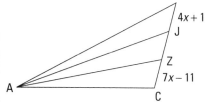

Figure 3-7: A trisected segment.

Okay, here's how you solve it: $\overline{BC}$ is trisected, so it's cut into three congruent parts; thus, $BJ = ZC$. Just set these equal to each other and solve for x:

$4x + 1 = 7x - 11$

$12 = 3x$

$x = 4$

Plugging $x = 4$ into $4x + 1$ and $7x - 11$ gives you 17 for each segment. JZ must also be 17, so BC must total $3 \cdot 17$, or 51.

Don't make the common mistake of thinking that because $\overline{BC}$ is trisected, $\angle BAC$ must be trisected as well. As it happens, this is *never* true.

WARNING! If a side of a triangle is trisected by rays from the opposite vertex, the vertex angle can't be trisected. The vertex angle often *looks* like it's trisected, and it's often divided into nearly-equal parts, but it's never an exact trisection.

In this particular problem, you might be especially likely to fall prey to this perilous pitfall because I have *rays $\overrightarrow{AJ}$ and $\overrightarrow{AZ}$* (rather than *points J and Z*) trisecting $\overline{BC}$, and rays often *do* trisect angles. But the givens in this problem say nothing about ∠BAC, and the way the rays cut up segment $\overline{BC}$ just doesn't tell you anything about the way they divide ∠BAC.

Bisecting and trisecting angles

REMEMBER Brace yourself for a real shocker: The terms *bisecting* and *trisecting* mean the same thing for angles as they do for segments! (Their definitions are often used in proofs. Check out Chapter 4.)

- **Angle bisection:** A ray that cuts an angle into two congruent angles *bisects* the angle. The ray is called the *angle bisector*.

- **Angle trisection:** Two rays that divide an angle into three congruent angles *trisect* the angle. These rays are called *angle trisectors*.

Take a stab at this problem: In Figure 3-8, $\overrightarrow{TP}$ bisects ∠STL, which equals $(12x - 24)°$; $\overrightarrow{TL}$ bisects ∠PTI, which equals $(8x)°$. Is ∠STI trisected, and what is its measure?

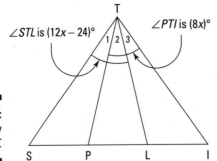

Figure 3-8: A three-way SPLIT.

Nothing to it. First, yes, ∠STI is trisected. You know this because ∠STL is bisected, so ∠1 must equal ∠2. And because ∠PTI is bisected, ∠2 equals ∠3. Thus, all three angles must be equal, and that means ∠STI is trisected.

Now find the measure of ∠STI. Because ∠STL — which measures $(12x - 24)°$ — is bisected, ∠2 must be half its size, or $(6x - 12)°$. And because ∠PTI is bisected, ∠2 must also be half the size of ∠PTI — that's half of $(8x)°$, or $(4x)°$. Because ∠2 equals both $(6x - 12)°$ and $(4x)°$, you set those expressions equal to each other and solve for x:

$$6x - 12 = 4x$$
$$2x = 12$$
$$x = 6$$

Then just plug $x = 6$ into, say, $(4x)°$, which gives you $4 \cdot 6$, or 24° for ∠2. Angle STI is three times that, or 72°. That does it.

When rays trisect an angle of a triangle, the opposite side of the triangle is *never* trisected by these rays.

In Figure 3-8, for instance, because ∠STI is trisected, $\overline{SI}$ is definitely *not* trisected. Note that this is the reverse of the warning in the previous section, which tells you that if a side of a triangle is trisected, the angle is not trisected.

Proving (Not Jumping to) Conclusions about Figures

Here's something that's unusual about the study of geometry: In geometry diagrams, you're *not* allowed to assume that everything that looks true is true.

Consider the triangle in Figure 3-9. Now, if this figure were to appear in a non-geometry context (for example, the figure could be a roof with a horizontal beam and a vertical support), it'd be perfectly sensible to conclude that the two sides of the roof are equal, that the beam is perfectly horizontal, that the support is perfectly vertical, and therefore, that the support and the crossbeam are perpendicular. If you see this figure in a geometry problem, however, you can't assume any of these things. These things might certainly turn out to be true (actually, it's extremely likely that they are true), but you can't assume that they're true. Instead, you have to prove they're true by airtight, mathematical logic. This way of dealing with figures gives you practice in proving things using rigorous deductive reasoning.

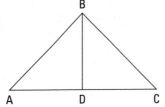

Figure 3-9: A typical geometry diagram.

One way to understand why figures are treated this way in geometry courses is to consider that just because two lines in a figure look perpendicular, that doesn't guarantee that they are precisely perpendicular. In Figure 3-9, for example, the two angles on either side of $\overline{BD}$ could be 89.99° and 90.01° instead of two 90° angles. Even if you had the most accurate instrument in the world to measure the angles, you could never be perfectly accurate. No instrument can measure the difference between a 90° angle and, say, a 90.00000000001° angle. So if you want to know for sure that two lines are perpendicular, you have to use pure logic, not measurement.

Here are the lists of things you can and cannot assume about geometry diagrams. Refer again to Figure 3-9.

In geometry diagrams, you *can* assume four things; all of them have to do with *straight lines*. Here's an example of each type of valid assumption using △*ABC*:

- $\overline{AC}$ is straight.
- ∠*ADC* is a straight angle.
- *A*, *D*, and *C* are collinear.
- *D* is between *A* and *C*.

In geometry diagrams, you *can't* assume things that concern the size of segments or angles. You can't assume that segments and angles that look congruent are congruent or that segments and angles that look unequal are unequal; nor can you assume anything about the relative sizes of segments and angles. For instance, in Figure 3-9, the following aren't necessarily true:

- $\overline{AB} \cong \overline{CB}$; $\overline{AD} \cong \overline{CD}$.
- *D* is the midpoint of $\overline{AC}$.
- ∠*A* ≅ ∠*C*; ∠*ABD* ≅ ∠*CBD*.
- $\overrightarrow{BD}$ bisects ∠*ABC*.
- $\overline{AC} \perp \overline{BD}$.

- ∠*ADB* is a right angle.
- *AB* (that's the length of $\overline{AB}$) is greater than *AD*.
- ∠*ADB* is larger than ∠*A*.

Now, I don't want to suggest that the way figures look isn't important. Especially when doing geometry proofs, it's a good idea to check out the proof diagram and pay attention to whether segments, angles, and triangles look congruent. If they look congruent, they probably are, so the appearance of the diagram is a valuable *hint* about the truth of the diagram. But to establish that something is in fact true, you have to *prove* it.

Hold onto your hat, because in this somewhat peculiar discussion about the treatment of figures, I've saved the worst for last. Occasionally, geometry teachers and authors will throw you a curveball and draw figures that are warped out of their proper shape. This may seem weird, but it's allowed under the rules of the geometry game. Luckily, warped figures like this are fairly rare.

Consider Figure 3-10. The given triangle on the left is a triangle you might see in a geometry problem. In this particular problem, you'd be asked to determine *x* and *y*, the lengths of the two unknown sides of the triangle. The key to this problem is that the three angles are marked congruent. As you may know, the only triangle with three equal angles is the *equilateral* triangle, which has three 60° angles and three equal sides. Thus, *x* and *y* must both equal 5. The figure on the right shows what this triangle really looks like. So Figure 3-10 shows an example of equal segments and angles that are drawn to look unequal.

Figure 3-10: A triangle as drawn and the real thing.

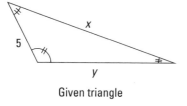
Given triangle

What the triangle actually looks like

Now check out Figure 3-11. It illustrates the opposite type of warping: Segments and angles that are unequal in reality appear equal in the geometry diagram. If the quadrilateral on the left were to appear in a non-geometry context, you would, of course, refer to it as a rectangle (and you could safely assume that its four angles were right angles). But in this geometry diagram, you'd be wrong to call this shape a rectangle. Despite its appearance, it's *not* a rectangle; it's a no-name quadrilateral. Check out its actual shape on the right.

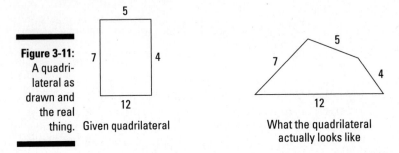

Figure 3-11: A quadrilateral as drawn and the real thing.

If you're having trouble wrapping your head around this strange treatment of geometry diagrams, don't sweat it. It'll become clear in subsequent chapters when you see it in action.

Part II
Introducing Proofs

In this part . . .

Here you get your first taste of two-column geometry proofs. Chapter 4 gives you the basic structure of two-column proofs, explaining their basic parts: the diagram, the *givens* (your starting point), the *prove* statement (your goal), and the statements and reasons that take you from the beginning to the end. In Chapter 5, you get your first dozen or so theorems, along with several examples of shorter proofs that show you how to use the theorems. Chapter 6 walks you though the entire thought process involved in solving a longer proof. Many students find two-column proofs extremely challenging, but if you pay attention to the tips and strategies in this part, you'll soon find that proofs really aren't so tough after all.

Chapter 4

Prelude to Proofs

In This Chapter
- Getting geared up for geometry proofs
- Introducing if-then logic
- Theorizing about theorems and defining definitions
- Proving that horses don't talk

Traditional two-column geometry proofs are arguably the most important topic in standard high school geometry courses. And — sorry to be the bearer of bad news — geometry proofs give many students more difficulty than anything else in the entire high school mathematics curriculum. But before you consider dropping geometry for underwater basket weaving, here's the good news: Over the course of several chapters, I give you ten fantastically helpful strategies that make proofs much easier than they seem at first (you can also find summaries of the strategies on the Cheat Sheet). Practice these strategies, and you'll become a proof-writing whiz in no time.

In this chapter, I lay the groundwork for the two-column proofs you do in subsequent chapters. First, I give you a schematic drawing that shows you all the elements of a two-column proof and where they go. Next, I explain how you prove something using a deductive argument. Finally, I show you how to use deductive reasoning to prove that Clyde the Clydesdale will not give your high school commencement address. Big surprise!

Getting the Lay of the Land: The Components of a Formal Geometry Proof

Basically, a two-column geometry proof is a problem involving a geometric diagram of some sort. You're told one or more things that are true about the diagram (the *givens*), and you're asked to prove that something else is true about the diagram (the *prove* statement). Every proof proceeds like this:

1. You begin with one or more of the given facts about the diagram.
2. You then state something that follows from the given fact or facts; then you state something that follows from that; then, something that follows from that; and so on. Each deduction leads to the next.
3. You end by making your final deduction — the fact you're trying to prove.

Every standard, two-column geometry proof contains the following elements. The proof mockup in Figure 4-1 shows how these elements all fit together.

- **The diagram:** The shape or shapes in the diagram are the subject matter of the proof. Your goal is to prove some fact about the diagram (for example, that two triangles or two angles in the diagram are congruent). The proof diagrams are usually but not always drawn accurately. Don't forget, however, that you can't assume that things that look true *are* true. For instance, just because two angles look congruent doesn't mean they are. (See Chapter 3 for more on making assumptions.)

- **The givens:** The givens are true facts about the diagram that you build upon to reach your goal, the *prove* statement. You always begin a proof with one of the givens, putting it in line 1 of the statement column.

   Most people like to mark the diagram to show the information from the givens. For example, if one of the givens were $\overline{AB} \cong \overline{CB}$, you'd put little tick marks on both segments so that when you glance at the diagram, the congruence is immediately apparent.

- **The *prove* statement:** The *prove* statement is the fact about the diagram that you must establish with your chain of logical deductions. It always goes in the last line of the statement column.

- **The statement column:** In the statement column, you put all the given facts, the facts that you deduce, and in the final line, the *prove* statement. In this column, you put *specific* facts about *specific* geometric objects, such as $\angle ABD \cong \angle CBD$.

- **The reason column:** In the reason column, you put the justification for each statement that you make. In this column, you write *general* rules about things in *general*, such as *If an angle is bisected, then it's divided into two congruent parts*. You do not give the names of specific objects.

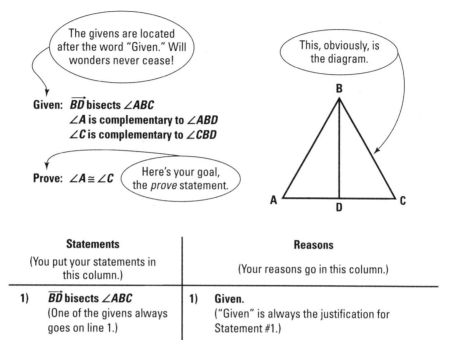

Figure 4-1:
Anatomy of a geometry proof.

Reasoning with If-Then Logic

Every geometry proof is a sequence of logical deductions. You write one of the given facts as statement 1. Then, for statement 2, you put something that follows from statement 1 and write your justification for that in the reason column. Then you proceed to statement 3, and so on, till you get to the *prove* statement. The way you get from statement 1 to statement 2, from statement 2 to statement 3, and so on is by using *if-then* logic.

By the way, the ideas in the next few sections are huge, so if by any chance you've been dozing off a bit, wake up and pay attention!

If-then chains of logic

A two-column geometry proof is in essence a logical argument or a chain of logical deductions, like

1. If I study, then I'll get good grades.
2. If I get good grades, then I'll get into a good college.
3. If I get into a good college, then I'll become a babe/guy magnet.
4. (And so on . . .)

(Except that geometry proofs are about geometric figures, naturally.) Note that each of these steps is a sentence with an *if* clause and a *then* clause.

Here's an example of a two-column proof from everyday life. Say you've got a Dalmatian named Spot, and you want to prove that he's a mammal. Figure 4-2 shows the proof.

On the first line of the statement column, you put down the given fact that Spot is a Dalmatian, and you write *Given* in the reason column. Then, in statement 2, you put down a new fact that you deduce from statement 1 — namely, *Spot is a dog*. In reason 2, you justify or defend that claim with the reason *If something is a Dalmatian, then it's a dog*.

Statements (or Conclusions)	Reasons (or Justifications)
1) Spot is a Dalmatian	1) Given.
2) Spot is a dog	2) If something is a Dalmatian, then it's a dog.
3) Spot is a mammal	3) If something is a dog, then it's a mammal.

Figure 4-2: Proving that Spot is a mammal.

Here are a couple of ways of looking at how reasons work:

- Imagine I know that Spot is a Dalmatian and then you say to me, "Spot is a dog." I ask you, "How do you know?" Your response to me is what you'd write in the reason column.

- When you write a reason like *If something is a Dalmatian, then it's a dog,* you can think of the word *if* as meaning *because I already know,* and you can think of the word *then* as meaning *I can now deduce.* So basically, the second reason in Figure 4-2 means that because you already know that Spot is a Dalmatian, you can deduce or conclude that Spot is a dog.

Continuing with the proof, in statement 3, you write something that you can deduce from statement 2, namely that Spot is a mammal. Finally, for reason 3, you write your justification for statement 3: *If something is a dog, then it's a mammal.* Every geometry proof solution has this same, basic structure.

You've got your reasons: Definitions, theorems, and postulates

Definitions, theorems, and postulates are the building blocks of geometry proofs. With very few exceptions, every justification in the reason column is one of these three things. Look back at Figure 4-2. If that had been a geometry proof instead of a dog proof, the reason column would contain *if-then* definitions, theorems, and postulates about geometry instead of *if-then* ideas about dogs. Here's the lowdown on definitions, theorems, and postulates.

Using definitions in the reason column

Definition: I'm sure you know what a definition is — it defines or explains what a term means. Here's an example: "A *midpoint* divides a segment into two congruent parts."

You can write all definitions in *if-then* form in either direction: "If a point is a midpoint of a segment, then it divides that segment into two congruent parts" or "If a point divides a segment into two congruent parts, then it's the midpoint of that segment."

Figure 4-3 shows you how to use both versions of the midpoint definition in a two-column proof.

54 Part II: Introducing Proofs

The following mini proofs use this figure:

A———M———B

For the first mini proof, it's *given* that M is the midpoint of $\overline{AB}$, and you need to *prove* that $\overline{AM} \cong \overline{MB}$.

Statements (or Conclusions)	Reasons (or Justifications)
1) M is the midpoint of $\overline{AB}$	1) Given.
2) $\overline{AM} \cong \overline{MB}$	2) *If* a point is the midpoint of a segment, *then* it divides the segment into two congruent parts.

Figure 4-3: Double duty — using both versions of the midpoint definition in the reason column.

For the second mini proof, it's *given* $\overline{AM} \cong \overline{MB}$, and you need to *prove* that M is the midpoint of $\overline{AB}$.

Statements (or Conclusions)	Reasons (or Justifications)
1) $\overline{AM} \cong \overline{MB}$	1) Given.
2) M is the midpoint of $\overline{AB}$	2) *If* a point divides a segment into two congruent parts, *then* it's the midpoint of the segment.

When you have to choose between these two versions of the midpoint definition, remember that you can think of the word *if* as meaning *because I already know* and the word *then* as meaning *I can now deduce*. For example, for reason 2 in the first proof in Figure 4-3, you choose the version that goes, "*If* a point is the midpoint of a segment, *then* it divides the segment into two congruent parts," because you already know that M is the midpoint of $\overline{AB}$ (because it's given) and from that given fact you can deduce that $\overline{AM} \cong \overline{MB}$.

Using theorems and postulates in the reason column

Theorem and postulate: Both theorems and postulates are statements of geometrical truth, such as *All right angles are congruent* or *All radii of a circle are congruent*. The difference between postulates and theorems is that postulates are assumed to be true, but theorems must be proven to be true based on postulates and/or already-proven theorems. This distinction isn't something you have to care a great deal about unless you happen to be writing your Ph.D. dissertation on the deductive structure of geometry. However, because I suspect that you're *not* currently working on your Ph.D. in geometry, I wouldn't sweat this fine point.

Written in *if-then* form, the theorem *All right angles are congruent* would read, "If two angles are right angles, then they're congruent." Unlike definitions, theorems are generally *not* reversible. For example, if you reverse this right-angle theorem, you get a false statement: "If two angles are congruent, then they're right angles." (If a theorem works in both directions, you'll get a separate theorem for each version. The two isosceles-triangle theorems — *If sides, then angles* and *If angles, then sides* — are an example. See Chapter 9.) Figure 4-4 shows you the right-angle theorem in a proof.

For this mini proof, it's *given* that $\angle A$ and $\angle B$ are right angles, and you need to *prove* that $\angle A \cong \angle B$.

Figure 4-4: Using a theorem in the reason column of a proof.

Statements (or Conclusions)	Reasons (or Justifications)
1) $\angle A$ is a right angle $\angle B$ is a right angle	1) Given.
2) $\angle A \cong \angle B$	2) If two angles are right angles, then they're congruent.

When you're doing your first proofs, or later if you're struggling with a difficult one, it's very helpful to write your reasons (definitions, theorems, and postulates) in *if-then* form. When you use *if-then* form, the logical structure of the proof is easier to follow. After you become a proof expert, you can abbreviate your reasons in non-*if-then* form or simply list the name of the definition, theorem, or postulate.

Bubble logic for two-column proofs

I like to add bubbles and arrows to a proof solution to show the connections between the statements and the reasons. You won't be asked to do this when you solve a proof; it's just a way to help you understand how proofs work. Figure 4-5 shows the Spot-the-dog proof from Figure 4-2, this time with bubbles and arrows that show how the logic flows through the proof.

Figure 4-5: Follow the arrows from bubble to bubble.

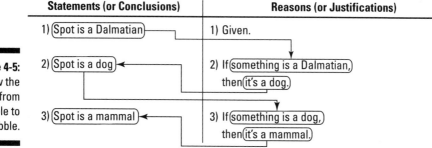

Follow the arrows from bubble to bubble, and follow my tips to stay out of trouble! (It's because of high poetry like this that they pay me the big bucks.) The next tip is really huge, so take heed.

In a two-column proof,

- The idea in the *if* clause of each reason must come from the statement column somewhere *above* the reason.
- The idea in the *then* clause of each reason must match the idea in the statement on the *same line* as the reason.

The arrows and bubbles in Figure 4-5 show how this incredibly important logical structure works.

Horsing Around with a Two-Column Proof

To wrap up this geometry proof prelude, I want to give you one more non-geometry proof to show you how a deductive argument all hangs together. In the following proof, I brilliantly establish that Clyde the Clydesdale won't be giving an address at your high school commencement. Here's the basic argument:

1. Clyde is a Clydesdale.
2. Therefore, Clyde is a horse (because all Clydesdales are horses).
3. Therefore, Clyde can't talk (because horses can't talk).
4. Therefore, Clyde can't give a commencement address (because something that doesn't talk can't give a commencement address).
5. Therefore, Clyde won't be giving an address at your high school commencement (because something that can't give a commencement address won't be giving one at your high school commencement).

Here's the argument in a nutshell: Clydesdale → horse → can't talk → can't give a commencement address → won't give an address at your high school commencement.

Now take a look at what this argument or proof would look like in the standard two-column geometry proof format with the reasons written in *if-then* form. When reasons are written this way, you can see how the chain of logic flows.

Given: Clyde is a Clydesdale.

Prove: Clyde won't be giving an address at your high school commencement.

Chapter 4: Prelude to Proofs 57

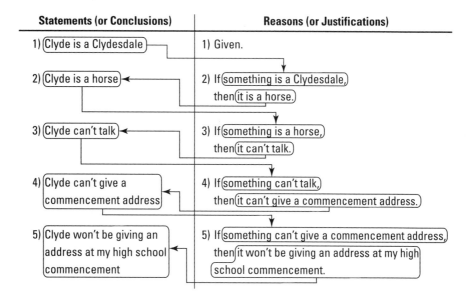

Follow the arrows from bubble to bubble. Note again that the idea in the *if* clause of each reason connects to the same idea in the statement column *above* the line of the reason; the idea in the *then* clause of each reason connects to the same idea in the statement column on the *same line* as the reason.

Notice the difference between the things you put in the statement column and the things you put in the reason column: In all proofs, the statement column contains *specific facts* (things about a particular horse, like *Clyde is a Clydesdale*), and the reason column contains *general principles* (ideas about horses in general, such as *If something is a horse, then it can't talk*).

Chapter 5
Your Starter Kit of Easy Theorems and Short Proofs

In This Chapter
- Understanding the complementary and supplementary angle theorems
- Summing up addition and subtraction theorems
- Using doubles, triples, halves, and thirds
- Identifying congruent angles with the vertical angle theorem
- Swapping places with the Transitive and Substitution Properties

In this chapter, you move past the warm-up material in previous chapters and get to work for real on some honest-to-goodness geometry proofs. (If you're not ready to take this leap, Chapter 4 goes over the components of a two-column geometry proof and its logical structure.) Here, I give you a starter kit that contains 18 theorems along with some proofs that illustrate how those theorems are used. On your mark, get set, go!

Doing Right and Going Straight: Complementary and Supplementary Angles

Ready for your first theorems? Okay, then dig these groovy ones, man. This section introduces you to theorems about complementary and supplementary angles. *Complementary angles* are two angles that add up to 90°, or a right angle; two *supplementary angles* add up to 180°, or a straight angle. These angles aren't the most exciting things in geometry, but you have to be able to spot them in a diagram and know how to use the related theorems.

You use the theorems I list here for complementary angles:

- **Complements of the same angle are congruent.** If two angles are each complementary to a third angle, then they're congruent to each other. (Note that this theorem involves three total angles.)
- **Complements of congruent angles are congruent.** If two angles are complementary to two other congruent angles, then they're congruent. (This theorem involves four total angles.)

The following examples show how incredibly simple the logic of these two theorems is.

Complements of the Same Angle

Given: Diagram as shown

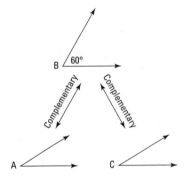

Conclusion: ∠A ≅ ∠C because they'd both have to be 30° angles.

Complements of Congruent Angles

Given: Diagram as shown

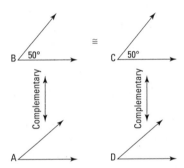

Conclusion: ∠A ≅ ∠D because they'd both have to be 40° angles.

Note: The logic shown in these two figures works the same, of course, when you don't know the size of the given angles (∠B on the left and ∠B and ∠C on the right).

And here are the two theorems about supplementary angles that work exactly the same way as the two complementary angle theorems:

- **Supplements of the same angle are congruent.** If two angles are each supplementary to a third angle, then they're congruent to each other. (This is the three-angle version.)
- **Supplements of congruent angles are congruent.** If two angles are supplementary to two other congruent angles, then they're congruent. (This is the four-angle version.)

Chapter 5: Your Starter Kit of Easy Theorems and Short Proofs

The previous four theorems about complementary and supplementary angles, as well as the addition and subtraction theorems and the transitivity theorems (which you see later in this chapter), come in pairs: One of the theorems involves *three* segments or angles, and the other, which is based on the same idea, involves *four* segments or angles. When doing a proof, note whether the relevant part of the proof diagram contains three or four segments or angles to determine whether to use the three- or four-object version of the appropriate theorem.

Take a look at one of the complementary-angle theorems and one of the supplementary-angle theorems in action:

Given: $\overline{TD} \perp \overline{DC}$
$\overline{QC} \perp \overline{DC}$
$\angle TDQ \cong \angle QCT$
Prove: $\angle UDC \cong \angle SCD$

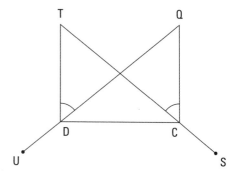

Extra credit: What does *UDTQCS* stand for?

Before trying to write out a formal, two-column proof, it's often a good idea to think through a seat-of-the-pants argument about why the *prove* statement has to be true. I call this argument a *game plan* (I give you more details on making a game plan in Chapter 6). Game plans are especially helpful for longer proofs, because without a plan, you might get lost in the middle of the proof. Throughout this book, I include example game plans for many of the proofs; when I don't, you can try coming up with your own before you read the formal two-column solution for the proof.

When working through a game plan, you may find it helpful to make up arbitrary sizes for segments and angles in the proof. You can do this for segments and angles in the givens and, sometimes, for unmentioned segments and angles. You should not, however, make up sizes for things that you're trying to show are congruent.

Game plan: In this proof, for example, you might say to yourself, "Let's see.... Because of the given perpendicular segments, I have two right angles. Next, the other given tells me that $\angle TDQ \cong \angle QCT$. If they were both 50°, $\angle QDC$ and $\angle TCD$ would both be 40°, and then $\angle UDC$ and $\angle SCD$ would both have to be 140° (because a straight line is 180°)." That's it.

Here's the formal proof:

Statements	Reasons
1) $\overline{TD} \perp \overline{DC}$ $\overline{QC} \perp \overline{DC}$	1) Given. (Why would they tell you this? See reason 2.)
2) $\angle TDC$ is a right angle $\angle QCD$ is a right angle	2) If segments are perpendicular, then they form right angles (definition of perpendicular).
3) $\angle CDQ$ is complementary to $\angle TDQ$, $\angle DCT$ is complementary to $\angle QCT$	3) If two angles form a right angle, then they're complementary (definition of complementary angles).
4) $\angle TDQ \cong \angle QCT$	4) Given.
5) $\angle CDQ \cong \angle DCT$	5) If two angles are complementary to two other congruent angles, then they're congruent.
6) $\angle UDQ$ is a straight angle $\angle SCT$ is a straight angle	6) Assumed from diagram.
7) $\angle UDC$ is supplementary to $\angle CDQ$, $\angle SCD$ is supplementary to $\angle DCT$	7) If two angles form a straight angle, then they're supplementary (definition of supplementary angles).
8) $\angle UDC \cong \angle SCD$	8) If two angles are supplementary to two other congruent angles, then they're congruent.

Note: Depending on where your geometry teacher falls on the loose-to-rigorous scale, he or she might allow you to omit a step like step 6 in this proof because it's so simple and obvious. Many teachers begin the first semester insisting that every little step be included, but then, as the semester progresses, they loosen up a bit and let you skip some of the simplest steps.

The answer to the extra credit question is as easy as one, two, three: *UDTQCS* stands for *Un, Due, Tre, Quattro, Cinque, Sei.* That's counting to six in Italian. (All of you multilingual folks out there may know that counting to six in French will also give you the same six letters.)

Chapter 5: Your Starter Kit of Easy Theorems and Short Proofs

Addition and Subtraction: Eight No-Big-Deal Theorems

In this section, I give you eight simple theorems: four about adding or subtracting segments and four (that work exactly the same way) about adding or subtracting angles. I'm sure you'll have no trouble with these theorems because they all involve ideas that you would've easily understood — and I'm not exaggerating — when you were about 7 or 8 years old.

Addition theorems

In this section, I go over the four addition theorems: two for segments and two for angles . . . as easy as 2 + 2 = 4.

Use these two addition theorems for proofs involving three segments or three angles:

- **Segment addition (three total segments):** If a segment is added to two congruent segments, then the sums are congruent.
- **Angle addition (three total angles):** If an angle is added to two congruent angles, then the sums are congruent.

After you're comfortable with proofs and know your theorems well, you can abbreviate these theorems as *segment addition* or *angle addition* or simply *addition;* however, when you're starting out, writing the theorems out in full is a good idea.

Figure 5-1 shows you how these two theorems work.

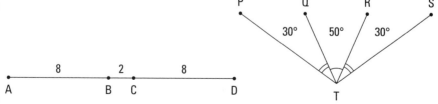

Figure 5-1: Adding one thing to two congruent things.

If you add $\overline{BC}$ to the congruent segments $\overline{AB}$ and $\overline{CD}$, the sums, namely $\overline{AC}$ and $\overline{BD}$, are congruent. In other words, 8 + 2 = 8 + 2. Extraordinary!

And if you add ∠QTR to congruent angles ∠PTQ and ∠RTS, the sums, ∠PTR and ∠QTS, will be congruent: 30° + 50° = 30° + 50°. Brilliant!

Note: In proofs, you won't be given segment lengths and angle measures like the ones in Figure 5-1. I put them in the figure so you can more easily see what's going on.

As you come across theorems in this book, look carefully at the figures that accompany them. The figures show the logic of the theorems in a visual way that can help you remember the wording of the theorems. Try quizzing yourself by reading a theorem and seeing whether you can draw the figure or by looking at a figure and trying to state the theorem.

Use these addition theorems for proofs involving four segments or four angles (also abbreviated as *segment addition, angle addition,* or just *addition*):

- **Segment addition (four total segments):** If two congruent segments are added to two other congruent segments, then the sums are congruent.
- **Angle addition (four total angles):** If two congruent angles are added to two other congruent angles, then the sums are congruent.

Check out Figure 5-2, which illustrates these theorems.

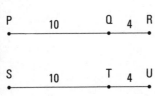
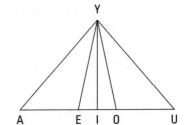

Figure 5-2: Adding congruent things to congruent things.

If $\overline{PQ}$ and $\overline{ST}$ are congruent and $\overline{QR}$ and $\overline{TU}$ are congruent, then $\overline{PR}$ is obviously congruent to $\overline{SU}$, right?

And if ∠AYE ≅ ∠UYO (say they're both 40°) and ∠EYI ≅ ∠OYI (say they're both 20°), then ∠AYI ≅ ∠UYI (they'd both be 60°).

Now for a proof that uses segment addition:

Given: $\overline{MD} \cong \overline{VI}$

$\overline{DX} \cong \overline{CV}$

Prove: $\overline{MC} \cong \overline{XI}$

Impress me: What year is MDXCVI?

Chapter 5: Your Starter Kit of Easy Theorems and Short Proofs

Really impress me: What famous mathematician (who made a major breakthrough in geometry) was born in this year?

I've put what amounts to a game plan for this proof inside the following two-column solution, between the numbered lines.

Statements	Reasons
1) $\overline{MD} \cong \overline{VI}$	1) Given.
2) $\overline{DX} \cong \overline{CV}$	2) Given.

I expect you know what comes next, but for the sake of argument, pretend you don't. Statement 3 has to use one or both of the givens. To see how you can use the four segments from the givens, make up arbitrary lengths for the segments: say $\overline{MD}$ and $\overline{VI}$ both have a length of 5, and $\overline{DX}$ and $\overline{CV}$ are both 2. Obviously, that makes both $\overline{MX}$ and $\overline{CI}$ equal to 7, and that's called addition, of course. So now you've got line 3.

3) $\overline{MX} \cong \overline{CI}$	3) If two congruent segments are added to two other congruent segments, then the sums are congruent.

Now imagine that $\overline{XC}$ is 10. That would make both $\overline{MC}$ and $\overline{XI}$ equal to 17, and thus they're congruent. This is the three-segment version of segment addition, and that's a wrap.

4) $\overline{MC} \cong \overline{XI}$	4) If a segment is added to two congruent segments, then the sums are congruent.

By the way, did you see the other way of doing this proof? It uses the three-segment addition theorem in line 3 and the four-segment addition theorem in line 4.

For the trivia question, did you come up with René Descartes, born in 1596? You can see his famous Cartesian plane in Chapter 18.

Before looking at the next example, check out these two tips — they're huge! They can often make a tricky problem much easier and get you unstuck when you're stuck:

✔ **Use every given.** You have to do something with every given in a proof. So if you're not sure how to do a proof, don't give up until you've asked yourself, "Why did they give me this given?" for every single one of the givens. If you then write down what follows from each given (even if you don't know how that information will help you), you might see how to proceed. You may have a geometry teacher who likes to throw you the occasional curveball, but in every geometry book that I know, the authors don't give you irrelevant givens. And that means that *every given is a built-in hint.*

✓ **Work backward.** Thinking about how a proof will end — what the last and second-to-last lines will look like — is often very helpful. In some proofs, you may be able to work backward from the final statement to the second-to-last statement and then to the third-to-last statement and maybe even to the fourth-to-last. This makes the proof easier to finish because you no longer have to "see" all the way from the *given* to the *prove statement*. The proof has, in a sense, been shortened. You can use this process when you get stuck somewhere in the middle of a proof, or sometimes it's a good thing to try as you begin to tackle a proof.

The following proof shows how you use angle addition:

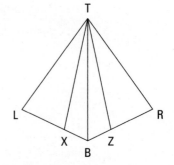

Given: $\overrightarrow{TB}$ bisects $\angle XTZ$
$\overrightarrow{TX}$ and $\overrightarrow{TZ}$ trisect $\angle LTR$

Prove: $\overrightarrow{TB}$ bisects $\angle LTR$

In this proof, I've added a partial game plan that deals with the part of the proof where people might get stuck. The only ideas missing from this game plan are the things (which you see in lines 2 and 4) that follow immediately from the two givens.

Statements	Reasons
1) $\overrightarrow{TB}$ bisects $\angle XTZ$	1) Given. (Why would they tell you this? See statement 2.)
2) $\angle XTB \cong \angle ZTB$	2) If an angle is bisected, then it's divided into two congruent angles (definition of bisect).
3) $\overrightarrow{TX}$ and $\overrightarrow{TZ}$ trisect $\angle LTR$	3) Given. (And why would they tell you that?)
4) $\angle LTX \cong \angle RTZ$	4) If an angle is trisected, then it's divided into three congruent angles (definition of trisect).

Statements	Reasons
Say you're stuck here. Try jumping to the end of the proof and working backward. You know that the final statement must be the *prove* conclusion, $\overrightarrow{TB}$ bisects ∠LTR. Now ask yourself what you'd need to know in order to draw that final conclusion. To conclude that a ray bisects an angle, you need to know that the ray cuts the angle into two equal angles. So the second-to-last statement must be ∠LTB ≅ ∠RTB. And how do you deduce that? Well, with angle addition. The congruent angles from statements 2 and 4 add up to ∠LTB and ∠RTB. That does it.	
5) ∠LTB ≅ ∠RTB	5) If two congruent angles are added to two other congruent angles, then the sums are congruent.
6) $\overrightarrow{TB}$ bisects ∠LTR	6) If a ray divides an angle into two congruent angles, then it bisects the angle (definition of bisect).

Subtraction theorems

In this section, I introduce you to the four subtraction theorems: two for segments and two for angles. Each of these corresponds to one of the addition theorems.

Here are the subtraction theorems for three segments and three angles (abbreviated as *segment subtraction, angle subtraction,* or just *subtraction*):

- **Segment subtraction (three total segments):** If a segment is subtracted from two congruent segments, then the differences are congruent.

- **Angle subtraction (three total angles):** If an angle is subtracted from two congruent angles, then the differences are congruent.

Check out Figure 5-3, which provides the visual aids for these two theorems. If $\overline{JL} \cong \overline{KM}$, then $\overline{JK}$ must be congruent to $\overline{LM}$. (Say $\overline{KL}$ has a length of 3 and $\overline{JL}$ and $\overline{KM}$ are both 10. Then $\overline{JK}$ and $\overline{LM}$ are both 10 – 3, or 7.) For the angles, if ∠EFB ≅ ∠DFG and you subtract ∠GFB from both, you end up with congruent differences, ∠EFG and ∠DFB.

Figure 5-3: The three-thing versions of the segment- and angle-subtraction theorems.

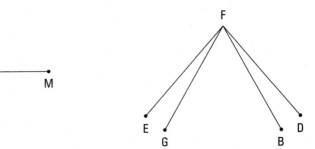

Last but not least, I give you the subtraction theorems for four segments and for four angles (abbreviated just like the subtraction theorems for three things):

- **Segment subtraction (four total segments):** If two congruent segments are subtracted from two other congruent segments, then the differences are congruent.
- **Angle subtraction (four total angles):** If two congruent angles are subtracted from two other congruent angles, then the differences are congruent.

Figure 5-4 illustrates these two theorems.

Figure 5-4: The four-thing versions of the segment- and angle-subtraction theorems.

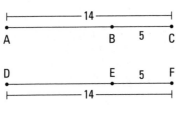 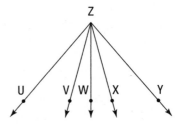

Because $\overline{AC}$ and $\overline{DF}$ are congruent and $\overline{BC}$ and $\overline{EF}$ are congruent, $\overline{AB}$ and $\overline{DE}$ would have to be congruent as well (both would equal 14 – 5, or 9). It works the same for the angles: If ∠UZW and ∠YZW are congruent and ∠VZW and ∠XZW are also congruent, then subtracting the small pair of angles from the big pair would leave congruent angles ∠UZV and ∠YZX.

Chapter 5: Your Starter Kit of Easy Theorems and Short Proofs

Before reading the formal, two-column solution of the next proof, try to think through your own game plan or commonsense argument about why the *prove* statement has to be true. **_Hint:_** Making up angle measures for the two congruent angles in the given and for ∠PUS and ∠QUR may help you see how everything works.

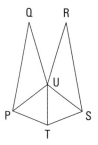

Given: ∠PUR ≅ ∠SUQ
$\overrightarrow{UT}$ bisects ∠PUS

Prove: ∠QUT ≅ ∠RUT

Statements	Reasons
1) ∠PUR ≅ ∠SUQ	1) Given.
2) ∠PUQ ≅ ∠SUR	2) If an angle (∠QUR) is subtracted from two congruent angles (∠PUR and ∠SUQ), then the differences are congruent.
3) $\overrightarrow{UT}$ bisects ∠PUS	3) Given.
4) ∠PUT ≅ ∠SUT	4) If a ray bisects an angle, then it divides it into two congruent angles (definition of bisect).
5) ∠QUT ≅ ∠RUT	5) If two congruent angles (the angles from statement 2) are added to two other congruent angles (the ones from statement 4), then the sums are congruent.

Piece o' cake, right? Now, before moving on to the next section, check out the following. You may have noticed that each of the addition theorems corresponds to one of the subtraction theorems and that a similar diagram is used to illustrate each corresponding pair. Figure 5-1, about addition theorems, pairs up with Figure 5-3, about subtraction theorems; and Figures 5-2 and 5-4 pair up the same way. Because of the similarity of these figures and the ideas that underlie them, people sometimes mix up addition theorems and subtraction theorems. Here's how to keep them straight.

TIP: In a proof, you use one of the *addition theorems* when you add *small* segments (or angles) and conclude that two *big* segments (or angles) are congruent. You use one of the *subtraction theorems* when you subtract segments (or angles) from *big* segments (or angles) to conclude that two *small* segments (or angles) are congruent. In short, *addition* theorems take you from small to big; *subtraction* theorems take you from big to small.

Like Multiples and Like Divisions? Then These Theorems Are for You!

The two theorems in this section are based on very simple ideas (multiplication and division), but they do trip people up from time to time, so make sure to pay careful attention to how these theorems are used in the example proofs. And note my oh-so-helpful tips. They'll keep you from getting the Like Multiples and Like Divisions Theorems confused with the definitions of midpoint, bisect, and trisect (which you find in Chapter 3).

Like Multiples: If two segments (or angles) are congruent, then their *like multiples* are congruent. For example, if you have two congruent angles, then three times one will equal three times the other.

See Figure 5-5. If $\overline{AB} \cong \overline{WX}$ and $\overline{AD}$ and $\overline{WZ}$ are both trisected, then the Like Multiples Theorem tells you that $\overline{AD} \cong \overline{WZ}$.

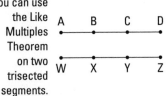

Figure 5-5: You can use the Like Multiples Theorem on two trisected segments.

Like Divisions: If two segments (or angles) are congruent, then their *like divisions* are congruent. If you have, say, two congruent segments, then $\frac{1}{4}$ of one equals $\frac{1}{4}$ of the other, or $\frac{1}{10}$ of one equals $\frac{1}{10}$ of the other, and so on.

Look at Figure 5-6. If ∠BAC ≅ ∠YXZ and both angles are bisected, then the Like Divisions Theorem tells you that ∠1 ≅ ∠3 and that ∠2 ≅ ∠4. And you could also use the theorem to deduce that ∠1 ≅ ∠4 and that ∠2 ≅ ∠3. But note that you *cannot* use the Like Divisions Theorem to conclude that ∠1 ≅ ∠2 or ∠3 ≅ ∠4. Those congruencies follow from the definition of bisect.

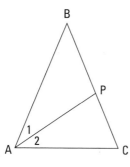
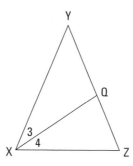

Figure 5-6: Congruent angles divided into congruent parts.

People sometimes get the Like Multiples and Like Divisions Theorems mixed up. Here's a tip that'll help you keep them straight: In a proof, you use the Like Multiples Theorem when you use congruent *small* segments (or angles) to conclude that two *big* segments (or angles) are congruent. You use the Like Divisions Theorem when you use congruent *big* things to conclude that two *small* things are congruent. In short, *Like Multiples* takes you from small to big; *Like Divisions* takes you from big to small.

When you look at the givens in a proof and you see one of the terms *midpoint*, *bisect*, or *trisect* mentioned *twice*, then you'll probably use either the Like Multiples Theorem or the Like Divisions Theorem. But if the term is used only once, you'll likely use the definition of that term instead.

You see how to use the Like Multiples Theorem in the next proof.

Given: ∠EHM ≅ ∠JMH

∠NHM ≅ ∠IMH

$\overrightarrow{HE}$ and $\overrightarrow{HF}$ trisect ∠GHN

$\overrightarrow{MJ}$ and $\overrightarrow{MK}$ trisect ∠LMI

Prove: ∠GHN ≅ ∠LMI

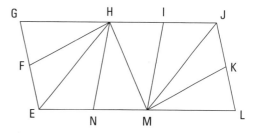

Game plan: Here's how your thought process for this proof might go: Ask yourself how you can use the givens. In this proof, can you see what you can deduce from the two pairs of congruent angles in the given? If not, make up arbitrary measures for the angles. Say ∠EHM and ∠JMH are each 65° and ∠NHM and ∠IMH are each 40°. What would follow from that? You subtract 40° from 65° and get 25° for both ∠EHN and ∠JMI. Then, when you see *trisect* mentioned twice in the other givens, that should ring a bell and make you think *Like Multiples* or *Like Divisions*. Because you use small things (∠EHN and ∠JMI) to deduce the congruence of bigger things (∠GHN and ∠LMI), *Like Multiples* is the ticket.

Statements	Reasons
1) ∠EHM ≅ ∠JMH	1) Given.
2) ∠NHM ≅ ∠IMH	2) Given.
3) ∠EHN ≅ ∠JMI	3) If two congruent angles are subtracted from two other congruent angles, then the differences are congruent.
4) $\overrightarrow{HE}$ and $\overrightarrow{HF}$ trisect ∠GHN	4) Given.
5) $\overrightarrow{MJ}$ and $\overrightarrow{MK}$ trisect ∠LMI	5) Given.
6) ∠GHN ≅ ∠LMI	6) If two angles are congruent (angles *EHN* and *JMI*), then their like multiples are congruent (three times one equals three times the other).

Now for a proof that uses *Like Divisions*:

Given: $\overline{ND} \cong \overline{EL}$

   O is the midpoint of $\overline{NE}$

   A is the midpoint of $\overline{DL}$

Prove: $\overline{NO} \cong \overline{AL}$

Here's a possible game plan: What can you do with the first given? If you can't figure that out right away, make up lengths for $\overline{ND}$, $\overline{EL}$, and $\overline{DE}$. Say that $\overline{ND}$ and $\overline{EL}$ are both 12 and that $\overline{DE}$ is 6. That would make both $\overline{NE}$ and $\overline{DL}$ 18 units long. Then, because both of these segments are bisected by their mid points, $\overline{NO}$ and $\overline{AL}$ must both be 9. That's a wrap.

Statements	Reasons
1) $\overline{ND} \cong \overline{EL}$	1) Given.
2) $\overline{NE} \cong \overline{DL}$	2) If a segment is added to two congruent segments, then the sums are congruent.
3) O is the midpoint of $\overline{NE}$ A is the midpoint of $\overline{DL}$	3) Given.
4) $\overline{NO} \cong \overline{AL}$	4) If two segments are congruent ($\overline{NE}$ and $\overline{DL}$), then their like divisions are congruent (half of one equals half of the other).

TIP The Like Divisions Theorem is particularly easy to get confused with the definitions of *midpoint, bisect,* and *trisect* (see Chapter 3), so remember this: Use the definition of *midpoint, bisect,* or *trisect* when you want to show that parts of *one* bisected or trisected segment or angle are equal to each other. Use the Like Divisions Theorem when *two* objects are bisected or trisected (like $\overline{NE}$ and $\overline{DL}$ in the preceding proof) and you want to show that a part of one ($\overline{NO}$) is equal to a part of the other ($\overline{AL}$).

The X-Files: Congruent Vertical Angles Are Out There

When two lines intersect to make an X, angles on opposite sides of the X are called *vertical* angles (more on that in Chapter 2). These angles are equal, and here's the official theorem that tells you so.

Vertical angles are congruent: If two angles are vertical angles, then they're congruent (see Figure 5-7).

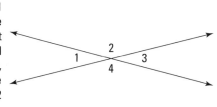

Figure 5-7: Angles 1 and 3 are congruent vertical angles, as are angles 2 and 4.

74 Part II: Introducing Proofs

TIP

Vertical angles are one of the most frequently used things in proofs and other types of geometry problems, and they're one of the easiest things to spot in a diagram. Don't neglect to check for them!

Here's an algebraic geometry problem that illustrates this simple concept: Determine the measure of the six angles in the following figure.

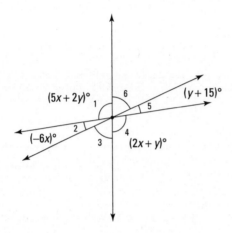

Vertical angles are congruent, so $\angle 1 \cong \angle 4$ and $\angle 2 \cong \angle 5$; and thus you can set their measures equal to each other:

$\angle 1 \cong \angle 4$ and $\angle 2 \cong \angle 5$

$5x + 2y = 2x + y$ $-6x = y + 15$

Now you have a system of two equations and two unknowns. To solve the system, first solve each equation for y:

$y = -3x$ $y = -6x - 15$

Next, because both equations are solved for y, you can set the two x-expressions equal to each other and solve for x:

$-3x = -6x - 15$

$3x = -15$

$x = -5$

To get y, plug in –5 for x in the first simplified equation:

$y = -3x$

$y = -3(-5)$

$y = 15$

Now plug –5 and 15 into the angle expressions to get four of the six angles:

$\angle 4 \cong \angle 1 = 5x + 2y = 5(-5) + 2(15) = 5°$

$\angle 5 \cong \angle 2 = -6x = -6(-5) = 30°$

To get angle 3, note that angles 1, 2, and 3 make a straight line, so they must sum to 180°:

$\angle 1 + \angle 2 + \angle 3 = 180°$

$5° + 30° + \angle 3 = 180°$

$\angle 3 = 145°$

Finally, ∠3 and ∠6 are congruent vertical angles, so ∠6 must be 145° as well. Did you notice that the angles in the figure are absurdly out of scale? Don't forget that you can't assume anything about the relative sizes of angles or segments in a diagram (see Chapter 3).

Pulling the Switch with the Transitive and Substitution Properties

The Transitive Property and the Substitution Property are two principles that you should understand right off the bat. If $a = b$ and $b = c$, then $a = c$, right? That's transitivity. And if $a = b$ and $b < c$, then $a < c$. That's substitution. Easy enough. In the following list, you see these theorems in greater detail:

- **Transitive Property (for three segments or angles):** If two segments (or angles) are each congruent to a third segment (or angle), then they're congruent to each other. For example, if $\angle A \cong \angle B$ and $\angle B \cong \angle C$, then $\angle A \cong \angle C$ ($\angle A$ and $\angle C$ are each congruent to $\angle B$, so they're congruent to each other). See Figure 5-8.

Figure 5-8: The Transitive Property tells you that ∠A ≅ ∠C.

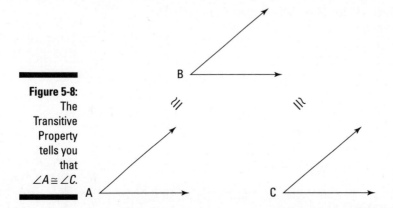

✔ **Transitive Property (for four segments or angles):** If two segments (or angles) are congruent to congruent segments (or angles), then they're congruent to each other. For example, if $\overline{AB} \cong \overline{CD}$, $\overline{CD} \cong \overline{EF}$, and $\overline{EF} \cong \overline{GH}$, then $\overline{AB} \cong \overline{GH}$. ($\overline{AB}$ and $\overline{GH}$ are congruent to the congruent segments $\overline{CD}$ and $\overline{EF}$, so they're congruent to each other.) See Figure 5-9.

Figure 5-9: Three congruency connections make $\overline{AB}$ and $\overline{GH}$ congruent to each other.

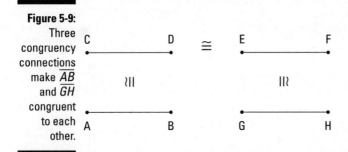

✔ **Substitution Property:** If two geometric objects (segments, angles, triangles, or whatever) are congruent and you have a statement involving one of them, you can pull the switcheroo and replace the one with the other. For example, if ∠X ≅ ∠Y and ∠Y is supplementary to ∠Z, then ∠X is supplementary to ∠Z. A figure isn't especially helpful for this property, so I'm skipping it.

Chapter 5: Your Starter Kit of Easy Theorems and Short Proofs

TIP

To avoid getting the Transitive and Substitution Properties mixed up, just follow these guidelines:

- Use the *Transitive Property* as the reason in a proof when the statement on the same line involves congruent things.
- Use the *Substitution Property* when the statement does not involve a congruence. *Note:* The Substitution Property is the only theorem in this chapter that doesn't involve a congruence in the statements column.

Check out this *TGIF* rectangle proof, which deals with angles:

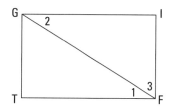

Given: ∠TFI is a right angle

∠1 ≅ ∠2

Prove: ∠2 is complementary to ∠3

No need for a game plan here because the proof is so short — take a look:

Statements	Reasons
1) ∠TFI is a right angle	1) Given.
2) ∠1 is complementary to ∠3	2) If two angles form a right angle, then they're complementary (definition of complementary).
3) ∠1 is congruent to ∠2	3) Given.
4) ∠2 is complementary to ∠3	4) Substitution Property (statements 2 and 3; ∠2 replaces ∠1).

And for the final segment of the program, here's a related proof, *OSIM* (Oh Shoot, It's Monday):

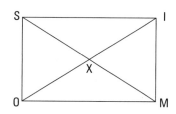

Given: X is the midpoint of $\overline{MS}$ and $\overline{OI}$

$\overline{SX} \cong \overline{IX}$

Prove: $\overline{MX} \cong \overline{OX}$

Part II: Introducing Proofs

This is another incredibly short proof that doesn't call for a game plan.

Statements	Reasons
1) X is the midpoint of $\overline{MS}$ and $\overline{OI}$	1) Given.
2) $\overline{SX} \cong \overline{MX}$ $\overline{IX} \cong \overline{OX}$	2) A midpoint divides a segment into two congruent segments.
3) $\overline{SX} \cong \overline{IX}$	3) Given.
4) $\overline{MX} \cong \overline{OX}$	4) Transitive Property (for four segments; statements 2 and 3).

Chapter 6
The Ultimate Guide to Tackling a Longer Proof

In This Chapter
- Making a game plan
- Starting at the start, working from the end, and meeting in the middle
- Making sure your logic holds

Chapters 4 and 5 start you off with short proofs and a couple dozen basic theorems. Here, I go through a single, longer proof in great detail, carefully analyzing each step. Throughout the chapter, I walk you through the entire thought process that goes into solving a proof, reviewing and expanding on the half dozen or so proof strategies from Chapters 4 and 5. When you're working on a proof and you get stuck, this chapter is a good one to come back to for tips on how to get moving again.

The proof I've created for this chapter isn't so terribly gnarly; it's just a bit longer than the ones in Chapter 5. Here it is:

Given: $\overline{BD} \perp \overline{DE}$
$\overline{BF} \perp \overline{FE}$
$\angle 1 \cong \angle 2$
$\angle 5$ is complementary to $\angle 3$
$\angle 6$ is complementary to $\angle 4$

Prove: $\overrightarrow{BX}$ bisects $\angle ABC$

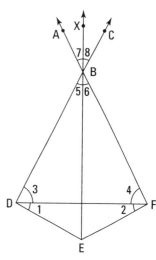

Making a Game Plan

A good way to begin any proof is to make a *game plan,* or rough outline, of how you'd do the proof. The formal way of writing out a two-column proof can be difficult, especially at first — almost like learning a foreign language. Writing a proof is easier if you break it into two shorter, more-manageable pieces.

First, you jot down or simply think through a game plan, in which you go through the logic of the proof with your common sense without being burdened by getting the technical language right. Then the second step of translating that logic into the two-column format isn't so hard.

As you see in Chapter 5, when you're working through a game plan, it's sometimes a good idea to make up arbitrary numbers for the segments and angles in the givens and for unmentioned segments and angles. You should not, however, make up numbers for segments and angles that you're trying to show are congruent. This optional step makes the proof diagram more concrete and makes it easier for you to get a handle on how the proof works.

Here's one possible game plan for the proof we're working on: The givens provide you with two pairs of perpendicular segments; that gives you 90° for $\angle BDE$ and $\angle BFE$. Then, say congruent angles $\angle 1$ and $\angle 2$ are both 30°. That would make $\angle 3$ and $\angle 4$ both equal to 90° – 30°, or 60°. Next, because $\angle 3$ and $\angle 5$ are complementary, as are $\angle 4$ and $\angle 6$, $\angle 5$ and $\angle 6$ would both be 30°. Angles 5 and 8 are congruent vertical angles, as are $\angle 6$ and $\angle 7$, so $\angle 7$ and $\angle 8$ would also have to be 30° — and thus they're congruent. Finally, because $\angle 7 \cong \angle 8$, $\angle ABC$ is bisected. That does it.

If you have trouble keeping track of the chain of logic as you work through a game plan, you might want to put marks on the proof diagram as you go through each logical step. For example, whenever you deduce that a pair of segments or angles is congruent, you could show that by putting tick marks on the diagram. Marking the diagram gives you a quick visual way to keep track of your reasoning.

When doing a proof, thinking through a rough sketch of the proof argument like the preceding game plan is always a good idea. However, there'll likely be occasions when you can't figure out the entire argument right away. If this happens to you, you can use the strategies presented in the rest of this chapter to help you think through the proof. The upcoming sections also provide some tips that can help you turn a bare-bones game plan into a fleshed-out two-column proof.

Using All the Givens

Perhaps you don't follow the game plan in the previous section — or you get it but don't think you would've been able to come up with it on your own in one shot — and so you're staring at the proof and just don't know where to begin. My advice: Check all the givens in the proof and ask yourself *why* they'd tell you each given.

Every given is a built-in hint.

Look at the five givens in this proof (see the chapter intro). It's not immediately clear how the third, fourth, and fifth givens can help you, but what about the first two about the perpendicular segments? Why would they tell you this? What do perpendicular lines give you? Right angles, of course. Okay, so you're on your way — you know the first two lines of the proof (see Figure 6-1).

Statements	Reasons
1) $\overline{BD} \perp \overline{DE}$ $\overline{BF} \perp \overline{FE}$	1) Given.
2) $\angle BDE$ is a right angle $\angle BFE$ is a right angle	2) If segments are perpendicular, then they form right angles.

Figure 6-1: The first two lines of the proof.

Note that the second reason just about writes itself if you remember how the if-then structure of reasons works (see the next section and Chapter 4 for more on if-then logic).

Making Sure You Use If-Then Logic

Moving from the givens to the final conclusion in a two-column proof is like knocking over a row of dominoes: Just as each domino knocks over the next domino, each proof statement leads to the next statement. The if-then sentence structure of each reason in a two-column proof shows you how each statement "knocks over" the next statement. In Figure 6-1, for example, you see the reason "*if* two segments are perpendicular, *then* they form a right angle." The perpendicular domino (statement 1) knocks over the right-angle domino (statement 2). This process continues throughout the whole proof.

Focusing on the if-then logic of a proof helps you see how the whole proof fits together.

Make sure that the if-then structure of your reasons is correct (I cover if-then logic in more depth in Chapter 4):

- The idea or ideas in the *if* clause of a reason must appear in the statement column somewhere *above* the line of that reason.
- The single idea in the *then* clause of a reason must be the same idea that's in the statement *directly across from* the reason.

Look back at Figure 6-1. Because statement 1 is the only statement above reason 2, it's the only place you can look for the ideas that go in the *if* clause of reason 2. So if you begin this proof by putting the two pairs of perpendicular segments in statement 1, then you have to use that information in reason 2, which must therefore begin "if segments are perpendicular, then . . ."

Now say you didn't know what to put in statement 2. The if-then structure of reason 2 helps you out. Because reason 2 begins "if two segments are perpendicular . . ." you'd ask yourself, "Well, what happens when two segments are perpendicular?" The answer, of course, is that right angles are formed. The right-angle idea must therefore go in the *then* clause of reason 2 and right across from it in statement 2.

Okay, now what? Well, think about reason 3. One way it could begin is with the right angles from statement 2. The *if* clause of reason 3 might be "if two angles are right angles . . ." Can you finish that? Of course: If two angles are right angles, then they're congruent. So that's it: You've got reason 3, and statement 3 must contain the idea from the *then* clause of reason 3, the congruence of right angles. Figure 6-2 shows you the proof so far.

Figure 6-2: The first three lines of the proof.

Statements	Reasons
1) $\overline{BD} \perp \overline{DE}$ $\overline{BF} \perp \overline{FE}$	1) Given.
2) $\angle BDE$ is a right angle $\angle BFE$ is a right angle	2) If segments are perpendicular, then they form right angles.
3) $\angle BDE \cong \angle BFE$	3) If two angles are right angles, then they're congruent.

Chapter 6: The Ultimate Guide to Tackling a Longer Proof

When writing proofs, you need to spell out every little step as if you had to make the logic clear to a computer. For example, it may seem obvious that if you have two pairs of perpendicular segments, you've got congruent right angles, but this simple deduction takes three steps in a two-column proof. You have to go from perpendicular segments to right angles and then to congruent right angles — you can't jump straight to the congruent right angles. That's the way computers "think": A leads to B, B leads to C, and so on.

Chipping Away at the Problem

Face it: You're going to get stuck at one point or another, or heaven forbid, at several points in one proof! Wondering what you should do when you get stuck?

Try something. When doing geometry proofs, you need to be willing to experiment with ideas using trial and error. Doing proofs isn't as black and white as the math you've done before. You often can't know for sure what'll work. Just try something, and if it doesn't work, try something else. Sooner or later, the whole proof should fall into place.

So far in the proof in this chapter, you have the two congruent angles in statement 3, but you can't make more progress with that idea alone. So check out the givens again. Which of the three unused givens might build on statement 3? There's no way to answer that with certainty, so you need to trust your instincts, pick a given, and try it (or if you're thinking you don't have instincts for this, then just try something, anything).

The third given says $\angle 1 \cong \angle 2$. That looks promising because angles 1 and 2 are part of the right angles from statement 3. You should ask yourself, "What would follow if $\angle 1$ and $\angle 2$ were, say, 35°?" You know the right angles are 90°, so if $\angle 1$ and $\angle 2$ were 35°, then $\angle 3$ and $\angle 4$ would both have to be 55° and thus, obviously, they'd be congruent. That's it. You're making progress. You can use that third given in statement 4 and then state that $\angle 3 \cong \angle 4$ in statement 5.

Figure 6-3 shows the proof up to statement 5. The bubbles and arrows show you how the statements and reasons connect to each other. You can see that the *if* clause of each reason connects to a statement from above the reason and that the *then* clause connects to the statement on the same line as the reason. Because I haven't gone over reason 5 yet, it's not in the figure. See whether you can figure out reason 5 before reading the explanation that follows. **Hint:** The *then* clause for reason 5 must connect to statement 5 as shown in the figure.

Part II: Introducing Proofs

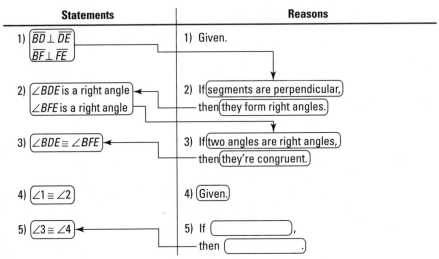

Figure 6-3: The first five lines of the proof (minus reason 5).

So, did you figure out reason 5? It's angle *subtraction* because — using 35° for ∠1 and ∠2 — ∠3 and ∠4 in statement 5 ended up being 55° angles, and you get the answer of 55° by doing a subtraction problem, 90° – 35° = 55°. (Don't make the mistake of thinking that this is angle *addition* because 35° + 55° = 90°.) You're subtracting two angles from two other angles, so you use the four-angle version of angle subtraction (see Chapter 5). Reason 5 is, therefore, "If two congruent angles (∠1 and ∠2) are subtracted from two other congruent angles (the right angles), then the differences (∠3 and ∠4) are congruent."

At this stage, you may feel a bit (or more than a bit) disconcerted if you don't know where these five lines are taking you or whether they're correct or not. "What good is it," you might ask, "to get five lines done when I don't know where I'm going?" That's an understandable reaction and question. Here's the answer.

If you're in the middle of solving a proof and can't see how to get to the end, remember that taking steps is a good thing. If you're able to deduce more and more facts and can begin filling in the statement column, you're very likely on the right path. Don't worry about the possibility that you're going the wrong way. (Although such detours do happen from time to time, don't sweat it. If you hit a dead end, just go back and try a different tack.)

Don't feel like you've got to score a touchdown (that is, see how the whole proof fits together). Instead, be content with just making a first down (getting one more statement), then another first down, then another, and so on.

Sooner or later, you'll make it into the end zone. I once heard about a student who went from getting Ds and Fs in geometry to As and Bs by merely changing his focus from scoring touchdowns to just making yardage.

Jumping Ahead and Working Backward

Assume that you're in the middle of a proof and you can't see how to get to the finish line from where you are now. No worries — just jump to the end of the proof and *work backward*.

Okay, so picking up where I left off on this chapter's proof: You've completed five lines of the proof, and you're up to $\angle 3 \cong \angle 4$. Where to now? Going forward from here might be a bit tricky, so work backward. You know that the final line of the proof has to be the *prove* statement: $\overrightarrow{BX}$ bisects $\angle ABC$. Now, if you think about what the final reason has to be or what the second-to-last statement should be, it shouldn't be hard to see that you need to have two congruent angles (the two half-angles) to conclude that a larger angle is bisected. Figure 6-4 shows you what the end of the proof looks like. Note the if-then logic bubbles (*if* clauses in reasons connect to statements above; *then* clauses in reasons connect to statements on the same line).

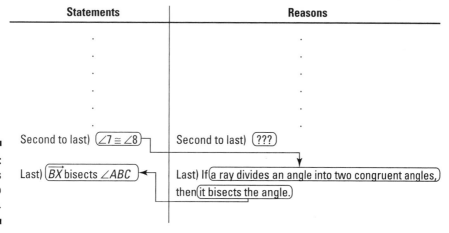

Figure 6-4: The proof's last two lines.

Try to continue going backward to the third-to-last statement, the fourth-to-last statement, and so on. (Working backward through a proof always involves some guesswork, but don't let that stop you.) Why might $\angle 7$ be congruent to $\angle 8$? Well, you probably don't have to look too hard to spot the pair of congruent vertical angles $\angle 5$ and $\angle 8$ and the other pair, $\angle 6$ and $\angle 7$.

Okay, so you want to show that ∠7 is congruent to ∠8, and you know that ∠6 equals ∠7 and ∠5 equals ∠8. So if you were to know that ∠5 and ∠6 are congruent, you'd be home free.

Now that you've worked backward a number of steps, here's the argument in the forward direction: The proof could end by stating in the fourth-to-last statement that ∠5 ≅ ∠6, then in the third-to-last that ∠5 ≅ ∠8 and ∠6 ≅ ∠7 (because vertical angles are congruent), and then in the second-to-last that ∠7 ≅ ∠8 by the Transitive Property (for four angles — see Chapter 5). Figure 6-5 shows how this all looks written out in the two-column format.

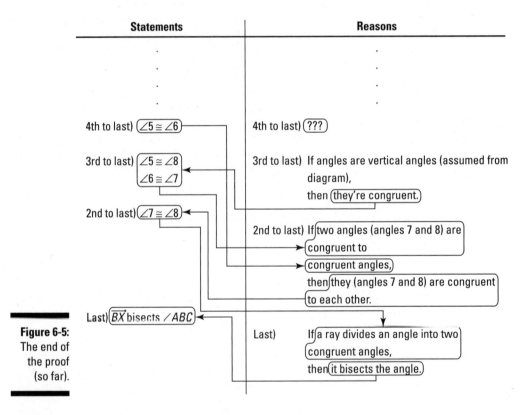

Figure 6-5: The end of the proof (so far).

Filling In the Gaps

As I explain in the preceding section, working backward from the end of a proof is a great strategy. You can't always work as far backward as I did in this proof — sometimes you can only get to the second-to-last statement or maybe to the third-to-last. But even if you fill in only one or two statements (in addition to the automatic final statement), those additions can be very helpful. After making the additions, the proof is easier to finish because your new "final" destination (say the third-to-last statement) is fewer steps away from the beginning of the proof and is thus an easier goal to aim for. It's kind of like solving one of those mazes you see in a magazine or newspaper: You can work from the Start; then, if you get stuck, you can work from the Finish. Finally, you can go back to where you left off on the path from the Start and simply connect the ends. Figure 6-6 shows the process.

Figure 6-6: Work from both ends, and then bridge the gap.

1. Work from Start. 2. Work from Finish. 3. Bridge the gap.

(This task is easier than the original problem because you no longer have to "see" all the way to the Finish point.)

Okay, what do you say I wrap up this proof? All that remains to be done is to bridge the gap between statement 5 ($\angle 3 \cong \angle 4$), and the fourth-to-last statement ($\angle 5 \cong \angle 6$). There are two givens you haven't used yet, so they must be the key to finishing the proof.

How can you use the givens about the two pairs of complementary angles? You might want to try the plugging-in-numbers idea again. Use the same numbers as before (from "Making a Game Plan") and say that congruent angles $\angle 3$ and $\angle 4$ are each 55°. Angle 5 is complementary to $\angle 3$, so if $\angle 3$ were 55°, $\angle 5$ would have to be 35°. Angle 6 is complementary to $\angle 4$, so $\angle 6$ also ends up being 35°. That does it — $\angle 5$ and $\angle 6$ are congruent, and you've connected the loose ends. All that's left is to finish writing out the formal proof, which I do in the next section.

By the way, using angle sizes like this is a great strategy, but it's often unnecessary. If you know your theorems well, you might simply realize that because $\angle 3$ and $\angle 4$ are congruent, their complements ($\angle 5$ and $\angle 6$) must also be congruent.

Writing Out the Finished Proof

Sound the trumpets! Here's the finished proof complete with the flow-of-logic bubbles (see Figure 6-7). (This time, I've put in only the arrows that connect to the *if* clause of each reason. You know that each reason's *then* clause must connect to the statement on the same line.) If you understand all the strategies and tips covered in this chapter and you can follow every step of this proof, you should be able to handle just about any proof they throw at you.

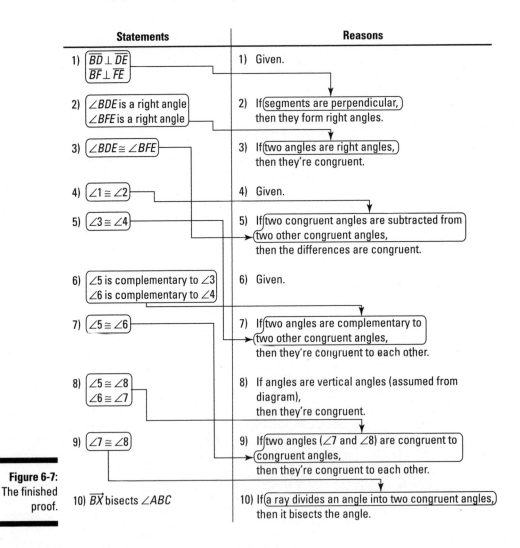

Figure 6-7: The finished proof.

Part III
Triangles: Polygons of the Three-Sided Variety

In this part . . .

Part III is all about triangles. You might think that the triangle is a pretty simple shape, but there's much more to triangles than meets the eye (I'll let you decide whether that's good news or bad news). Chapters 7 and 8 cover triangle basics such as acute, obtuse, and right triangles; scalene, isosceles, and equilateral triangles; and how to compute the area of a triangle. But get a load of these wild and crazy advanced concepts you'll also see: Pythagorean triples, 30°- 60°- 90° triangles, orthocenters, incenters, circumcenters, and centroids.

Then, in Chapter 9, you get back to proofs. Triangle proofs are the meat and potatoes of high school geometry, the warp and woof, the be-all and end-all, the alpha and omega, the Fred and Ginger . . . well, you get my drift.

Chapter 7

Grasping Triangle Fundamentals

In This Chapter

▶ Looking at a triangle's sides: Equal or unequal
▶ Uncovering the triangle inequality principle
▶ Classifying triangles by their angles
▶ Calculating the area of a triangle
▶ Finding the four "centers" of a triangle

Considering that it's the runt of the polygon family, the triangle sure does play a big role in geometry. Triangles are one of the most important components of geometry proofs (you see triangle proofs in Chapter 9). They also have a great number of interesting properties that you might not expect from the simplest possible polygon. Maybe Leonardo da Vinci (1452–1519) was on to something when he said, "Simplicity is the ultimate sophistication."

In this chapter, I take you through the triangle basics — their names, sides, angles, and area. I also show you how to find the four "centers" of a triangle.

Taking In a Triangle's Sides

Triangles are classified according to the length of their sides or the measure of their angles. These classifications come in threes, just like the sides and angles themselves. That is, a triangle has three sides, and three terms describe triangles based on their sides; a triangle also has three angles, and three classifications of triangles are based on their angles. I talk about classifications based on angles in the upcoming section "Getting to Know Triangles by Their Angles."

The following are triangle classifications based on sides:

  ✓ **Scalene triangle:** A triangle with no congruent sides
  ✓ **Isosceles triangle:** A triangle with at least two congruent sides
  ✓ **Equilateral triangle:** A triangle with three congruent sides

Because an equilateral triangle is also isosceles, all triangles are either scalene or isosceles. But when people call a triangle *isosceles,* they're usually referring to a triangle with only two equal sides, because if the triangle had three equal sides, they'd call it *equilateral.* So is this three types of triangles or only two? You be the judge.

Scalene triangles: Akilter, awry, and askew

In addition to having three unequal sides, scalene triangles have three unequal angles. The shortest side is across from the smallest angle, the medium side is across from the medium angle, and — surprise, surprise — the longest side is across from the largest angle. Figure 7-1 shows you what I mean.

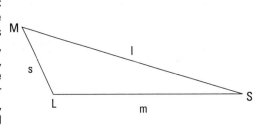

Figure 7-1: The Goldilocks rule: Small, medium, and large sides mirror small, medium, and large angles.

Scalene triangles aplenty

This fact may surprise you: In contrast to what you see in geometry books, which are loaded with isosceles and equilateral triangles (and right triangles), 99.999... percent of triangles in the mathematical universe are non-right-angle, scalene triangles. All the special triangles (isosceles, equilateral, and right triangles) are sort of like infinitesimal needles in the haystack of all triangles. So if you were to pick a triangle at random from all possible triangles, the probability of picking an isosceles triangle, an equilateral triangle, or a right triangle is pretty much a big, fat 0 percent! This is surprising to most people because special triangles (isosceles, right, and equilateral) do pop up all over the place in the real world (in buildings, everyday products, and so on) — and that's why we study them.

The ratio of sides doesn't equal the ratio of angles. Don't assume that if one side of a triangle is, say, twice as long as another side that the angles opposite those sides are also in a 2 : 1 ratio. The ratio of the sides may be close to the ratio of the angles, but these ratios are *never* exactly equal (except when the sides are equal).

If you're trying to figure out something about triangles — such as whether an angle bisector also bisects (cuts in half) the opposite side — you can sketch a triangle and see whether it looks true. But the triangle you sketch should be a non-right-angle, scalene triangle (as opposed to an isosceles, equilateral, or right triangle). This is because scalene triangles, by definition, lack special properties such as congruent sides or right angles. If you sketch, say, an isosceles triangle instead, any conclusion you reach may be true only for triangles of this special type. In general, in any area of mathematics, when you want to investigate some idea, you shouldn't make things more special than they have to be.

Isosceles triangles: Nice pair o' legs

An isosceles triangle has two equal sides and two equal angles. The equal sides are called *legs*, and the third side is the *base*. The two angles touching the base (which are congruent, or equal) are called *base angles*. The angle between the two legs is called the *vertex angle*. See Figure 7-2.

Figure 7-2:
Two run-of-the-mill isosceles triangles.

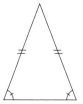 

Equilateral triangles: All parts are created equal

An equilateral triangle has three equal sides and three equal angles (which are each 60°). Its equal angles make it *equiangular* as well as equilateral. You don't often hear the expression *equiangular triangle*, however, because the only triangle that's equiangular is the equilateral triangle, and everyone calls this triangle *equilateral*. (With quadrilaterals and other polygons, however, you need both terms, because an equiangular figure, such as a rectangle, can have sides of different lengths, and an equilateral figure, such as a rhombus, can have angles of different sizes. See Chapter 12 for details.)

If you cut an equilateral triangle in half right down the middle, you get two 30°- 60°- 90° triangles. You see the incredibly important 30°- 60°- 90° triangle in the next chapter.

Introducing the Triangle Inequality Principle

The triangle inequality principle: The sum of the lengths of any two sides of a triangle must be greater than the length of the third side. This principle comes up in a fair number of problems, so don't forget it! It's based on the simple fact that the shortest distance between two points is a straight line. Check out Figure 7-3 and the explanation that follows to see what I mean.

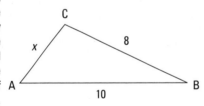

Figure 7-3: The triangle inequality principle lets you find the possible lengths of side $\overline{AC}$.

In △ABC, what's the shortest route from A to B? Naturally, going straight across from A to B is shorter than taking a detour by traveling from A to C and then on to B. That's the triangle inequality principle in a nutshell.

In △ABC, because you know that AB must be less than AC plus CB, $x + 8$ must be greater than 10; therefore,

$x + 8 > 10$

$x > 2$

But don't forget that the same principle applies to the path from A to C; thus, $8 + 10$ must be greater than x:

$8 + 10 > x$

$18 > x$

You can write both of these answers as a single inequality:

2 < x < 18

These are the possible lengths of side $\overline{AC}$. Figure 7-4 shows this range of lengths. Think of vertex B as a hinge. As the hinge opens more and more, the length of $\overline{AC}$ grows.

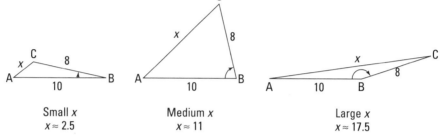

Figure 7-4: Triangle ABC changes as side $\overline{AC}$ grows.

Small x
x ≈ 2.5

Medium x
x ≈ 11

Large x
x ≈ 17.5

Note: Give yourself a pat on the back if you're wondering why I didn't mention the third path, from B to C. Here's why: In the first two inequalities, note that I put the longer known side (the 10) and the unknown side (the x) on the right sides of the inequalities. That's all you need to do to get your answer. You don't have to do a third inequality with the shorter of the two known sides (the 8) on the right side of the inequality, because that won't add anything to your answer — you'd simply find that x has to be greater than –2, and side lengths have to be positive, anyway.

By the way, if this problem had been about three towns A, B, and C instead of △ABC, then the possible distances between towns A and C would look the same except that the less-than symbols would be less-than-or-equal-to symbols:

2 ≤ x ≤ 18

This is because — unlike vertices A, B, and C of △ABC — towns A, B, and C can lie in a straight line. Look again at Figure 7-4. If ∠B goes down to 0°, the towns would be in a line, and the distance from A to C would be exactly 2; if ∠B opens all the way to 180°, the towns would again be in a line, and the distance from A to C would be exactly 18. You can't do this with the triangle problem, however, because when A, B, and C are in a line, there's no triangle left.

Getting to Know Triangles by Their Angles

As I mention in the earlier section titled "Taking In a Triangle's Sides," you can classify triangles by their angles as well as by their sides. Classifications by angles are as follows:

- **Acute triangle:** A triangle with three acute angles (less than 90°).
- **Obtuse triangle:** A triangle with one obtuse angle (greater than 90°). The other two angles are acute. If a triangle were to have two obtuse angles (or three), two of its sides would go out in opposite directions and never come together to form a triangle.
- **Right triangle:** A triangle with a single right angle (90°) and two acute angles. The *legs* of a right triangle are the sides touching the right angle, and the *hypotenuse* is the side across from the right angle. I devote Chapter 8 to right triangles.

The angles of a triangle add up to 180°. That's another reason why if one of the angles of a triangle is 90° or larger, the other two angles have to be acute.

I show you one example of this 180° total in the section titled "Equilateral triangles." The angles of an equilateral triangle are 60°, 60°, and 60°. In Chapter 8, you see two other important examples: the 30°- 60°- 90° triangle and the 45°- 45°- 90° triangle.

Sizing Up Triangle Area

In this section, I run through everything you need to know to determine a triangle's area (as you probably know, *area* is the amount of space inside a figure). I show you what an altitude is and how you use it in the standard triangle area formula. I also let you in on a shortcut that you can use when you know all three sides of a triangle and want to find the area directly, without bothering to calculate the length of an altitude.

Scaling altitudes

Altitude (of a triangle): A segment from a vertex of a triangle to the opposite side (or to the extension of the opposite side if necessary) that's perpendicular to the opposite side; the opposite side is called the *base*. (You use the definition of altitude in some triangle proofs. See Chapter 9.)

Imagine that you have a cardboard triangle standing straight up on a table. The altitude of the triangle tells you exactly what you'd expect — the triangle's height *(h)* measured from its peak straight down to the table. This height goes down to the base of the triangle that's flat on the table. Figure 7-5 shows you an example of an altitude.

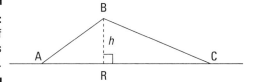

Figure 7-5: $\overline{BR}$ is one of the altitudes of △*ABC*.

Every triangle has three altitudes, one for each side. Figure 7-6 shows the same triangle from Figure 7-5 standing up on a table in the other two possible positions: with $\overline{CB}$ as the base and with $\overline{BA}$ as the base.

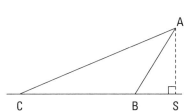 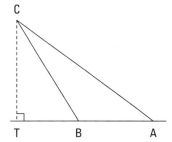

Figure 7-6: $\overline{AS}$ and $\overline{CT}$ are the other two altitudes of △*ABC*.

Every triangle has three altitudes whether or not the triangle is standing up on a table. And you can use any side of a triangle as a base, regardless of whether that side is on the bottom. Figure 7-7 shows △*ABC* again with all three of its altitudes.

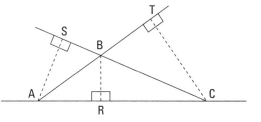

Figure 7-7: Triangle *ABC* with its three altitudes: $\overline{BR}$, $\overline{AS}$, and $\overline{CT}$.

The following points tell you about the length and location of the altitudes of the different types of triangles (see "Taking In a Triangle's Sides" and "Getting to Know Triangles by Their Angles" for more on naming triangles):

- **Scalene:** None of the altitudes has the same length.
- **Isosceles:** Two altitudes have the same length.
- **Equilateral:** All three altitudes have the same length.
- **Acute:** All three altitudes are inside the triangle.
- **Right:** One altitude is inside the triangle, and the other two altitudes are the legs of the triangle (remember this when figuring the area of a right triangle).
- **Obtuse:** One altitude is inside the triangle, and two altitudes are outside the triangle.

Determining a triangle's area

In this section, I give you three methods for calculating a triangle's area: the well-known standard formula, a little-known but very useful 2,000-year-old fancy-pants formula, and the formula for the area of an equilateral triangle.

Tried and true: The triangle area formula everyone knows

Triangle area formula: You likely first ran into the basic triangle area formula in about sixth or seventh grade. If you've forgotten it, no worries — I have it right here:

$$\text{Area}_\triangle = \frac{1}{2} \, base \cdot height$$

Assume for the sake of argument that you have trouble remembering this formula. Well, you won't forget it if you focus on why it's true — which brings me to one of the most important tips in this book.

Whenever possible, don't just memorize math concepts, formulas, and so on by rote. Try to understand *why* they're true. When you grasp the *whys* underlying the ideas, you remember them better and develop a deeper appreciation of the interconnections among mathematical ideas. That appreciation makes you a more successful math student.

So why does the area of a triangle equal $\frac{1}{2} \, base \cdot height$? Because the area of a rectangle is *base · height* (which is the same thing as *length · width*), and a triangle is half of a rectangle.

Check out Figure 7-8, which shows two triangles inscribed in rectangles *HALF* and *PINT*.

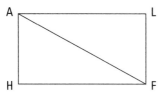
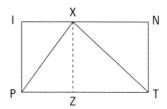

Figure 7-8: A triangle takes up half the area of a rectangle.

It should be really obvious that △*HAF* has half the area of rectangle *HALF*. And it shouldn't exactly give you a brain hemorrhage to see that △*PXT* also has half the area of the rectangle around it. (Triangle *PXZ* is half of rectangle *PIXZ*, and △*ZXT* is half of rectangle *ZXNT*.) Because every possible triangle (including △*HAF*, by the way, if you use $\overline{AF}$ for its base) fits in some rectangle just like △*PXT* fits in rectangle *PINT*, every triangle is half a rectangle.

Now for a problem that involves finding the area of a triangle: What's the length of altitude $\overline{XT}$ in △*WXR* in Figure 7-9?

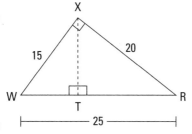

Figure 7-9: Right triangle WXR with its three altitudes.

The trick here is to note that because △*WXR* is a right triangle, legs $\overline{WX}$ and $\overline{RX}$ are also altitudes. So you can use either one as the altitude, and then the other leg automatically becomes the base. Plug their lengths into the formula to determine the triangle's area:

$$\text{Area}_{\triangle WXR} = \frac{1}{2} \text{ base} \cdot \text{height}$$
$$= \frac{1}{2}(RX)(WX)$$
$$= \frac{1}{2}(20)(15)$$
$$= 150$$

Now you can use the area formula again, using this area of 150, base $\overline{WR}$, and altitude $\overline{XT}$:

$$\text{Area}_{\triangle WXR} = \frac{1}{2} \text{base} \cdot \text{height}$$
$$150 = \frac{1}{2}(WR)(XT)$$
$$150 = \frac{1}{2}(25)(XT)$$
$$12 = XT$$

Bingo.

A Heroic trick: The area formula almost no one knows

Hero's Formula: When you know the length of a triangle's three sides and you don't know an altitude, Hero's formula works like a charm. Check it out:

$$\text{Area}_\triangle = \sqrt{S(S-a)(S-b)(S-c)},$$

where a, b, and c are the lengths of the triangle's sides and S is the triangle's *semiperimeter* (that's half the perimeter: $S = \frac{a+b+c}{2}$).

Let's use Hero's formula to calculate the area of a triangle with sides of length 5, 6, and 7. First, you need the triangle's perimeter (the sum of the lengths of its sides), and from that you get the semiperimeter. The perimeter is $5 + 6 + 7 = 18$, so you get 9 for the semiperimeter. Now just plug 9, 5, 6, and 7 into the formula:

$$\text{Area}_\triangle = \sqrt{9(9-5)(9-6)(9-7)}$$
$$= \sqrt{9 \cdot 4 \cdot 3 \cdot 2}$$
$$= \sqrt{36 \cdot 6}$$
$$= 6\sqrt{6}, \text{ or about } 14.7$$

Third time's the charm: The area of an equilateral triangle

You can get by without the formula for the area of an equilateral triangle because you can use the length of a side to calculate the altitude and then use the regular area formula (see the discussion of 30°- 60°- 90° triangles in Chapter 8). But this formula is nice to know because it gives you the answer in one fell swoop.

Area of an equilateral triangle (with side s):

$$\text{Area}_{\text{Equilateral}\triangle} = \frac{s^2\sqrt{3}}{4}$$

You get a chance to see this formula in action in Chapters 12 and 14.

Now you see it, now you don't

Put your thinking cap on — here's a tough brainteaser for you. The first figure here is made up of four pieces. Below it, the same four pieces have been rearranged. But mysteriously, you get that extra little white square of area. How can the identical four pieces from the first figure not completely fill up the second figure? (Stop reading here if you want to work out the solution on your own.)

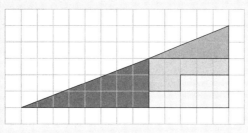

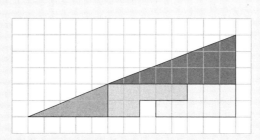

Few people can solve this problem without any hints. (If you did, I'm very impressed. If not, read on for some hints and give it another try before reading the entire solution.) Take a look at the four pieces: the two triangles and those two sort of L-shaped pieces. The black triangle has a base of 8 and a height of 3, so its area is $\frac{1}{2} \cdot 8 \cdot 3$, or 12. The area of the dark gray triangle is $\frac{1}{2} \cdot 5 \cdot 2$, or 5. The two L-shaped pieces have areas of 7 and 8. That gives you a total of 32. But both big triangles have bases of 13 and heights of 5, so their areas are 32.5. What gives?

Here's what gives: The two big "triangles" aren't triangles at all — they're quadrilaterals! Take the book in your hands, making sure this page is totally flat, close one eye, and turn the book so you can look along the "hypotenuse" of the first "triangle" — you know, so that the "hypotenuse" is sort of pointing right into your eye. If you look carefully, you should see that this "hypotenuse" has a very slight downward bend in it. This little depression explains why the four pieces add up to only 32. If not for this indentation (if the hypotenuse were straight), the total of the four pieces would have to be 32.5 — the area of a triangle with a base of 13 and height of 5. (If you can't see the bend, try taking a ruler and lining it up with the two ends of the hypotenuse. If you do this very carefully, you should be able to see the extremely slight downward bend.)

The "hypotenuse" of the second "triangle" bends slightly upward. This upward bend creates the little extra room needed for the four pieces that total 32 plus the empty square that has an area of 1. This grand total of 33 square units fits in the bulging-out triangle that would have an area of 32.5 without the bulge. Pretty sneaky, eh?

102 Part III: Triangles: Polygons of the Three-Sided Variety

Locating the "Centers" of a Triangle

In this section, you look at four points associated with every triangle. One of these points is called the *centroid,* and the other three are called *centers,* but none of them is the "real" center of a triangle. Unlike circles, squares, and rectangles, triangles (except for equilateral triangles) don't really have a true center.

Balancing on the centroid

Before I discuss centroids, you need the definition of another triangle term.

Median: A median of a triangle is a segment that goes from one of the triangle's vertices to the midpoint of the opposite side. Every triangle has three medians. (You use the definition of median in some triangle proofs. See Chapter 9.)

Centroid: The three medians of a triangle intersect at its centroid. The centroid is the triangle's balance point, or center of gravity.

On each median, the distance from the vertex to the centroid is twice as long as the distance from the centroid to the midpoint. Take a look at Figure 7-10.

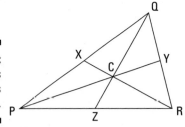

Figure 7-10: Point C is $\triangle PQR$'s centroid.

X, Y, and Z are the midpoints of the sides of $\triangle PQR$; $\overline{RX}$, $\overline{PY}$, and $\overline{QZ}$ are the medians; and the medians intersect at point C, the centroid. If you take out a ruler (or just use your fingers), you can verify that the centroid is at the $\frac{1}{3}$ mark along, say, median $\overline{PY}$ — in other words, $\overline{CY}$ is $\frac{1}{3}$ as long as $\overline{PY}$ (and therefore, $\overline{CY}$ is half as long as $\overline{CP}$).

If you're from Missouri (the Show-Me State), you might want to actually see how a triangle balances on its centroid. Cut a triangle of any shape out of a fairly stiff piece of cardboard. Carefully find the midpoints of two of the sides, and then draw the two medians to those midpoints. The centroid is where these medians cross. (You can draw in the third median if you like, but you don't need it to find the centroid.) Now, using something with a small, flat top

An infinite series of triangles

Check this out (this just dawned on me as I was writing this chapter): The triangle here is $\triangle PQR$ from Figure 7-10 with a new triangle added, $\triangle XYZ$. To get the new triangle, I simply connected the midpoints of the three sides.

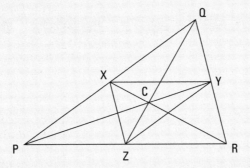

It turns out that $\triangle XYZ$ is the same shape as $\triangle PQR$ and has exactly $\frac{1}{4}$ its area. And C, the centroid of $\triangle PQR$, is also the centroid of $\triangle XYZ$.

Now look at the same triangle again but with two more triangles added:

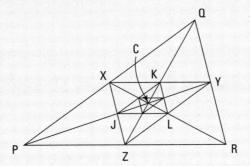

The next triangle, $\triangle JKL$, works the same way: it's the same shape as $\triangle XYZ$ takes up $\frac{1}{4}$ of $\triangle XYZ$'s area, and C is its centroid.

This pattern continues indefinitely. You end up with an infinite number of nested triangles, all of the same shape and all having C as their centroid. Finally, despite the fact that you have an infinite number of triangles, their total area is not infinite; the total area of all the triangles is a mere $1\frac{1}{3}$ times the area of $\triangle PQR$. Thus, if the area of $\triangle PQR$ is 3, the total area of the infinite series of triangles is only 4. (In case you're wondering, finding that answer involves calculus. Something to look forward to, right? Check out *Calculus For Dummies* if you just can't wait.)

such as an unsharpened pencil, the triangle will balance if you place the centroid right in the center of the pencil's tip.

A triangle's centroid is probably as good a point as any to give you a rough idea of where its center is. The centroid is definitely a better candidate for a triangle's center than the three "centers" I discuss in the next section.

Finding three more "centers" of a triangle

In addition to a centroid, every triangle has three more "centers" that are located at the intersection of rays, lines, and segments associated with the triangle:

- **Incenter:** Where a triangle's three angle bisectors intersect (an *angle bisector* is a ray that cuts an angle in half); the incenter is the center of a circle *inscribed* in (drawn inside) the triangle.

- **Circumcenter:** Where the three perpendicular bisectors of the sides of a triangle intersect (a *perpendicular bisector* is a line that forms a 90° angle with a segment and cuts the segment in half); the circumcenter is the center of a circle *circumscribed* about (drawn around) the triangle.

- **Orthocenter:** Where the triangle's three altitudes intersect (see the earlier "Scaling altitudes" section for more on altitudes).

Keeping the centers straight

Each of the following four "centers" is paired up with the lines, rays, or segments that intersect at that center:

- Centroid — Medians
- Circumcenter — Perpendicular bisectors
- Incenter — Angle bisectors
- Orthocenter — Altitudes

The two "centers" that begin with a consonant pair up with terms that also begin with a consonant. And ditto for the two "centers" that begin with a vowel. Sweet, eh? Also, "centroid" and "medians" are the only two words containing a double vowel (*oi* and *ia*); and "orthocenter" and "altitudes" are the only two terms with two *t*'s. This mnemonic may be a bit lame, but it's better than nothing. If you can come up with a better one, use it! With this incredibly important information at your disposal, you'll have something to talk about if you come to an awkward silence while out on a date. (And speaking of dates, another way to remember the list is to think about going out to the movies. When you alphabetize the four centers on the left, the initial letters of the terms on the right form the acronym for the Motion Picture Association of America.)

Investigating the incenter

You find a triangle's *incenter* at the intersection of the triangle's three angle bisectors. This location gives the incenter an interesting property: The incenter is equally far away from the triangle's three sides. No other point has this quality. Incenters, like centroids, are always inside their triangles.

Figure 7-11 shows two triangles with their incenters and *inscribed circles,* or *incircles* (circles drawn inside the triangles so the circles barely touch the sides of each triangle). The incenters are the centers of the incircles. (Don't talk about this stuff too much if you want to be in with the in-crowd.)

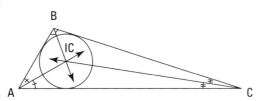
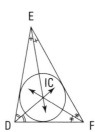

Figure 7-11: Two triangles with their incenters.

Spotting the circumcenter

You find a triangle's *circumcenter* at the intersection of the perpendicular bisectors of the triangle's sides. This location gives the circumcenter an interesting property: The circumcenter is equally far away from the triangle's three vertices.

Figure 7-12 shows two triangles with their circumcenters and *circumscribed circles,* or *circumcircles* (circles drawn around the triangles so that the circles go through each triangle's vertices). The circumcenters are the centers of the circumcircles. (***Note:*** In case you're curious, a circumcircle, in addition to being circumscribed about a triangle, is also circumambient and circumjacent to the triangle; but maybe I'm getting a bit carried away with this circumlocutory circumlocution.)

You can see in Figure 7-12 that, unlike centroids and incenters, a circumcenter is sometimes outside the triangle. The circumcenter is

- Inside all acute triangles
- Outside all obtuse triangles
- On all right triangles (at the midpoint of the hypotenuse)

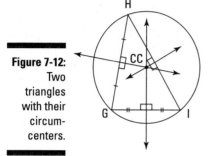
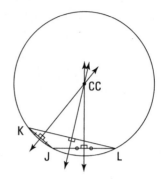

Figure 7-12: Two triangles with their circumcenters.

Obtaining the orthocenter

Check out Figure 7-13 to see a couple of orthocenters. You find a triangle's *orthocenter* at the intersection of its altitudes. Unlike the centroid, incenter, and circumcenter — all of which are located at an interesting point of the triangle (the triangle's center of gravity, the point equidistant from the triangle's sides, and the point equidistant from the triangle's vertices, respectively), a triangle's orthocenter doesn't lie at a point with any such nice characteristics. Well, three out of four ain't bad.

But get a load of this: Look again at the triangles in Figure 7-13. Take the four labeled points of either triangle (the three vertices plus the orthocenter). If you make a triangle out of any three of those four points, the fourth point is the orthocenter of that triangle. Pretty sweet, eh?

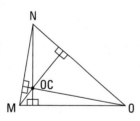
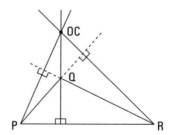

Figure 7-13: Two triangles with their orthocenters.

Orthocenters follow the same rule as circumcenters (note that both orthocenters and circumcenters involve perpendicular lines — altitudes and perpendicular bisectors): The orthocenter is

- ✓ Inside all acute triangles
- ✓ Outside all obtuse triangles
- ✓ On all right triangles (at the right angle vertex)

Chapter 8
Regarding Right Triangles

In This Chapter
▶ Poring over the Pythagorean Theorem
▶ Playing with Pythagorean triples and families
▶ Reasoning about angle ratios: 45°- 45°- 90° and 30°- 60°- 90° triangles

*I*n the mathematical universe of all possible triangles, right triangles are extremely rare (see Chapter 7). But in the so-called *real* world, right angles — and therefore right triangles — are extremely common. Right angles are everywhere: the corners of almost every wall, floor, ceiling, door, window, and wall hanging; the corners of every book, table, box, and piece of paper; the intersection of most streets; the angle between the height of anything (a building, tree, or mountain) and the ground — not to mention the angle between the height and base of any two- or three-dimensional geometrical figure. The list is endless. And everywhere you see a right angle, you potentially have a right triangle. Right triangles abound in navigation, surveying, carpentry, and architecture — even the builders of the Great Pyramids in Egypt used right-triangle mathematics.

Another reason for the abundance of right triangles between the covers of geometry books is the simple connection between the lengths of their sides. Because of this connection, right triangles are a great source of geometry problems. In this chapter, I show you how right triangles pull their weight.

Applying the Pythagorean Theorem

The Pythagorean Theorem has been known for at least 2,500 years (I say *at least* because no one really knows whether someone else discovered it before Pythagoras did).

You use the Pythagorean Theorem when you know the lengths of two sides of a right triangle and you want to figure out the length of the third side.

> ### Pythagoras and the mathematikoi gang
>
> By all accounts, Pythagoras (born on the Greek island of Samos in about 575 B.C.; died circa 500 B.C.) was a great mathematician and thinker. He did original work in mathematics, philosophy, and music theory. However, he and his followers, the *mathematikoi,* were more than a bit on the strange side. Unlike his famous theorem, some of the rules that the members of his society followed haven't exactly stood the test of time: not to eat beans, not to stir a fire with an iron poker, not to step over a crossbar, not to pick up what has fallen, not to look in a mirror next to a light, and not to touch a white rooster.

The Pythagorean Theorem: The sum of the squares of the legs of a right triangle is equal to the square of the hypotenuse:

$$a^2 + b^2 = c^2$$

Here, a and b are the lengths of the legs and c is the length of the hypotenuse. The *legs* are the two short sides that touch the right angle, and the *hypotenuse* (the longest side) is opposite the right angle.

Figure 8-1 shows how the Pythagorean Theorem works for a right triangle with legs of 3 and 4 and a hypotenuse of 5.

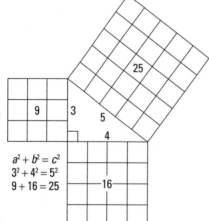

Figure 8-1: The Pythagorean Theorem is as easy as $9 + 16 = 25$.

$a^2 + b^2 = c^2$
$3^2 + 4^2 = 5^2$
$9 + 16 = 25$

Try your hand at the following three problems, which use the Pythagorean Theorem — they get harder as you go along.

Here's the first (Figure 8-2): On your walk to work, you can walk around a park or diagonally across it. If the park is 2 blocks by 3 blocks, how much shorter is your walk if you take the shortcut through the park?

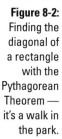

Figure 8-2: Finding the diagonal of a rectangle with the Pythagorean Theorem — it's a walk in the park.

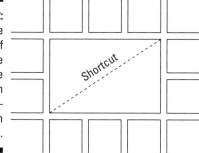

You have a right triangle with legs of 2 and 3 blocks. Plug those numbers into the Pythagorean Theorem to calculate the length of the shortcut that runs along the triangle's hypotenuse:

$$2^2 + 3^2 = c^2$$
$$4 + 9 = c^2$$
$$13 = c^2$$
$$c = \sqrt{13} \approx 3.6 \text{ blocks}$$

That's the length of the shortcut. Going around the park is 2 + 3 = 5 blocks, so the shortcut saves you about 1.4 blocks.

Here's a problem of medium difficulty. It's a multi-stage problem in which you have to use the Pythagorean Theorem more than once: In Figure 8-3, find x and the area of hexagon *ABCDEF*.

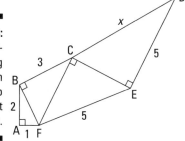

Figure 8-3: A funny-looking hexagon made up of right triangles.

ABCDEF is made up of four connected right triangles, each of which shares at least one side with another triangle. To get *x*, you set up a chain reaction in which you solve for the unknown side of one triangle and then use that answer to find the unknown side of the next triangle — just use the Pythagorean Theorem four times. You already know the lengths of two sides of △*BAF*, so start there to find *BF*:

$$(BF)^2 = (AF)^2 + (AB)^2$$
$$(BF)^2 = 1^2 + 2^2$$
$$(BF)^2 = 5$$
$$BF = \sqrt{5}$$

Now that you have *BF*, you know two of the sides of △*CBF*. Use the Pythagorean Theorem to find *CF*:

$$(CF)^2 = (BF)^2 + (BC)^2$$
$$(CF)^2 = \sqrt{5}^2 + 3^2$$
$$(CF)^2 = 5 + 9$$
$$CF = \sqrt{14}$$

With *CF* filled in, you can find the short leg of △*ECF*:

$$(CE)^2 + (CF)^2 = (FE)^2$$
$$(CE)^2 + \sqrt{14}^2 = 5^2$$
$$(CE)^2 + 14 = 25$$
$$(CE)^2 = 11$$
$$CE = \sqrt{11}$$

And now that you know *CE*, you can solve for *x*:

$$x^2 = (CE)^2 + (ED)^2$$
$$x^2 = \sqrt{11}^2 + 5^2$$
$$x^2 = 11 + 25$$
$$x^2 = 36$$
$$x = 6$$

Okay, on to the second half of the problem. To get the area of *ABCDEF*, just add up the areas of the four right triangles. The area of a triangle is $\frac{1}{2} \cdot$ base $\cdot$ height. For a right triangle, you can use the two legs for the base and the height. Solving for *x* has already given you the lengths of all the sides of the triangles, so just plug the numbers into the area formula:

$$\text{Area}_{\triangle BAF} = \frac{1}{2} \cdot 1 \cdot 2 \qquad\qquad \text{Area}_{\triangle CBF} = \frac{1}{2} \cdot \sqrt{5} \cdot 3$$
$$= 1 \qquad\qquad\qquad\qquad = 1.5\sqrt{5}$$

$$\text{Area}_{\triangle ECF} = \frac{1}{2} \cdot \sqrt{11} \cdot \sqrt{14} \qquad \text{Area}_{\triangle DCE} = \frac{1}{2} \cdot \sqrt{11} \cdot 5$$
$$= 0.5\sqrt{154} \qquad\qquad\qquad = 2.5\sqrt{11}$$

Thus, the area of hexagon *ABCDEF* is $1 + 1.5\sqrt{5} + 0.5\sqrt{154} + 2.5\sqrt{11}$, or about 18.9 units^2.

And now for a more challenging problem. For this one, you need to solve a system of two equations in two unknowns. Dust off your algebra and get ready to go! Here's the problem: Find the area of △*FAC* in Figure 8-4 using the standard triangle area formula, not Hero's formula, which would make this challenge problem much easier; then use Hero's formula to confirm your answer (see Chapter 7 for both formulas).

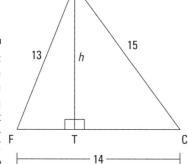

Figure 8-4: The altitude is also the shared leg of two right triangles — it's a *FACT.*

You already know the length of the base of △*FAC*, so to find its area, you need to know its altitude. The altitude forms right angles with the base of △*FAC*, so you have two smaller right triangles — △*FAT* and △*CAT*. Therefore, if you can find the length of the bottom leg of one of these triangles, you can use the Pythagorean Theorem to find the height.

You know that FT and TC add up to 14. So if you let FT equal x, TC becomes $14 - x$. You now have two variables in the problem, h and x. If you use the Pythagorean Theorem for both triangles, you get a system of two equations in two unknowns:

$$\triangle FAT: \begin{aligned} h^2 + x^2 &= 13^2 \\ h^2 + x^2 &= 169 \end{aligned}$$

$$\triangle CAT: \begin{aligned} h^2 + (14 - x)^2 &= 15^2 \\ h^2 + 196 - 28x + x^2 &= 225 \\ h^2 + x^2 - 28x &= 29 \end{aligned}$$

To solve this system, you first need to come up with a single equation in one unknown. I do this with the substitution method. Take the final FAT equation and solve it for h^2:

$$\triangle FAT: \begin{aligned} h^2 + x^2 &= 169 \\ h^2 &= 169 - x^2 \end{aligned}$$

Now take the right side of that equation and plug it in for the h^2 in the final CAT equation. This substitution gives you a single equation in x which you can then solve:

$$\triangle CAT: \begin{aligned} h^2 + x^2 - 28x &= 29 \\ (169 - x^2) + x^2 - 28x &= 29 \\ 169 - 28x &= 29 \\ -28x &= -140 \\ x &= 5 \end{aligned}$$

Did you notice the nifty shortcut for solving this system? (I went through the longer, standard solution because this shortcut rarely works.) The FAT equation tells you that $h^2 + x^2 = 169$. Because the final CAT equation also contains an $h^2 + x^2$, you can just replace that expression with 169, giving you $169 - 28x = 29$. You finish from there like I just did. (By the way, ideally you'd like to solve for h instead of x because h is the thing you need to finish the problem; here, however, that'd involve square roots and get too complicated, so solving for x first is your best bet.)

So now just plug 5 into the x in the first FAT equation (or the first CAT equation, though the FAT equation is simpler) and solve for h:

$$\begin{aligned} h^2 + x^2 &= 169 \\ h^2 + 5^2 &= 169 \\ h^2 &= 144 \\ h &= \pm 12 \text{ (you can reject the } -12) \end{aligned}$$

Finally, finish with the area formula:

$$\text{Area}_{\triangle FAC} = \frac{1}{2}bh$$
$$= \frac{1}{2} \cdot 14 \cdot 12$$
$$= 84 \text{ units}^2$$

Now confirm this result with Hero's formula: $\text{Area}_{\triangle} = \sqrt{S(S-a)(S-b)(S-c)}$. You can say that $a = 13$, $b = 14$, and $c = 15$ (it doesn't matter which is which), and then $S = \frac{13+14+15}{2} = 21$. Thus,

$$\text{Area}_{\triangle FAC} = \sqrt{21(21-13)(21-14)(21-15)}$$
$$= \sqrt{21 \cdot 8 \cdot 7 \cdot 6}$$
$$= \sqrt{7,056}$$
$$= 84 \text{ units}^2$$

The answer checks.

Perusing Pythagorean Triple Triangles

If you use any old numbers for two sides of a right triangle, the Pythagorean Theorem almost always gives you the square root of something for the third side. For example, a right triangle with legs of 5 and 6 has a hypotenuse of $\sqrt{61}$; if the legs are 3 and 8, the hypotenuse is $\sqrt{73}$; and if one of the legs is 6 and the hypotenuse is 9, the other leg works out to $\sqrt{81-36}$, which is $\sqrt{45}$, or $3\sqrt{5}$.

A *Pythagorean triple triangle* is a right triangle with sides whose lengths are all whole numbers, such as 3, 4, and 5 or 5, 12, and 13. People like to use these triangles in problems because they don't contain those pesky square roots. Despite there being an infinite number of such triangles, they're few and far between (like the fact that multiples of 100 are few and far between the other integers, even though there are an infinite number of these multiples).

The Fab Four Pythagorean triple triangles

The first four Pythagorean triple triangles are the favorites of geometry problem-makers. These triples — especially the first and second in the list that follows — pop up all over the place in geometry books. (*Note:* The first two numbers in each of the triple triangles are the lengths of the legs, and the third, largest number is the length of the hypotenuse).

Here are the first four Pythagorean triple triangles:

- The 3-4-5 triangle
- The 5-12-13 triangle
- The 7-24-25 triangle
- The 8-15-17 triangle

You'd do well to memorize these Fab Four so you can quickly recognize them on tests.

Forming irreducible Pythagorean triple triangles

As an alternative to counting sheep some night, you may want to see how many other Pythagorean triple triangles you can come up with.

The first three on the previous list follow a pattern. Consider the 5-12-13 triangle, for example. The square of the smaller, odd leg ($5^2 = 25$) is the sum of the longer leg and the hypotenuse ($12 + 13 = 25$). And the longer leg and the hypotenuse are always consecutive numbers. This pattern makes it easy to generate as many more triangles as you want. Here's what you do:

1. **Take any odd number and square it.**

    $9^2 = 81$, for example

2. **Find the two consecutive numbers that add up to this value.**

    $40 + 41 = 81$

    You can often just come up with the two numbers off the top of your head, but if you don't see them right away, just subtract 1 from the result in Step 1 and then divide that answer by 2:

    $$\frac{81 - 1}{2} = 40$$

    That result and the next larger number are your two numbers.

3. **Write the number you squared and the two numbers from Step 2 in consecutive order to name your triple.**

    You now have another Pythagorean triple triangle: 9-40-41.

Here are the next few Pythagorean triple triangles that follow this pattern:

- 11-60-61   ($11^2 = 121$; $60 + 61 = 121$)
- 13-84-85   ($13^2 = 169$; $84 + 85 = 169$)
- 15-112-113   ($15^2 = 225$; $112 + 113 = 225$)

This list is endless — capable of dealing with the worst possible case of insomnia. And note that each triangle on this list is irreducible; that is, it's not a multiple of some smaller Pythagorean triple triangle (in contrast to the 6-8-10 triangle, for example, which is *not* irreducible because it's the 3-4-5 triangle doubled).

When you make a new Pythagorean triple triangle (like the 6-8-10) by blowing up a smaller one (the 3-4-5), you get triangles with the exact same shape. But every *irreducible* Pythagorean triple triangle has a shape different from all the other irreducible triangles.

A new pattern: Forming further Pythagorean triple triangles

The 8-15-17 triangle is the first Pythagorean triple triangle that doesn't follow the pattern I mention in the preceding section. Here's how you generate triples that follow the 8-15-17 pattern:

1. **Take any multiple of 4.**

    Say you choose 12.

2. **Square half of it.**

    $(12 \div 2)^2 = 6^2 = 36$

3. **Take the number from Step 1 and the two odd numbers on either side of the result in Step 2 to get a Pythagorean triple triangle.**

    12-35-37

The next few triples in this infinite set are

- 16-63-65 $(16 \div 2 = 8; 8^2 = 64; 63, 65)$
- 20-99-101 $(20 \div 2 = 10; 10^2 = 100; 99, 101)$
- 24-143-145 $(24 \div 2 = 12; 12^2 = 144; 143, 145)$

By the way, you can use this process for the other even numbers (the non-multiples of 4) such as 10, 14, 18, and so on. But you get a triangle such as the 10-24-26 triangle, which is the 5-12-13 Pythagorean triple triangle blown up to twice its size, rather than an irreducible, uniquely-shaped triangle.

Families of Pythagorean triple triangles

Each irreducible Pythagorean triple triangle such as the 5-12-13 triangle is the matriarch of a family with an infinite number of children. The 3 : 4 : 5 *family* (note the colons), for example, consists of the 3-4-5 triangle and all her offspring. Offspring are created by blowing up or shrinking the 3-4-5 triangle:

They include the 3/100-4/100-5/100 triangle, the 6-8-10 triangle, the 21-28-35 triangle (3-4-5 times 7), and their eccentric siblings such as the $3\sqrt{11} - 4\sqrt{11} - 5\sqrt{11}$ triangle and the 3π-4π-5π triangle. All members of the 3 : 4 : 5 family — or any other triangle family — have the same shape (they're *similar* — see Chapter 13 for more on similarity).

When you know only two of the three sides of a right triangle, you can compute the third side with the Pythagorean Theorem. But if the triangle happens to be a member of one of the Fab Four Pythagorean triple triangle families — and you're able to recognize that fact — you can often save yourself some time and effort (see "The Fab Four Pythagorean triple triangles"). All you need to do is figure out the blow-up or shrink factor that converts the main Fab Four triangle into the given triangle and use that factor to compute the missing side of the given triangle.

No-brainer cases

You can often just see that you have one of the Fab Four families and figure out the blow-up or shrink factor in your head. Check out Figure 8-5.

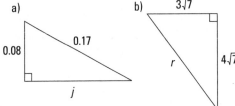

Figure 8-5: Two triangles from famous families.

In Figure 8-5a, the digits 8 and 17 in the 0.08 and 0.17 should give you a big hint that this triangle is a member of the 8 : 15 : 17 family. Because 8 divided by 100 is 0.08 and 17 divided by 100 is 0.17, this triangle is an 8-15-17 triangle shrunk down 100 times. Side *j* is thus 15 divided by 100, or 0.15. Bingo. This shortcut is definitely easier than using the Pythagorean Theorem.

Likewise, the digits 3 and 4 should make it a dead giveaway that the triangle in Figure 8-5b is a member of the 3 : 4 : 5 family. Because $3\sqrt{7}$ is $\sqrt{7}$ times 3 and $4\sqrt{7}$ is $\sqrt{7}$ times 4, you can see that this triangle is a 3-4-5 triangle blown up by a factor of $\sqrt{7}$. Thus, side *r* is simply $\sqrt{7}$ times 5, or $5\sqrt{7}$.

Make sure the sides of the given triangle match up correctly with the sides of the Fab Four triangle family you're using. In a 3 : 4 : 5 triangle, for example, the legs must be the 3 and the 4, and the hypotenuse must be the 5. So a triangle with legs of 30 and 50 (despite the 3 and the 5) is not in the 3 : 4 : 5 family because the 50 (the 5) is one of the legs instead of the hypotenuse.

The step-by-step triple triangle method

If you can't immediately see what Fab Four family a triangle belongs to, you can always use the following step-by-step method to pick the family and find the missing side. Don't be put off by the length of the method; it's easier to do than to explain. I use the triangle in Figure 8-6 to illustrate this process.

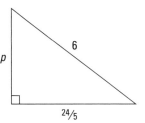

Figure 8-6: Use a ratio to figure out what family this triangle belongs to.

1. **Take the two known sides and make a ratio (either in fraction form or colon form) of the smaller to the larger side.**

   Take the $\frac{24}{5}$ and the 6 and make the ratio of $\frac{24/5}{6}$.

2. **Reduce this ratio to whole numbers in lowest terms.**

   If you multiply the top and bottom of $\frac{24/5}{6}$ by 5, you get $\frac{24}{30}$; that reduces to $\frac{4}{5}$. (With a calculator, this step is a snap because many calculators have a function that reduces fractions to lowest terms.)

3. **Look at the fraction from Step 2 to spot the particular triangle family.**

   The numbers 4 and 5 are part of the 3-4-5 triangle, so you're dealing with the 3 : 4 : 5 family.

4. **Divide the length of a side from the given triangle by the corresponding number from the family ratio to get your multiplier (which tells you how much the basic triangle has been blown-up or shrunk).**

   Use the length of the hypotenuse from the given triangle (because working with a whole number is easier) and divide it by the 5 from the 3 : 4 : 5 ratio. You should get $\frac{6}{5}$ for your multiplier.

5. **Multiply the third family number (the number you don't see in the reduced fraction in Step 2) by the result from Step 4 to find the missing side of your triangle.**

   Three times $\frac{6}{5}$ is $\frac{18}{5}$. That's the length of side *p*; and that's a wrap.

You may be wondering why you should go through all this trouble when you could just use the Pythagorean Theorem. Good point. The Pythagorean Theorem is easier for some triangles (especially if you're allowed to use your calculator). But — take my word for it — this triple triangle technique can come in handy. Take your pick.

Getting to Know Two Special Right Triangles

Make sure you know the two right triangles in this section: the 45°- 45°- 90° triangle and the 30°- 60°- 90° triangle. They come up in many, many geometry problems, not to mention their frequent appearance in trigonometry, pre-calculus, and calculus. Despite the pesky irrational (square-root) lengths they have for some of their sides, they're both more basic and more important than the Pythagorean triple triangles I discuss earlier. They're more basic because they're the progeny of the square and equilateral triangle, and they're more important because their angles are nice fractions of a right angle.

The 45°- 45°- 90° triangle — half a square

The 45°-45°-90° triangle (or isosceles right triangle): A triangle with angles of 45°, 45°, and 90° and sides in the ratio of $1 : 1 : \sqrt{2}$. Note that it's the shape of half a square, cut along the square's diagonal, and that it's also an isosceles triangle (both legs have the same length). See Figure 8-7.

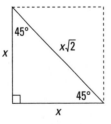

Figure 8-7: The 45°-45°- 90° triangle.

Try a couple of problems. Find the lengths of the unknown sides in triangles *BAT* and *BOY* shown in Figure 8-8.

Chapter 8: Regarding Right Triangles 119

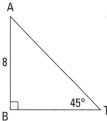
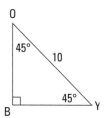

Figure 8-8: Find the missing lengths.

You can solve 45°- 45°- 90° triangle problems in two ways: the formal book method and the street-smart method. Try 'em both and take your pick. The formal method uses the ratio of the sides from Figure 8-7.

$$\text{leg} : \text{leg} : \text{hypotenuse}$$
$$x : x : x\sqrt{2}$$

For △BAT, because one of the legs is 8, the x in the ratio is 8. Plugging 8 into the three x's gives you

$$\text{leg} : \text{leg} : \text{hypotenuse}$$
$$8 : 8 : 8\sqrt{2}$$

And for △BOY, the hypotenuse is 10, so you set the $x\sqrt{2}$ from the ratio equal to 10 and solve for x:

$$x\sqrt{2} = 10$$
$$x = \frac{10}{\sqrt{2}} = \frac{10}{\sqrt{2}} \cdot \frac{\sqrt{2}}{\sqrt{2}} = \frac{10\sqrt{2}}{2} = 5\sqrt{2}$$

That does it:

$$\text{leg} : \text{leg} : \text{hypotenuse}$$
$$5\sqrt{2} : 5\sqrt{2} : 10$$

Now for *the street-smart method* for working with the 45°- 45°- 90° triangle (the street-smart method is based on the same math as the formal method, but it involves fewer steps): Remember the 45°- 45°- 90° triangle as the "$\sqrt{2}$ triangle." Using that tidbit, do one of the following:

- If you know a leg and want to compute the hypotenuse (a *longer* thing), you *multiply* by $\sqrt{2}$. In Figure 8-8, one of the legs in △BAT is 8, so you multiply that by $\sqrt{2}$ to get the *longer* hypotenuse — $8\sqrt{2}$.

- If you know the hypotenuse and want to compute the length of a leg (a *shorter* thing), you *divide* by $\sqrt{2}$. In Figure 8-8, the hypotenuse in △BOY is 10, so you *divide* that by $\sqrt{2}$ to get the *shorter* legs; they're each $\frac{10}{\sqrt{2}}$.

Look back at the lengths of the sides in △BAT and △BOY. In △BAT, the hypotenuse is the only side that contains a radical. In △BOY, the hypotenuse is the only side without a radical. These two cases are by far the most common ones, but in unusual cases, all three sides may contain a radical symbol.

However, it's impossible for none of the sides to contain a radical symbol. This situation would be possible only if the 45°- 45°- 90° triangle were a member of one of the Pythagorean triple families — which it isn't (see the earlier "Perusing Pythagorean Triple Triangles" section). The sidebar "Close but no cigar: Special right triangles and Pythagorean triples" in the following section tells you more about this interesting fact.

The 30°- 60°- 90° triangle — half of an equilateral triangle

The 30°- 60°- 90° triangle: A triangle with angles of 30°, 60°, and 90° and sides in the ratio of $1 : \sqrt{3} : 2$. Note that it's the shape of half an equilateral triangle, cut straight down the middle along its altitude. Check out Figure 8-9.

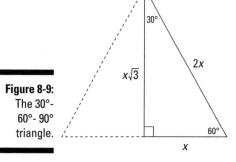

Figure 8-9: The 30°- 60°- 90° triangle.

Get acquainted with this triangle by doing a couple of problems. Find the lengths of the unknown sides in △UMP and △IRE in Figure 8-10.

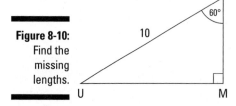
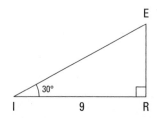

Figure 8-10: Find the missing lengths.

You can solve 30°-60°-90° triangles with the textbook method or the street-smart method. The textbook method begins with the ratio of the sides from Figure 8-9:

short leg : long leg : hypotenuse
x : $x\sqrt{3}$: $2x$

In △UMP, the hypotenuse is 10, so you set $2x$ equal to 10 and solve for x, getting $x = 5$. Now just plug 5 in for the x's, and you have △UMP:

short leg : long leg : hypotenuse
5 : $5\sqrt{3}$: 10

In △IRE, the long leg is 9, so set $x\sqrt{3}$ equal to 9 and solve:

$x\sqrt{3} = 9$

$x = \dfrac{9}{\sqrt{3}} = \dfrac{9}{\sqrt{3}} \cdot \dfrac{\sqrt{3}}{\sqrt{3}} = \dfrac{9\sqrt{3}}{3} = 3\sqrt{3}$

Plug in the value of x, and you're done:

short leg : long leg : hypotenuse
$(3\sqrt{3})$: $(3\sqrt{3})\sqrt{3}$: $2(3\sqrt{3})$
$3\sqrt{3}$: 9 : $6\sqrt{3}$

Here's the *street-smart method* for the 30°-60°-90° triangle. (It's slightly more involved than the method for the 45°-45°-90° triangle.) Think of the 30°-60°-90° triangle as the "$\sqrt{3}$ triangle." Using that fact, do the following:

- The relationship between the short leg and the hypotenuse is a no-brainer: The hypotenuse is twice as long as the short leg. So if you know one of them, you can get the other in your head. The $\sqrt{3}$ method mainly concerns the connection between the short and long legs.

- If you know the short leg and want to compute the long leg (a *longer* thing), you *multiply* by $\sqrt{3}$. If you know the long leg and want to compute the length of the short leg (a *shorter* thing), you *divide* by $\sqrt{3}$.

Try out the street-smart method with the triangles in Figure 8-10. The hypotenuse in △UMP is 10, so first you cut that in half to get the length of the short leg, which is thus 5. Then to get the *longer* leg, you *multiply* that by $\sqrt{3}$, which gives you $5\sqrt{3}$. In △IRE, the long leg is 9, so to get the *shorter* leg, you *divide* that by $\sqrt{3}$, which gives you $\dfrac{9}{\sqrt{3}}$, or $3\sqrt{3}$. The hypotenuse is twice that, $6\sqrt{3}$.

Just like with 45°-45°-90° triangles, 30°-60°-90° triangles almost always have one or two sides whose lengths contain a square root. But with 30°-60°-90° triangles, the *long leg* is the odd one out. And also like 45°-45°-90° triangles, all three sides of a 30°-60°-90° triangle could contain square roots, but it's impossible that none of the sides would — which brings me to the following warning.

Because at least one side of a 30°-60°-90° triangle must contain a square root, a 30°-60°-90° triangle *cannot* belong to any of the Pythagorean triple triangle families. So don't make the mistake of thinking that a 30°-60°-90° triangle is in, say, the 8 : 15 : 17 family or that any triangle that *is* in one of the Pythagorean triple triangle families is also a 30°-60°-90° triangle. There's no overlap between the 30°-60°-90° triangle (or the 45°-45°-90° triangle) and any of the Pythagorean triple triangles and their families. The sidebar "Close but no cigar: Special right triangles and Pythagorean triples" tells you more about this important idea.

Close but no cigar: Special right triangles and Pythagorean triples

You can find a Pythagorean triple triangle that's as close as you like to the shape of a 30°-60°-90° triangle, but you'll never find a perfect match.

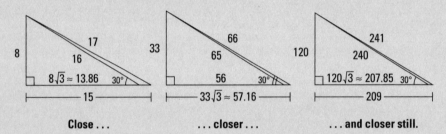

Close . . .   . . . closer . . .   . . . and closer still.

There's no limit to how close you can get. The same thing is true for the 45°-45°-90° triangle. No Pythagorean triple triangle matches its shape exactly, but you can get awfully darn close. There's the oh-so-familiar 803,760-803,761-1,136,689 right triangle, for example, whose legs are in the ratio of about 1 : 1.0000012. That makes it *almost* isosceles but, of course, not a 45°-45°-90° isosceles right triangle.

Chapter 9

Completing Congruent Triangle Proofs

In This Chapter:

▶ Proving triangles congruent with SSS, SAS, ASA, AAS, and HLR

▶ CPCTC: Focusing on parts of congruent triangles

▶ Addressing the two isosceles triangle theorems

▶ Finding perpendicular bisectors and congruent segments with the equidistance theorems

▶ Taking a different tack with indirect proofs

*Y*ou've arrived at high school geometry's main event: triangle proofs. The proofs in Chapters 4, 5, and 6 are complete proofs that show you how proofs work, and they illustrate many of the most important proof strategies. But on the other hand, they're sort of just warm-up or preliminary proofs that lay the groundwork for the real, full-fledged triangle proofs you see in this chapter. Here, I show you how to prove triangles congruent, work with congruent parts of triangles, and use the incredibly important isosceles triangle theorems. I also explain the somewhat peculiar logic involved in indirect proofs.

Introducing Three Ways to Prove Triangles Congruent

Actually, you can prove triangles congruent in five ways, but I think overindulging should be confined to holidays such as Thanksgiving and Pi Day (March 14). So that you can enjoy your proofs in moderation, I give you just the first three ways here and the final two in the brilliantly titled section that follows later: "Trying Out Two More Ways to Prove Triangles Congruent."

Congruent triangles: Triangles in which all pairs of corresponding sides and angles are congruent.

Maybe the best way to think about what it means for two triangles (or any other shapes) to be congruent is that you could move them around (by shifting, rotating, and/or flipping them) so that they'd stack perfectly on top of one another.

You indicate that triangles are congruent with a statement such as $\triangle ABC \cong \triangle XYZ$, which means that vertex A (the first letter) corresponds with and would stack on vertex X (the first letter), B would stack on Y, and C would stack on Z. Side $\overline{AB}$ would stack on side $\overline{XY}$, $\angle B$ would stack on $\angle Y$, and so on.

Figure 9-1 shows two congruent triangles in any old configuration (on the left) and then aligned. The triangles on the left are congruent, but the statement $\triangle ABC \cong \triangle PQR$ is false. Visualize how you'd have to move $\triangle PQR$ to align it with $\triangle ABC$ — you'd have to flip it over and then rotate it. On the right, I've moved $\triangle PQR$ so that it lines up perfectly with $\triangle ABC$. And there you have it: $\triangle ABC \cong \triangle RQP$. All corresponding parts of the triangles are congruent: $\overline{AB} \cong \overline{RQ}$, $\overline{BC} \cong \overline{QP}$, $\angle C \cong \angle P$, and so on.

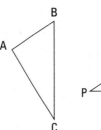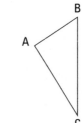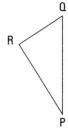

Figure 9-1: Check out how these two congruent triangles stack up.

SSS: Using the side-side-side method

SSS (Side-Side-Side): If the three sides of one triangle are congruent to the three sides of another triangle, then the triangles are congruent. Figure 9-2 illustrates this idea.

Figure 9-2: Triangles with congruent sides are congruent.

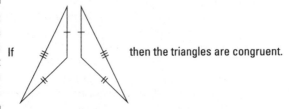

If ... then the triangles are congruent.

Chapter 9: Completing Congruent Triangle Proofs 125

You can use the SSS postulate in the following "*TRIANGLE*" proof:

Given: $\overline{AG} \cong \overline{EG}$
$\overline{NG} \cong \overline{LG}$
$\overline{AR} \cong \overline{ET}$
$\overline{NI} \cong \overline{LI}$
T is the midpoint of $\overline{NI}$
R is the midpoint of $\overline{LI}$

Prove: $\triangle ANT \cong \triangle ELR$

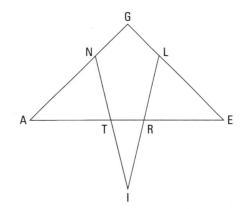

Before you begin writing a formal proof, figure out your game plan. Here's how that might work.

You know you've got to prove the triangles congruent, so your first question should be "Can you show that the three pairs of corresponding sides are congruent?" Sure, you can do that:

- Subtract $\overline{NG}$ and $\overline{LG}$ from $\overline{AG}$ and $\overline{EG}$ to get the first pair of congruent sides, $\overline{AN}$ and $\overline{EL}$.
- Subtract $\overline{TR}$ from $\overline{AR}$ and $\overline{ET}$ to get the second pair of congruent sides, $\overline{AT}$ and $\overline{ER}$.
- Cut congruent segments $\overline{NI}$ and $\overline{LI}$ in half to get the third pair, $\overline{NT}$ and $\overline{LR}$. That's it.

To make the game plan more tangible, you may want to make up lengths for the various segments. For instance, say *AG* and *EG* are 9, *NG* and *LG* are 3, *AR* and *ET* are 8, *TR* is 3, and *NI* and *LI* are 8. When you do the math, you see that $\triangle ANT$ and $\triangle ELR$ both end up with sides of 4, 5, and 6, which means, of course, that they're congruent.

Here's how the formal proof shapes up:

Statements	Reasons
1) $\overline{AG} \cong \overline{EG}$ $\overline{NG} \cong \overline{LG}$	1) Given.
2) $\overline{AN} \cong \overline{EL}$	2) If two congruent segments are subtracted from two other congruent segments, then the differences are congruent.

(continued)

Statements	Reasons
3) $\overline{AR} \cong \overline{ET}$	3) Given.
4) $\overline{AT} \cong \overline{ER}$	4) If a segment is subtracted from two congruent segments, then the differences are congruent.
5) $\overline{NI} \cong \overline{LI}$ T is the midpoint of $\overline{NI}$ R is the midpoint of $\overline{LI}$	5) Given.
6) $\overline{NT} \cong \overline{LR}$	6) If segments are congruent, then their Like Divisions are congruent (half of one equals half of the other — see Chapter 5).
7) $\triangle ANT \cong \triangle ELR$	7) SSS (2, 4, 6).

Note: After SSS in the final step, I indicate the three lines from the statement column where I've shown the three pairs of sides to be congruent. You don't have to do this, but it's a good idea. It can help you avoid some careless mistakes. Remember that each of the three lines you list must show a *congruence* of segments (or angles, if you're using one of the other approaches to proving triangles congruent).

SAS: Taking the side-angle-side approach

SAS (Side-Angle-Side): If two sides and the included angle of one triangle are congruent to two sides and the included angle of another triangle, then the triangles are congruent. (The *included angle* is the angle formed by the two sides.) Figure 9-3 illustrates this method.

Figure 9-3:
Two sides and the angle between them make these triangles congruent.

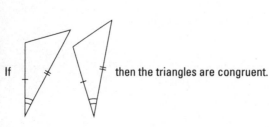

If   then the triangles are congruent.

Check out the SAS postulate in action:

Given: △QZX is isosceles with base $\overline{QX}$
$\overline{JQ} \cong \overline{XF}$
∠1 ≅ ∠2

Prove: △JZX ≅ △FZQ

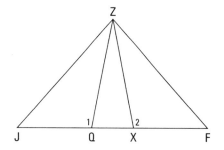

When overlapping triangles muddy your understanding of a proof diagram, try redrawing the diagram with the triangles separated. Doing so can give you a clearer idea of how the triangles' sides and angles relate to each other. Focusing on your new diagram may make it easier to figure out what you need to prove the triangles congruent. However, you still need to use the original diagram to understand some parts of the proof, so use the second diagram as a sort of aid to get a better handle on the original diagram.

Figure 9-4 shows you what this proof diagram looks like with the triangles separated.

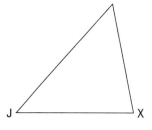 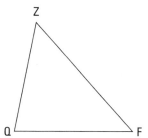

Figure 9-4: An amicable separation of triangles.

Looking at Figure 9-4, you can easily see that the triangles are congruent (they're mirror images of each other). You also see that, for example, side $\overline{ZX}$ corresponds to side $\overline{ZQ}$ and that ∠X corresponds to ∠Q.

So using both diagrams, here's a possible game plan:

- **Determine which congruent triangle postulate is likely to be the ticket for proving the triangles congruent.** You know you have to prove the triangles congruent, and one of the givens is about angles, so SAS looks like a better candidate than SSS for the final reason. (You don't have to figure this out now, but it's not a bad idea to at least have a guess about the final reason.)

- **Look at the givens and think about what they tell you about the triangles.** Triangle QZX is isosceles, so that tells you $\overline{ZQ} \cong \overline{ZX}$. Look at these sides in both figures. Put tick marks on $\overline{ZQ}$ and $\overline{ZX}$ in Figure 9-4 to show that you know they're congruent. Now consider why they'd tell you the next given, $\overline{JQ} \cong \overline{XF}$. Well, what if they were both 6 and $\overline{QX}$ were 2? $\overline{JX}$ and $\overline{QF}$ would both be 8, so you have a second pair of congruent sides. Put tick marks on Figure 9-4 to show this congruence.

- **Find the pair of congruent angles.** Look at Figure 9-4 again. If you can show that $\angle X$ is congruent to $\angle Q$, you'll have SAS. Do you see where $\angle X$ and $\angle Q$ fit into the original diagram? Note that they're the supplements of $\angle 1$ and $\angle 2$. That does it. Angles 1 and 2 are congruent, so their supplements are congruent as well. (If you fill in numbers, you can see that if $\angle 1$ and $\angle 2$ are both 100°, $\angle Q$ and $\angle X$ would both be 80°.)

Here's the formal proof:

Statements	Reasons
1) △QZX is isosceles with base $\overline{QX}$	1) Given.
2) $\overline{ZX} \cong \overline{ZQ}$	2) Definition of isosceles triangle.
3) $\overline{JQ} \cong \overline{XF}$	3) Given.
4) $\overline{JX} \cong \overline{FQ}$	4) If a segment is added to two congruent segments, then the sums are congruent.
5) $\angle 1 \cong \angle 2$	5) Given.
6) $\angle ZXJ \cong \angle ZQF$	6) If two angles are supplementary to two other congruent angles, then they're congruent.
7) △JZX $\cong$ △FZQ	7) SAS (2, 6, 4).

ASA: Taking the angle-side-angle tack

ASA (Angle-Side-Angle): If two angles and the included side of one triangle are congruent to two angles and the included side of another triangle, then the triangles are congruent. (The *included side* is the side between the vertices of the two angles.) See Figure 9-5.

Figure 9-5:
Two angles and their shared side make these triangles congruent.

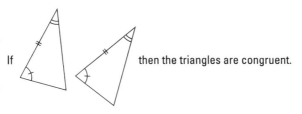

Here's a congruent-triangle proof with the ASA postulate:

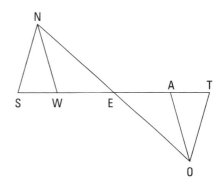

Given: E is the midpoint of $\overline{NO}$

$\angle SNW \cong \angle TOA$

$\overrightarrow{NW}$ bisects $\angle SNE$

$\overrightarrow{OA}$ bisects $\angle TOE$

Prove: $\triangle SNE \cong \triangle TOE$

Here's my game plan:

- **Note any congruent sides and angles in the diagram.** First and foremost, notice the congruent vertical angles (I introduce vertical angles in Chapter 2). Vertical angles are important in many proofs, so you can't afford to miss them. Next, midpoint E gives you $\overline{NE} \cong \overline{OE}$. So now you have a pair of congruent angles and a pair of congruent sides.

- **Determine which triangle postulate you need to use.** To finish with SAS, you'd need to show $\overline{SE} \cong \overline{TE}$; and to finish with ASA, you'd need $\angle SNE \cong \angle TOE$. A quick glance at the bisected angles in the givens (or the title of this section, but that's cheating!) makes the second alternative much more likely. Sure enough, you can get $\angle SNE \cong \angle TOE$ because a given says that half of $\angle SNE$ ($\angle SNW$) is congruent to half of $\angle TOE$ ($\angle TOA$). That's a wrap.

Here's how the formal proof plays out:

Statements	Reasons
1) $\angle SEN \cong \angle TEO$	1) Vertical angles are congruent.
2) E is the midpoint of $\overline{NO}$	2) Given.
3) $\overline{NE} \cong \overline{OE}$	3) Definition of midpoint.
4) $\angle SNW \cong \angle TOA$	4) Given.
5) $\overrightarrow{NW}$ bisects $\angle SNE$ $\overrightarrow{OA}$ bisects $\angle TOE$	5) Given.
6) $\angle SNE \cong \angle TOE$	6) If two angles are congruent (angles SNW and TOA), then their Like Multiples are congruent (twice one equals twice the other).
7) $\triangle SNE \cong \triangle TOE$	7) ASA (1, 3, 6).

Sticking to the basics: The workings behind SSS, SAS, and ASA

The idea behind the SSS postulate is pretty simple. Say you have three sticks of given lengths (how about 5, 7, and 9 inches?) and then you make a triangle out of them. Now you take three more sticks of the same lengths and make a second triangle. No matter how you connect the sticks, you end up with two triangles of the exact same size and shape — in other words, two *congruent* triangles.

Maybe you're thinking, "Of course if you make two triangles using the same-length sticks for both, you'll end up with two triangles of the same size and shape. What's the point?" Well, maybe it is sort of obvious, but this principle doesn't hold for polygons of four or more sides. If you take, for example, four sticks of 4, 5, 6, and 7 inches, there's no limit to the number of differently shaped quadrilaterals you can make. Try it.

Now consider the SAS postulate. If you start with two sticks that connect to form a given angle, you're again locked into a single triangle of a definite shape. And lastly, the ASA postulate works because if you start with one stick and two angles of given sizes that must go on the ends of the stick, there's also only a single triangle you can make (this one's a little harder to picture).

By the way, you don't really need this stick idea to understand the three postulates. I just want explain *why* the postulates work. In a nutshell, they all work because the three given things (three sides, or two sides and an angle, or two angles and a side) lock you into one definite triangle.

CPCTC: Taking Congruent Triangle Proofs a Step Further

In the preceding section, the relatively short proofs end with showing that two triangles are congruent. But in more-advanced proofs, showing triangles congruent is just a stepping stone for going on to prove other things. In this section, you take proofs a step further.

Proving triangles congruent is often the focal point of a proof, so always check the proof diagram for *all* pairs of triangles that look like they're the same shape and size. If you find any, you'll very likely have to prove one (or more) of the pairs of triangles congruent.

Defining CPCTC

CPCTC: An acronym for *corresponding parts of congruent triangles are congruent*. This idea sort of has the feel of a theorem, but it's really just the definition of congruent triangles.

Because congruent triangles have six pairs of congruent parts (three pairs of segments and three pairs of angles) and you need three of the pairs for SSS, SAS, or ASA, there will always be three remaining pairs that you didn't use. The purpose of CPCTC is to show one or more of these remaining pairs congruent.

CPCTC is very easy to use. After you show that two triangles are congruent, you can state that two of their sides or angles are congruent on the next line of the proof, using CPCTC as the justification for that statement. This group of two consecutive lines makes up the core or heart of many proofs.

Say you're in the middle of some proof (shown in Figure 9-6), and by line 6, you're able to show with ASA that $\triangle PQR$ is congruent to $\triangle XYZ$. The tick marks in the diagram show the pair of congruent sides and the two pairs of congruent angles that were used for ASA. Now that you know that the triangles are congruent, you can state on line 7 that $\overline{QR} \cong \overline{YZ}$ and use CPCTC for the reason (you could also use CPCTC to justify that $\overline{PR} \cong \overline{XZ}$ or that $\angle QRP \cong \angle YZX$).

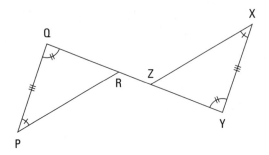

Statements	Reasons
...	...
...	...
...	...
6) △PQR ≅ △XYZ	6) ASA.
7) $\overline{QR} \cong \overline{YZ}$	7) CPCTC.
...	...
...	...
...	...

Figure 9-6: A critical pair of proof lines: Congruent triangles and CPCTC.

Tackling a CPCTC proof

You can check out CPCTC in action in the proof that follows. But before I get there, here's a property you need to do the problem. It's an incredibly simple concept that comes up in many proofs.

The Reflexive Property: Any segment or angle is congruent to itself. (Who would've thought?)

Whenever you see two triangles that share a side or an angle, that side or angle belongs to both triangles. With the Reflexive Property, the shared side or angle becomes a pair of congruent sides or angles that you can use as one of the three pairs of congruent things that you need to prove the triangles congruent. Check out Figure 9-7.

Chapter 9: Completing Congruent Triangle Proofs **133**

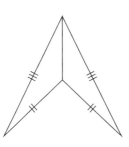

Figure 9-7: Using the Reflexive Property for the shared side, these triangles are congruent by SSS.

Here's your CPCTC proof:

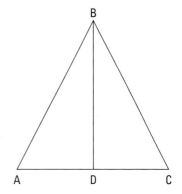

Given: $\overline{BD}$ is a median and an altitude of △ABC

Prove: $\overrightarrow{BD}$ bisects ∠ABC

Before you write out the formal proof, come up with a game plan. Here's one possibility:

- **Look for congruent triangles.** The congruent triangles should just about jump out at you from this diagram. Think about how you'll show that they're congruent. The triangles share side $\overline{BD}$, giving you one pair of congruent sides. $\overline{BD}$ is an altitude, so that gives you congruent right angles. And because $\overline{BD}$ is a median, $\overline{AD} \cong \overline{CD}$ (see Chapter 7 for more on medians and altitudes). That does it; you have SAS.

- **Now think about what you have to prove and what you'd need to know to get there.** To conclude that $\overrightarrow{BD}$ bisects ∠ABC, you need ∠ABD ≅ ∠CBD in the second-to-last line. And how will you get that? Why, with CPCTC, of course!

Here's the two-column proof:

Statements	Reasons
1) $\overline{BD}$ is a median of $\triangle ABC$	1) Given.
2) D is the midpoint of $\overline{AC}$	2) Definition of median.
3) $\overline{AD} \cong \overline{CD}$	3) Definition of midpoint.
4) $\overline{BD}$ is an altitude of $\triangle ABC$	4) Given.
5) $\overline{BD} \perp \overline{AC}$	5) Definition of altitude (if a segment is an altitude [Statement 4], then it is perpendicular to the triangle's base [Statement 5]).
6) $\angle ADB$ is a right angle $\angle CDB$ is a right angle	6) Definition of perpendicular.
7) $\angle ADB \cong \angle CDB$	7) All right angles are congruent.
8) $\overline{BD} \cong \overline{BD}$	8) Reflexive Property.
9) $\triangle ABD \cong \triangle CBD$	9) SAS (3, 7, 8).
10) $\angle ABD \cong \angle CBD$	10) CPCTC.
11) $\overrightarrow{BD}$ bisects $\angle ABC$	11) Definition of bisect.

Remember: Every little step in a proof must be spelled out. For instance, in the preceding proof, you can't go from the idea of a median (line 1) to congruent segments (line 3) in one step — even though it's obvious — because the definition of median says nothing about congruent segments. By the same token, you can't go from the idea of an altitude (line 4) to congruent right angles (line 7) in one step or even two steps. You need three steps to connect the links in this chain of logic: Altitude → perpendicular → right angles → congruent angles.

Eying the Isosceles Triangle Theorems

The earlier sections in this chapter involve *pairs* of congruent triangles. Here, you get two theorems that involve a *single* isosceles triangle. Although you often need these theorems for proofs in which you show that two triangles are congruent, the theorems themselves concern only one triangle.

The following two theorems are based on one simple idea about isosceles triangles that happens to work in both directions:

- **If sides, then angles:** If two sides of a triangle are congruent, then the angles opposite those sides are congruent. Figure 9-8 shows you how this works.

Figure 9-8: The congruent sides tell you that the angles are congruent.

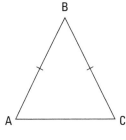 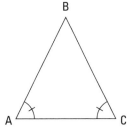

If you know this . . .   . . . you can conclude this.

- **If angles, then sides:** If two angles of a triangle are congruent, then the sides opposite those angles are congruent. Take a look at Figure 9-9.

Figure 9-9: The congruent angles tell you that the sides are congruent.

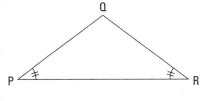 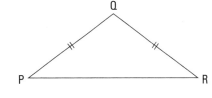

If you know this . . .   . . . you can conclude this.

Look for isosceles triangles. The two angle-side theorems are critical for solving many proofs, so when you start doing a proof, look at the diagram and identify all triangles that look isosceles. Then make a mental note that you may have to use one of the theorems for one or more of the isosceles triangles. These theorems are incredibly easy to use if you spot all the isosceles triangles (which shouldn't be too hard). But if you fail to notice them, the proof may become impossible. And note that your goal here is to spot single isosceles triangles because unlike SSS, SAS, and ASA, the isosceles-triangle theorems do not involve pairs of triangles.

Here's a proof. Try to work through a game plan and/or a formal proof on your own before reading the ones I present here.

Given: ∠P ≅ ∠T
$\overline{PX} \cong \overline{TY}$
$\overline{RX} \cong \overline{RY}$

Prove: ∠Q ≅ ∠S

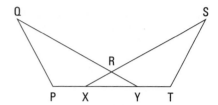

Here's a game plan:

- **Check the proof diagram for isosceles triangles and pairs of congruent triangles.** This proof's diagram has an isosceles triangle, which is a huge hint that you'll likely use one of the isosceles triangle theorems. You also have a pair of triangles that look congruent (the overlapping ones), which is another huge hint that you'll want to show that they're congruent.

- **Think about how to finish the proof with a triangle congruence theorem and CPCTC.** You're given the sides of the isosceles triangle, so that gives you congruent angles. You're also given ∠P ≅ ∠T, so that gives you a second pair of congruent angles. If you can get $\overline{PY} \cong \overline{TX}$, you'd have ASA. And you can get that by adding $\overline{XY}$ to the given congruent segments, $\overline{PX}$ and $\overline{TY}$. You finish with CPCTC.

Check out the formal proof:

Statements	Reasons
1) $\overline{RX} \cong \overline{RY}$	1) Given.
2) ∠RYX ≅ ∠RXY	2) If two sides of a triangle are congruent, then the angles opposite those sides are congruent.
3) $\overline{PX} \cong \overline{TY}$	3) Given.
4) $\overline{PY} \cong \overline{TX}$	4) If a segment is added to two congruent segments, then the sums are congruent.
5) ∠P ≅ ∠T	5) Given.
6) △PQY ≅ △TSX	6) ASA (2, 4, 5).
7) ∠Q ≅ ∠S	7) CPCTC.

Trying Out Two More Ways to Prove Triangles Congruent

Back in the "Introducing Three Ways to Prove Triangles Congruent" section, I promised you that I'd give you two more ways to prove triangles congruent, and because I'm a man of my word, here they are.

Don't try to find some nice connection between these two additional methods. They're together simply because I didn't want to give you all five methods in the first section and risk giving you a case of triangle-congruence-theorem overload.

AAS: Using the angle-angle-side theorem

AAS (Angle-Angle-Side): If two angles and a nonincluded side of one triangle are congruent to the corresponding parts of another triangle, then the triangles are congruent. Figure 9-10 shows you how AAS works.

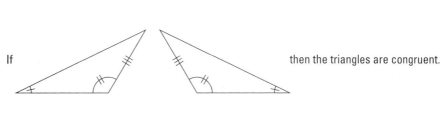

Figure 9-10: Two congruent angles and a side not between them make these triangles congruent.

Like ASA (see the earlier section), to use AAS, you need two pairs of congruent angles and one pair of congruent sides to prove two triangles congruent. But for AAS, the two angles and one side in each triangle must go in the order angle-angle-side (going around the triangle either clockwise or counterclockwise).

138 Part III: Triangles: Polygons of the Three-Sided Variety

ASS and SSA don't prove anything, so don't try using ASS (or its backward twin, SSA) to prove triangles congruent. You can use SSS, SAS, ASA, and AAS (or SAA, the backward twin of AAS) to prove triangles congruent, but not ASS. In short, every three-letter combination of *A*'s and *S*'s proves something unless it spells *ass* or is *ass* backward. (You work with AAA in Chapter 13, but it shows that triangles are similar, not congruent.)

Try to solve the following proof by first looking for all isosceles triangles (with the two isosceles triangle theorems in mind) and for all pairs of congruent triangles (with CPCTC in mind). I may sound like a broken record, but I can't tell you how much easier some proofs become when you remember to check for these things!

Given: $\angle QRT \cong \angle UTR$

$\angle VRT \cong \angle VTR$

$\overline{SQ} \cong \overline{SU}$

Prove: *V* is the midpoint of $\overline{QU}$

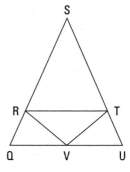

Here's a game plan that shows how you might think through this proof:

- ✓ **Take note of isosceles triangles and pairs of congruent triangles.** You should notice three isosceles triangles ($\triangle QSU$, $\triangle RST$, and $\triangle RVT$). The given congruent sides of $\triangle QSU$ give you $\angle Q \cong \angle U$, and the given congruent angles of $\triangle RVT$ give you $\overline{RV} \cong \overline{TV}$.

  You should also notice the two congruent-looking triangles ($\triangle QRV$ and $\triangle UTV$) and then realize that showing them congruent and using CPCTC is very likely the ticket.

- ✓ **Look at the *prove* statement and plan your next move.** To prove the midpoint, you need $\overline{QV} \cong \overline{UV}$ on the second-to-last line, and you could get that by CPCTC if you knew that $\triangle QRV$ and $\triangle UTV$ were congruent.

- ✓ **Figure out how to prove the triangles congruent.** You already have (from the first bullet) a pair of congruent angles ($\angle Q$ and $\angle U$) and a pair of congruent sides ($\overline{RV}$ and $\overline{TV}$). Because of where these angles and sides are, SAS and ASA won't work, so the key has to be AAS. To use AAS, you'd need $\angle QRV \cong \angle UTV$. Can you get that? Sure. Check out the givens: You subtract congruent angles *VRT* and *VTR* from congruent angles *QRT* and *UTR*. Checkmate.

Here's the formal proof:

Statements	Reasons
1) $\angle VRT \cong \angle VTR$	1) Given.
2) $\overline{RV} \cong \overline{TV}$	2) If angles, then sides.
3) $\angle QRT \cong \angle UTR$	3) Given.
4) $\angle QRV \cong \angle UTV$	4) If two congruent angles ($\angle VRT$ and $\angle VTR$) are subtracted from two other congruent angles ($\angle QRT$ and $\angle UTR$), then the differences ($\angle QRV$ and $\angle UTV$) are congruent.
5) $\overline{SQ} \cong \overline{SU}$	5) Given.
6) $\angle RQV \cong \angle TUV$	6) If sides, then angles.
7) $\triangle QRV \cong \triangle UTV$	7) AAS (6, 4, 2).
8) $\overline{QV} \cong \overline{UV}$	8) CPCTC.
9) V is the midpoint of $\overline{QU}$	9) Definition of midpoint.

HLR: The right approach for right triangles

HLR (Hypotenuse-Leg-Right angle): If the hypotenuse and a leg of one right triangle are congruent to the hypotenuse and a leg of another right triangle, then the triangles are congruent. Figure 9-11 shows you an example. HLR is different from the other four ways of proving triangles congruent because it works only for right triangles.

Figure 9-11: Congruent legs and hypotenuses make these right triangles congruent.

If 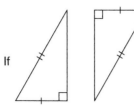 then the triangles are congruent.

In other books, HLR is usually called HL. Rebel that I am, I'm boldly renaming it HLR because its three letters emphasize that — as with SSS, SAS, ASA, and AAS — before you can use it in a proof, you need to have three things in the statement column (congruent hypotenuses, congruent legs, and right angles).

Note: When you use HLR, listing the pair of right angles in the statement column is sufficient for that part of the theorem. If you want to use a pair of right angles with SAS, ASA, and AAS, you have to state that the right angles are congruent, but with HLR, you don't have to do that.

Ready for an HLR proof? Well, ready or not, here you go.

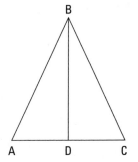

Given: △ABC is isosceles with base $\overline{AC}$
$\overline{BD}$ is an altitude

Prove: $\overline{BD}$ is a median

Here's a possible game plan. You see the pair of congruent triangles and then ask yourself how you can prove them congruent. You know you have a pair of congruent sides because the triangle is isosceles. You have another pair of congruent sides because of the Reflexive Property for $\overline{BD}$. And you have the right angles because of the altitude. Voilà, that's HLR. You then get $\overline{AD} \cong \overline{CD}$ with CPCTC, and you're home free. Here it is in two-column format:

Statements	Reasons
1) △ABC is isosceles with base $\overline{AC}$	1) Given.
2) $\overline{AB} \cong \overline{CB}$	2) Definition of isosceles triangle.
3) $\overline{BD} \cong \overline{BD}$	3) Reflexive Property.
4) $\overline{BD}$ is an altitude	4) Given.
5) $\overline{BD} \perp \overline{AC}$	5) Definition of altitude.
6) ∠ADB is a right angle ∠CDB is a right angle	6) Definition of perpendicular.

Statements	Reasons
7) △ABD ≅ △CBD	7) HLR (2, 3, 6).
8) $\overline{AD} \cong \overline{CD}$	8) CPCTC.
9) D is the midpoint of $\overline{AC}$	9) Definition of midpoint.
10) $\overline{BD}$ is a median of △ABC	10) Definition of median.

Going the Distance with the Two Equidistance Theorems

Although congruent triangles are the focus of this chapter, in this section, I give you two theorems that you can often use *instead* of proving triangles congruent. Even though you see congruent triangles in this section's proof diagrams, you don't have to prove the triangles congruent — one of the *equidistance* theorems gives you a shortcut to the *prove* statement.

Be on your toes for the equidistance shortcut. When doing triangle proofs, be alert for two possibilities: Look for congruent triangles and think about ways to prove them congruent, but at the same time, try to see whether one of the equidistance theorems can get you around the congruent triangle issue.

Determining a perpendicular bisector

The first equidistance theorem tells you that two points determine the perpendicular bisector of a segment. (To "determine" something means to fix or lock in its position, basically to show you where something is.) Here's the theorem.

Two equidistant points determine the perpendicular bisector: If two points are each (one at a time) equidistant from the endpoints of a segment, then those points determine the perpendicular bisector of the segment. (Here's an easy way to think about it: If you have *two* pairs of congruent segments, then there's a perpendicular bisector.)

This theorem is a royal mouthful, so the best way to understand it is visually. Consider the kite-shaped diagram in Figure 9-12.

Figure 9-12: The first equidistance theorem.

If you know that $\overline{XW} \cong \overline{XY}$ and $\overline{ZW} \cong \overline{ZY}$, then you can conclude that $\overleftrightarrow{XZ}$ is the perpendicular bisector of $\overline{WY}$.

The theorem works like this: If you have one point (like *X*) that's equally distant from the endpoints of a segment (*W* and *Y*) and another point (like *Z*) that's also equally distant from the endpoints, then the two points (*X* and *Z*) determine the perpendicular bisector of that segment ($\overline{WY}$). You can also see the meaning of the short form of the theorem in this diagram: If you have *two* pairs of congruent segments ($\overline{XW} \cong \overline{XY}$ and $\overline{ZW} \cong \overline{ZY}$), then there's a perpendicular bisector ($\overleftrightarrow{XZ}$ is the perpendicular bisector of $\overline{WY}$).

Here's a *"SHORT"* proof that shows how to use the first equidistance theorem as a shortcut so you can skip showing that triangles are congruent.

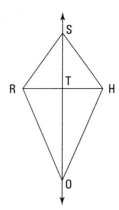

Given: $\overline{SR} \cong \overline{SH}$

$\angle ORT \cong \angle OHT$

Prove: *T* is the midpoint of $\overline{RH}$

You can do this proof using congruent triangles, but it'd take you about nine steps and you'd have to use two different pairs of congruent triangles.

Statements	Reasons
1) $\angle ORT \cong \angle OHT$	1) Given.
2) $\overline{OR} \cong \overline{OH}$	2) If angles, then sides.
3) $\overline{SR} \cong \overline{SH}$	3) Given.
4) $\overleftrightarrow{SO}$ is the perpendicular bisector of $\overline{RH}$	4) If two points (S and O) are each equidistant from the endpoints of a segment ($\overline{RH}$), then they determine the perpendicular bisector of that segment.
5) $\overline{RT} \cong \overline{TH}$	5) Definition of bisect.
6) T is the midpoint of $\overline{RH}$	6) Definition of midpoint.

Using a perpendicular bisector

With the second equidistance theorem, you use a point on a perpendicular bisector to prove two segments congruent.

A point on the perpendicular bisector is equidistant from the segment's endpoints: If a point is on the perpendicular bisector of a segment, then it's equidistant from the endpoints of the segment. (Here's my abbreviated version: If you have a perpendicular bisector, then there's *one* pair of congruent segments.)

Figure 9-13 shows you how the second equidistance theorem works.

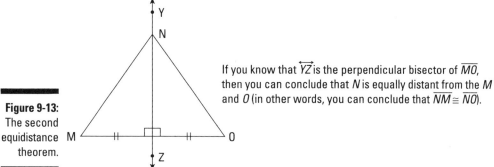

Figure 9-13: The second equidistance theorem.

If you know that $\overleftrightarrow{YZ}$ is the perpendicular bisector of $\overline{MO}$, then you can conclude that N is equally distant from the M and O (in other words, you can conclude that $\overline{NM} \cong \overline{NO}$).

This theorem tells you that if you begin with a segment (like $\overline{MO}$) and its perpendicular bisector (like $\overleftrightarrow{YZ}$) and you have a point on the perpendicular bisector (like N), then that point is equally distant from the endpoints of the segment. Note that you can see the reasoning behind the short form of the theorem in this diagram: If you have a perpendicular bisector (line $\overleftrightarrow{YZ}$ is the perpendicular bisector of $\overline{MO}$), then there's *one* pair of congruent segments ($\overline{NM} \cong \overline{NO}$).

Here's a proof that uses the second equidistance theorem:

Given: $\angle 1 \cong \angle 4$
$\overline{LQ} \cong \overline{NQ}$

Prove: $\overline{LP} \cong \overline{NP}$

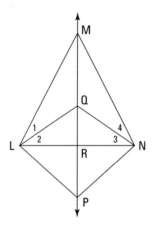

Statements	Reasons
1) $\overline{LQ} \cong \overline{NQ}$	1) Given.
2) $\angle 2 \cong \angle 3$	2) If sides, then angles.
3) $\angle 1 \cong \angle 4$	3) Given.
4) $\angle MLR \cong \angle MNR$	4) If two congruent angles ($\angle 2$ and $\angle 3$) are added to two other congruent angles ($\angle 1$ and $\angle 4$), then the sums are congruent.
5) $\overline{ML} \cong \overline{MN}$	5) If angles, then sides.
6) $\overleftrightarrow{MQ}$ is the perpendicular bisector of $\overline{LN}$	6) If two points (M and Q) are equidistant from the endpoints of a segment ($\overline{LN}$; see statements 1 and 5), then they determine the perpendicular bisector of that segment.
7) $\overline{LP} \cong \overline{NP}$	7) If a point (point P) is on the perpendicular bisector of a segment, then it is equidistant from the endpoints of that segment.

Making a Game Plan for a Longer Proof

In previous sections, I show you some typical examples of triangle proofs and all the theorems you need for them. Here, I walk through a game plan for a longer, slightly gnarlier proof. This section gives you the opportunity to use some of the most important proof-solving strategies. Because the point of this section is to show you how to think through the commonsense reasoning for a longer proof, I skip the proof itself.

Here's the setup for the proof. Try working through your own game plan before reading the one that follows.

Given: $\overline{EU} \perp \overline{LH}$, $\overline{EU} \perp \overline{SR}$
$\overline{EL} \cong \overline{RU}$, $\overline{ES} \cong \overline{HU}$

Prove: $\overline{EH} \cong \overline{SU}$

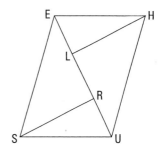

Here's a game plan that illustrates how your thought process might play out:

- **Look for congruent triangles.** You should see three pairs of congruent triangles: the two small ones, $\triangle ELH$ and $\triangle URS$; the two medium ones, $\triangle ERS$ and $\triangle ULH$; and the two large ones, $\triangle EUS$ and $\triangle UEH$. Then you should ask yourself how showing that one or more of these pairs of triangles are congruent and then using CPCTC could play a part in this proof.

- **Work backward.** The final statement must be $\overline{EH} \cong \overline{SU}$. CPCTC is the likely final reason. To use CPCTC, you'd have to show the congruence of either the small triangles or the large ones. Can you do that?

- **Use every given to see whether you can prove the triangles congruent.** The two pairs of perpendicular segments give you congruent right angles in the small and medium triangles. For the small triangles, you also have $\overline{EL} \cong \overline{RU}$. So to show that the small triangles are congruent, you'd need to get $\overline{LH} \cong \overline{SR}$ and use SAS or get $\angle LEH \cong \angle RUS$ and use AAS. Unfortunately, there doesn't seem to be any way to get either of these two congruences.

  If you could get $\overline{ES} \cong \overline{SU}$, you could show the small triangles congruent by HLR, but $\overline{EH} \cong \overline{SU}$ is what you're trying to prove, so that nixes that option.

  As for showing the two large triangles congruent, you have $\overline{ES} \cong \overline{HU}$ and $\overline{EU} \cong \overline{EU}$ by the Reflexive Property. To finish with SAS, you'd need to get $\angle SEU \cong \angle HUE$, but that looks like another dead end.

So there doesn't seem to be any direct way of showing either the small triangles congruent or the large triangles congruent and then using CPCTC to finish the proof. Well, if there isn't a direct way, then there must be a roundabout way.

- **Try the third set of triangles.** If you could prove the two medium triangles congruent, you could use CPCTC on those triangles to get $\overline{LH} \cong \overline{SR}$ (the thing you needed to prove the small triangles congruent with SAS). Then, as described in the preceding bullet, you'd show the small triangles congruent and finish with CPCTC. So now all you have to do is show the medium triangles congruent, and after you do that, you know how to get to the end.

- **Use the givens again.** Try to use the givens to prove $\triangle ERS \cong \triangle ULH$ (the medium triangles). You already have two pairs of parts — the right angles and $\overline{ES} \cong \overline{HU}$ — so all you need is the third pair of congruent parts; $\angle ESR \cong \angle UHL$ or $\angle SER \cong \angle HUL$ would give you AAS, and $\overline{ER} \cong \overline{UL}$ would give you HLR. Can you get any of these four pairs? Sure. The givens include $\overline{EL} \cong \overline{RU}$. Adding $\overline{LR}$ to both of those gives you what you need, $\overline{ER} \cong \overline{UL}$. (If you can't see this, put in numbers: If EL and RU were both 3 and LR were 4, ER and UL would both be 7.) That does it. You use HLR for the medium triangles, and you know how to finish from there. You're done.

So taking it from the top, you add $\overline{LR}$ to $\overline{EL}$ and $\overline{RU}$, giving you $\overline{ER} \cong \overline{UL}$. You use that plus $\overline{ES} \cong \overline{HU}$ (given) and the right angles from the given perpendicular segments to get $\triangle ERS \cong \triangle ULH$ by HLR. Then you use CPCTC to get $\overline{LH} \cong \overline{SR}$. Using these sides, $\overline{EL} \cong \overline{RU}$ (given), and another pair of congruent right angles, you get $\triangle ELH \cong \triangle URS$ with SAS. CPCTC then gives you $\overline{EH} \cong \overline{SU}$ to finish the proof.

(Note that there's at least one other good way to do this proof. So if you saw a method different from what I described in this game plan, your method may be just as good.)

Running a Reverse with Indirect Proofs

To wrap up this chapter, I want to discuss indirect proofs — a different type of proof that's sort of a weird uncle of your regular two-column proofs. With an *indirect* proof, instead of proving that something must be true, you prove it *indirectly* by showing that it can't be false.

Note the *not*. When your task in a proof is to prove that things are *not* congruent, *not* perpendicular, and so on, it's a dead giveaway that you're dealing with an indirect proof.

Chapter 9: Completing Congruent Triangle Proofs

For the most part, an indirect proof is very similar to a regular two-column proof. What makes it different is the way it begins and ends. And except for the beginning and end, to solve an indirect proof, you use the same techniques and theorems that you've been using on regular proofs.

The best way to explain indirect proofs is by showing you an example. Here you go.

Given: $\overrightarrow{SQ}$ bisects ∠PSR

∠PQS ≇ ∠RQS

Prove: $\overline{PS}$ ≇ $\overline{RS}$

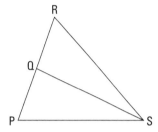

Note two peculiar things about this odd duck of a proof: the *not*-congruent symbols in the givens and the *prove* statement. The one in the *prove* statement is sort of what makes this an indirect proof.

Here's a game plan showing how you can tackle this indirect proof. You assume that the *prove* statement is false, namely that $\overline{PS}$ is congruent to $\overline{RS}$, and then your goal is to arrive at a contradiction of some known true thing (usually a given fact about things that are *not* congruent, *not* perpendicular, and so on). In this problem, your goal is to show that ∠PQS *is* congruent to ∠RQS, which contradicts the given.

One last thing before showing you the solution — you can write out indirect proofs in the regular two-column format, but many geometry textbooks and teachers present indirect proofs in paragraph form, like this:

1. **Assume the opposite of the *prove* statement, treating this opposite statement as a given.**

   Assume $\overline{PS} \cong \overline{RS}$.

2. **Work through the problem as usual, trying to prove the opposite of one of the givens (usually the one that states something is *not* perpendicular, congruent, or the like).**

   Because $\overrightarrow{SQ}$ bisects ∠PSR, you know that ∠PSQ ≅ ∠RSQ. You also know that $\overline{QS} \cong \overline{QS}$ by the Reflexive Property. Using these two congruences plus the one from Step 1, you can conclude that △PSQ ≅ △RSQ by SAS, and hence ∠PQS ≅ ∠RQS by CPCTC.

3. Finish by stating that you've reached a contradiction and that, therefore, the *prove* statement must be true.

This last statement is impossible because it contradicts the given fact that ∠PQS ≇ ∠RQS. Consequently, the assumption ($\overline{PS} \cong \overline{RS}$) must be false, and thus its opposite ($\overline{PS} \not\cong \overline{RS}$) must be true. Q.E.D. (*Quod erat demonstrandum* — "which was to be demonstrated" — for all you Latin-speakers out there; the rest of you can just say, "We're done!")

Note: After you assume that $\overline{PS} \cong \overline{RS}$, it works just like a given. And after you identify your goal of showing ∠PQS ≅ ∠RQS, this goal now works like an ordinary *prove* statement. In fact, after you do these two indirect proof steps, the rest of the proof, which starts with the givens (including the new "given" $\overline{PS} \cong \overline{RS}$) and ends with ∠PQS ≅ ∠RQS, is exactly like a regular proof (although it looks different because it's in paragraph form).

Understanding why indirect proofs work

With indirect proofs, you enter the realm of the double negative. Just as multiplying two negative numbers gives you a positive answer, using two negatives in the English language gives you a positive statement. For example, something that *isn't false* is true, and if the statement that two angles *aren't congruent* is false, then the angles are congruent.

Say that you have to prove that idea *P* is true. A regular proof might go like this: *A* and *B* are given, and from that you can deduce *C;* and if *C* is true, *P* must be true. With an indirect proof, you take a different tack. You prove that *P* can't be false. To do that, you assume that *P is* false and then show that that leads to an impossible conclusion (like the conclusion that *A* is false, which is impossible because *A* is given and therefore true). Finally, because assuming *P* was false led to an impossibility, *P* must be true. Like 2 and 2 is 4, right?

Unfortunately, there's one more little twist. In a typical indirect proof, the thing you're asked to prove is a negative-sounding statement that something is *not* congruent or *not* perpendicular (you could have a *prove* statement like ∠M ≇ ∠N, for example). So when you assume that that's false, you're assuming some regular-sounding, positive thing (like ∠M ≅ ∠N). So just note that a *true* statement can be *negative* (like $\overline{AB} \not\cong \overline{CD}$) and a *false* statement can be *positive* (like $\overline{AB} \cong \overline{CD}$). This little twist doesn't affect the basic logic, but I point it out just because all this stuff about assuming that something is *false* and the statements about angles *not* being congruent can be a bit confusing. (I hope that none of this hasn't not been too unclear.)

Part IV
Polygons of the Four-or-More Sided Variety

In this part . . .

In Part IV, you move on up to polygons that have more than three sides. First, you check out the properties of quadrilaterals: four-sided polygons such as rectangles, parallelograms, kites, and rhombuses (or *rhombi,* if you're a stickler for Latin plurals). Next, you find out how to prove that a shape with four sides qualifies as one of these special quadrilaterals. Then you take a look at a number of perfectly pleasing polygon principles that apply to polygons of any number of sides, from three to almost ∞ (which means *infinity,* in case you're wondering). You also see how to compute a polygon's area and the number of its diagonals. Lastly, you find out what it means for two polygons to be *similar* (they're the same shape), and then you return to the simplest polygon, the triangle, to discover how to prove that two triangles are similar.

Chapter 10

The Seven Wonders of the Quadrilateral World

In This Chapter
▶ Crossing the road to get to the other side: Parallel lines and transversals
▶ Tracing the family tree of quadrilaterals
▶ Flying high with kites and trapezoids
▶ Plumbing the depths of parallelograms, rhombuses, rectangles, and squares

*I*n Chapters 7, 8, and 9, you deal with three-sided polygons — triangles. In this chapter and the next, you check out *quadrilaterals,* polygons with four sides. Then, in Chapter 12, you see polygons up to a gazillion sides. Totally exciting, right?

The most familiar quadrilateral, the rectangle, is by far the most common shape in the everyday world around you. Look around. Wherever you are, there are surely rectangular shapes in sight: books, tabletops, picture frames, walls, ceilings, floors, laptops, and so on.

Mathematicians have been studying quadrilaterals for over 2,000 years. All sorts of fascinating things have been discovered about these four-sided figures, and that's why I've devoted this chapter to their definitions, properties, and classifications. Most of these quadrilaterals have parallel sides, so I introduce you to some parallel-line properties as well.

Getting Started with Parallel-Line Properties

Parallel lines are important when you study quadrilaterals because six of the seven types of quadrilaterals (all of them except the kite) contain parallel lines. In this section, I show you some interesting parallel-line properties.

Crossing the line with transversals: Definitions and theorems

Check out Figure 10-1, which shows three lines that kind of resemble a giant not-equal sign. The two horizontal lines are parallel, and the third line that crosses them is called a *transversal*. As you can see, the three lines form eight angles.

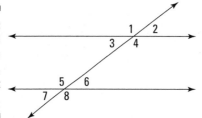

Figure 10-1: Two parallel lines, one transversal, and eight angles.

The eight angles formed by parallel lines and a transversal are either congruent or supplementary. The following theorems tell you how various pairs of angles relate to each other.

Proving that angles are congruent: If a transversal intersects two parallel lines, then the following angles are congruent (refer to Figure 10-1):

- **Alternate interior angles:** The pair of angles 3 and 6 (as well as 4 and 5) are *alternate interior angles*. These angle pairs are on opposite (alternate) sides of the transversal and are in between (in the interior of) the parallel lines.

- **Alternate exterior angles:** Angles 1 and 8 (and angles 2 and 7) are called *alternate exterior angles*. They're on opposite sides of the transversal, and they're outside the parallel lines.

- **Corresponding angles:** The pair of angles 1 and 5 (also 2 and 6, 3 and 7, and 4 and 8) are *corresponding angles*. Angles 1 and 5 are corresponding because each is in the same position (the upper left-hand corner) in its group of four angles.

Also notice that angles 1 and 4, 2 and 3, 5 and 8, and 6 and 7 are across from each other, forming vertical angles, which are also congruent (see Chapter 5 for details).

Proving that angles are supplementary: If a transversal intersects two parallel lines, then the following angles are supplementary (see Figure 10-1):

- **Same-side interior angles:** Angles 3 and 5 (and 4 and 6) are on the same side of the transversal and are in the interior of the parallel lines, so they're called (ready for a shock?) *same-side interior angles.*

- **Same-side exterior angles:** Angles 1 and 7 (and 2 and 8) are called *same-side exterior angles* — they're on the same side of the transversal, and they're outside the parallel lines.

Any two of the eight angles are either *congruent* or *supplementary.* You can sum up the definitions and theorems about transversals in this simple, concise idea. When you have two parallel lines cut by a transversal, you get four acute angles and four obtuse angles (except when you get eight right angles). All the acute angles are congruent, all the obtuse angles are congruent, and each acute angle is supplementary to each obtuse angle.

Proving that lines are parallel: All the theorems in this section work in reverse. You can use the following theorems to prove that lines are parallel. That is, two lines are parallel if they're cut by a transversal such that

- Two corresponding angles are congruent.
- Two alternate interior angles are congruent.
- Two alternate exterior angles are congruent.
- Two same-side interior angles are supplementary.
- Two same-side exterior angles are supplementary.

Applying the transversal theorems

Here's a problem that lets you take a look at some of the theorems in action: Given that lines *m* and *n* are parallel, find the measure of ∠1.

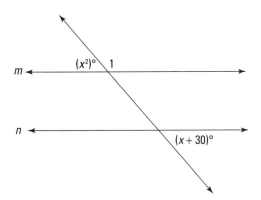

Here's the solution: The $(x^2)°$ angle and the $(x + 30)°$ angle are alternate exterior angles and are therefore congruent. (Or you can use the preceding section's tip about transversals: Because the two angles are both obviously acute, they must be congruent.) Set them equal to each other and solve for x:

$$x^2 = x + 30$$

Set equal to zero: $\quad x^2 - x - 30 = 0$

Factor: $\quad (x - 6)(x + 5) = 0$

Use Zero Product Property: $\quad x - 6 = 0 \quad \text{or} \quad x + 5 = 0$

$$x = 6 \quad \text{or} \quad x = -5$$

This equation has two solutions, so take them one at a time and plug them into the x's in the alternate exterior angles. Plugging $x = 6$ into x^2 gives you 36° for that angle. And because ∠1 is its supplement, ∠1 must be $180° - 36°$, or 144°. The $x = -5$ solution gives you 25° for the x^2 angle and 155° for ∠1. So 144° and 155° are your answers for ∠1.

When you get two solutions (such as $x = 6$ and $x = -5$) in a problem like this, you *do not* plug one of them into one of the x's (like $6^2 = 36$) and the other solution into the other x (like $-5 + 30 = 25$). You have to plug one of the solutions into *all* x's, giving you one result for both angles ($6^2 = 36$ and $6 + 30 = 36$); then you have to separately plug the other solution into *all* x's, giving you a second result for both angles ($[-5]^2 = 25$ and $-5 + 30 = 25$).

Angles and segments can't have negative measures or lengths. Make sure that each solution for x produces *positive* answers for *all* the angles or segments in a problem (in the preceding problem, you should check both the $(x^2)°$ angle and the $(x + 30)°$ angle with each solution for x). If a solution makes any angle or segment in the diagram negative, it must be rejected even if the angles or segments you care about end up being positive. However, *do not* reject a solution just because x is negative: x can be negative as long as the angles and segments are positive ($x = -5$, for example, works just fine in the example problem).

Now here's a proof that uses some of the transversal theorems:

Given: $\overline{LK} \cong \overline{HI}$
$\overline{LK} \parallel \overline{HI}$
$\overline{GK} \cong \overline{JH}$

Prove: $\overline{LJ} \parallel \overline{GI}$

Check out the formal proof:

Statements	Reasons
1) $\overline{LK} \cong \overline{HI}$	1) Given.
2) $\overline{LK} \parallel \overline{HI}$	2) Given.
3) $\angle K \cong \angle H$	3) If lines are parallel, then alternate interior angles are congruent.
4) $\overline{GK} \cong \overline{JH}$	4) Given.
5) $\overline{JK} \cong \overline{GH}$	5) If a segment ($\overline{GJ}$) is subtracted from two congruent segments, then the differences are congruent.
6) $\triangle JKL \cong \triangle GHI$	6) SAS (1, 3, 5).
7) $\angle LJK \cong \angle IGH$	7) CPCTC.
8) $\overline{LJ} \parallel \overline{GI}$	8) If alternate exterior angles are congruent, then lines are parallel.

Extend the lines in transversal problems. Extending the parallel lines and transversals may help you see how the angles are related.

For instance, if you have a hard time seeing that $\angle K$ and $\angle H$ are indeed alternate interior angles (for step 3 of the proof), rotate the figure (or tilt your head) until the parallel segments $\overline{LK}$ and $\overline{HI}$ are horizontal; then extend $\overline{LK}$, $\overline{HI}$, and $\overline{HK}$ in both directions, turning them into lines (you know, with arrows). After doing that, you're looking at the familiar parallel-line scheme shown in Figure 10-1. You can do the same thing for $\angle LJK$ and $\angle IGH$ by extending $\overline{LJ}$ and $\overline{GI}$.

Working with more than one transversal

When a parallel-lines-with-transversal drawing contains more than three lines, identifying congruent and supplementary angles can be kind of challenging. Figure 10-2 shows you parallel lines with two transversals.

Part IV: Polygons of the Four-or-More Sided Variety

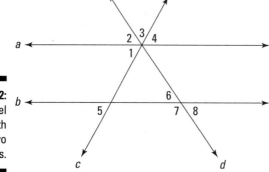

Figure 10-2: Parallel lines with two transversals.

If you get a figure that has more than three lines and you want to use any of the transversal ideas, make sure you're using only three of the lines at a time (two parallel lines and one transversal). If you aren't using a set of three lines like this, the theorems just don't work. With Figure 10-2, you can use lines *a*, *b*, and *c*, or you can use lines *a*, *b*, and *d*, but you *can't* use both transversals *c* and *d* at the same time. Thus, you can't, for example, conclude anything about ∠1 and ∠6 because ∠1 is on transversal *c* and ∠6 is on transversal *d*.

Table 10-1 shows what you can say about several pairs of angles in Figure 10-2, noting whether you can conclude that they're congruent or supplementary. As you read through this table, remember the warning about using only two parallel lines and one transversal.

Table 10-1		Sorting Out Angles and Transversals
Angle Pair	*Conclusion*	*Reason*
2 and 8	Congruent	∠2 and ∠8 are alternate exterior angles on transversal *d*
3 and 6	Nothing	To make ∠3 you need to use both transversals, *c* and *d*
4 and 5	Congruent	∠4 and ∠5 are alternate exterior angles on *c*
4 and 6	Nothing	∠4 is on transversal *c*; ∠6 is on transversal *d*
2 and 7	Supplementary	∠2 and ∠7 are same-side exterior angles on *d*
1 and 8	Nothing	∠1 is on transversal *c*; ∠8 is on transversal *d*
4 and 8	Nothing	∠4 is on transversal *c*; ∠8 is on transversal *d*

If you get a figure with more than one transversal or more than one set of parallel lines, you may want to do the following: Trace the figure from your book to a sheet of paper and then highlight one pair of parallel lines and one transversal. (Or you can just trace the three lines you want to work with.) Then you can use the transversal ideas on the highlighted lines. After that, you can highlight a different group of three lines and work with those.

Of course, instead of tracing and highlighting, you can just make sure that the two angles you're analyzing use only three lines (one ray of each angle should be the common transversal, and the other ray should be one of two parallel lines).

Meeting the Seven Members of the Quadrilateral Family

A *quadrilateral* is a shape with four straight sides. In this section and the next, you find out about the seven quadrilaterals. Some are surely familiar to you, and some may not be so familiar. Check out the following definitions and the quadrilateral family tree in Figure 10-3.

If you know what the quadrilaterals look like, their definitions should make sense and be pretty easy to understand (though the first one is a bit of a mouthful). Here are the seven quadrilaterals:

- **Kite:** A quadrilateral in which two disjoint pairs of consecutive sides are congruent ("disjoint pairs" means that one side can't be used in both pairs)
- **Parallelogram:** A quadrilateral that has two pairs of parallel sides
- **Rhombus:** A quadrilateral with four congruent sides; a rhombus is both a kite and a parallelogram
- **Rectangle:** A quadrilateral with four right angles; a rectangle is a type of parallelogram
- **Square:** A quadrilateral with four congruent sides and four right angles; a square is both a rhombus and a rectangle
- **Trapezoid:** A quadrilateral with exactly one pair of parallel sides (the parallel sides are called *bases*)
- **Isosceles trapezoid:** A trapezoid in which the nonparallel sides (the *legs*) are congruent

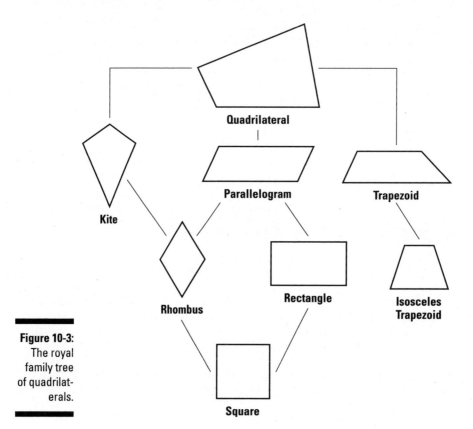

Figure 10-3: The royal family tree of quadrilaterals.

In the hierarchy of quadrilaterals shown in Figure 10-3, a quadrilateral below another on the family tree is a special case of the one above it. A rectangle, for example, is a special case of a parallelogram. Thus, you can say that a rectangle is a parallelogram but not that a parallelogram is a rectangle (a parallelogram is only *sometimes* a rectangle).

Looking at quadrilateral relationships

The quadrilateral family tree shows you the relationships among the various quadrilaterals. Table 10-2 gives you a taste of some of these relationships. (You can test yourself by supplying the answers — *always, sometimes,* or *never* — before you check the answer column.)

Chapter 10: The Seven Wonders of the Quadrilateral World

Table 10-2	How Are These Quadrilaterals Related?
Assertion	**Answer**
A rectangle is a rhombus.	Sometimes (when it's a square)
A kite is a parallelogram.	Sometimes (when it's a rhombus)
A rhombus is a parallelogram.	Always
A kite is a rectangle.	Sometimes (when it's a square)
A trapezoid is a kite.	Never
A parallelogram is a square.	Sometimes
An isosceles trapezoid is a rectangle.	Never
A square is a kite.	Always
A rectangle is a square.	Sometimes

Keep your family tree (Figure 10-3) handy when you're doing *always, sometimes, never* problems because you can use the quadrilaterals' positions on the tree to figure out the answer. Here's how:

- If you go *up* from the first figure to the second, the answer is *always*.
- If you go *down* from the first figure to the second, then the answer is *sometimes*.
- If you can make the connection by going *down and then up* (like from a rectangle to a kite or vice versa), the answer's *sometimes*.
- If the only way to get from one figure to the other is by going *up and then down* (like from a parallelogram to an isosceles trapezoid), the answer is *never*.

Working with auxiliary lines

The following proof introduces you to a new idea: adding a line or segment (called an *auxiliary line*) to a proof diagram to help you do the proof. Some proofs are impossible to solve until you add a line to the diagram.

Auxiliary lines often create congruent triangles, or they intersect existing lines at right angles. So if you're stumped by a proof, see whether drawing an auxiliary line (or lines) could get you one of those things.

160 Part IV: Polygons of the Four-or-More Sided Variety

Two points determine a line: When you draw in an auxiliary line, just write something like "Draw $\overline{AB}$" in the statement column; then use the following postulate in the reason column: Two points determine a line (or ray or segment).

Here's an example proof:

Given: *GRAM* is a parallelogram
Prove: $\overline{GR} \cong \overline{AM}$

You might come up with a game plan like the following:

- **Take a look at the givens.** The only thing you can conclude from the single given is that the sides of *GRAM* are parallel (using the definition of a parallelogram). But it doesn't seem like you can go anywhere from there.

- **Jump to the end of the proof.** What could be the justification for the final statement, $\overline{GR} \cong \overline{AM}$? At this point, no justification seems possible, so put on your thinking cap.

- **Consider drawing an auxiliary line.** If you draw $\overline{RM}$, as shown in Figure 10-4, you get triangles that look congruent. And if you could show that they're congruent, the proof could then end with CPCTC. (*Note:* You can do the proof in a similar way by drawing $\overline{GA}$ instead of $\overline{RM}$.)

- **Show the triangles congruent.** To show that the triangles are congruent, you use $\overline{RM}$ as a transversal. First use it with parallel sides $\overline{RA}$ and $\overline{GM}$; that gives you congruent, alternate interior angles *GMR* and *ARM* (see the earlier "Getting Started with Parallel-Line Properties" section). Then use $\overline{RM}$ with parallel sides $\overline{GR}$ and $\overline{MA}$; that gives you two more congruent, alternate interior angles, *GRM* and *AMR*. These two pairs of congruent angles, along with side $\overline{RM}$ (which is congruent to itself by the Reflexive Property), prove the triangles congruent with ASA. That does it.

Figure 10-4: Connecting two points on the figure creates triangles you can use in your proof.

Here's the formal proof:

Statements	Reasons
1) *GRAM* is a parallelogram	1) Given.
2) Draw $\overline{RM}$	2) Two points determine a segment.
3) $\overline{RA} \parallel \overline{GM}$	3) Definition of parallelogram.
4) $\angle GMR \cong \angle ARM$	4) If two parallel lines ($\overleftrightarrow{RA}$ and $\overleftrightarrow{GM}$) are cut by a transversal ($\overleftrightarrow{RM}$), then alternate interior angles are congruent.
5) $\overline{GR} \parallel \overline{MA}$	5) Definition of parallelogram.
6) $\angle GRM \cong \angle AMR$	6) Same as Reason 4, but this time $\overleftrightarrow{GR}$ and $\overleftrightarrow{MA}$ are the parallel lines.
7) $\overline{RM} \cong \overline{MR}$	7) Reflexive Property.
8) $\triangle GRM \cong \triangle AMR$	8) ASA (4, 7, 6).
9) $\overline{GR} \cong \overline{AM}$	9) CPCTC.

A good way to spot congruent alternate interior angles in a diagram is to look for pairs of so-called *Z-angles*. Look for a Z or backward Z — or a stretched-out Z or backward Z — as shown in Figures 10-5 and 10-6. The angles in the crooks of the Z are congruent.

Figure 10-5: Four pairs of Z-angles.

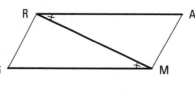
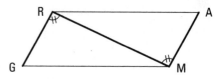

Figure 10-6: The two pairs of Z-angles from the preceding proof — a backward Z and a tipped Z.

Backward "Z"
∠GMR ≅ ∠ARM

Turn figure sideways to see this "Z"
∠GRM ≅ ∠AMR

Giving Props to Quads: The Properties of Quadrilaterals

The *properties* of the quadrilaterals are simply the things that are true about them. The properties of a particular quadrilateral concern its

- ✓ **Sides:** Are they congruent? Parallel?
- ✓ **Angles:** Are they congruent? Supplementary? Right?
- ✓ **Diagonals:** Are they congruent? Perpendicular? Do they bisect each other? Do they bisect the angles whose vertices they meet?

I present a total of about 30 quadrilateral properties, which may seem like a lot to memorize. No worries. You don't have to rely solely on memorization. Here's a great tip that makes learning the properties a snap.

If you can't remember whether something is a property of some quadrilateral, just sketch the quadrilateral in question. If the thing looks true, it's probably a property; if it doesn't look true, it's not a property. (This method is almost foolproof, but it's a bit un-math-teacherly of me to say it — so don't quote me, or I might get in trouble with the math police.)

Properties of the parallelogram

I have a feeling you can guess what's in this section. You got it — the properties of parallelograms.

The parallelogram has the following properties:

- Opposite sides are parallel by definition.
- Opposite sides are congruent.
- Opposite angles are congruent.
- Consecutive angles are supplementary.
- The diagonals bisect each other.

If you just look at a parallelogram, the things that look true (namely, the things on this list) *are* true and are thus properties, and the things that don't look like they're true aren't properties.

If you draw a picture to help you figure out a quadrilateral's properties, make your sketch as general as possible. For instance, as you sketch your parallelogram, make sure it's not almost a rhombus (with four sides that are almost congruent) or almost a rectangle (with four angles close to right angles). If your parallelogram sketch is close to, say, a rectangle, something that's true for all rectangles but not true for all parallelograms (such as congruent diagonals) may look true and thus cause you to mistakenly conclude that it's a property of parallelograms. Capiche?

Imagine that you can't remember the properties of a parallelogram. You could just sketch one (as in Figure 10-7) and run through all things that might be properties.

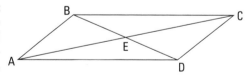

Figure 10-7:
A run-of-the-mill parallelogram.

Table 10-3 concerns questions about the sides of a parallelogram (refer to Figure 10-7).

Table 10-3   Asking Questions about Sides of Parallelograms

Do Any Sides Appear to Be...	Answer
Congruent?	Yes, opposite sides look congruent, and that's a property. But adjacent sides don't look congruent, and that's *not* a property.
Parallel?	Yes, opposite sides look parallel (and of course, you know this property if you know the definition of a parallelogram).

Table 10-4 explores the angles of a parallelogram (see Figure 10-7 again).

Table 10-4   Asking Questions about Angles of Parallelograms

Do Any Angles Appear to Be...	Answer
Congruent?	Yes, opposite angles look congruent, and that's a property. (Angles A and C appear to be about 45°, and angles B and D look like about 135°).
Supplementary?	Yes, consecutive angles (like angles A and B) look like they're supplementary, and that's a property. (Using parallel lines $\overleftrightarrow{BC}$ and $\overleftrightarrow{AD}$ and transversal $\overleftrightarrow{AB}$, angles A and B are same-side interior angles and are therefore supplementary.)
Right angles?	Obviously not, and that's not a property.

Table 10-5 addresses statements about the diagonals of a parallelogram (see Figure 10-7).

Table 10-5   Asking Questions about Diagonals of Parallelograms

Do the Diagonals Appear to Be...	Answer
Congruent?	Not even close (in Figure 10-7, one is roughly twice as long as the other, which surprises most people; measure them if you don't believe me!) — not a property.
Perpendicular?	Not even close; not a property.

Chapter 10: The Seven Wonders of the Quadrilateral World

Do the Diagonals Appear to Be...	Answer
Bisecting each other?	Yes, each one seems to cut the other in half, and that's a property.
Bisecting the angles whose vertices they meet?	No. At a quick glance, you might think that $\angle A$ (or $\angle C$) is bisected by diagonal $\overline{AC}$, but if you look closely, you see that $\angle BAC$ is actually about twice as big as $\angle DAC$. And of course, diagonal $\overline{BD}$ doesn't come close to bisecting $\angle B$ or $\angle D$. Not a property.

Look at your sketch carefully. When I show students a parallelogram like the one in Figure 10-7 and ask them whether the diagonals look congruent, they often tell me that they do despite the fact that one is literally twice as long as the other! So when asking yourself whether a potential property looks true, don't just take a quick glance at the quadrilateral, and don't let your eyes play tricks on you. Look at the segments or angles in question very carefully.

The sketching-quadrilaterals method and the questions in the preceding three tables brings me to an important tip about mathematics in general.

Whenever possible, bolster your memorization of rules, formulas, concepts, and so on by trying to see *why* they're true or *why* they seem to make sense. Not only does this make the ideas easier to memorize, but it also helps you see connections to other ideas, and that fosters a deeper understanding of mathematics.

And now for a parallelogram proof:

Given: *JKLM* is a parallelogram

T is the midpoint of $\overline{JM}$

S is the midpoint of $\overline{KL}$

Prove: $\angle 1 \cong \angle 2$

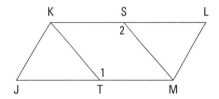

Use quadrilateral properties in quadrilateral proofs. If one of the givens in a proof is that a shape is a particular quadrilateral, you can be sure that you'll need to use one or more of the properties of that quadrilateral in the proof — usually to show triangles congruent.

Your thought process might go like this:

- **Note the congruent triangles.** If you could prove the triangles congruent, you could get $\angle JTK \cong \angle LSM$ by CPCTC and then get $\angle 1 \cong \angle 2$ by supplements of congruent angles.
- **Prove the triangles congruent.** Can you use some of the properties of a parallelogram? Sure: *Opposite sides are congruent* gives you $\overline{JK} \cong \overline{LM}$, and *opposite angles are congruent* gives you $\angle J \cong \angle L$. Two congruent pairs down, one to go. The two midpoints in the given is a clue that you need to use Like Divisions (see Chapter 5). That's it — you cut congruent sides $\overline{JM}$ and $\overline{KL}$ in half to give you $\overline{JT} \cong \overline{LS}$, which gives you congruent triangles by SAS.

Here's the formal proof:

Statements	Reasons
1) JKLM is a parallelogram	1) Given.
2) $\overline{JK} \cong \overline{LM}$	2) Opposite sides of a parallelogram are congruent.
3) $\angle J \cong \angle L$	3) Opposite angles of a parallelogram are congruent.
4) $\overline{JM} \cong \overline{KL}$	4) Opposite sides of a parallelogram are congruent.
5) T is the midpoint of $\overline{JM}$ S is the midpoint of $\overline{KL}$	5) Given.
6) $\overline{JT} \cong \overline{LS}$	6) Like Divisions.
7) $\triangle JTK \cong \triangle LSM$	7) SAS (2, 3, 6).
8) $\angle JTK \cong \angle LSM$	8) CPCTC.
9) $\angle 1 \cong \angle 2$	9) Supplements of congruent angles are congruent.

Properties of the three special parallelograms

Figure 10-8 shows you the three *special* parallelograms, so-called because they're, as mathematicians say, *special cases* of the parallelogram. (In addition, the square is a special case or type of both the rectangle and the rhombus.)

The three-level hierarchy you see with *parallelogram → rectangle → square* or *parallelogram → rhombus → square* in the quadrilateral family tree (Figure 10-3) works just like *mammal → dog → Dalmatian*. A dog is a special type of a mammal, and a Dalmatian is a special type of a dog.

Before reading the properties that follow, try figuring them out on your own. Using the shapes in Figure 10-8, run down the list of possible properties from the beginning of "Giving Props to Quads: The Properties of Quadrilaterals," asking yourself whether they look like they're true for the rhombus, the rectangle, and the square. (Note that both the rhombus and the rectangle in Figure 10-8 are drawn as generally as possible; in other words, neither one resembles a square. Also, note that the rhombus is vertical rather than on its side like parallelograms are usually drawn; this is the easier and better way to draw a rhombus because you can more easily see its symmetry and the fact that its diagonals are perpendicular.)

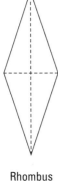
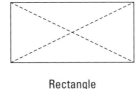

Figure 10-8: The two kids and one grandkid (the square) of the parallelogram.

Rhombus   Rectangle   Square

Here are the properties of the rhombus, rectangle, and square. Note that because these three quadrilaterals are all parallelograms, their properties include the parallelogram properties.

✓ **The rhombus has the following properties:**

- All the properties of a parallelogram apply (the ones that matter here are parallel sides, opposite angles are congruent, and consecutive angles are supplementary).
- All sides are congruent by definition.
- The diagonals bisect the angles.
- The diagonals are perpendicular bisectors of each other.

✔ **The rectangle has the following properties:**

- All the properties of a parallelogram apply (the ones that matter here are parallel sides, opposite sides are congruent, and diagonals bisect each other).
- All angles are right angles by definition.
- The diagonals are congruent.

✔ **The square has the following properties:**

- All the properties of a rhombus apply (the ones that matter here are parallel sides, diagonals are perpendicular bisectors of each other, and diagonals bisect the angles).
- All the properties of a rectangle apply (the only one that matters here is diagonals are congruent).
- All sides are congruent by definition.
- All angles are right angles by definition.

Now try working through a couple of problems: Given the rectangle as shown, find the measures of ∠1 and ∠2:

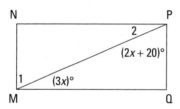

Here's the solution: *MNPQ* is a rectangle, so ∠*Q* = 90°. Thus, because there are 180° in a triangle, you can say the following:

$$90° + (3x)° + (2x + 20)° = 180°$$
$$3x + 2x + 20 = 90$$
$$5x = 70$$
$$x = 14$$

Now plug in 14 for all the *x*'s. Angle *QMP*, (3*x*)°, is 3 · 14, or 42°, and because you have a rectangle, ∠1 is its complement and is therefore 90° – 42°, or 48°. Angle *QPM*, (2*x* + 20)°, is 2 · 14 + 20, or 48°, and ∠2, its complement, is therefore 42°.

Chapter 10: The Seven Wonders of the Quadrilateral World

Now find the perimeter of rhombus *RHOM*.

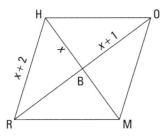

Here's the solution: All the sides of a rhombus are congruent, so *HO* equals $x + 2$. And because the diagonals of a rhombus are perpendicular, $\triangle HBO$ is a right triangle. You finish with the Pythagorean Theorem:

$$a^2 + b^2 = c^2$$
$$(HB)^2 + (BO)^2 = (HO)^2$$
$$x^2 + (x+1)^2 = (x+2)^2$$
$$x^2 + x^2 + 2x + 1 = x^2 + 4x + 4$$

Combine like terms and set equal to zero: $\quad x^2 - 2x - 3 = 0$

Factor: $\quad (x-3)(x+1) = 0$

Use Zero Product Property: $\quad x - 3 = 0 \quad \text{or} \quad x + 1 = 0$
$$x = 3 \quad \text{or} \quad x = -1$$

You can reject $x = -1$ because that would result in $\triangle HBO$ having legs with lengths of -1 and 0. So x equals 3, which gives $\overline{HR}$ a length of 5. Because rhombuses have four congruent sides, *RHOM* has a perimeter of $4 \cdot 5$, or 20 units.

Properties of the kite

Check out the kite in Figure 10-9 and try to figure out its properties before reading the list that follows.

Part IV: Polygons of the Four-or-More Sided Variety

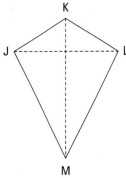

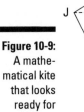

Figure 10-9: A mathematical kite that looks ready for flying.

The properties of the kite are as follows:

- Two disjoint pairs of consecutive sides are congruent by definition ($\overline{JK} \cong \overline{LK}$ and $\overline{JM} \cong \overline{LM}$). *Note: Disjoint* means that one side can't be used in both pairs — the two pairs are totally separate.
- The diagonals are perpendicular.
- One diagonal ($\overline{KM}$, the *main diagonal*) is the perpendicular bisector of the other diagonal ($\overline{JL}$, the *cross diagonal*). (The terms "main diagonal" and "cross diagonal" are quite useful, but don't look for them in other geometry books because I made them up.)
- The main diagonal bisects a pair of opposite angles ($\angle K$ and $\angle M$).
- The opposite angles at the endpoints of the cross diagonal are congruent ($\angle J$ and $\angle L$).

The last three properties are called the *half properties* of the kite.

Grab an energy drink and get ready for another proof. Due to space considerations, I'm going to skip the game plan this time. You're on your own — egad!

Given: $RSTV$ is a kite with $\overline{RS} \cong \overline{TS}$
$\overline{WV} \cong \overline{UV}$

Prove: $\overline{WS} \cong \overline{US}$

Chapter 10: The Seven Wonders of the Quadrilateral World   **171**

Statements	Reasons
1) $RSTV$ is a kite with $\overline{RS} \cong \overline{TS}$	1) Given.
2) $\overline{RV} \cong \overline{TV}$	2) A kite has two disjoint pairs of congruent sides.
3) $\overline{WV} \cong \overline{UV}$	3) Given.
4) $\overline{RW} \cong \overline{TU}$	4) If two congruent segments ($\overline{WV}$ and $\overline{UV}$) are subtracted from two other congruent segments ($\overline{RV}$ and $\overline{TV}$), then the differences are congruent.
5) $\angle R \cong \angle T$	5) The opposite angles at the endpoints of the cross diagonal are congruent.
6) $\triangle SRW \cong \triangle STU$	6) SAS (1, 5, 4).
7) $\overline{WS} \cong \overline{US}$	7) CPCTC.

Properties of the trapezoid and the isosceles trapezoid

Practice your picking-out-properties proficiency one more time with the trapezoid and isosceles trapezoid in Figure 10-10. *Remember:* What looks true is likely true, and what doesn't, isn't.

Figure 10-10:
A trapezoid (on the left) and an isosceles trapezoid (on the right).

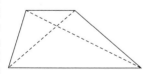

✓ **The properties of the trapezoid are as follows:**

- The bases are parallel by definition.
- Each lower base angle is supplementary to the upper base angle on the same side.

The properties of the isosceles trapezoids are as follows:

- The properties of trapezoids apply by definition (parallel bases).
- The legs are congruent by definition.
- The lower base angles are congruent.
- The upper base angles are congruent.
- Any lower base angle is supplementary to any upper base angle.
- The diagonals are congruent.

Perhaps the hardest property to spot in both diagrams is the one about supplementary angles. Because of the parallel sides, consecutive angles are same-side interior angles and are thus supplementary. (All the quadrilaterals except the kite, by the way, contain consecutive supplementary angles.)

Here's an isosceles trapezoid proof for you. I trust you to handle the game plan this time:

Given: *ZOID* is an isosceles trapezoid with bases $\overline{OI}$ and $\overline{ZD}$

Prove: $\overline{TO} \cong \overline{TI}$

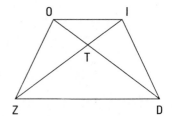

Statements	Reasons
1) *ZOID* is an isosceles trapezoid with bases $\overline{OI}$ and $\overline{ZD}$	1) Given.
2) $\overline{ZO} \cong \overline{DI}$	2) The legs of an isosceles trapezoid are congruent.
3) $\angle ZOI \cong \angle DIO$	3) The upper base angles of an isosceles trapezoid are congruent.
4) $\overline{OI} \cong \overline{IO}$	4) Reflexive Property.
5) $\triangle ZOI \cong \triangle DIO$	5) SAS (2, 3, 4).
6) $\angle ZIO \cong \angle DOI$	6) CPCTC.
7) $\overline{TO} \cong \overline{TI}$	7) If angles, then sides.

Chapter 11

Proving That You've Got a Particular Quadrilateral

In This Chapter

▶ Noting the property-proof connection
▶ Proving a figure is a parallelogram or other quadrilateral

Chapter 10 tells you all about seven different quadrilaterals — their definitions, their properties, what they look like, and where they fit on the family tree. Here, I fill you in on proving that a given quadrilateral qualifies as one of those particular types.

Throughout this chapter, you work on proofs in which you have to show that the quadrilateral in the diagram is, say, a parallelogram or a rectangle or a kite (the final line of the proof — or a line near the end — might be something like "*ABCD* is a rectangle"). Now, maybe you're thinking that'll be easy, that all you have to do is show that the quadrilateral has, say, one of the rectangle properties to prove it's a rectangle. Unfortunately, it's not that simple because some quadrilaterals share the same traits. But don't worry — this chapter tells you how to look beyond any family resemblance to get a positive ID.

Putting Properties and Proof Methods Together

Before getting into the many ways you can prove that a figure is a parallelogram, a rectangle, a kite, and so on, I want to talk about how these proof methods relate to the quadrilateral properties that I cover in Chapter 10. (If, like Sergeant Joe Friday, your motto is "Just the facts, Ma'am," you can skip this discussion and just memorize the proof methods, which you find in the upcoming sections. But if you want a fuller understanding of this topic, read on.)

To begin our discussion of the connection between proof methods and properties, let's consider one of the parallelogram properties from Chapter 10: *Opposite sides of a parallelogram are congruent.* Here's this property in if-then form so you can see its logical structure: *If a quadrilateral is a parallelogram, then its opposite sides are congruent.*

It turns out that the converse (reverse) of this property is also true: *If opposite sides of a quadrilateral are congruent, then it's a parallelogram.* Because the converse of the property is true, you can use the converse as a method of proof. If you're doing a two-column proof in which you have to prove that a quadrilateral is a parallelogram, the final statement might be something like, "*ABCD* is a parallelogram," and the final reason might be, "If opposite sides of a quadrilateral are congruent, then it's a parallelogram."

Some quadrilateral properties are reversible; others are not. *Reversibile or not* is the key. If the converse of a property is also a true statement, then you can use it as a method of proof. But if a property isn't reversible (in other words, its converse is false), you can't use it as a method of proof. The relationship between properties and methods of proof is a bit complicated, but the following guidelines and examples should help clear things up:

- **Definitions always work as a method of proof.** One of the properties of the rhombus is that all its sides are congruent. An abbreviated if-then statement would be *if rhombus, then all sides congruent.*

  Because this property follows from the definition of a rhombus, its converse is also true: *if all sides congruent, then rhombus.* (As I show you in Chapter 4, *all* definitions are reversible; but only some theorems and postulates are reversible.) Because this property is reversible, it's one of the ways of proving that a quadrilateral is a rhombus.

- **When a "child" quadrilateral shares a property with a "parent" quadrilateral, you can't use the converse of that property to prove that you have the child quadrilateral.** (A "child" quadrilateral connects to its "parent" above it on the quadrilateral family tree.) One rhombus property is that both pairs of opposite sides are parallel. In short, *if rhombus, then two pairs of parallel sides.* This property also belongs to the parallelogram, and a rhombus has this property by virtue of the fact that it's a special type of a parallelogram.

  The converse of this statement — *if two pairs of parallel sides, then rhombus* — is clearly false because not all parallelograms are rhombuses; thus, you can't use the converse to prove that you've got a rhombus. If you have two pairs of parallel sides, all you can conclude is that you've got a parallelogram.

- **Some other properties of quadrilaterals are reversible — but you can't always count on it.** Out of the properties that don't follow from definitions and that aren't shared with parent quadrilaterals, most are reversible, but a few aren't. One of the properties of a parallelogram is that its diagonals bisect each other: *if parallelogram, then diagonals bisect each other.* The converse of this — *if diagonals bisect each other, then parallelogram* — is true and is thus one of the ways of proving that a figure is a parallelogram.

  Now consider this property of a rectangle: *if rectangle, then diagonals are congruent.* The converse of this — *if diagonals are congruent, then rectangle* — is false, and thus you can't use it as a method of proving that you have a rectangle. The converse is false because all isosceles trapezoids, some kites, and some no-name quadrilaterals have congruent diagonals. (You can see this by taking two pens or pencils of equal length, crossing them over each other to make them work like diagonals, and moving them around to create different shapes.) And finally, to make matters even more complicated . . .

- **Some methods of proof aren't properties in reverse.** For example, one of the methods of proving that a quadrilateral is a parallelogram is to show that one pair of sides is *both* parallel and congruent. Although this proof method is related to the parallelogram's properties, it isn't the reverse of any single property. For these oddball methods of proof, you may have to resort to just memorizing them.

In case you're curious, this is the end of the theoretical mumbo-jumbo I said you could skip. Now let's come back down to earth and go through the basic proof methods and how to use them.

Before attempting to prove that a figure is a certain quadrilateral, make sure you know all the proof methods well. You want to have them all at your fingertips, and then you should remain flexible — ready to use any of them. After considering the givens, pick the proof method that seems most likely to do the trick. If it works, great; but if after working on it for a little while, it looks like it's not going to work or that it'd take too many steps, switch course and try a different proof method and then maybe a third. After you get really familiar with all the methods, you can sort of consider them all simultaneously.

Proving That a Quadrilateral Is a Parallelogram

The five methods for proving that a quadrilateral is a parallelogram are among the most important proof methods in this chapter. One reason they're important is that you often have to prove that a quadrilateral is a parallelogram before going on to prove that it's one of the special parallelograms (a rectangle, a rhombus, or a square). Parallelogram proofs are the most common type of quadrilateral proof in geometry textbooks, so you'll use these methods over and over again.

Surefire ways of ID-ing a parallelogram

Five ways to prove that a quadrilateral is a parallelogram: You can choose from five ways of proving that a quadrilateral is a parallelogram. The first four are the converses of parallelogram properties (including the definition of a parallelogram). Make sure you remember the oddball fifth one — which isn't the converse of a property — because it often comes in handy:

- If both pairs of opposite sides of a quadrilateral are parallel, then it's a parallelogram (reverse of the definition). Because this is the reverse of the definition, it's technically a definition, not a theorem or postulate, but it works exactly like a theorem, so don't sweat this distinction.

- If both pairs of opposite sides of a quadrilateral are congruent, then it's a parallelogram (converse of a property).

    Tip: To get a feel for why this proof method works, take two toothpicks and two pens or pencils of the same length and put them all together tip-to-tip; create a closed figure, with the toothpicks opposite each other. The only shape you can make is a parallelogram.

- If both pairs of opposite angles of a quadrilateral are congruent, then it's a parallelogram (converse of a property).

- If the diagonals of a quadrilateral bisect each other, then it's a parallelogram (converse of a property).

    Tip: Take, say, a pencil and a toothpick (or two pens or pencils of different lengths) and make them cross each other at their midpoints. No matter how you change the angle they make, their tips form a parallelogram.

- If one pair of opposite sides of a quadrilateral are both parallel and congruent, then it's a parallelogram (neither the reverse of the definition nor the converse of a property).

    Tip: Take two pens or pencils of the same length, holding one in each hand. If you keep them parallel, no matter how you move them around, you can see that their four ends form a parallelogram.

The preceding list contains the converses of four of the five parallelogram properties. If you're wondering why the converse of the fifth property (*consecutive angles are supplementary*) isn't on the list, you have a good mind for details. Essentially, the converse of this property, while true, is difficult to use, and you can always use one of the other methods instead.

Trying some parallelogram proofs

Here's your first parallelogram proof:

Given: ∠UQV ≅ ∠RVQ
∠TUQ ≅ ∠SRV

Prove: QRVU is a parallelogram

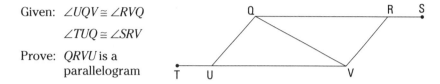

Here's a game plan outlining how your thinking might go:

- **Notice the congruent triangles.** Always check for triangles that look congruent!

- **Jump to the end of the proof and ask yourself whether you could prove that *QRVU* is a parallelogram if you knew that the triangles were congruent.** Using CPCTC, you could show that *QRVU* has two pairs of congruent sides, and that would make it a parallelogram. So . . .

- **Show that the triangles are congruent.** You already have $\overline{QV}$ congruent to itself by the Reflexive Property and one pair of congruent angles (given), and you can get the other angle for AAS with *supplements of congruent angles* (see Chapter 5). That does it.

There are two other good ways to do this proof. If you noticed that the given congruent angles, *UQV* and *RVQ*, are alternate interior angles, you could've correctly concluded that $\overline{UQ}$ and $\overline{VR}$ are parallel. (This is a good thing to notice, so congratulations if you did.) You might then have had the good idea to try to prove the other pair of sides parallel so you could use the first parallelogram proof method. You can do this by proving the triangles congruent, using CPCTC, and then using alternate interior angles *VQR* and *QVU*, but assume, for the sake of argument, that you didn't realize this. It would seem like you're at a dead end. Don't let this frustrate you. When doing proofs, it's not uncommon for good ideas and good plans to lead to dead ends. When this happens, just go back to the drawing board. A third way to do the proof is to get that first pair of parallel lines and then show that they're also congruent — with congruent triangles and CPCTC — and then finish with the fifth parallelogram proof method. These proof methods are perfectly good alternatives; I just thought my method was a bit more straightforward.

Take a look at the formal proof:

Statements	Reasons
1) $\angle UQV \cong \angle RVQ$	1) Given.
2) $\angle TUQ \cong \angle SRV$	2) Given.
3) $\angle VUQ \cong \angle QRV$	3) If two angles are supplementary to two other congruent angles, then they're congruent.
4) $\overline{QV} \cong \overline{VQ}$	4) Reflexive Property.
5) $\triangle VUQ \cong \triangle QRV$	5) AAS (3, 1, 4).
6) $\overline{QU} \cong \overline{RV}$	6) CPCTC.
7) $\overline{UV} \cong \overline{QR}$	7) CPCTC.
8) $QRVU$ is a parallelogram	8) If both pairs of opposite sides of a quadrilateral are congruent, then the quadrilateral is a parallelogram.

Here's another proof — with a pair of parallelograms. This problem gives you more practice with parallelogram proof methods, and because it's a bit longer than the first proof, it'll give you a chance to think through a longer game plan.

Given: *HEJG* is a parallelogram

$\angle DGH \cong \angle FEJ$

Prove: *DEFG* is a parallelogram

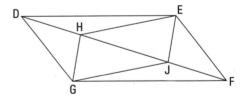

Because all quadrilaterals (except for the kite) contain parallel lines, be on the lookout for opportunities to use the parallel-line theorems from Chapter 10. And always keep your eyes peeled for congruent triangles.

Your game plan might go something like this:

- **Look for congruent triangles.** This diagram takes the cake for containing congruent triangles — it has six pairs of them! Don't spend much time thinking about them — except the ones that might help you — but at least make a quick mental note that they're there.

✔ **Consider the givens.** The given congruent angles, which are parts of △DGH and △FEJ, are a huge hint that you should try to show these triangles congruent. You have those congruent angles and the congruent sides $\overline{HG}$ and $\overline{EJ}$ from parallelogram HEJG, so you need only one more pair of congruent sides or angles to use SAS or ASA (see Chapter 9).

✔ **Think about the end of the proof.** To prove that DEFG is a parallelogram, it would help to know that $\overline{DG} \cong \overline{EF}$, so you'd like to be able to prove the triangles congruent and then get $\overline{DG} \cong \overline{EF}$ by CPCTC. That eliminates the SAS option for proving the triangles congruent because to use SAS, you'd need to know that $\overline{DG} \cong \overline{EF}$ — the very thing you're trying to get with CPCTC. (And if you knew $\overline{DG} \cong \overline{EF}$, there'd be no point to showing that the triangles are congruent, anyway.) So you should try the other option: proving the triangles congruent with ASA.

The second angle pair you'd need for ASA consists of ∠DHG and ∠FJE. They're congruent because they're alternate exterior angles using parallel lines $\overleftrightarrow{HG}$ and $\overleftrightarrow{EJ}$ and transversal $\overleftrightarrow{DF}$. Okay, so the triangles are congruent by ASA, and then you get $\overline{DG} \cong \overline{EF}$ by CPCTC. You're on your way.

✔ **Consider parallelogram proof methods.** You now have one pair of congruent sides of DEFG. Two of the parallelogram proof methods use a pair of congruent sides. To complete one of these methods, you need to show one of the following:

- That the other pair of opposite sides are congruent
- That $\overline{DG}$ and $\overline{EF}$ are parallel as well as congruent

Ask yourself which approach looks easier or quicker. Showing $\overline{DE} \cong \overline{GF}$ would probably require showing a second pair of triangles congruent, and that looks like it'd take a few more steps, so try the other tack.

Can you show $\overline{DG} \parallel \overline{EF}$? Sure, with one of the parallel-line theorems from Chapter 10. Because angles GDH and EFJ are congruent (by CPCTC), you can finish by using those angles as congruent alternate interior angles, or Z-angles, to give you $\overline{DG} \parallel \overline{EF}$. That's a wrap!

Now take a look at the formal proof:

Statements	Reasons
1) HEJG is a parallelogram	1) Given.
2) $\overline{HG} \cong \overline{EJ}$	2) Opposite sides of a parallelogram are congruent.
3) $\overline{HG} \parallel \overline{EJ}$	3) Opposite sides of a parallelogram are parallel.
4) ∠DHG ≅ ∠FJE	4) If lines are parallel, then alternate exterior angles are congruent.

(continued)

Statements	Reasons
5) $\angle DGH \cong \angle FEJ$	5) Given.
6) $\triangle DGH \cong \triangle FEJ$	6) ASA (4, 2, 5).
7) $\overline{DG} \cong \overline{EF}$	7) CPCTC.
8) $\angle GDH \cong \angle EFJ$	8) CPCTC.
9) $\overline{DG} \parallel \overline{EF}$	9) If alternate interior angles are congruent ($\angle GDH$ and $\angle EFJ$), then lines are parallel.
10) DEFG is a parallelogram	10) If one pair of opposite sides of a quadrilateral are both parallel and congruent, then the quadrilateral is a parallelogram (lines 9 and 7).

Note: As I mention in the game plan, you can prove that *DEFG* is a parallelogram by showing that both pairs of opposite sides are congruent. Your first eight steps would be the same, and then you'd go on to show that $\triangle DEF \cong \triangle FGD$ and then use CPCTC. This proof method would take you about 12 steps. Or you could prove that both pairs of opposite sides of *DEFG* are parallel (if you have some strange urge to make the proof really long). This proof method would take you about 15 steps.

Proving That a Quadrilateral Is a Rectangle, Rhombus, or Square

Some of the ways to prove that a quadrilateral is a rectangle or a rhombus are directly related to the rectangle or rhombus properties (including their definitions). Other methods require you to first show (or be given) that the quadrilateral is a parallelogram and then go on to prove that the parallelogram is a rectangle or rhombus. The same thing goes for proving that a quadrilateral is a square, except that instead of showing that the quadrilateral is a parallelogram, you have to show that it's both a rectangle and a rhombus. I introduce you to these proofs in the following sections.

Revving up for rectangle proofs

Three ways to prove that a quadrilateral is a rectangle: Note that the second and third methods require that you first show (or be given) that the quadrilateral in question is a parallelogram:

- If all angles in a quadrilateral are right angles, then it's a rectangle (reverse of the rectangle definition). (Actually, you only need to show that three angles are right angles — if they are, the fourth one is automatically a right angle as well.) This is a definition, not a theorem or postulate, but it works exactly like a theorem, so don't sweat it.

- If the diagonals of a parallelogram are congruent, then it's a rectangle (neither the reverse of the definition nor the converse of a property).

- If a parallelogram contains a right angle, then it's a rectangle (neither the reverse of the definition nor the converse of a property).

Tip: Do the following to visualize why this method works: Take an empty cereal box and push in the top flaps. If you then look into the empty box, the top of the box makes a rectangular shape, right? Now, start to crush the top of the box — you know, like you want to make it flat before putting it in the trash (I hope you get what I mean so you can do this highly scientific experiment). As you start to crush the top of the box, you see a parallelogram shape. Now, after you've crushed it a bit, if you take this parallelogram and make one of the angles a right angle, the whole top has to become a rectangle again. You can't make one of the angles a right angle without the other three also becoming right angles.

Before I show any of these proof methods in action, here's a useful little theorem that you need to do the upcoming proof.

Congruent supplementary angles are right angles: If two angles are both supplementary and congruent, then they're right angles. This idea makes sense because $90° + 90° = 180°$.

Okay, so here's the proof. The game plan's up to you.

Given: ∠1 is supplementary to ∠2

∠2 is supplementary to ∠3

∠1 is supplementary to ∠3

Prove: $\overline{NL} \cong \overline{EG}$

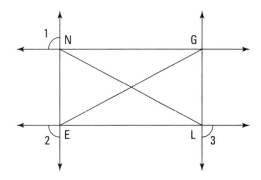

Statements	Reasons
1) ∠1 is a supplementary to ∠2 ∠2 is a supplementary to ∠3	1) Given.
2) $\overleftrightarrow{NG} \parallel \overleftrightarrow{EL}$ $\overleftrightarrow{NE} \parallel \overleftrightarrow{GL}$	2) If same-side exterior angles are supplementary, then lines are parallel.
3) NGLE is a parallelogram	3) If both pairs of opposite sides of a quadrilateral are parallel, then the quadrilateral is a parallelogram.
4) ∠1 ≅ ∠3	4) If two angles are supplementary to the same angle, then they're congruent.
5) ∠1 is a supplementary to ∠3	5) Given.
6) ∠1 is a right angle ∠3 is a right angle	6) If two angles are both supplementary and congruent, then they're right angles.
7) $\overleftrightarrow{NG} \perp \overleftrightarrow{NE}$	7) If lines form a right angle, then they're perpendicular.
8) ∠ENG is a right angle	8) If lines are perpendicular, then they form right angles.
9) NGLE is a rectangle	9) If a parallelogram contains a right angle, then it's a rectangle.
10) $\overline{NL} \cong \overline{EG}$	10) The diagonals of a rectangle are congruent.

Waxing rhapsodic about rhombus proofs

Six ways to prove that a quadrilateral is a rhombus: You can use the following six methods to prove that a quadrilateral is a rhombus. The last three methods on this list require that you first show (or be given) that the quadrilateral in question is a parallelogram:

- If all sides of a quadrilateral are congruent, then it's a rhombus (reverse of the definition). This is a definition, not a theorem or postulate.

- If the diagonals of a quadrilateral bisect all the angles, then it's a rhombus (converse of a property).

- If the diagonals of a quadrilateral are perpendicular bisectors of each other, then it's a rhombus (converse of a property).

Chapter 11: Proving That You've Got a Particular Quadrilateral

Tip: To visualize this one, take two pens or pencils of different lengths and make them cross each other at right angles and at their midpoints. Their four ends must form a diamond shape — a rhombus.

- If two consecutive sides of a parallelogram are congruent, then it's a rhombus (neither the reverse of the definition nor the converse of a property).

- If either diagonal of a parallelogram bisects two angles, then it's a rhombus (neither the reverse of the definition nor the converse of a property).

- If the diagonals of a parallelogram are perpendicular, then it's a rhombus (neither the reverse of the definition nor the converse of a property).

Here's a rhombus proof for you. Try to come up with your own game plan before reading the two-column proof.

Given: *ABCD* is a rectangle

W, *X*, and *Z* are the midpoints of $\overline{AD}$, $\overline{AB}$, and $\overline{DC}$

$\triangle WXY$ and $\triangle WZY$ are isosceles triangles with shared base $\overline{WY}$

Prove: *WXYZ* is a rhombus

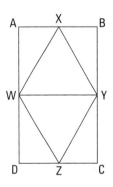

Statements	Reasons
1) *ABCD* is a rectangle	1) Given.
2) $\overline{AB} \cong \overline{DC}$	2) Opposite sides of a rectangle are congruent.
3) *X* is the midpoint of $\overline{AB}$ *Z* is the midpoint of $\overline{DC}$	3) Given.
4) $\overline{AX} \cong \overline{DZ}$	4) Like Divisions Theorem.
5) $\angle WAX$ is a right angle $\angle WDZ$ is a right angle	5) All angles of a rectangle are right angles.
6) $\angle WAX \cong \angle WDZ$	6) All right angles are congruent.
7) *W* is the midpoint of $\overline{AD}$	7) Given.
8) $\overline{AW} \cong \overline{DW}$	8) A midpoint divides a segment into two congruent segments.

(continued)

Statements	Reasons
9) △WAX ≅ △WDZ	9) SAS (4, 6, 8).
10) $\overline{WX} \cong \overline{WZ}$	10) CPCTC.
11) △WXY is an isosceles triangle with base $\overline{WY}$, △WZY is an isosceles triangle with base $\overline{WY}$	11) Given.
12) $\overline{WX} \cong \overline{YX}$ $\overline{WZ} \cong \overline{YZ}$	12) If a triangle is isosceles, then its two legs are congruent.
13) $\overline{YX} \cong \overline{WX} \cong \overline{WZ} \cong \overline{YZ}$	13) Transitivity (10 and 12).
14) WXYZ is a rhombus	14) If a quadrilateral has four congruent sides, then it's a rhombus.

Squaring off with square proofs

Four methods to prove that a quadrilateral is a square: In the last three of these methods, you first have to prove (or be given) that the quadrilateral is a rectangle, rhombus, or both:

- If a quadrilateral has four congruent sides and four right angles, then it's a square (reverse of the square definition). This is a definition, not a theorem or postulate.

- If two consecutive sides of a rectangle are congruent, then it's a square (neither the reverse of the definition nor the converse of a property).

- If a rhombus contains a right angle, then it's a square (neither the reverse of the definition nor the converse of a property).

- If a quadrilateral is both a rectangle and a rhombus, then it's a square (neither the reverse of the definition nor the converse of a property).

You should know these four ways to prove that you've got a square, but I'll skip the example proof this time. If you understand the preceding rectangle and rhombus proofs, you should have no trouble with any square proofs you run across.

Proving That a Quadrilateral Is a Kite

Two ways to prove that a quadrilateral is a kite: Proving that a quadrilateral is a kite is pretty easy. Usually, all you have to do is use congruent triangles or isosceles triangles. Here are the two methods:

- If two disjoint pairs of consecutive sides of a quadrilateral are congruent, then it's a kite (reverse of the kite definition). This is a definition, not a theorem or postulate.
- If one of the diagonals of a quadrilateral is the perpendicular bisector of the other, then it's a kite (converse of a property).

When you're trying to prove that a quadrilateral is a kite, the following tips may come in handy:

- **Check the diagram for congruent triangles.** Don't fail to spot triangles that look congruent and to consider how using CPCTC might help you.
- **Keep the first equidistance theorem in mind (see Chapter 9).** When you have to prove that a quadrilateral is a kite, you might have to use the equidistance theorem in which two points determine the perpendicular bisector.
- **Draw in diagonals.** One of the methods for proving that a quadrilateral is a kite involves diagonals, so if the diagram lacks either of the two diagonals, try drawing in one or both of them.

Now get ready for a proof:

Given: $\overleftrightarrow{RS}$ bisects $\angle CRA$ and $\angle CHA$

Prove: $CRAS$ is a kite

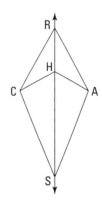

Game plan: Here's how your plan of attack might work for this proof.

- **Note that one of the kite's diagonals is missing.** Draw in the missing diagonal, $\overline{CA}$.
- **Check the diagram for congruent triangles.** After drawing in $\overline{CA}$, there are six pairs of congruent triangles. The two triangles most likely to help you are $\triangle CRH$ and $\triangle ARH$.

- **Prove the triangles congruent.** You can use ASA (see Chapter 9).

- **Use the equidistance theorem.** Use CPCTC (Chapter 9) with $\triangle CRH$ and $\triangle ARH$ to get $\overline{CR} \cong \overline{AR}$ and $\overline{CH} \cong \overline{AH}$. Then, using the equidistance theorem, those two pairs of congruent sides determine the perpendicular bisector of the diagonal you drew in. Over and out.

Check out the formal proof:

Statements	Reasons
1) Draw $\overline{CA}$	1) Two points determine a line.
2) $\overleftrightarrow{RS}$ bisects $\angle CRA$	2) Given.
3) $\angle CRH \cong \angle ARH$	3) Definition of bisect.
4) $\overline{RH} \cong \overline{RH}$	4) Reflexive Property.
5) $\overleftrightarrow{RS}$ bisects $\angle CHA$	5) Given.
6) $\angle CHS \cong \angle AHS$	6) Definition of bisect.
7) $\angle CHR \cong \angle AHR$	7) If two angles are supplementary to two other congruent angles ($\angle CHS$ and $\angle AHS$), then they're congruent.
8) $\triangle CRH \cong \triangle ARH$	8) ASA (3, 4, 7).
9) $\overline{CR} \cong \overline{AR}$	9) CPCTC.
10) $\overline{CH} \cong \overline{AH}$	10) CPCTC.
11) $\overleftrightarrow{RS}$ is the perpendicular bisector of $\overline{CA}$	11) If two points (R and H) are each equidistant from the endpoints of a segment ($\overline{CA}$), then they determine the perpendicular bisector of that segment.
12) $CRAS$ is a kite	12) If one of the diagonals of a quadrilateral ($\overleftrightarrow{RS}$) is the perpendicular bisector of the other ($\overline{CA}$), then the quadrilateral is a kite.

If you're wondering why I don't include a section about proving that a quadrilateral is a trapezoid or isosceles trapezoid, that's good — you're on your toes. I left these proofs out because there's nothing particularly interesting about them, and they're easier than the proofs in this chapter. On top of that, it's very unlikely that you'll ever be asked to do one. For info on trapezoid and isosceles trapezoid properties, flip to Chapter 10.

Chapter 12

Polygon Formulas: Area, Angles, and Diagonals

In This Chapter
- Finding the area of quadrilaterals
- Computing the area of regular polygons
- Determining the number of diagonals in a polygon
- Heating things up with the number of degrees in a polygon

In this chapter, you take a break from proofs and move on to problems that have a *little* more to do with the real world. I emphasize *little* because the shapes you deal with here — such as trapezoids, hexagons, octagons, and yep, even pentadecagons (15 sides) — aren't exactly things you encounter outside of math class on a regular basis. But at least the concepts you work with here — the length and size and shape of polygons — are fairly ordinary things. For nearly everyone, relating to visual, real-world things like this is easier than relating to proofs, which are more in the realm of pure mathematics.

Calculating the Area of Quadrilaterals

I'm sure you've had to calculate the area of a square or rectangle before, whether it was in a math class or in some more practical situation, such as when you wanted to know the area of a room in your house. In this section, you see the square and rectangle formulas again, and you also get some new, gnarlier, quadrilateral area formulas you may not have seen before.

Setting forth the quadrilateral area formulas

Here are the five area formulas for the seven special quadrilaterals. There are only five formulas because some of them do double duty — for example, you can calculate the area of a rhombus with the kite formula.

Quadrilateral area formulas (for info on types of quadrilaterals, see Chapter 10):

- $\text{Area}_{\text{Rectangle}} = \text{base} \cdot \text{height}$ (or length · width, which is the same thing)
- $\text{Area}_{\text{Parallelogram}} = \text{base} \cdot \text{height}$ (because a rhombus is a type of parallelogram, you can use this formula for a rhombus)
- $\text{Area}_{\text{Kite}} = \frac{1}{2} \text{diagonal}_1 \cdot \text{diagonal}_2$, or $\frac{1}{2} d_1 d_2$ (a rhombus is also a type of kite, so you can use the kite formula for a rhombus as well)
- $\text{Area}_{\text{Square}} = \text{side}^2$, or $\frac{1}{2} \text{diagonal}^2$ (this second formula works because a square is a type of kite)
- $\text{Area}_{\text{Trapezoid}} = \frac{\text{base}_1 + \text{base}_2}{2} \cdot \text{height}$
  $= \text{median} \cdot \text{height}$

Note: The *median* of a trapezoid is the segment that connects the midpoints of the legs. Its length equals the average of the lengths of the bases. You use this formula for all trapezoids, including isosceles trapezoids.

Because the square is a special type of four quadrilaterals — a parallelogram, a rectangle, a kite, and a rhombus — it doesn't really need its own area formula. You can find the area of a square by using the parallelogram/rectangle/rhombus formula (base · height) or the kite/rhombus formula ($\frac{1}{2} d_1 d_2$). The simple formula $A = s^2$ is nice to know, however, and because it's so well known, I thought it'd look a bit weird to leave it off the bulleted list.

Getting behind the scenes of the formulas

The area formulas for the parallelogram, kite, and trapezoid are based on the area of a rectangle. The following figures show you how each of these three quadrilaterals relates to a rectangle, and the following list gives you the details:

- **Parallelogram:** In Figure 12-1, if you cut off the little triangle on the left and fill it in on the right, the parallelogram becomes a rectangle (and the area obviously hasn't changed). This rectangle has the same base and height as the original parallelogram. The area of the rectangle is *base · height*, so that formula gives you the area of the parallelogram as well. If you don't believe me (even though you should by now) you can try this yourself by cutting out a paper parallelogram and snipping off the triangle as shown in Figure 12-1.

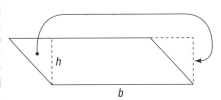

Figure 12-1:
The relationship between a parallelogram and a rectangle.

- **Kite:** Figure 12-2 shows that the kite has half the area of the rectangle drawn around it (this follows from the fact that $\triangle 1 \cong \triangle 2$, $\triangle 3 \cong \triangle 4$, and so on). You can see that the length and width of the large rectangle are the same as the lengths of the diagonals of the kite. The area of the rectangle (length · width) thus equals $d_1 d_2$, and because the kite has half that area, its area is $\frac{1}{2} d_1 d_2$.

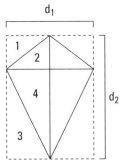

Figure 12-2:
The kite takes up half of each of the four small rectangles and thus is half the area of the large rectangle.

- **Trapezoid:** If you cut off the two triangles and move them as I show you in Figure 12-3, the trapezoid becomes a rectangle. This rectangle has the same height as the trapezoid, and its base equals the median *(m)* of the trapezoid. Thus, the area of the rectangle (and therefore the trapezoid as well) is *median · height*.

Figure 12-3: The relationship between a trapezoid and a rectangle.

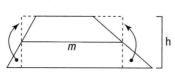

Trying a few area problems

This section lets you try your hand at some example problems.

The key for many quadrilateral area problems is to draw altitudes and other perpendicular segments on the diagram. Doing so creates one or more right triangles, which allows you to use the Pythagorean Theorem or your knowledge of special right triangles, such as the 45°- 45°- 90° and 30°- 60°- 90° triangles (see Chapter 8).

Identifying special right triangles in a parallelogram problem

Find the area of parallelogram *ABCD* in Figure 12-4.

Figure 12-4: Use a 30°- 60°- 90° triangle to find the area of this parallelogram.

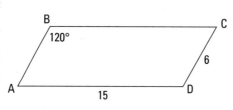

When you see a 120° angle in a problem, a 30°- 60°- 90° triangle is likely lurking somewhere in the problem. (Of course, a 30° or 60° angle is a dead giveaway of a 30°- 60°- 90° triangle.) And if you see a 135° angle, a 45°- 45°- 90° triangle is likely lurking.

To get started, draw in the height of the parallelogram straight down from *B* to base $\overline{AD}$ to form a right triangle as shown in Figure 12-5.

Figure 12-5: Drawing in the height creates a right triangle.

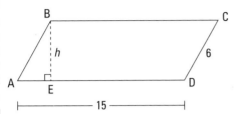

Consecutive angles in a parallelogram are supplementary. Angle *ABC* is 120°, so ∠*A* is 60° and △*ABE* is thus a 30°- 60°- 90° triangle. Now, if you know the ratio of the lengths of the sides in a 30°- 60°- 90° triangle, $x : x\sqrt{3} : 2x$ (see Chapter 8), the rest is a snap. *AB* (the $2x$ side) equals *CD* and is thus 6. Then *AE* (the x side) is half of that, or 3; *BE* (the $x\sqrt{3}$ side) is therefore $3\sqrt{3}$. Here's the finish with the area formula:

$$\text{Area}_{\text{Parallelogram}} = b \cdot h$$
$$= 15 \cdot 3\sqrt{3}$$
$$= 45\sqrt{3} \approx 77.9 \text{ units}^2$$

Using triangles and ratios in a rhombus problem

Now for a rhombus problem: Find the area of rhombus *RHOM* given that *MB* is 6 and that the ratio of *RB* to *BH* is 4 : 1 (see Figure 12-6).

Figure 12-6: Find the area of this rhombus.

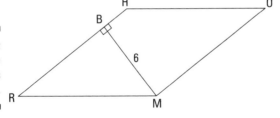

This one's a bit tricky. You might feel that you're not given enough information to solve it or that you just don't know how to begin. If you ever feel this way when you're in the middle of a problem, I have a great tip for you.

If you get stuck when doing a geometry problem — or any kind of math problem, for that matter — *do something, anything!* Begin anywhere you can: Use the given information or any ideas you have (try simple ideas before more-advanced ones) and write something down. Maybe draw a diagram if you don't have one. *Put something down on paper.* One idea may trigger another, and before you know it, you've solved the problem. This tip is surprisingly effective.

Because the ratio of RB to BH is 4 : 1, you can give $\overline{RB}$ a length of $4x$ and $\overline{BH}$ a length of x. Then, because all sides of a rhombus are congruent, RM must equal RH, which is $4x + x$, or $5x$. Now you have a right triangle ($\triangle RBM$) with legs of $4x$ and 6 and a hypotenuse of $5x$, so you can use the Pythagorean Theorem:

$$a^2 + b^2 = c^2$$
$$(4x)^2 + 6^2 = (5x)^2$$
$$16x^2 + 36 = 25x^2$$
$$36 = 9x^2$$
$$4 = x^2$$
$$x = 2 \text{ or } -2$$

Because side lengths must be positive, you reject the answer $x = -2$. The length of the base, $\overline{RH}$, is thus $5(2)$, or 10. (Triangle RBM is your old, familiar friend, a 3-4-5 triangle blown up by a factor of 2 — see Chapter 8.) Now use the parallelogram/rhombus area formula:

$$\text{Area}_{RHOM} = b \cdot h$$
$$= 10 \cdot 6$$
$$= 60 \text{ units}^2$$

Drawing in diagonals to find a kite's area

What's the area of kite *KITE* in Figure 12-7?

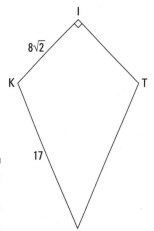

Figure 12-7: A kite with a funky side length.

Draw in diagonals if necessary. For kite and rhombus area problems (and sometimes other quadrilateral problems), the diagonals are almost always necessary for the solution (because they form right triangles). You may have to add them to the figure.

Draw in $\overline{KT}$ and $\overline{IE}$ as shown in Figure 12-8.

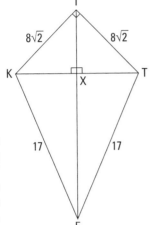

Figure 12-8: Kite *KITE* with its diagonals drawn in.

Triangle *KIT* is a right triangle with congruent legs, so it's a 45°- 45°- 90° triangle with sides in the ratio of $x : x : x\sqrt{2}$ (see Chapter 8). The length of the hypotenuse, $\overline{KT}$, thus equals one of the legs times $\sqrt{2}$; that's $8\sqrt{2} \cdot \sqrt{2}$, or 16. *KX* is half of that, or 8.

Triangle *KIX* is another 45°- 45°- 90° triangle ($\overline{IE}$, the kite's main diagonal, bisects opposite angles *KIT* and *KET*, and half of ∠*KIT* is 45° — see Chapter 10 for other kite properties); therefore, *IX*, like *KX*, is 8. You've got another right triangle, △*KXE*, with a side of 8 and a hypotenuse of 17. I hope that rings a bell! You're looking at an 8-15-17 triangle (see Chapter 8), so without any work, you see that *XE* is 15. (No bells? No worries. You can get *XE* with the Pythagorean Theorem instead.) Add *XE* to *IX* and you get 8 + 15 = 23 for diagonal $\overline{IE}$.

Now that you know the diagonal lengths, you have what you need to finish. The length of diagonal $\overline{KT}$ is 16, and diagonal $\overline{IE}$ is 23. Plug these numbers into the kite area formula for your final answer:

$$\text{Area}_{KITE} = \frac{1}{2} d_1 d_2$$
$$= \frac{1}{2} \cdot 16 \cdot 23$$
$$= 184 \text{ units}^2$$

Using the right-triangle trick for trapezoids

What's the area of trapezoid *TRAP* in Figure 12-9? It looks like an isosceles trapezoid, doesn't it? Don't forget — looks can be deceiving.

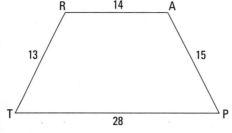

Figure 12-9: A trapezoid with given side lengths.

You should be thinking, *right triangles, right triangles, right triangles.* So draw in two heights straight down from *R* and *A* as shown in Figure 12-10.

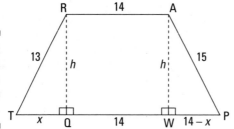

Figure 12-10: Trapezoid *TRAP* with two heights drawn in.

You can see that *QW*, like *RA*, is 14. Then, because *TP* is 28, that leaves 28 – 14, or 14, for the sum of *TQ* and *WP*. Next, you can assign $\overline{TQ}$ a length of *x*, which gives $\overline{WP}$ a length of 14 – *x*. Now you're all set to use — what else? — the Pythagorean Theorem. You have two unknowns, *x* and *h*, so to solve, you need two equations:

$\triangle PAW$: $(14 - x)^2 + h^2 = 15^2$

$\triangle TRQ$: $x^2 \quad\ \ + h^2 = 13^2$

Now solve the system of equations. First, you subtract the second equation from the first, column by column: Subtract the x^2 from the $(14 - x)^2$, the h^2 from the h^2 (canceling it out), and the 13^2 from the 15^2. Then you solve for *x*.

$$(14 - x)^2 - x^2 = 15^2 - 13^2$$

$$(196 - 28x + x^2) - x^2 = 225 - 169$$
$$196 - 28x = 56$$
$$-28x = -140$$
$$x = 5$$

So TQ is 5, and $\triangle TRQ$ is yet another Pythagorean triple triangle — the 5-12-13 triangle (see Chapter 8). The height of $TRAP$ is thus 12. (You can also get h, of course, by plugging $x = 5$ into the $\triangle TRQ$ or $\triangle PAW$ equations.) Now finish with the trapezoid area formula:

$$\text{Area}_{TRAP} = \frac{b_1 + b_2}{2} \cdot h$$
$$= \frac{14 + 28}{2} \cdot 12$$
$$= 252 \text{ units}^2$$

Finding the Area of Regular Polygons

In case you've been dying to know how to figure the area of your ordinary, octagonal stop sign, you've come to the right place. (By the way, did you know that each of the eight sides of a regular-size stop sign is about 12.5 inches long? Hard to believe but true.) In this section, you discover how to find the area of equilateral triangles, hexagons, octagons, and other shapes that have equal sides and angles.

Presenting polygon area formulas

A *regular polygon* is equilateral (it has equal sides) and equiangular (it has equal angles). To find the area of a regular polygon, you use an *apothem* — a segment that joins the polygon's center to the midpoint of any side and that is perpendicular to that side ($\overline{HM}$ in Figure 12-11 is an apothem).

Area of a regular polygon: Use the following formula to find the area of a regular polygon.

$$\text{Area}_{\text{Regular Polygon}} = \tfrac{1}{2} \text{ perimeter} \cdot \text{apothem, or } \tfrac{1}{2} pa$$

Note: This formula is usually written as $\tfrac{1}{2} ap$, but if I do say so myself, the way I've written it, $\tfrac{1}{2} pa$, is better. I like this way of writing it because the formula is based on the triangle area formula, $\tfrac{1}{2} bh$: The polygon's perimeter (p) is related to the triangle's base (b), and the apothem (a) is related to the height (h).

An equilateral triangle is the regular polygon with the fewest possible number of sides. To figure its area, you can use the regular polygon formula; however, it also has its own area formula (which you may remember from Chapter 7):

Area of an equilateral triangle: Here's the area formula for an equilateral triangle.

$$\text{Area}_{\text{Equilateral}\triangle} = \frac{s^2\sqrt{3}}{4}$$ (where s is the length of each of the triangle's sides)

Tackling more area problems

Don't tell me about your problems; I've got problems of my own — and here they are.

Lifting the hex on hexagon area problems

Here's your first regular polygon problem: What's the area of a regular hexagon with an apothem of $10\sqrt{3}$?

For hexagons, use 30°- 60°- 90° and equilateral triangles. A regular hexagon can be cut into six equilateral triangles, and an equilateral triangle can be divided into two 30°- 60°- 90° triangles. So if you're doing a hexagon problem, you may want to cut up the figure and use equilateral triangles or 30°- 60°- 90° triangles to help you find the apothem, perimeter, or area.

First, sketch the hexagon with its three diagonals, creating six equilateral triangles. Then draw in an apothem, which goes from the center to the midpoint of a side. Figure 12-11 shows hexagon *EXAGON*.

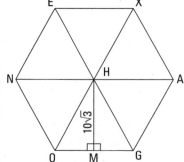

Figure 12-11: A regular hexagon cut into six congruent, equilateral triangles.

Chapter 12: Polygon Formulas: Area, Angles, and Diagonals

Note that the apothem divides △OHG into two 30°- 60°- 90° triangles (halves of an equilateral triangle — see Chapter 8). The apothem is the long leg (the $x\sqrt{3}$ side) of a 30°- 60°- 90° triangle, so

$$x\sqrt{3} = 10\sqrt{3}$$
$$x = 10$$

$\overline{OM}$ is the short leg (the x side), so its length is 10. $\overline{OG}$ is twice as long, so it's 20. And the perimeter is six times that, or 120.

Now you can finish with either the regular polygon formula or the equilateral triangle formula (multiplied by 6). They're equally easy. Take your pick. Here's what it looks like with the regular polygon formula:

$$\text{Area}_{EXAGON} = \frac{1}{2}pa$$
$$= \frac{1}{2} \cdot 120 \cdot 10\sqrt{3}$$
$$= 600\sqrt{3} \text{ units}^2$$

And here's how to do it with the handy equilateral triangle formula:

$$\text{Area}_{\triangle HOG} = \frac{s^2\sqrt{3}}{4}$$
$$= \frac{20^2\sqrt{3}}{4}$$
$$= 100\sqrt{3} \text{ units}^2$$

EXAGON is six times as big as △HOG, so it's $6 \cdot 100\sqrt{3}$, or $600\sqrt{3}$ units².

Picking out squares and triangles in an octagon problem

Check out this nifty octagon problem: Given that *EIGHTPLU* in Figure 12-12 is a regular octagon with sides of length 6 and that △SUE is a right triangle,

Use a paragraph proof to show that △SUE is a 45°- 45°- 90° triangle.

Find the area of octagon *EIGHTPLU*.

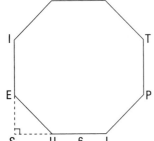

Figure 12-12: A regular octagon and a triangle named *SUE*.

Here's the paragraph proof: *EIGHTPLU* is a *regular* octagon, so all its angles are congruent; therefore, ∠*IEU* ≅ ∠*LUE*. Because supplements of congruent angles are congruent, ∠*SEU* ≅ ∠*SUE*, and thus △*SUE* is isosceles. Finally, angle *S* is a right angle, so △*SUE* is an isosceles right triangle and therefore a 45°- 45°- 90° triangle.

Now for the solution to the area part of the problem. But first, here are two great tips.

- For octagons, use 45°- 45°- 90° triangles. If a problem involves a regular octagon, add segments to the diagram to get one or more 45°- 45°- 90° triangles and some squares and rectangles to help you solve the problem.

- Think outside the box. It's easy to get into the habit of looking only inside a figure because that suffices for the vast majority of problems. But occasionally, you need to break out of that rut and look outside the perimeter of the figure.

Okay, so here's what you do. Draw three more 45°- 45°- 90° triangles to fill out the corners of a square as shown in Figure 12-13.

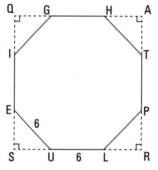

Figure 12-13: Drawing extra lines outside the octagon creates a square.

To find the area of the octagon, you just subtract the area of the four little triangles from the area of the square. *EU* is 6. It's the length of the hypotenuse (the $x\sqrt{2}$ side) of 45°- 45°- 90° triangle △*SUE*, so go ahead and solve for *x*:

$$x\sqrt{2} = 6$$
$$x = \frac{6}{\sqrt{2}} = 3\sqrt{2}$$

$\overline{SU}$ is a leg of the 45°- 45°- 90° triangle, so it's an *x* side and thus has a length of $3\sqrt{2}$. $\overline{LR}$ also has a length of $3\sqrt{2}$, so square *SQAR* has a side of $3\sqrt{2} + 6 + 3\sqrt{2}$, or $6 + 6\sqrt{2}$ units.

You're in the home stretch. First calculate the area of the square and the area of a single corner triangle:

$$\text{Area}_{SQAR} = s^2$$
$$= \left(6 + 6\sqrt{2}\right)^2$$
$$= 36 + 72\sqrt{2} + 72$$
$$= 108 + 72\sqrt{2}$$
$$\text{Area}_{\triangle SUE} = \frac{1}{2}bh$$
$$= \frac{1}{2} \cdot 3\sqrt{2} \cdot 3\sqrt{2}$$
$$= 9$$

To finish, subtract the total area of the four corner triangles from the area of the square:

$$\text{Area}_{EIGHTPLU} = \text{area}_{SQAR} - 4 \cdot \text{area}_{\triangle SUE}$$
$$= \left(108 + 72\sqrt{2}\right) - (4 \cdot 9)$$
$$= 72 + 72\sqrt{2} \approx 173.8 \text{ units}^2$$

Using Polygon Angle and Diagonal Formulas

In this section, you get polygon formulas involving — hold onto your hat! — angles and diagonals. You can use these formulas to answer a couple of questions that I bet have been keeping you awake at night: 1) How many diagonals does a 100-sided polygon have? *Answer:* 4,850; and 2) What's the sum of the measures of all the angles in an icosagon (a 20-sided polygon)? *Answer:* 3,240°.

Interior and exterior design: Exploring polygon angles

Everything you need to know about a polygon doesn't necessarily fall within its sides. You use angles that are outside the polygon as well.

You use two kinds of angles when working with polygons (see Figure 12-14):

- **Interior angle:** An interior angle of a polygon is an angle inside the polygon at one of its vertices. Angle *Q* is an interior angle of quadrilateral *QUAD*.

- **Exterior angle:** An exterior angle of a polygon is an angle outside the polygon formed by one of its sides and the extension of an adjacent side. Angle *ADZ*, ∠*XUQ*, and ∠*YUA* are exterior angles of *QUAD*; vertical angle ∠*XUY* is *not* an exterior angle of *QUAD*.

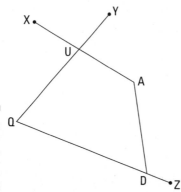

Figure 12-14: Interior and exterior angles.

Interior and exterior angle formulas:

- The *sum* of the measures of the *interior angles* of a polygon with n sides is $(n-2)180$.

- The measure of *each interior angle* of an equiangular n-gon is $\frac{(n-2)180}{n}$ or $180 - \frac{360}{n}$ (the supplement of an exterior angle).

- If you count one exterior angle at each vertex, the *sum* of the measures of the *exterior angles* of a polygon is always 360°.

- The measure of *each exterior angle* of an equiangular n-gon is $\frac{360}{n}$.

Handling the ins and outs of a polygon angle problem

You can practice the interior and exterior angle formulas in the following three-part problem: Given a regular dodecagon (12 sides),

1. **Find the sum of the measures of its interior angles.**

    Just plug the number of sides (12) into the formula for the sum of the interior angles of a polygon:

    Sum of interior angles $= (n-2)180$

    $= (12-2)180$

    $= 1{,}800°$

2. Find the measure of a single interior angle.

This polygon has 12 sides, so it has 12 angles; and because you're dealing with a *regular* polygon, all its angles are congruent. So to find the measure of a single angle, just divide your answer from the first part of the problem by 12. (Note that this is basically the same as using the first formula for a single interior angle.)

$$\text{Measure of a single interior angle} = \frac{1,800}{12}$$
$$= 150°$$

3. Find the measure of a single exterior angle with the exterior angle formula; then check that its supplement, an interior angle, equals the answer you got from part 2 of the problem.

First, plug 12 into the oh-so-simple exterior angle formula:

$$\text{Measure of a single exterior angle} = \frac{360}{12}$$
$$= 30°$$

Now take the supplement of your answer to find the measure of a single interior angle, and check that it's the same as your answer from part 2:

$$\text{Measure of a single interior angle} = 180 - 30$$
$$= 150°$$

It checks. (And note that this final computation is basically the same thing as using the second formula for a single interior angle.)

Criss-crossing with diagonals

Number of diagonals in an n-gon: The number of diagonals that you can draw in an *n*-gon is $\frac{n(n-3)}{2}$.

This formula looks like it came outta nowhere, doesn't it? I promise it makes sense, but you might have to think about it a little first. (Of course, just memorizing it is okay, but what's the fun in that?)

Here's where the diagonal formula comes from and why it works. Each diagonal connects one point to another point in the polygon that isn't its next-door neighbor. In an *n*-sided polygon, you have *n* starting points for diagonals. And each diagonal can go to (*n* – 3) ending points because a diagonal can't end at its own starting point or at either of the two neighboring points. So the first step is to multiply *n* by (*n* – 3). Then, because each diagonal's ending point can be used as a starting point as well, the product *n*(*n* – 3) counts each diagonal twice. That's why you divide by 2.

A round-robin tennis tourney

Here's a nifty real-world application of the formula for the number of diagonals in a polygon. Say there's a small tennis tournament with six people in which everyone has to play everyone else. How many total matches will there be? The following figure shows the six tennis players with segments connecting each pair of players.

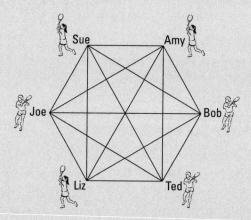

Each segment represents a match between two contestants. So to get the total number of matches, you just have to count up all the segments in the figure: the number of sides of the hexagon (6) plus the number of diagonals in the hexagon $\left(\dfrac{6(6-3)}{2} = 9\right)$. The total is therefore 15 matches. For the general case, the total number of matches in a round-robin tournament with n players would be $n + \dfrac{n(n-3)}{2}$, which simplifies to $\dfrac{n(n-1)}{2}$. Game, set, match.

Here's one last problem for you: If a polygon has 90 diagonals, how many sides does it have?

You know what the formula for the number of diagonals in a polygon is, and you know that the polygon has 90 diagonals, so plug 90 in for the answer and solve for n:

$$\dfrac{n(n-3)}{2} = 90$$
$$n^2 - 3n = 180$$
$$n^2 - 3n - 180 = 0$$
$$(n-15)(n+12) = 0$$

Thus, n equals 15 or –12. But because a polygon can't have a negative number of sides, n must be 15. So you have a 15-sided polygon (a pentadecagon, in case you're curious).

Chapter 13

Similarity: Same Shape, Different Size

In This Chapter
▸ Sizing up similar figures
▸ Doing similar triangle proofs and using CASTC and CSSTP
▸ Scoping out theorems about proportionality

You know the meaning of the word *similar* in everyday speech. In geometry, it has a related but more technical meaning. Two figures are *similar* if they have exactly the same shape. You get similar figures, for example, when you use a photocopy machine to blow up some image. The result is bigger but exactly the same shape as the original. And photographs show shapes that are smaller than but geometrically similar to the original objects.

You witness similarity in action virtually every minute of the day. As you see things (people, objects, anything) moving toward or away from you, you see them appear to get bigger or smaller, but they retain their same shape. The shape of this book is a rectangle. If you hold it close to you, you see a rectangular shape of a certain size. When you hold it farther away, you see a smaller but *similar* rectangle. In this chapter, I show you all sorts of interesting things about similarity and how it's used in geometry.

Note: Congruent figures are automatically similar, but when you have two congruent figures, you call them *congruent*, naturally; you don't have much reason to point out that they're also similar (the same shape) because it's so obvious. So even though congruent figures qualify as similar figures, problems about similarity usually deal with figures of the same shape but different size.

204 Part IV: Polygons of the Four-or-More Sided Variety

Getting Started with Similar Figures

In this section, I cover the formal definition of similarity, how similar figures are named, and how they're positioned. Then you get to practice these ideas by working though a problem.

Defining and naming similar polygons

As you see in Figure 13-1, quadrilateral *WXYZ* is the same shape as quadrilateral *ABCD*, but it's ten times larger (though not drawn to scale, of course). These quadrilaterals are therefore similar.

Figure 13-1: These quadrilaterals are *similar* because they're exactly the same shape; note that their angles are congruent.

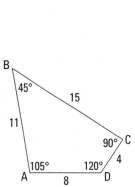 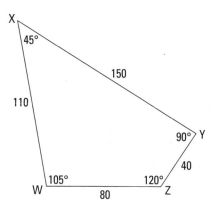

Similar polygons: For two polygons to be similar, both of the following must be true:

- ✓ Corresponding angles are congruent.
- ✓ Corresponding sides are proportional.

To fully understand this definition, you have to know what *corresponding angles* and *corresponding sides* mean. (Maybe you've already figured this out by just looking at the figure.) Here's the lowdown on *corresponding*. In Figure 13-1, if you expand *ABCD* to the same size as *WXYZ* and slide it to the right, it'd stack perfectly on top of *WXYZ*. *A* would stack on *W, B* on *X, C* on *Y,* and *D* on *Z*. These vertices are thus *corresponding*. And therefore, you say that ∠*A* corresponds to ∠*W,* ∠*B* corresponds to ∠*X,* and so on. Also, side $\overline{AB}$

corresponds to side $\overline{WX}$, $\overline{BC}$ to $\overline{XY}$, and so on. In short, if one of two similar figures is expanded or shrunk to the size of the other, angles and sides that would stack on each other are called *corresponding*.

When you name similar polygons, pay attention to how the vertices pair up. For the quadrilaterals in Figure 13-1, you write that *ABCD ~ WXYZ* (the squiggle symbol means *is similar to*) because *A* and *W* (the first letters) are corresponding vertices, *B* and *X* (the second letters) are corresponding, and so on. You can also write *BCDA ~ XYZW* (because corresponding vertices pair up) but *not ABCD ~ YZWX*.

Now I'll use quadrilaterals *ABCD* and *WXYZ* to explore the definition of similar polygons in greater depth:

- **Corresponding angles are congruent.** You can see that $\angle A$ and $\angle W$ are both 105° and therefore congruent, $\angle B \cong \angle X$, and so on. When you blow up or shrink a figure, the angles don't change.

- **Corresponding sides are proportional.** The ratios of corresponding sides are equal, like this:

$$\frac{\text{Left side}_{WXYZ}}{\text{Left side}_{ABCD}} = \frac{\text{top}_{WXYZ}}{\text{top}_{ABCD}} = \frac{\text{right side}_{WXYZ}}{\text{right side}_{ABCD}} = \frac{\text{base}_{WXYZ}}{\text{base}_{ABCD}}$$

$$\frac{WX}{AB} = \frac{XY}{BC} = \frac{YZ}{CD} = \frac{ZW}{DA}$$

$$\frac{110}{11} = \frac{150}{15} = \frac{40}{4} = \frac{80}{8} = 10$$

Each ratio equals 10, the expansion factor. (If the ratios were flipped upside down — which is equally valid — each would equal $\frac{1}{10}$, the shrink factor.) And not only do these ratios all equal 10, but the ratio of the perimeters of *ABCD* and *WXYZ* also equals 10.

Perimeters of similar polygons: The ratio of the perimeters of two similar polygons equals the ratio of any pair of their corresponding sides.

How similar figures line up

Two similar figures can be positioned so that they either line up or don't line up. You can see that figures *ABCD* and *WXYZ* in Figure 13-1 are positioned in the same way in the sense that if you were to blow up *ABCD* to the size of *WXYZ* and then slide *ABCD* over, it'd match up perfectly with *WXYZ*. Now check out Figure 13-2, which shows *ABCD* again with another similar quadrilateral. You can easily see that, unlike the quadrilaterals in Figure 13-1, *ABCD* and *PQRS* are *not* positioned in the same way.

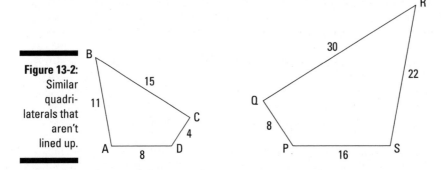

Figure 13-2: Similar quadrilaterals that aren't lined up.

In the preceding section, you see how to set up a proportion for similar figures using the positions of their sides, which I've labeled *left side, right side, top,* and *base* — for example, one valid proportion is $\frac{\text{Left side}}{\text{Left side}} = \frac{\text{top}}{\text{top}}$ (note that both numerators must come from the same figure; ditto for both denominators). This is a good way to think about how proportions work with similar figures, but proportions like this work only if the figures are drawn like *ABCD* and *WXYZ* are. When similar figures are drawn facing different ways, as in Figure 13-2, the left side doesn't necessarily correspond to the left side, and so on, and you have to take greater care that you're pairing up the proper vertices and sides.

Quadrilaterals *ABCD* and *PQRS* are similar, but you *can't* say that *ABCD ~ PQRS* because the vertices don't pair up in this order. Ignoring its size, *PQRS* is the mirror image of *ABCD* (or you can say it's flipped over compared with *ABCD*). If you flip *PQRS* over in the left-right direction, you get the image in Figure 13-3.

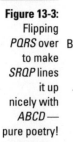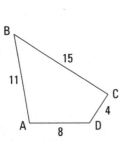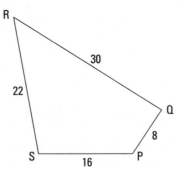

Figure 13-3: Flipping *PQRS* over to make *SRQP* lines it up nicely with *ABCD* — pure poetry!

Now it's easier to see how the vertices pair up. *A* corresponds to *S*, *B* with *R*, and so on, so you write the similarity like this: *ABCD ~ SRQP*.

Chapter 13: Similarity: Same Shape, Different Size

Align similar polygons. If you get a problem with a diagram of similar polygons that aren't lined up, consider redrawing one of them so that they're both positioned in the same way. This may make the problem easier to solve.

And here are a few more things you can do to help you see how the vertices of similar polygons match up when the polygons are positioned differently:

- You can often tell how the vertices correspond just by looking at the polygons, which is actually a pretty good way of seeing whether one polygon has been flipped over or spun around.

- If the similarity is given to you and written out like $\triangle JKL \sim \triangle TUV$, you know that the first letters, J and T, correspond, K and U correspond, and L and V correspond. The order of the letters also tells you that $\overline{KL}$ corresponds to $\overline{UV}$, and so on.

- If you know the measures of the angles or which angles are congruent to which, that information tells you how the vertices correspond because corresponding angles are congruent.

- If you're given (or you figure out) which sides are proportional, that info tells you how the sides would stack up, and from that you can see how the vertices correspond.

Solving a similarity problem

Enough of this general, schmeneral stuff — time to see these ideas in action:

Given: *ROTFL ~ SUBAG*

Perimeter of *ROTFL* is 52

Find:
1. The lengths of $\overline{AG}$ and $\overline{GS}$
2. The perimeter of *SUBAG*
3. The measures of $\angle S$, $\angle G$, and $\angle A$

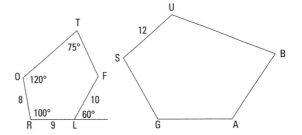

You can see that the *ROTFL* and *SUBAG* aren't positioned the same way just by looking at the figure (and noting that their first letters, *R* and *S*, aren't in the same place). So you need to figure out how their vertices correspond. Let's use one of the methods from the bulleted list. The letters in *ROTFL* ~ *SUBAG* show you what corresponds to what. *R* corresponds to *S*, *O* corresponds to *U*, and so on. (By the way, do you see what you'd have to do to line up *SUBAG* with *ROTFL*? *SUBAG* has sort of been tipped over to the right, so you'd have to rotate it counterclockwise a bit and stand it up on base $\overline{GS}$. You may want to redraw *SUBAG* like that, which can really help you see how all the parts of the two pentagons correspond.)

1. **Find the lengths of $\overline{AG}$ and $\overline{GS}$.**

   The order of the vertices in *ROTFL* ~ *SUBAG* tells you that $\overline{SU}$ corresponds to $\overline{RO}$ and that $\overline{AG}$ corresponds to $\overline{FL}$; thus, you can set up the following proportion to find missing length *AG*:

   $$\frac{AG}{FL} = \frac{SU}{RO}$$

   $$\frac{AG}{10} = \frac{12}{8}$$

   $$\frac{AG}{10} = 1.5 \text{ (or you could have cross-multiplied)}$$

   $$AG = 15$$

   This method of setting up a proportion and solving for the unknown length is the standard way of solving this type of problem. It's often useful, and you should know how to do it (including knowing how to cross-multiply).

   But another method can come in handy. Here's how to use it to find *GS*: Divide the lengths of two known sides of the figures like this: $\frac{SU}{RO} = \frac{12}{8}$, which equals 1.5. That answer tells you that all the sides of *SUBAG* (and its perimeter) are 1.5 times as long as their counterparts in *ROTFL*. The order of the vertices in *ROTFL* ~ *SUBAG* tells you that $\overline{GS}$ corresponds to $\overline{LR}$; thus, $\overline{GS}$ is 1.5 times as long as $\overline{LR}$:

   $$GS = 1.5 \cdot LR$$
   $$= 1.5 \cdot 9$$
   $$= 13.5$$

2. **Find the perimeter of *SUBAG*.**

   The method I just introduced tells you immediately that

   $$\text{Perimeter}_{SUBAG} = 1.5 \cdot \text{perimeter}_{ROTFL}$$
   $$= 1.5 \cdot 52$$
   $$= 78$$

But for math teachers and other fans of formality, here's the standard method using cross-multiplication:

$$\frac{\text{Perimeter}_{SUBAG}}{\text{Perimeter}_{ROTFL}} = \frac{SU}{RO}$$

$$\frac{P}{52} = \frac{12}{8}$$

$$8 \cdot P = 52 \cdot 12$$

$$P = 78$$

3. **Find the measures of ∠S, ∠G, and ∠A.**

   S corresponds to R, G corresponds to L, and A corresponds to F, so

   - Angle S is the same as ∠R, or 100°.
   - Angle G is the same as ∠RLF, which is 120° (the supplement of the 60° angle).

   To get ∠A, you first have to find ∠F with the sum-of-angles formula from Chapter 12:

   $$\text{Sum of angles}_{\text{Pentagon } ROTFL} = (n-2)180$$
   $$= (5-2)180$$
   $$= 540°$$

   Because the other four angles of ROTFL (clockwise from L) add up to 120° + 100° + 120° + 75° = 415°, ∠F, and therefore ∠A, must equal 540° − 415°, or 125°.

Proving Triangles Similar

Chapter 9 explains five ways to prove triangles congruent: SSS, SAS, ASA, AAS, and HLR. In this section, I show you something related — the three ways to prove triangles *similar:* AA, SSS~, and SAS~.

Use the following methods to prove triangles similar:

- ✔ **AA:** If two angles of one triangle are congruent to two angles of another triangle, then the triangles are similar.
- ✔ **SSS~:** If the ratios of the three pairs of corresponding sides of two triangles are equal, then the triangles are similar.
- ✔ **SAS~:** If the ratios of two pairs of corresponding sides of two triangles are equal and the included angles are congruent, then the triangles are similar.

In the sections that follow, I dive into some problems so you can see how these methods work.

Tackling an AA proof

The AA method is the most frequently used and is therefore the most important. Luckily, it's also the easiest of the three methods to use. Give it a whirl with the following proof:

Given: ∠1 is supplementary to ∠2

$\overline{AY} \parallel \overline{LR}$

Prove: △CYA ~ △LTR

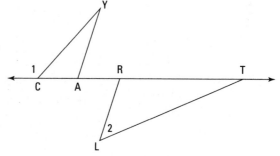

Whenever you see parallel lines in a similar-triangle problem, look for ways to use the parallel-line theorems from Chapter 10 to get congruent angles.

Here's a game plan describing how your thought process might go (this hypothetical thought process assumes that you don't know that this is an AA proof from the title of this section): The first given is about angles, and the second given is about parallel lines, which will probably tell you something about congruent angles. Therefore, this proof is almost certainly an AA proof. So all you have to do is think about the givens and figure out which two pairs of angles you can prove congruent to use for AA. Duck soup.

Take a look at how the proof plays out:

Statements	Reasons
1) ∠1 is supplementary to ∠2	1) Given.
2) ∠1 is supplementary to ∠YCA	2) Two angles that form a straight angle (assumed from diagram) are supplementary.
3) ∠YCA ≅ ∠2	3) Supplements of the same angle are congruent.
4) $\overline{AY} \parallel \overline{LR}$	4) Given.
5) ∠CAY ≅ ∠LRT	5) Alternate exterior angles are congruent (using parallel segments $\overline{AY}$ and $\overline{LR}$ and transversal $\overrightarrow{CT}$).
6) △CYA ≅ △LTR	6) AA. (If two angles of one triangle are congruent to two angles of another triangle, then the triangles are similar; lines 3 and 5.)

Using SSS~ to prove triangles similar

The upcoming SSS~ proof incorporates the Midline Theorem, which I present to you here.

The Midline Theorem: According to the *Midline Theorem,* a segment joining the midpoints of two sides of a triangle is

- One-half the length of the third side, and
- Parallel to the third side.

Figure 13-4 provides the visual for the theorem.

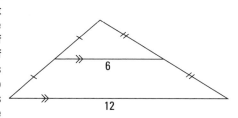

Figure 13-4: A segment joining the midpoints of two sides of a triangle is parallel to and half as long as the third side.

Check out this theorem in action with an SSS~ proof:

Given: *A*, *W*, and *Y* are the midpoints of $\overline{KN}$, $\overline{KE}$, and $\overline{NE}$ respectively

Prove: In a paragraph proof, show that △*WAY* ~ △*NEK* using

1. The first part of the Midline Theorem
2. The second part of the Midline Theorem

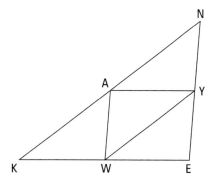

1. **Use the first part of the Midline Theorem to prove that △WAY ~ △NEK.**

   Here's the solution: The first part of the Midline Theorem says that a segment connecting the midpoints of two sides of a triangle is half the length of the third side. You have three such segments: $\overline{AY}$ is half the length of $\overline{KE}$, $\overline{WY}$ is half the length of $\overline{KN}$, and $\overline{AW}$ is half the length of $\overline{NE}$. That gives you the proportionality you need: $\frac{AY}{KE} = \frac{WY}{KN} = \frac{AW}{NE} = \frac{1}{2}$. Thus, the triangles are similar by SSS~.

2. **Use part two of the Midline Theorem to prove that △WAY ~ △NEK.**

   Solve this one as follows: The second part of the Midline Theorem tells you that a segment connecting the midpoints of two sides of a triangle is parallel to the third side. You have three segments like this in the diagram, $\overline{AY}$, $\overline{WY}$, and $\overline{AW}$, each of which is parallel to a side of △NEK. The pairs of parallel segments should make you think about using the parallel-line theorems (from Chapter 10), which could give you the congruent angles you need to prove the triangles similar with AA.

   Look at parallel segments $\overline{AY}$ and $\overline{KE}$, with transversal $\overline{NE}$. You can see that ∠E is congruent to ∠AYN because corresponding angles (the parallel-line kind of *corresponding*) are congruent.

   Now look at parallel segments $\overline{AW}$ and $\overline{NE}$, with transversal $\overline{AY}$. Angle AYN is congruent to ∠WAY because they're alternate interior angles. So by the Transitive Property, ∠E ≅ ∠WAY.

   With identical reasoning, you next show that ∠K ≅ ∠WYA or that ∠N ≅ ∠AWY. And that does it. The triangles are similar by AA.

Working through an SAS~ proof

Try using the SAS~ method to solve the following proof:

Given: △BOA ~ △BYT

Prove: △BAT ~ △BOY
(paragraph proof)

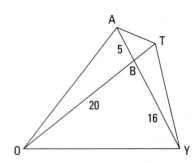

Game Plan: Your thinking might go like this. You have one pair of congruent angles, the vertical angles ∠ABT and ∠OBY. But because it doesn't look like you can get another pair of congruent angles, the AA approach is out. What other method can you try? You're given side lengths in the figure, so the combination of angles and sides should make you think of SAS~. To prove △BAT ~ △BOY with SAS~, you need to find the length of $\overline{BT}$ so you can show that $\overline{BA}$ and $\overline{BT}$ (the sides that make up ∠ABT) are proportional to $\overline{BO}$ and $\overline{BY}$ (the sides that make up ∠OBY). To find BT, you can use the similarity in the given.

So you begin solving the problem by figuring out the length of $\overline{BT}$. △BOA ~ △BYT, so — paying attention to the order of the letters — you see that $\overline{BO}$ corresponds to $\overline{BY}$ and that $\overline{BA}$ corresponds to $\overline{BT}$. Thus, you can set up this proportion:

$$\frac{BO}{BY} = \frac{BA}{BT}$$
$$\frac{20}{16} = \frac{5}{BT}$$
$$20 \cdot BT = 16 \cdot 5$$
$$BT = 4$$

Now, to prove △BAT ~ △BOY with SAS~, you use the congruent vertical angles and then check that the following proportion works:

$$\frac{BA}{BO} \stackrel{?}{=} \frac{BT}{BY}$$
$$\frac{5}{20} \stackrel{?}{=} \frac{4}{16}$$

This checks. You're done. (By the way, these fractions both reduce to $\frac{1}{4}$, so △BAT is $\frac{1}{4}$ as big as △BOY.)

CASTC and CSSTP, the Cousins of CPCTC

In this section, you prove triangles similar (as in the preceding section) and then go a step further to prove other things about the triangles using CASTC and CSSTP (which are just acronyms for the parts of the definition of similar polygons, as applied to triangles).

Similar triangles have the following two characteristics:

- **CASTC:** Corresponding angles of similar triangles are congruent.
- **CSSTP:** Corresponding sides of similar triangles are proportional.

This definition of similar triangles follows from the definition of similar polygons, and thus it isn't really a new idea, so you might wonder why it deserves an icon. Well, what's new here isn't the definition itself; it's how you use the definition in two-column proofs.

CASTC and CSSTP work just like CPCTC. In a two-column proof, you use CASTC or CSSTP on the very next line after showing triangles similar, just like you use CPCTC (see Chapter 9) on the line after you show triangles congruent.

Working through a CASTC proof

The following proof shows you how CASTC works:

Given: Diagram as shown

Prove: $\overleftrightarrow{AB} \parallel \overleftrightarrow{DE}$ (paragraph proof)

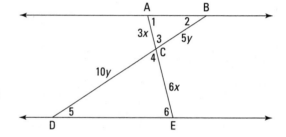

Here's how your game plan might go: When you see the two triangles in this proof diagram and you're asked to prove that the lines are parallel, you should be thinking about proving the triangles similar. Then, using CASTC, you've got congruent angles that you can use with the parallel-line theorems (from Chapter 10) to finish.

So here's the solution. You've got the pair of congruent vertical angles, $\angle 3$ and $\angle 4$, so if you could show that the sides that make up those angles are proportional, the triangles would be similar by SAS~. So check that the sides are proportional:

$$\frac{AC}{EC} \stackrel{?}{=} \frac{BC}{DC}$$

$$\frac{3x}{6x} \stackrel{?}{=} \frac{5y}{10y}$$

$$\frac{1}{2} = \frac{1}{2}$$

Check. Thus, $\triangle ABC \sim \triangle EDC$ by SAS~. (Note that the similarity is written so that corresponding vertices pair up.) Vertices B and D correspond, so $\angle 2 \cong \angle 5$ by CASTC. Because $\angle 2$ and $\angle 5$ are alternate interior angles that are congruent, $\overleftrightarrow{AB} \parallel \overleftrightarrow{DE}$. (Note that you could instead show that $\angle 1$ and $\angle 6$ are congruent by CASTC and then use those angles as the alternate interior angles.)

Taking on a CSSTP proof

CSSTP proofs can be a bit trickier than CASTC proofs because they often involve an odd step at the end in which you have to prove that one product of sides equals another product of sides. You'll see what I mean in the following problem:

Given: △LMN is isosceles with base $\overline{LN}$

∠1 ≅ ∠8

Prove: JL · NP = QN · LK

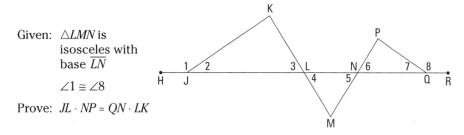

You can often use a proportion to prove that two products are equal; therefore, if you're asked to prove that a product equals another product (as with JL · NP = QN · LK), the proof probably involves a proportion related to similar triangles (or maybe, though less likely, a proportion related to one of the theorems in the upcoming sections). So look for similar triangles that contain the four segments in the *prove* statement. You can then set up a proportion using those four segments and finally cross-multiply to arrive at the desired product.

Here's the formal proof:

Statements	Reasons
1) △LMN is an isosceles with base $\overline{LN}$	1) Given.
2) $\overline{ML} ≅ \overline{MN}$	2) Definition of isosceles triangle.
3) ∠4 ≅ ∠5	3) If sides, then angles.
4) ∠3 ≅ ∠4 ∠5 ≅ ∠6	4) Vertical angles are congruent.
5) ∠3 ≅ ∠6	5) Transitive Property for four angles. (If two angles are congruent to two other congruent angles, then they're congruent.)
6) ∠1 ≅ ∠8	6) Given.
7) ∠2 ≅ ∠7	7) Supplements of congruent angles are congruent.
8) △JKL ~ △QPN	8) AA (lines 5 and 7).

(continued)

Statements	Reasons
9) $\dfrac{JL}{QN} = \dfrac{LK}{NP}$	9) CSSTP.
10) $JL \cdot NP = QN \cdot LK$	10) Cross-multiplication.

Splitting Right Triangles with the Altitude-on-Hypotenuse Theorem

In a right triangle, the altitude that's perpendicular to the hypotenuse has a special property: It creates two smaller right triangles that are both similar to the original right triangle.

Altitude-on-Hypotenuse Theorem: If an altitude is drawn to the hypotenuse of a right triangle as shown in Figure 13-5, then

- The two triangles formed are similar to the given triangle and to each other:

    $\triangle ACB \sim \triangle ADC \sim \triangle CDB$,

- $h^2 = xy$, and
- $a^2 = yc$ and $b^2 = xc$.

Note that the two equations in this third bullet are really just one idea, not two. It works exactly the same way on both sides of the big triangle:

(leg of big $\triangle$)² = (part of hypotenuse below it) · (whole hypotenuse)

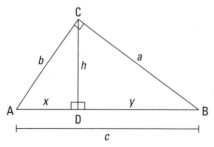

Figure 13-5: Three similar right triangles: small, medium, and large.

Here's a problem for you: Use Figure 13-6 to answer the following questions.

Figure 13-6: Altitude $\overline{KM}$ lets you apply the Altitude-on-Hypotenuse Theorem.

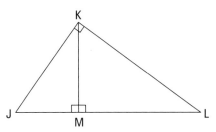

1. **If *JL* = 17 and *KL* = 15, what are *JK*, *JM*, *ML*, and *KM*?**

    Here's how you do this one: *JK* is 8 because you have an 8-15-17 triangle (or you can get *JK* with the Pythagorean Theorem; see Chapter 8 for more info). Now you can find *JM* and *ML* using part three of the Altitude-on-Hypotenuse Theorem:

    $$(JK)^2 = (JM)(JL) \quad \text{and} \quad (KL)^2 = (ML)(JL)$$
    $$8^2 = JM \cdot 17 \qquad\qquad 15^2 = ML \cdot 17$$
    $$JM = \frac{64}{17} \approx 3.8 \qquad\qquad ML = \frac{225}{17} \approx 13.2$$

    (I included the *ML* solution just to show you another example of the theorem, but obviously, it would've been easier to get *ML* by just subtracting *JM* from *JL*.)

    Finally, use the second part of the theorem (or the Pythagorean Theorem, if you prefer) to get *KM*:

    $$(KM)^2 = (JM)(ML)$$
    $$(KM)^2 = \left(\frac{64}{17}\right)\left(\frac{225}{17}\right)$$
    $$KM = \sqrt{\frac{14,400}{289}} = \frac{120}{17} \approx 7.1$$

2. **If *ML* = 16 and *JK* = 15, what's *JM*?**

    Set *JM* equal to *x*; then use part three of the theorem.

    $$(JK)^2 = (JM)(JL)$$
    $$15^2 = x(x + 16)$$
    $$225 = x^2 + 16x$$
    $$x^2 + 16x - 225 = 0$$
    $$(x - 9)(x + 25) = 0$$
    $$x - 9 = 0 \quad \text{or} \quad x + 25 = 0$$
    $$x = 9 \quad \text{or} \quad x = -25$$

You know that a length can't be –25, so JM = 9. (If you have a hard time seeing how to factor this one, you can use the quadratic formula to get the values of *x* instead.)

When doing a problem involving an altitude-on-hypotenuse diagram, don't assume that you must use the second or third part of the Altitude-on-Hypotenuse Theorem. Sometimes, the easiest way to solve the problem is with the Pythagorean Theorem. And at other times, you can use ordinary similar-triangle proportions to solve the problem.

The next problem illustrates this tip: Use the following figure to find *h*, the altitude of △*ABC*.

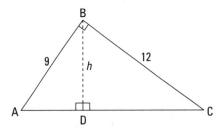

On your mark, get set, go. First get *AC* with the Pythagorean Theorem or by noticing that you have a triangle in the 3 : 4 : 5 family — namely a 9-12-15 triangle. So *AC* = 15. Then, though you could finish with the Altitude-on-Hypotenuse Theorem, that approach is a bit complicated and would take some work. Instead, just use an ordinary similar-triangle proportion:

$$\frac{\text{Long leg}_{\triangle ABD}}{\text{Long leg}_{\triangle ACB}} = \frac{\text{hypotenuse}_{\triangle ABD}}{\text{hypotenuse}_{\triangle ACB}}$$

$$\frac{h}{12} = \frac{9}{15}$$

$$15h = 108$$

$$h = 7.2$$

Finito.

Getting Proportional with Three More Theorems

In this section, you get three theorems that involve proportions in one way or another. The first of these theorems is a close relative of CSSTP, and the

Chapter 13: Similarity: Same Shape, Different Size   219

second is a distant relative (see the earlier section titled "CASTC and CSSTP, the Cousins of CPCTC" for details on similar-triangle proportions). The third is no kin at all (but two out of three ain't bad).

The Side-Splitter Theorem: It'll make you split your sides

The Side-Splitter Theorem isn't really necessary because the problems in which you use it involve similar triangles, so you can solve them with the ordinary similar-triangle proportions I present earlier in this chapter. The Side-Splitter Theorem just gives you an alternative, shortcut solution method.

Side-Splitter Theorem: If a line is parallel to a side of a triangle and it intersects the other two sides, it divides those sides proportionally. See Figure 13-7.

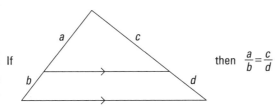

Figure 13-7: A line parallel to a side cuts the other two sides proportionally.

Check out the following problem, which shows this theorem in action:

Given: $\overline{PQ} \parallel \overline{TR}$

Prove: $\triangle PQS \sim \triangle TRS$
(paragraph proof)

Find: x and y

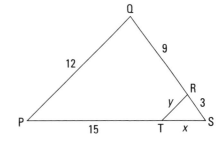

Here's the proof: Because $\overline{PQ} \parallel \overline{TR}$, $\angle Q$ and $\angle TRS$ are congruent corresponding angles (that's *corresponding* in the parallel-lines sense — see Chapter 10 — but these angles also turn out to be *corresponding* in the similar-triangle sense. Get it?) Then, because both triangles contain $\angle S$, the triangles are similar by AA.

Now find x and y. Because $\overline{PQ} \parallel \overline{TR}$, you use the Side-Splitter Theorem to get x:

$$\frac{x}{15} = \frac{3}{9}$$
$$9x = 45$$
$$x = 5$$

And here's the solution for y: First, don't fall for the trap and conclude that $y = 4$. This is a doubly sneaky trap that I'm especially proud of. Side y looks like it should equal 4 for two reasons: First, you could jump to the erroneous conclusion that $\triangle TRS$ is a 3-4-5 right triangle. But nothing tells you that $\angle TRS$ is a right angle, so you can't conclude that.

Second, when you see the ratios of $9:3$ (along $\overline{QS}$) and $15:5$ (along $\overline{PS}$, after solving for x), both of which reduce to $3:1$, it looks like PQ and y should be in the same $3:1$ ratio. That would make $PQ:y$ a $12:4$ ratio, which again leads to the wrong answer that y is 4. The answer comes out wrong because this thought process amounts to using the Side-Splitter Theorem for the sides that aren't split — which you aren't allowed to do.

Don't use the Side-Splitter Theorem on sides that aren't split. You can use the Side-Splitter Theorem *only* for the four segments on the split sides of the triangle. Do *not* use it for the parallel sides, which are in a different ratio. For the parallel sides, use similar-triangle proportions. (Whenever a triangle is divided by a line parallel to one of its sides, the triangle created is similar to the original, large triangle.)

So finally, the correct way to get y is to use an ordinary similar-triangle proportion. The triangles in this problem are positioned the same way, so you can write the following:

$$\frac{\text{Left side}_{\triangle TRS}}{\text{Left side}_{\triangle PQS}} = \frac{\text{base}_{\triangle TRS}}{\text{base}_{\triangle PQS}}$$

$$\frac{y}{12} = \frac{5}{20}$$
$$20y = 60$$
$$y = 3$$

That's a wrap.

Crossroads: The Side-Splitter Theorem extended

This next theorem takes the side-splitter principle and generalizes it, giving it a broader context. With the Side-Splitter Theorem, you draw one parallel line that divides a triangle's sides proportionally. With this next theorem, you can draw any number of parallel lines that cut any lines (not just a triangle's sides) proportionally.

Extension of the Side-Splitter Theorem: If three or more parallel lines are intersected by two or more transversals, the parallel lines divide the transversals proportionally.

See Figure 13-8. Given that the horizontal lines are parallel, the following proportions (among others) follow from the theorem:

$$\frac{AB}{BC} = \frac{PQ}{QR}, \quad \frac{PQ}{QR} = \frac{WX}{XY}, \quad \frac{PR}{RS} = \frac{WY}{YZ}, \quad \frac{AD}{BC} = \frac{WZ}{XY}, \quad \frac{QS}{PQ} = \frac{XZ}{WX}$$

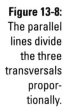

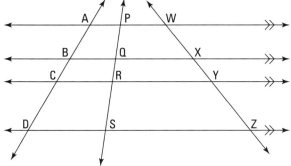

Figure 13-8: The parallel lines divide the three transversals proportionally.

Ready for a problem? Here goes nothing:

Given: $AB = 12$, $BD = 32$, $FJ = 33$, $KM = 45$, $MN = 10$

Find: CD, BC, FG, GH, HJ, KL, and LM

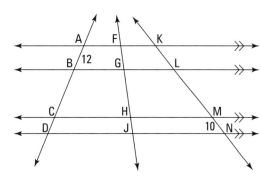

This is a long process, so I go through the unknown lengths one by one.

1. **Set up a proportion to get *CD*.**

   $$\frac{CD}{AD} = \frac{MN}{KN}$$

   $$\frac{CD}{44} = \frac{10}{55}$$

   $$55 \cdot CD = 44 \cdot 10$$

   $$CD = 8$$

2. **Now just subtract *CD* from *BD* to get *BC*.**

   $$BD - CD = BC$$

   $$32 - 8 = BC$$

   $$24 = BC$$

3. **Skip over the segments that make up $\overline{FJ}$ for a minute and use a proportion to find *KL*.**

   $$\frac{AB}{CD} = \frac{KL}{MN}$$

   $$\frac{12}{8} = \frac{KL}{10}$$

   $$8 \cdot KL = 12 \cdot 10$$

   $$KL = 15$$

4. **Subtract to get *LM*.**

   $$KM - KL = LM$$

   $$45 - 15 = LM$$

   $$30 = LM$$

5. **To solve for the parts of $\overline{FJ}$, use the total length of $\overline{FJ}$ and the lengths along $\overline{AD}$.**

   To get *FG*, *GH*, and *HJ*, note that because the ratio *AB* : *BC* : *CD* is 12 : 24 : 8, which reduces to 3 : 6 : 2, the ratio of *FG* : *GH* : *HJ* must also equal 3 : 6 : 2. So let $FG = 3x$, $GH = 6x$, and $HJ = 2x$. Because you're given the length of $\overline{FJ}$, you know that these three segments must add up to 33:

   $$3x + 6x + 2x = 33$$

   $$11x = 33$$

   $$x = 3$$

   So $FG = 3 \cdot 3 = 9$, $GH = 6 \cdot 3 = 18$, and $HJ = 2 \cdot 3 = 6$. A veritable walk in the park.

Chapter 13: Similarity: Same Shape, Different Size **223**

The Angle-Bisector Theorem

In this final section, you get another theorem involving a proportion; but unlike everything else in this chapter, this theorem has nothing to do with similarity. (The extension of the Side-Splitter Theorem in the preceding section might not look like it involves similarity, but it is subtly related.)

Angle-Bisector Theorem: If a ray bisects an angle of a triangle, then it divides the opposite side into segments that are proportional to the other two sides. See Figure 13-9.

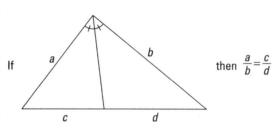

Figure 13-9: Because the angle is bisected, segments c and d are proportional to sides a and b.

If ... then $\frac{a}{b} = \frac{c}{d}$

When you bisect an angle of a triangle, you *never* get similar triangles (unless you bisect the vertex angle of an isosceles triangle, in which case the angle bisector divides the triangle into two congruent triangles).

Don't forget the Angle-Bisector Theorem. (For some reason, students often do forget this theorem.) So whenever you see a triangle with one of its angles bisected, consider using the theorem.

How about an angle-bisector problem? Why? Oh, just *BCUZ*.

Given: Diagram as shown
Find: 1. BZ, CU, UZ, and BU
2. The area of $\triangle BCU$ and $\triangle BUZ$

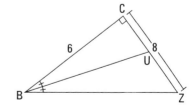

1. **Find BZ, CU, UZ, and BU.**

    You get BZ with the Pythagorean Theorem ($6^2 + 8^2 = c^2$) or by noticing that $\triangle BCZ$ is in the 3 : 4 : 5 family. It's a 6-8-10 triangle, so BZ is 10.

    Next, set CU equal to x and UZ equal to $8 - x$. Set up the angle-bisector proportion and solve for x:

    $$\frac{6}{10} = \frac{x}{8-x}$$
    $$48 - 6x = 10x$$
    $$48 = 16x$$
    $$3 = x$$

    So CU is 3 and UZ is 5.

    The Pythagorean Theorem then gives you BU:

    $$(BU)^2 = 6^2 + 3^2$$
    $$(BU)^2 = 45$$
    $$BU = \sqrt{45} = 3\sqrt{5} \approx 6.7$$

2. **Calculate the area of $\triangle BCU$ and $\triangle BUZ$.**

    Both triangles have a height of 6 (when you use $\overline{CU}$ and $\overline{UZ}$ as their bases), so just use the triangle area formula:

    $$\text{Area}_\triangle = \frac{1}{2}bh$$
    $$\text{Area}_{\triangle BCU} = \frac{1}{2} \cdot 3 \cdot 6 = 9 \text{ units}^2$$
    $$\text{Area}_{\triangle BUZ} = \frac{1}{2} \cdot 5 \cdot 6 = 15 \text{ units}^2$$

    Note that this ratio of triangle areas, 9 : 15, is equal to the ratio of the triangles' bases, 3 : 5. This equality holds whenever a triangle is divided into two triangles with a segment from one of its vertices to the opposite side (whether or not this segment cuts the vertex angle exactly in half).

Part V
Working with Not-So-Vicious Circles

The 5th Wave   By Rich Tennant

"We all know it's a pie, Helen. There's no need to pipe the number 3.141592653 on the top."

In this part . . .

In Part V, you get to ∞-sided polygons. (If the term ∞-*sided polygons* sounds a bit odd, sorry — that's the math geek in me rearing its ugly head. An ∞-sided polygon is just a circle. Get it? In all seriousness, a circle does work just like a polygon with an infinite number of sides, and using that fact is one way to compute the exact value of π.) Anyway, the circle is one of the most amazing shapes in geometry, and it has all kinds of interesting properties that you discover in Chapter 14. In the same chapter, you see how to use those properties in circle proofs. Next, you go on to Chapter 15, where you find the formulas for calculating the area and circumference of a circle as well as the area and length of different parts of a circle, such as chords, arcs, and sectors. Chapter 15 also gives you the angle-arc theorems and the power theorems. Lucky you!

Chapter 14

Coming Around to Circle Basics

In This Chapter
▶ Segments inside circles: Radii and chords
▶ The Three Musketeers: Arcs, central angles, and chords
▶ Common-tangent and walk-around problems

*I*n a sense, the circle is the simplest of all shapes — one smooth curve that's always the same distance from the circle's center: no corners, no irregularities, the same simple shape no matter how you turn it. On the other hand, that simple curve involves the number pi ($\pi \approx 3.14159...$), and nothing's simple about that. It goes on forever with no repeating pattern of digits. Despite the fact that mathematicians have been studying the circle and the number π for over 2,000 years, many unsolved mysteries about them remain.

The circle is also, perhaps, the most common shape in the natural world (if you count spheres, which are, of course, circular and whose surfaces contain an infinite collection of circles). The 10^{21} (or 1,000,000,000,000,000,000,000) stars in the universe are spherical. The tiny droplets of water in a cloud are spherical (one cloud can contain trillions of droplets). Throw a pebble in a pond, and the waves propagate outward in circular rings. The Earth travels around the sun in a circular orbit (okay, it's actually an ellipse for you astronomical nitpickers out there; though it's *extremely* close to a circle). And right this very minute, you're traveling in a circular path as the Earth rotates on its axis.

In this chapter, you investigate some of the circle's most fundamental properties. Time to get started.

The Straight Talk on Circles: Radii and Chords

I doubt you need a definition for *circle,* but for those of you who love math-speak, here's the fancy-pants definition.

Circle: A circle is a set of all points in a plane that are equidistant from a single point (the circle's center).

In the following sections, I talk about the three main types of line segments that you find inside a circle: radii, chords, and diameters. Although starting with all these *straight* things in a chapter on curving circles may seem a bit strange, some of the most interesting and important theorems about circles stem from these three segments. You later get to explore these theorems with some circle problems, so listen up.

Defining radii, chords, and diameters

Here are three terms that are fundamental to your investigation of the circle:

- **Radius:** A circle's radius — the distance from its center to a point on the circle — tells you the circle's size. In addition to being a measure of distance, a radius is also a segment that goes from a circle's center to a point on the circle.
- **Chord:** A segment that connects two points on a circle is called a chord.
- **Diameter:** A chord that passes through a circle's center is a diameter of the circle. A circle's diameter is twice as long as its radius.

Figure 14-1 shows circle O with diameter $\overline{AB}$ (which is also a chord), radii $\overline{OA}$, $\overline{OB}$, and $\overline{OC}$, and chord $\overline{PQ}$.

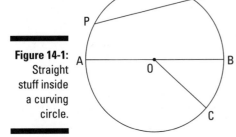

Figure 14-1: Straight stuff inside a curving circle.

How big is the full moon?

While I'm on the subject of circles, check this out: Take a penny and hold it out at arm's length. Now ask yourself how big that penny looks compared with the size of a full moon — you know, like if you went outside during a full moon and held the penny up at arm's length "next to" the moon. Which do you think would look bigger and by how much (is one of them twice as big, three times, or what)? Common answers are that the two are the same size or that the moon is two or three times as big as the penny. Well — hold on to your hat — the real answer is that the penny is three times as wide as the moon! (Give or take a bit, depending on the length of your arm.) Hard to believe but true. Try it some evening.

Introducing five circle theorems

I hope you have some available space on your mental hard drive for more theorems. (If not, maybe you can free up some room by deleting a few not-so-useful facts, such as the date of the Battle of Hastings, A.D. 1066) In this section, you get five important theorems about the properties of the segments inside a circle.

These theorems tell you about radii and chords (note that two of these theorems work in both directions):

- **Radii size:** All radii of a circle are congruent.
- **Perpendicularity and bisected chords:**
  - If a radius is perpendicular to a chord, then it bisects the chord.
  - If a radius bisects a chord (that isn't a diameter), then it's perpendicular to the chord.
- **Distance and chord size:**
  - If two chords of a circle are equidistant from the center of the circle, then they're congruent.
  - If two chords of a circle are congruent, then they're equidistant from its center.

Working through a proof

Here's a proof that uses three of the theorems from the preceding section:

Given: Circle G

F is the midpoint of $\overline{AE}$

$\overline{GB} \perp \overline{CA}$

$\overline{GD} \perp \overline{CE}$

Prove: $BCDG$ is a kite

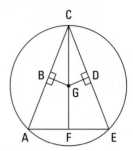

Before reading the proof, you may want to make your own game plan.

Statements	Reasons
1) Circle G F is the midpoint of $\overline{AE}$	1) Given.
2) $\overline{AF} \cong \overline{EF}$	2) Definition of midpoint.
3) $\overline{GF}$ bisects $\overline{AE}$	3) Definition of bisect.
4) $\overline{GF} \perp \overline{AE}$	4) If a radius bisects a chord (that's not a diameter), then it's perpendicular to the chord.
5) $\angle AFG$ and $\angle EFG$ are right angles	5) Definition of perpendicular.
6) $\angle AFG \cong \angle EFG$	6) All right angles are congruent.
7) $\overline{CF} \cong \overline{CF}$	7) Reflexive Property.
8) $\triangle AFC \cong \triangle EFC$	8) SAS (2, 6, 7).
9) $\overline{AC} \cong \overline{EC}$	9) CPCTC.
10) $\overline{GB} \cong \overline{GD}$	10) If two chords of a circle are congruent, then they're equidistant from its center.
11) $\overline{GB}$ bisects $\overline{AC}$ $\overline{GD}$ bisects $\overline{EC}$	11) If a radius is perpendicular to a chord, then it bisects the chord.
12) $\overline{CB} \cong \overline{CD}$	12) Like Divisions (statements 9 and 11).
13) $BCDG$ is a kite	13) Definition of a kite (two disjoint pairs of consecutive sides are congruent, statements 10 and 12).

Using extra radii to solve a problem

In real estate, the three most important factors are *location, location, location*. With circles, it's *radii, radii, radii*. In circle problems, you often add radii and partial radii to create right triangles or isosceles triangles that you can then use to solve the problem. Here's what you do in greater detail:

- **Draw additional radii on the figure.** You should draw radii to points where something else intersects or touches the circle, as opposed to just any old point on the circle.

- **Open your eyes and notice all the radii — including new ones you've drawn — and mark them all congruent.** For some reason — even though *all radii are congruent* is one of the simplest theorems of all — people frequently either fail to notice all the radii in a problem or fail to note that they're congruent.

- **Draw in the segment (part of a radius) that goes from the center of a circle to a chord and that's perpendicular to the chord.** This segment bisects the chord (I mention this theorem in the preceding section).

Now check out the following problem: Find the area of inscribed quadrilateral *GHJK* shown on the left. The circle has a radius of 2.

The tip that leads off this section gives you two hints for this problem. The first hint is to draw in the four radii to the four vertices of the quadrilateral as shown in the figure on the right.

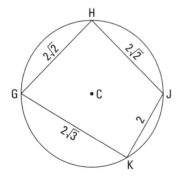
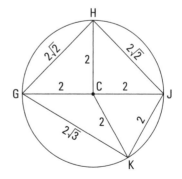

Now you simply need to find the area of the individual triangles. You can see that $\triangle JKC$ is equilateral, so you can use the equilateral triangle formula (in Chapter 7) for this one:

$$\text{Area} = \frac{s^2\sqrt{3}}{4} = \frac{2^2\sqrt{3}}{4} = \sqrt{3} \text{ units}^2$$

And if you're on the ball, you should recognize triangles *GHC* and *HJC*. Their sides are in the ratio of $2 : 2 : 2\sqrt{2}$, which reduces to $1 : 1 : \sqrt{2}$; thus, they're 45°- 45°- 90° triangles (see Chapter 8). You already know the base and height of these two triangles, so getting their areas should be a snap. For each triangle,

$$\text{Area} = \frac{1}{2} bh = \frac{1}{2} \cdot 2 \cdot 2 = 2 \text{ units}^2$$

Another hint from the tip helps you with $\triangle KGC$. Draw its altitude (a partial radius) from *C* to $\overline{GK}$. This radius is perpendicular to $\overline{GK}$ and thus bisects $\overline{GK}$ into two segments of length $\sqrt{3}$. You've divided $\triangle KGC$ into two right triangles; each has a hypotenuse of 2 and a leg of $\sqrt{3}$, so the other leg (the altitude) is 1 (by the Pythagorean Theorem or by recognizing that these are 30°- 60°- 90° triangles whose sides are in the ratio of $1 : \sqrt{3} : 2$ — see Chapter 8). So $\triangle KGC$ has an altitude of 1 and a base of $2\sqrt{3}$. Just use the regular area formula again:

$$\text{Area} = \frac{1}{2} bh = \frac{1}{2} \cdot 2\sqrt{3} \cdot 1 = \sqrt{3} \text{ units}^2$$

Now just add 'em up:

$$\begin{aligned} \text{Area}_{GHJK} &= \text{area}_{\triangle JKC} + \text{area}_{\triangle GHC} + \text{area}_{\triangle HJC} + \text{area}_{\triangle KGC} \\ &= \sqrt{3} + 2 + 2 + \sqrt{3} \\ &= 4 + 2\sqrt{3} \approx 7.46 \text{ units}^2 \end{aligned}$$

Pieces of the Pie: Arcs and Central Angles

In this section, I introduce you to arcs and central angles, and then you see six theorems about how arcs, central angles, and chords are all interrelated.

Three circle definitions for your mathematical pleasure

Okay, so maybe *pleasure* is a bit of a stretch. How about "more fun than sticking a hot poker in your eye"? These definitions may not rank up there with your greatest high school memories, but they are important in geometry. A circle's central angles and the arcs that they cut out are part of many circle proofs, as you can see in the following section. They also come up in many area problems, which you see in Chapter 15.

Keep these definitions in mind if you know what's good for you (and for a visual, see Figure 14-2):

Chapter 14: Coming Around to Circle Basics 233

- **Arc:** An arc is simply a curved piece of a circle. Any two points on a circle divide the circle into two arcs: a *minor arc* (the smaller piece) and a *major arc* (the larger) — unless the points are the endpoints of a diameter, in which case both arcs are semicircles. Figure 14-2 shows minor arc $\overset{\frown}{AB}$ (a 60° arc) and major arc $\overset{\frown}{ACB}$ (a 300° arc). Note that to name a minor arc, you use its two endpoints; to name a major arc, you use its two endpoints plus any point along the arc.

- **Central angle:** A central angle is an angle whose vertex is at the center of a circle. The two sides of a central angle are radii that hit the circle at the opposite ends of an arc — or as mathematicians say, the angle *intercepts* the arc.

  The measure of an arc is the same as the degree measure of the central angle that intercepts it. The figure shows central angle $\angle AQB$, which, like $\overset{\frown}{AB}$, measures 60°.

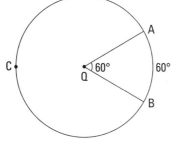

Figure 14-2:
A 60° central angle cuts out a 60° arc.

And here's one more definition that you need for the next section.

Congruent circles: Congruent circles are circles with congruent radii.

Six scintillating circle theorems

The next six theorems are all just variations on one basic idea about the interconnectedness of arcs, central angles, and chords (all six are illustrated in Figure 14-3):

- **Central angles and arcs:**
  - If two central angles of a circle (or of congruent circles) are congruent, then their intercepted arcs are congruent. (Short form: If central angles congruent, then arcs congruent.) In Figure 14-3, if $\angle WMX \cong \angle ZMY$, then $\overset{\frown}{WX} \cong \overset{\frown}{ZY}$.

- If two arcs of a circle (or of congruent circles) are congruent, then the corresponding central angles are congruent. (Short form: If arcs congruent, then central angles congruent.) If $\overset{\frown}{WX} \cong \overset{\frown}{ZY}$, then $\angle WMX \cong \angle ZMY$.

- **Central angles and chords:**
  - If two central angles of a circle (or of congruent circles) are congruent, then the corresponding chords are congruent. (Short form: If central angles congruent, then chords congruent.) In Figure 14-3, if $\angle WMX \cong \angle ZMY$, then $\overline{WX} \cong \overline{ZY}$.
  - If two chords of a circle (or of congruent circles) are congruent, then the corresponding central angles are congruent. (Short form: If chords congruent, then central angles congruent.) If $\overline{WX} \cong \overline{ZY}$, then $\angle WMX \cong \angle ZMY$.

- **Arcs and chords:**
  - If two arcs of a circle (or of congruent circles) are congruent, then the corresponding chords are congruent. (Short form: If arcs congruent, then chords congruent.) In Figure 14-3, if $\overset{\frown}{WX} \cong \overset{\frown}{ZY}$, then $\overline{WX} \cong \overline{ZY}$.
  - If two chords of a circle (or of congruent circles) are congruent, then the corresponding arcs are congruent. (Short form: If chords congruent, then arcs congruent.) If $\overline{WX} \cong \overline{ZY}$, then $\overset{\frown}{WX} \cong \overset{\frown}{ZY}$.

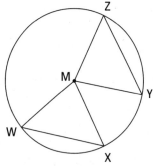

Figure 14-3: Arcs, chords, and central angles: All for one and one for all.

Here's a more condensed way of thinking about the six theorems:

- If the angles are congruent, both the chords and the arcs are congruent.
- If the chords are congruent, both the angles and the arcs are congruent.
- If the arcs are congruent, both the angles and the chords are congruent.

These three ideas condense further to one simple idea: If any pair (of central angles, chords, or arcs) is congruent, then the other two pairs are also congruent.

Trying your hand at some proofs

Time for a proof. Try to work out your own game plan before reading the solution:

Given: Circle U
$\overset{\frown}{DB} \cong \overset{\frown}{WE}$
$\overline{UO} \perp \overline{DW}$
$\overline{UL} \perp \overline{EB}$

Prove: $\triangle UOW \cong \triangle ULB$

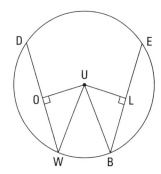

Hint: Arc addition and subtraction work just like segment addition and subtraction.

Behold the formal proof:

Statements	Reasons
1) Circle U $\overline{UO} \perp \overline{DW}$ $\overline{UL} \perp \overline{EB}$	1) Given.
2) $\angle UOW$ is a right angle $\angle ULB$ is a right angle	2) Definition of perpendicular.
3) $\overset{\frown}{DB} \cong \overset{\frown}{WE}$	3) Given.
4) $\overset{\frown}{DW} \cong \overset{\frown}{BE}$	4) Subtracting $\overset{\frown}{WB}$ from both $\overset{\frown}{DB}$ and $\overset{\frown}{WE}$.
5) $\overline{DW} \cong \overline{BE}$	5) If arcs are congruent, then chords are congruent.
6) $\overline{UO} \cong \overline{UL}$	6) If two chords of a circle are congruent, then they're equidistant from its center.
7) $\overline{UW} \cong \overline{UB}$	7) All radii are congruent.
8) $\triangle UOW \cong \triangle ULB$	8) HLR (7, 6, 2).

One proof down, one to go:

Given: Circle V
$\stackrel{\frown}{QR} \cong \stackrel{\frown}{ST}$

Prove: $\angle Q \cong \angle T$

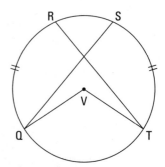

Here's a quick game plan: First, draw in radii to R and S, creating two triangles. Think about how you can prove the triangles congruent. With arc addition, you get $\stackrel{\frown}{QS} \cong \stackrel{\frown}{RT}$, and from that, you get $\angle QVS \cong \angle TVR$. You can then use those angles and four radii to get the triangles congruent with SAS and then finish with CPCTC.

Statements	Reasons
1) Circle V	1) Given.
2) Draw $\overline{VR}$ and $\overline{VS}$	2) Two points determine a line.
3) $\overline{VR} \cong \overline{VS}$	3) All radii of a circle are congruent.
4) $\overline{VT} \cong \overline{VQ}$	4) All radii of a circle are congruent.
5) $\stackrel{\frown}{QR} \cong \stackrel{\frown}{ST}$	5) Given.
6) $\stackrel{\frown}{QS} \cong \stackrel{\frown}{RT}$	6) Adding $\stackrel{\frown}{RS}$ to both $\stackrel{\frown}{QR}$ and $\stackrel{\frown}{ST}$.
7) $\angle QVS \cong \angle TVR$	7) If arcs are congruent, then central angles are congruent.
8) $\triangle QVS \cong \triangle TVR$	8) SAS (lines 3, 7, 4). (You could also have gotten congruent chords in line 7 and then used SSS in line 8.)
9) $\angle Q \cong \angle T$	9) CPCTC.

Going Off on a Tangent about Tangents

I hope you're enjoying *Geometry For Dummies* so far. I remember enjoying geometry in high school. Hey, that reminds me: I had this geometry teacher who had this old, run-down car, a real beater. He took it on a trip to the Ozarks, and on the way there he had to stop for gas. After filling up, he went into the station to buy some beef jerky, and when he was coming back, there was a bear trying to . . . hey, where was I? Oh, I guess I sort of went off on a *tangent* — get it? I really crack myself up.

Anyway, in this section, you look at lines that are tangent to circles. Tangents show up in a couple of interesting problems, the common-tangent problem and the walk-around problem. As you may have guessed, these problems have nothing to do with road trips to the Ozarks.

Introducing the tangent line

First, a definition: A line is *tangent* to a circle if it touches it at one and only one point.

Radius-tangent perpendicularity: If a line is tangent to a circle, then it is perpendicular to the radius drawn to the point of tangency. Check out the bicycle wheels in Figure 14-4.

Figure 14-4: The ground is tangent to the wheels.

In this figure, the wheels are, of course, circles, the spokes are radii, and the ground is a *tangent line*. The point where each wheel touches the ground is a *point of tangency*. And the most important thing — what the theorem tells you — is that the radius that goes to the point of tangency is *perpendicular* to the tangent line.

Don't neglect to check circle problems for tangent lines and the right angles that occur at points of tangency. You may have to draw in one or more radii to points of tangency to create the right angles. The right angles often become parts of right triangles (or sometimes rectangles).

Here's an example problem: Find the radius of circle C and the length of $\overline{DE}$ in the following figure.

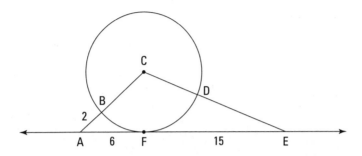

When you see a circle problem, you should be saying to yourself: *radii, radii, radii!* So draw in radius $\overline{CF}$, which, according to the theorem, is perpendicular to $\overleftrightarrow{AE}$. Set it equal to x, which gives $\overline{CB}$ a length of x as well. You now have right triangle $\triangle CFA$, so use the Pythagorean Theorem to find x:

$$x^2 + 6^2 = (x + 2)^2$$
$$x^2 + 36 = x^2 + 4x + 4$$
$$32 = 4x$$
$$8 = x$$

So the radius is 8. Then you can see that $\triangle CFE$ is an 8-15-17 triangle (see Chapter 8), so CE is 17. (Of course, you can also get CE with the Pythagorean Theorem.) CD is 8 (and it's the third radius in this problem; does *radii, radii, radii* ring a bell?). Therefore, DE is 17 – 8, or 9. That does it.

The common-tangent problem

The *common-tangent problem* is named for the single tangent line that's tangent to two circles. Your goal is to find the length of the tangent. These problems are a bit involved, but they should cause you little difficulty if you use the straightforward three-step solution method that follows.

The following example involves a common *external* tangent (where the tangent lies on the same side of both circles). You might also see a common-tangent problem that involves a common *internal* tangent (where the tangent lies between the circles). No worries: The solution technique is the same for both.

Chapter 14: Coming Around to Circle Basics 239

Given: The radius of circle A is 4

The radius of circle Z is 14

The distance between the circles is 8

Find: The length of the common tangent, $\overline{BY}$

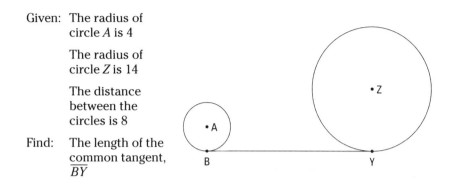

Here's how to solve it:

1. **Draw the segment connecting the centers of the two circles and draw the two radii to the points of tangency (if these segments haven't already been drawn for you).**

   Draw $\overline{AZ}$ and radii $\overline{AB}$ and $\overline{ZY}$. Figure 14-5 shows this step. Note that the given distance of 8 between the circles is the distance between the *outsides* of the circles along the segment that connects their centers.

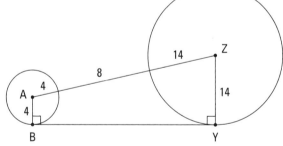

Figure 14-5: Here's the first step, which creates two right angles.

2. **From the center of the *smaller circle,* draw a segment parallel to the common tangent till it hits the radius of the larger circle (or the extension of the radius in a common-internal-tangent problem).**

   You end up with a right triangle and a rectangle; one of the rectangle's sides is the common tangent. Figure 14-6 illustrates this step.

Figure 14-6:
Here's the second step (and part of the third).

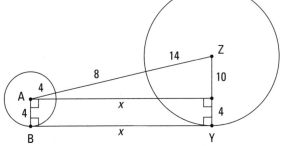

3. **You now have a right triangle and a rectangle and can finish the problem with the Pythagorean Theorem and the simple fact that opposite sides of a rectangle are congruent.**

    The triangle's hypotenuse is made up of the radius of circle *A*, the segment between the circles, and the radius of circle *Z*. Their lengths add up to $4 + 8 + 14 = 26$. You can see that the width of the rectangle equals the radius of circle *A*, which is 4; because opposite sides of a rectangle are congruent, you can then tell that one of the triangle's legs is the radius of circle *Z* minus 4, or $14 - 4 = 10$. You now know two sides of the triangle, and if you find the third side, that'll give you the length of the common tangent. You get the third side with the Pythagorean Theorem:

    $$x^2 + 10^2 = 26^2$$
    $$x^2 + 100 = 676$$
    $$x^2 = 576$$
    $$x = 24$$

    (Of course, if you recognize that the right triangle is in the 5 : 12 : 13 family, you can multiply 12 by 2 to get 24 instead of using the Pythagorean Theorem.)

    Because opposite sides of a rectangle are congruent, *BY* is also 24, and you're done.

    Now look back at Figure 14-6 and note where the right angles are and how the right triangle and the rectangle are situated; then make sure you heed the following tip and warning.

Note the location of the hypotenuse. In a common-tangent problem, the segment connecting the centers of the circles is *always* the hypotenuse of a right triangle. The common tangent is *always* the side of a rectangle, *not* a hypotenuse.

In a common-tangent problem, the segment connecting the centers of the circles is *never* one side of a right angle. Don't make this common mistake.

Taking a walk on the wild side with a walk-around problem

I think the way the next type of problem works out is really nifty. It's called a walk-around problem; you'll see why in a minute. But first, here's a theorem you need for the problem.

Dunce Cap Theorem: If two tangent segments are drawn to a circle from the same external point, then they're congruent. I call this the *Dunce Cap Theorem* because that's what the diagram looks like, but you won't have much luck if you try to find that name in another geometry book. See Figure 14-7.

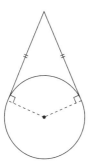

Figure 14-7:
The circle is wearing a dunce cap, which has congruent sides.

Now for the problem:

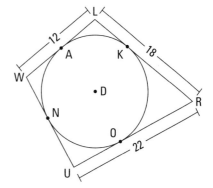

Given: Diagram as shown
$\overline{WL}$, $\overline{LR}$, $\overline{RU}$, and $\overline{UW}$ are tangent to circle D

Find: UW

The first thing to notice about a walk-around problem is exactly what the Dunce Cap Theorem says: Two tangent segments are congruent if they're drawn from the same point outside of the circle. So in this problem, you'd mark the following pairs of segments congruent: $\overline{WN}$ and $\overline{WA}$, $\overline{LA}$ and $\overline{LK}$, $\overline{RK}$ and $\overline{RO}$, and $\overline{UO}$ and $\overline{UN}$. (Are you starting to see why they call this beast the *walk-around?*)

Okay, let's finish taming this wild critter. Set WN equal to x. Then, by the Dunce Cap Theorem, WA is x as well. Next, because WL is 12 and WA is x, AL is $12 - x$. A to L to K is another dunce cap, so LK is also $12 - x$. LR is equal to 18, so KR is $LR - LK$, or $18 - (12 - x)$; this simplifies to $6 + x$. Continue walking around like this till you get back home to $\overline{NU}$, as I show you in Figure 14-8.

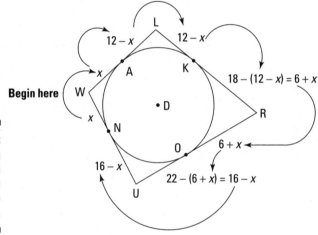

Figure 14-8: Walking around with the Dunce Cap Theorem.

Finally, UW equals $WN + NU$, or $x + (16 - x)$, which equals 16. That's it.

One of the things I find interesting about walk-around problems is that when you have an even number of sides in the figure (as in this example), you get a solution without ever solving for x. In the example problem, x can take on any value from 0 to 12 inclusive. Varying x changes the size of the circle and the shape of the quadrilateral, but the lengths of the four sides (including the solution) remain unchanged. When, on the other hand, a walk-around problem involves an odd number of sides, there's a single solution for x and the diagram has a fixed shape. Pretty cool, eh?

Chapter 15
Circle Formulas and Theorems

In This Chapter
- Analyzing arc length and sector area
- Theorizing about the angle-arc theorems
- Practicing products with the power theorems

Students are always asking, "When am I ever going to use this?" Well, I suppose this book contains many things you won't be using again very soon, but one thing I know for sure is that you'll use things with a circular shape thousands of times in your life. Every time you ride in a car, you encounter the four tires, the circular steering wheel, the circular knobs on the radio, the circular opening at the tailpipe, and so on; same goes for riding a bicycle — wheels, gears, the opening of the little air-filler thingamajig on the tires, and so on. And consider all the inventions and products that make use of this omnipresent shape: Ferris wheels, gyroscopes, iPod click wheels, lenses, manholes (and manhole covers), pipes, waterwheels, merry-go-rounds, and many, many more.

In this chapter, you discover fascinating things about the circle. You investigate many formulas and theorems about circles and the connections among circles, angles, arcs, and various segments associated with circles (chords, tangents, and secants). For the most part, these formulas involve the way these geometric objects cut up or divide each other: circles cutting up secants, angles cutting arcs out of circles, chords cutting up chords, and so on.

Chewing on the Pizza Slice Formulas

In this section, you begin with two basic formulas: the formulas for the area and the circumference of a circle. Then you use these formulas to compute lengths, perimeters, and areas of various parts of a circle: arcs, sectors, and segments (yes, *segment* is the name for a particular chunk of a circle, and it's completely different from a line segment — go figure). You could use these formulas when you're figuring out what size pizza to order, though I think it'd be more useful to simply calculate how hungry you are. So here you go.

Circumference and area of a circle: Along with the Pythagorean Theorem and a few other formulas, the two following circle formulas are among the most widely recognized formulas in geometry. In these formulas, *r* is a circle's radius and *d* is its diameter:

- Circumference = $2\pi r$ (or πd)
- $\text{Area}_{\text{Circle}} = \pi r^2$

Read on for info on finding arc length and the area of sectors and segments. If you understand the simple reasoning behind the formulas for these things, you should be able to solve arc, sector, and segment problems even if you forget the formulas.

Determining arc length

Before getting to the arc length formula, I want to mention a potential source of confusion about arcs and how you measure them. In Chapter 14, the *measure of an arc* is defined as the degree measure of the central angle that intercepts the arc. To say that the measure of an arc is 60° simply means that the associated central angle is a 60° angle. But now, in this section, I go over how you determine the length of an arc. An *arc's length* means the same common-sense thing length always means — you know, like the length of a piece of string (with an arc, of course, it'd be a curved piece of string). In a nutshell, the *measure* of an arc is the degree size of its central angle; the *length* of an arc is the regular length along the arc.

A circle is 360° all the way around; therefore, if you divide an arc's degree measure by 360°, you find the fraction of the circle's circumference that the arc makes up. Then, if you multiply the length all the way around the circle (the circle's circumference) by that fraction, you get the length along the arc. So finally, here's the formula you've been waiting for.

Arc length: The length of an *arc* (part of the circumference, like $\widehat{AB}$ in Figure 15-1) is equal to the circumference of the circle ($2\pi r$) times the fraction of the circle represented by the arc's measure (note that the degree measure of an arc is written like $m\widehat{AB}$):

$$\text{Length}_{\widehat{AB}} = \left(\frac{m\widehat{AB}}{360}\right)(2\pi r)$$

A circle is sort of an ∞-gon

A regular octagon, like a stop sign, has eight congruent sides and eight congruent angles. Now imagine what regular 12-gons, 20-gons, and 50-gons would look like. The more sides a polygon has, the closer it gets to a circle. Well, if you continue this to infinity, you sort of end up with an ∞-gon, which is exactly the same as a circle. (I say *sort of* because whenever you talk about infinity, you're on somewhat shaky ground.) The fact that you can think of a circle as an ∞-gon makes the following remarkable idea work:

> You can use the formula for the area of a regular polygon to compute the area of a circle!

Here's the regular-polygon formula from Chapter 12: $\text{Area}_{\text{Regular Polygon}} = \frac{1}{2} pa$ (where p is the polygon's perimeter and a is its apothem, the distance from the polygon's center to the midpoint of a side).

To explain how this formula works, I use an octagon as an example. Below is a regular octagon with its apothem drawn in and, to the right of it, what it would look like after being cut along its radii and unrolled.

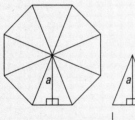
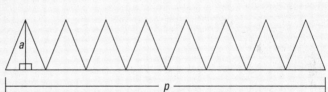

The octagon's perimeter has become the eight bases of the eight little triangles. You can see that the apothem is the same as the height of the triangles.

The polygon area formula is based on the triangle area formula $\left(\text{Area}_\triangle = \frac{1}{2} bh\right)$. All the polygon formula does is use the perimeter (the sum of all the triangle bases) instead of a single triangle base. In doing so, it just totals up the areas of all the little triangles in one fell swoop to give you the area of the polygon.

Now look at a circle cut into 16 thin sectors (or pizza slices) before and after being unrolled.

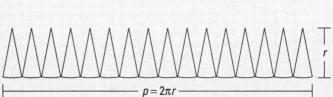

(continued)

(continued)

As you can see, a circle's "perimeter" is its circumference ($2\pi r$), and its "apothem" is its radius. Unlike the flat base of the unrolled octagon, the base of the unrolled circle is wavy because it's made up of little arcs of the circle. But if you were to cut up the circle into more and more sectors, this base would get flatter and flatter until — with an "infinite number" of "infinitely thin" sectors — it became perfectly flat and the sectors became infinitely thin triangles. Then the polygon formula would work for the same reason that it works for polygons:

$$\text{Area}_{Circle} = \frac{1}{2}pa$$
$$= \frac{1}{2}(2\pi r)(r)$$
$$= \pi r^2$$

Voilà! The old familiar πr^2.

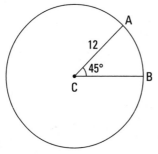

Figure 15-1: Arc $\widehat{AB}$ is $\frac{1}{8}$ of the circle's circumference.

Check out the calculations for $\widehat{AB}$. Its degree measure is 45° and the radius of the circle is 12, so here's the math for its length:

$$\text{Length}_{\widehat{AB}} = \left(\frac{m\widehat{AB}}{360}\right)(2\pi r)$$
$$= \frac{45}{360} \cdot 2 \cdot \pi \cdot 12$$
$$= \frac{1}{8} \cdot 24\pi$$
$$= 3\pi \approx 9.42 \text{ units}$$

As you can see, because 45° is $\frac{1}{8}$ of 360°, the length of arc $\widehat{AB}$ is $\frac{1}{8}$ of the circle's circumference. Pretty simple, eh?

Finding sector and segment area

Yes, you read that right. You can find the area of a segment — because the word *segment* also refers to a chunk of a circle. Mark off a section of a circle with an arc and a chord, and you have a segment. Throw a couple of radii around an arc, and you have a sector. In this section, I show you how to find the area of each of these regions. (You knew I wasn't just going to leave well enough alone at arcs, right?)

So here are the definitions of the two regions (Figure 15-2 shows you both):

- **Sector:** A region bounded by two radii and an arc of a circle (plain English definition: The shape of a piece of pizza)
- **Segment of a circle:** A region bounded by a chord and an arc of a circle

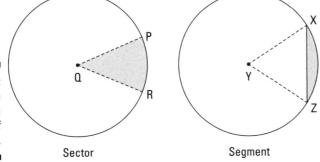

Figure 15-2: A pizza-slice sector and a segment of a circle.

Just as an arc is part of a circle's circumference, a sector is part of a circle's area; therefore, computing the area of a sector works like the arc-length formula in the preceding section.

Area of a sector: The area of a sector (such as sector *PQR* in Figure 15-2) is equal to the area of the circle (πr^2) times the fraction of the circle represented by the sector:

$$\text{Area}_{\text{Sector } PQR} = \left(\frac{m\widehat{PR}}{360}\right)(\pi r^2)$$

Use this formula to find the area of sector *ACB* from Figure 15-1:

$$\text{Area}_{\text{Sector } ACB} = \left(\frac{m\widehat{AB}}{360}\right)(\pi r^2)$$
$$= \frac{1}{8} \cdot \pi \cdot 12^2$$
$$= 18\pi \approx 56.55 \text{ units}^2$$

Because 45° is $\frac{1}{8}$ of 360°, the area of sector ACB is $\frac{1}{8}$ of the area of the circle (just like the length of $\overset{\frown}{AB}$ is $\frac{1}{8}$ of the circle's circumference).

Area of a segment: To compute the area of a segment like the one in Figure 15-2, just subtract the area of the triangle from the area of the sector (by the way, there's no technical way to name segments, but let's call this one *circle segment XZ*):

$$\text{Area}_{\text{Circle Segment } XZ} = \text{area}_{\text{sector } XYZ} - \text{area}_{\triangle XYZ}$$

You know how to compute the area of a sector from earlier in this section. To get the triangle's area, you draw an altitude that goes from the circle's center to the chord that makes up the triangle's base. This altitude then becomes a leg of a right triangle whose hypotenuse is a radius of the circle. You finish with right-triangle ideas such as the Pythagorean Theorem. I show you how to do all this in detail in the next section.

Pulling it all together in a problem

The following problem illustrates finding arc length, sector area, and segment area:

Given: Circle D with a radius of 6

Find: 1. Length of arc $\overset{\frown}{IK}$

2. Area of sector IDK

3. Area of circle segment IK

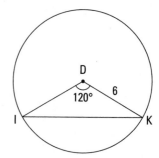

Here's the solution to this three-part problem:

1. **Find the length of arc $\overset{\frown}{IK}$.**

   You really don't need a formula for finding arc length if you understand the concepts: The measure of the arc is 120°, which is a third of 360°, so the length of $\overset{\frown}{IK}$ is a third of the circumference of circle D. That's all there is to it. Here's how all this looks when you plug it into the formula:

$$\text{Length}_{\widehat{IK}} = \left(\frac{m\widehat{IK}}{360}\right)(2\pi r)$$
$$= \frac{120}{360} \cdot 12\pi$$
$$= \frac{1}{3} \cdot 12\pi$$
$$= 4\pi \approx 12.6 \text{ units}$$

2. Find the area of sector *IDK*.

A sector is a portion of the circle's area. Because 120° takes up a third of the degrees in a circle, sector *IDK* occupies a third of the circle's area. Here's the formal solution:

$$\text{Area}_{\text{Sector }IDK} = \left(\frac{m\widehat{IK}}{360}\right)(\pi r^2)$$
$$= \frac{120}{360} \cdot 36\pi$$
$$= \frac{1}{3} \cdot 36\pi$$
$$= 12\pi \approx 37.7 \text{ units}^2$$

3. Find the area of circle segment *IK*.

To find the segment area, you need the area of △*IDK* so you can subtract it from the area of sector *IDK*. Draw an altitude straight down from *D* to $\overline{IK}$. That creates two 30°- 60°- 90° triangles. The sides of a 30°- 60°- 90° triangle are in the ratio of $x : x\sqrt{3} : 2x$ (see Chapter 8), where x is the short leg, $x\sqrt{3}$ the long leg, and $2x$ the hypotenuse. In this problem, the hypotenuse is 6, so the altitude (the short leg) is half of that, or 3, and the base (the long leg) is $3\sqrt{3}$. $\overline{IK}$ is twice as long as the base of the 30°- 60°- 90° triangle, so it's twice $3\sqrt{3}$, or $6\sqrt{3}$. You're all set to finish with the segment area formula:

$$\text{Area}_{\text{Segment }IDK} = \text{area}_{\text{Sector }IDK} - \text{area}_{\triangle IDK}$$
$$= 12\pi - \frac{1}{2}bh \quad \text{(You got the } 12\pi \text{ in part 2)}$$
$$= 12\pi - \frac{1}{2} \cdot 6\sqrt{3} \cdot 3$$
$$= 12\pi - 9\sqrt{3}$$
$$\approx 22.1 \text{ units}^2$$

Digesting the Angle-Arc Theorems and Formulas

In this section, you investigate angles that intersect a circle. The vertices of these angles can lie *inside* the circle, *on* the circle, or *outside* the circle. The formulas in this section tell you how each of these angles is related to the arcs they intercept. As with much of the other material in this book, Archimedes and other mathematicians from over two millennia ago knew these angle-arc relationships.

Angles on a circle

Of the three places an angle's vertex can be in relation to a circle, the angles whose vertices lie *on* a circle are the ones that come up in the most problems and are therefore the most important. These angles come in two flavors:

- **Inscribed angle:** An inscribed angle, like ∠BCD in Figure 15-3a, is an angle whose vertex lies on a circle and whose sides are two chords of the circle.

- **Tangent-chord angle:** A tangent-chord angle, like ∠JKL in Figure 15-3b, is an angle whose vertex lies on a circle and whose sides are a tangent and a chord of the circle (for more on tangents and chords, see Chapter 14).

Measure of an angle on a circle: The measure of an inscribed angle or a tangent-chord angle is *one-half* the measure of its intercepted arc.

For example, in Figure 15-3, $\angle BCD = \frac{1}{2}(m\widehat{BD})$ and $\angle JKL = \frac{1}{2}(m\widehat{JK})$.

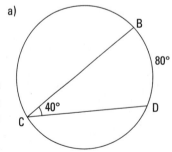
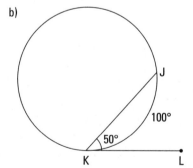

Figure 15-3: Angles with vertices *on* a circle.

Make sure you remember the simple idea that an angle on a circle is half the measure of the arc it intercepts (or if you look at it the other way around, the arc measure is double the angle). If you forget which is half of which, try this: Draw a quick sketch of a circle with a 90° arc (a quarter of the circle) and an inscribed angle that intercepts the 90° arc. You'll see right away that the angle is less than 90°, telling you that the angle is the thing that's half of the arc, not vice versa.

Congruent angles on a circle: The following theorems tell you about situations in which you get two congruent angles on a circle:

- If two inscribed or tangent-chord angles intercept the same arc, then they're congruent (see Figure 15-4a).

- If two inscribed or tangent-chord angles intercept congruent arcs, then they're congruent (see Figure 15-4b).

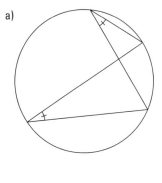
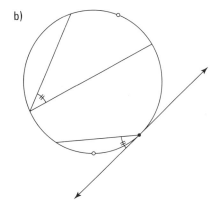

Figure 15-4: Congruent inscribed and tangent-chord angles.

Time to see these ideas in action — take a look at the following problem:

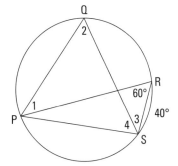

Given: Diagram as shown

∠QPS = 75°

Find: Angles 1, 2, 3, and 4 and the measures of arcs $\widehat{SP}, \widehat{PQ}$, and $\widehat{QR}$

Here's how you do this one. Just keep using the inscribed angle formula over and over. Remember — the angle is half the arc; the arc is twice the angle.

$\overset{\frown}{SP}$ is twice the 60° angle, so it's 120°. Angle 2 is half of that, so it's 60° (or by the first congruent-angle theorem in this section, ∠2 must equal ∠PRS because they both intercept $\overset{\frown}{SP}$).

$\overset{\frown}{RS}$ is 40°, so ∠RPS is half of that, or 20°. Subtracting that from ∠QPS (which the given says is 75°) gives you a measure of 55° for ∠1. $\overset{\frown}{QR}$ is twice that (110°); and because ∠3 intercepts $\overset{\frown}{QR}$, it's half of that, or 55°. (Again, you could've just figured out that ∠3 has to equal ∠1 because they both intercept $\overset{\frown}{QR}$.)

Now figure out the measure of $\overset{\frown}{PQ}$. Four arcs — $\overset{\frown}{QR}$, $\overset{\frown}{RS}$, $\overset{\frown}{SP}$, and $\overset{\frown}{PQ}$ — make up the entire circle, which is 360°. You've got the measures of the first three: 110°, 40°, and 120° respectively. That adds up to 270°. Thus, $\overset{\frown}{PQ}$ has to equal 360° − 270°, or 90°. Finally, ∠4 is half of that, or 45°. (You could also solve for ∠4 by instead using the fact that the angles in △PRS must add up to 180°. Just add up the measures of ∠R, ∠RPS, and ∠3, and subtract the total from 180°.)

Note: This triangle idea also gives you a good way to check your results — do the angles add up to 180°? Angle R is 60°, ∠RPS is 20°, ∠3 is 55°, and ∠4 is 45°. That does add up to 180°, so it checks, which brings me to the following tip.

Whenever possible, check your answers with a method that's different from your original solution method. This is a *much* more effective check of your results than simply going through your work a second time looking for mistakes.

Angles inside a circle

In this section, I discuss angles whose vertices are inside but not touching a circle.

Measure of an angle inside a circle: The measure of an angle whose vertex is *inside* a circle (a *chord-chord angle*) is one-half the sum of the measures of the arcs intercepted by the angle and its vertical angle. For example, check out Figure 15-5, which shows you chord-chord angle SVT. You find the measure of the angle like this:

$$\angle SVT = \tfrac{1}{2}\left(m\overset{\frown}{ST} + m\overset{\frown}{QR}\right)$$

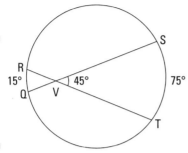

Figure 15-5: Chord-chord angles are inside a circle.

Here's a problem to show how this formula plays out:

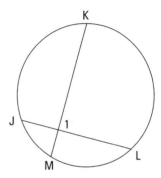

Given: $\widehat{MJ} : \widehat{JK} : \widehat{KL} : \widehat{LM} = 1 : 3 : 4 : 2$

Find: $\angle 1$

To use the formula to find $\angle 1$, you need the measures of arcs MJ and KL. You know the ratio of the arcs is $1 : 3 : 4 : 2$, so you can set their measures equal to $1x$, $3x$, $4x$, and $2x$. The four arcs make up an entire circle, so they must add up to $360°$. Thus,

$$1x + 3x + 4x + 2x = 360$$
$$10x = 360$$
$$x = 36$$

Plug 36 in for x to find the measures of $\widehat{MJ}$ and $\widehat{KL}$:

$$m\widehat{MJ} = x = 36°$$
$$m\widehat{KL} = 4x = 4 \cdot 36 = 144°$$

Now use the formula:

$$\angle 1 = \tfrac{1}{2}\left(m\widehat{KL} + m\widehat{MJ}\right)$$
$$= \tfrac{1}{2}(144 + 36)$$
$$= \tfrac{1}{2}(180)$$
$$= 90°$$

That does it. Take five.

Angles outside a circle

The preceding sections look at angles whose vertices are on a circle and whose vertices are inside a circle. There's only one other place an angle's vertex can be — outside a circle, of course. Three varieties of angles fall outside a circle, and all are made up of tangents and secants. You know what a tangent is (see Chapter 14), and here's the definition of *secant*.

Technically, a *secant* is a line that intersects a circle at two points. But the secants you use in this section and the section later in this chapter called "Powering Up with the Power Theorems" are segments that cut through a circle and that have one endpoint outside the circle and one endpoint on the circle.

So here are the three types of angles that are *outside* a circle:

- **Secant-secant angle:** A secant-secant angle, like ∠BDF in Figure 15-6a, is an angle whose vertex lies outside a circle and whose sides are two secants of the circle.
- **Secant-tangent angle:** A secant-tangent angle, like ∠GJK in Figure 15-6b, is an angle whose vertex lies outside a circle and whose sides are a secant and a tangent of the circle.
- **Tangent-tangent angle:** A tangent-tangent angle, like ∠LMN in Figure 15-6c, is an angle whose vertex lies outside a circle and whose sides are two tangents of the circle.

Measure of an angle outside a circle: The measure of a secant-secant angle, a secant-tangent angle, or a tangent-tangent angle is one-half the difference of the measures of the intercepted arcs. For example, in Figure 15-6,

$$\angle BDF = \tfrac{1}{2}\left(m\widehat{BF} - m\widehat{CE}\right)$$
$$\angle GJK = \tfrac{1}{2}\left(m\widehat{GK} - m\widehat{HK}\right)$$
$$\angle LMN = \tfrac{1}{2}\left(m\widehat{LPN} - m\widehat{LN}\right)$$

Chapter 15: Circle Formulas and Theorems

Note that you subtract the smaller arc from the larger (if you get a negative answer, you know you subtracted in the wrong order).

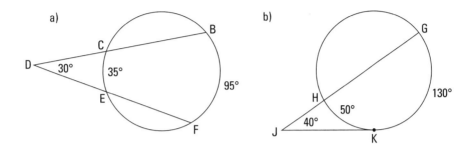

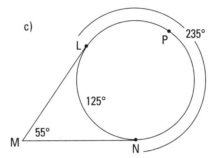

Figure 15-6: Three kinds of angles outside a circle.

Here's a problem that illustrates the angle-outside-a-circle formula:

Given: Diagram as shown

Find: $m\widehat{WT}$ and $m\widehat{TH}$

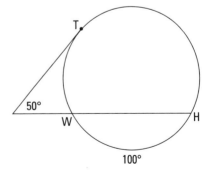

You know that arcs $\widehat{HW}$, $\widehat{WT}$, and $\widehat{TH}$ must add up to 360°, so because $\widehat{HW}$ is 100°, $\widehat{WT}$ and $\widehat{TH}$ add up to 260°. Thus, you can set $m\widehat{WT}$ equal to x and $m\widehat{TH}$ equal to $260 - x$. Plug these expressions into the formula, and you're home free:

$$\text{Measure of } \angle \text{ outside circle} = \tfrac{1}{2}(\text{arc} - \text{arc})$$
$$50 = \tfrac{1}{2}\left(m\widehat{TH} - m\widehat{WT}\right)$$
$$50 = \tfrac{1}{2}\left((260 - x) - x\right)$$
$$50 = \tfrac{1}{2}(260 - 2x)$$
$$50 = 130 - x$$
$$-80 = -x$$
$$x = 80$$

So $m\widehat{WT}$ is 80°, and $m\widehat{TH}$ is 180°.

Keeping your angle-arc formulas straight

I've got two great tips to help you remember when to use each of the three angle-arc formulas.

In the previous three sections, you see six types of angles made up of chords, secants, and tangents but only three angle-arc formulas. As you can tell from the titles of the three sections, to determine which of the three angle-arc formulas you need to use, all you need to pay attention to is where the angle's vertex is: inside, on, or outside the circle. You don't have to worry about whether the two sides of the angle are chords, tangents, secants, or some combination of these things.

The second tip can help you remember which formula goes with which category of angle. First, check out Figure 15-7.

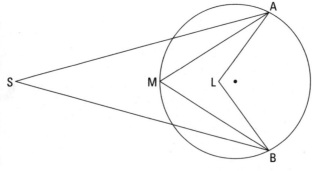

Figure 15-7: As the angle gets farther from the center of the circle, it gets smaller.

You can see that the *small* angle, ∠S (maybe about 35°) is *outside* the circle; the *medium* angle, ∠M (about 70°) is *on* the circle; and the *large* angle, ∠L (roughly 110°) is *inside* the circle. Here's one way to understand why the sizes of the angles go in this order. Say that the sides of ∠L are elastic. Picture grabbing ∠L at its vertex and pulling it to the left (as its ends remain attached to *A* and *B*). The farther you pull ∠L to the left, the smaller the angle would get.

Subtracting makes things smaller, and adding makes things larger, right? So here's how to remember which angle-arc formula to use (see Figure 15-7):

- To get the *small* angle, you *subtract*:

  $\angle S = \frac{1}{2}(arc - arc)$

- To get the *medium* angle, you *do nothing*:

  $\angle M = \frac{1}{2}(arc)$

- To get the *large* angle, you *add*:

  $\angle L = \frac{1}{2}(arc + arc)$

(**Note:** Whenever you use any of the angle-arc formulas, make sure you always use arcs that are in the *interior* of the angles.)

Powering Up with the Power Theorems

Like the preceding sections, this section takes a look at what happens when angles and circles intersect. But this time, instead of analyzing the size of angles and arcs, I analyze the lengths of the segments that make up the angles. The three power theorems that follow allow you to solve all sorts of interesting circle problems.

Striking a chord with the Chord-Chord Power Theorem

The Chord-Chord Power Theorem was brilliantly named for the fact that the theorem uses a chord and — can you guess? — another chord!

Chord-Chord Power Theorem: If two chords of a circle intersect, then the product of the measures of the parts of one chord is equal to the product of the measures of the parts of the other chord. (Whew, what a mouthful!)

For example, in Figure 15-8,

$5 \cdot 4 = 10 \cdot 2$

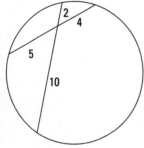

Figure 15-8: The Chord-Chord Power Theorem: (part) · (part) = (part) · (part).

Try out your power-theorem skills on this problem:

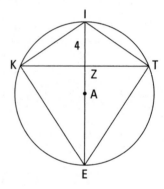

Given: Circle A has a radius of 6.5

   $KITE$ is a kite

   $IZ = 4$

Find: The area of $KITE$

To get the area of a kite, you need to know the lengths of its diagonals. The kite's diagonals are two chords that cross each other, so see whether you can apply the Chord-Chord Power Theorem.

To get diagonal $\overline{IE}$, note that $\overline{IE}$ is also the circle's diameter. Circle A has a radius of 6.5, so its diameter is twice as long, or 13, and thus that's the length of diagonal $\overline{IE}$. Then you see that ZE must be $13 - 4$, or 9. Now you have two of the lengths, $IZ = 4$ and $ZE = 9$, for the segments you use in the theorem:

   $(KZ)(ZT) = (IZ)(ZE)$

Because KITE is a kite, diagonal $\overline{IE}$ bisects diagonal $\overline{KT}$ (see Chapter 10 for kite properties). Thus, $\overline{KZ} \cong \overline{ZT}$, so you can set them both equal to x. Plug everything into the equation:

$$x \cdot x = 4 \cdot 9$$
$$x^2 = 36$$
$$x = 6 \text{ or } -6$$

You can obviously reject –6 as a length, so x is 6. KZ and ZT are thus both 6, and diagonal $\overline{KT}$ is therefore 12. You've already figured out that the length of the other diagonal is 13, so now you finish with the kite area formula:

$$\text{Area}_{KITE} = \frac{1}{2} d_1 d_2$$
$$= \frac{1}{2} \cdot 12 \cdot 13$$
$$= 78 \text{ units}^2$$

By the way, you can also do this problem with the Altitude-on-Hypotenuse Theorem, which I introduce in Chapter 13. Angles *IKE* and *ITE* intercept semicircles (180°), so they're both half of 180°, or right angles. The Altitude-on-Hypotenuse Theorem then gives you $(KZ)^2 = (IZ)(ZE)$ and $(TZ)^2 = (IZ)(ZE)$ for the two right triangles on the left and the right sides of the kite. After that, the math works out just like it does using the Chord-Chord Power Theorem.

Touching on the Tangent-Secant Power Theorem

In this section, I go through the Tangent-Secant Power Theorem — another absolutely awe-inspiring example of creative nomenclature.

Tangent-Secant Power Theorem: If a tangent and a secant are drawn from an external point to a circle, then the square of the measure of the tangent is equal to the product of the measures of the secant's external part and the entire secant. (Another mouthful!)

For example, in Figure 15-9,

$$8^2 = 4(4 + 12)$$

Dropping below the edge of the Earth

Here's a nifty application of the Tangent-Secant Power Theorem. Check out this figure of an adult of average height (say 5'7" or 5'8") standing at the ocean's shore.

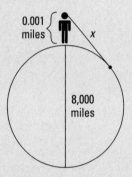

The *eyes* of someone of average height are about 5.3 feet above the ground, which is very close to $\frac{1}{1,000}$ of a mile. The Earth's diameter is about 8,000 miles. And *x* in the figure represents the distance to the horizon. You can plug everything into the Tangent-Secant Power Theorem and solve for *x*:

$$x^2 = 0.001(8,000 + 0.001)$$
$$x^2 \approx 0.001(8,000)$$
$$x^2 \approx 8$$
$$x \approx \sqrt{8} \approx 2.8 \text{ miles}$$

This short distance surprises most people. If you're standing on the shore, something floating on the water begins to drop below the horizon at a mere 2.8 miles from shore! (For yet another way of estimating distance to the horizon, see Chapter 22.)

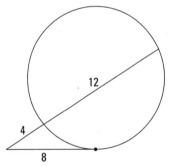

Figure 15-9: The Tangent-Secant Power Theorem: (tangent)² = (outside) · (whole).

Seeking out the Secant-Secant Power Theorem

Last but not least, I give you the Secant-Secant Power Theorem. Are you sitting down? This theorem involves two secants! (If you're trying to come up with a creative name for your child like Dweezil or Moon Unit, talk to Frank Zappa, not the guy who named the power theorems.)

Chapter 15: Circle Formulas and Theorems

Secant-Secant Power Theorem: If two secants are drawn from an external point to a circle, then the product of the measures of one secant's external part and that entire secant is equal to the product of the measures of the other secant's external part and that entire secant. (The biggest mouthful of all!)

For instance, in Figure 15-10,

$$4(4+2) = 3(3+5)$$

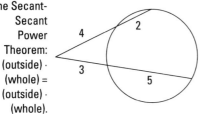

Figure 15-10: The Secant-Secant Power Theorem: (outside) · (whole) = (outside) · (whole).

This problem uses the last two power theorems:

Given: Diagram as shown

$\overline{BA}$ is tangent to circle H at A

Find: x and y

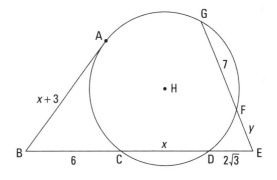

The figure includes a tangent and some secants, so look to your Tangent-Secant and Secant-Secant Power Theorems. First use the Tangent-Secant Power Theorem with tangent $\overline{AB}$ and secant $\overline{BD}$ to solve for x:

$$(x+3)^2 = 6(6+x)$$
$$x^2 + 6x + 9 = 36 + 6x$$
$$x^2 = 27$$
$$x = \pm\sqrt{27}$$
$$x = \pm 3\sqrt{3}$$

You can reject the negative answer, so x is $3\sqrt{3}$.

Now use the Secant-Secant Power Theorem with secants $\overline{EC}$ and $\overline{EG}$ to solve for y:

$$(2\sqrt{3})(2\sqrt{3}+3\sqrt{3}) = y(y+7)$$
$$2\sqrt{3} \cdot 5\sqrt{3} = y^2 + 7y$$
$$30 = y^2 + 7y$$
$$y^2 + 7y - 30 = 0$$
$$(y+10)(y-3) = 0$$
$$y+10 = 0 \quad \text{or} \quad y-3 = 0$$
$$y = -10 \qquad\qquad y = 3$$

A segment can't have a negative length, so $y = 3$. That does it.

Condensing the power theorems into a single idea

All three of the power theorems involve an equation with a product of two lengths (or one length squared) that equals another product of lengths. And each length is a distance from the vertex of an angle to the edge of the circle. Thus, all three theorems use the same scheme:

(vertex to circle) · (vertex to circle) = (vertex to circle) · (vertex to circle)

This unifying scheme can help you remember all three of the theorems I discuss in the preceding sections. And it'll help you avoid the common mistake of multiplying the external part of a secant by its internal part (instead of correctly multiplying the external part by the entire secant) when you're using the Tangent-Secant or Secant-Secant Power Theorem.

Part VI
Going Deep with 3-D Geometry

In this part . . .

This part takes you out of the flat 2-D world and into some 3-D geometry. In Chapter 16, you think about how lines (which are 1-D things) and triangles and planes (which are 2-D things) work when they're placed in the 3-D world. Then, in Chapter 17, you move on to real 3-D shapes such as cones, spheres, cylinders, and pyramids. You get formulas for the volumes of pointy-top solid objects and for flat-top objects, and you also find out how to compute their surface areas. Ready? Put on your 3-D glasses and dive in.

Chapter 16

3-D Space: Proofs in a Higher Plane of Existence

In This Chapter

▶ One-plane proofs: Lines that are perpendicular to planes

▶ Multi-plane proofs: Parallel lines, parallel planes, and much, much more

▶ Planes that cut through other planes

All the geometry in the chapters before this one involves two-dimensional shapes. In this chapter, you take your first look at three-dimensional diagrams and proofs, and you get a chance to check out lines and planes in 3-D space and investigate how they interact. But unlike the 3-D boxes, spheres, and cylinders you see in everyday life (and in Chapter 17), the 3-D things in this chapter simply boil down to the flat 2-D stuff from earlier chapters "standing up" in 3-D space.

Lines Perpendicular to Planes

A *plane* is just a flat thing, like a piece of paper, except that it's infinitely thin and goes on forever in all directions (Chapter 2 can tell you more about planes). In this section, you find out what it means for a line to be perpendicular to a plane and how to use this perpendicularity in two-column proofs.

Line-Plane perpendicularity definition: Saying that a line is perpendicular to a plane means that it's perpendicular to every line in the plane that passes through its foot. (A *foot* is the point where a line intersects a plane.)

Line-Plane Perpendicularity Theorem: If a line is perpendicular to two different lines that lie in a plane and pass through its foot, then it's perpendicular to the plane.

In two-column proofs, you use the preceding definition and theorem for different reasons:

- Use the definition when you already know that a line is perpendicular to a plane and you want to show that this line is perpendicular to a line that lies in the plane (in short, *if ⊥ to plane, then ⊥ to line*).

- Use the theorem when you already know that a line is perpendicular to two lines in a plane and you want to show that it's perpendicular to the plane itself (in short, *if ⊥ to two lines, then ⊥ to plane*). Note that this is almost exactly the reverse of the process in the first bullet.

Make sure you understand that a line must be perpendicular to *two* lines in a plane before you can conclude that it's perpendicular to the plane. Perpendicularity to one line in a plane isn't enough. Here's why: Imagine you have a big capital letter L made out of, say, plastic, and you're holding it on a table so it's pointing straight up. When it's pointing up, the vertical piece of the L is perpendicular to the tabletop. Now start to tip the L a bit (keeping its base on the table), so the top of the L is now on a slant. The top piece of the L is obviously still perpendicular to the bottom piece (which is a line that's on the plane of the table), but the top piece of the L is no longer perpendicular to the table. Thus, a line sticking out of a plane can make a right angle with a line in the plane and yet *not* be perpendicular to the plane.

Ready for a couple of problems? Here's the first one:

Given: $\overline{BD} \perp m$
$\angle EBC \cong \angle EDC$
Prove: $\overline{AB} \cong \overline{AD}$

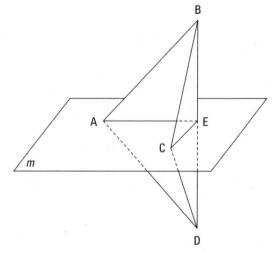

Statements	Reasons
1) $\overline{BD} \perp m$	1) Given.
2) $\overline{BD} \perp \overline{EC}$	2) If a line is perpendicular to a plane, then it's perpendicular to every line in the plane that passes through its foot (definition of perpendicularity of line to a plane).
3) $\angle BEC$ is a right angle $\angle DEC$ is a right angle	3) Definition of perpendicular.
4) $\angle BEC \cong \angle DEC$	4) All right angles are congruent.
5) $\angle EBC \cong \angle EDC$	5) Given.
6) $\overline{BC} \cong \overline{DC}$	6) If angles, then sides.
7) $\triangle BEC \cong \triangle DEC$	7) AAS (4, 5, 6).
8) $\overline{BE} \cong \overline{DE}$	8) CPCTC.
9) $\overline{AE} \cong \overline{AE}$	9) Reflexive Property.
10) $\overline{BD} \perp \overline{AE}$	10) If a line is perpendicular to a plane, then it's perpendicular to every line in the plane that passes through its foot.
11) $\angle BEA$ is a right angle $\angle DEA$ is a right angle	11) Definition of perpendicular.
12) $\angle BEA \cong \angle DEA$	12) All right angles are congruent.
13) $\triangle ABE \cong \triangle ADE$	13) SAS (8, 12, 9).
14) $\overline{AB} \cong \overline{AD}$	14) CPCTC.

Note: There are two other equally good ways to prove $\triangle BEC \cong \triangle DEC$, which you see in statement 7. Both use the Reflexive Property for $\overline{EC}$, and then one method finishes, like here, with AAS; the other finishes with HLR. All three methods take the same number of steps. I chose the method shown to reinforce the importance of the *if-angles-then-sides* theorem (reason 6).

The next example proof uses both the definition and the theorem about line-plane perpendicularity (for help deciding which to use where, see my explanation in the bulleted list in this section).

Given: Circle C

$\angle JCZ$ is a right angle

$\angle KCZ$ is a right angle

Prove: $\angle ZLM \cong \angle ZML$

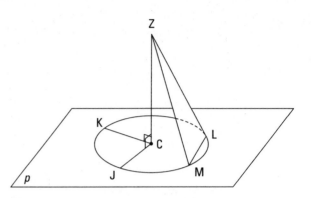

Here's the formal proof:

Statements	Reasons
1) Circle C $\angle JCZ$ is a right angle $\angle KCZ$ is a right angle	1) Given.
2) $\overline{ZC} \perp \overline{JC}$ $\overline{ZC} \perp \overline{KC}$	2) Definition of perpendicular.
3) $\overline{ZC} \perp p$	3) If a line is perpendicular to two lines that lie in a plane and pass through its foot, then it's perpendicular to the plane.
4) Draw $\overline{CL}$ Draw $\overline{CM}$	4) Two points determine a segment.
5) $\overline{CL} \cong \overline{CM}$	5) All radii of a circle are congruent.
6) $\overline{ZC} \perp \overline{CL}$ $\overline{ZC} \perp \overline{CM}$	6) If a line is perpendicular to a plane, then it's perpendicular to every line in the plane that passes through its foot.
7) $\angle ZCL$ is a right angle $\angle ZCM$ is a right angle	7) Definition of perpendicular.
8) $\angle ZCL \cong \angle ZCM$	8) All right angles are congruent.
9) $\overline{ZC} \cong \overline{ZC}$	9) Reflexive Property.
10) $\triangle ZCL \cong \triangle ZCM$	10) SAS (5, 8, 9).
11) $\overline{ZL} \cong \overline{ZM}$	11) CPCTC.
12) $\angle ZLM \cong \angle ZML$	12) If sides, then angles.

Parallel, Perpendicular, and Intersecting Lines and Planes

Fasten your seatbelt! In the preceding section, the proofs involve just a single plane, but in this section, you get on board with proofs and figures that really take off because they involve multiple planes at different altitudes. Ready for your flight?

The four ways to determine a plane

Before you get into multiple-plane proofs, you first have to know the several ways of determining a plane. *Determining a plane* is the fancy, mathematical way of saying "showing you where a plane is."

Here are the four ways to determine a plane:

- **Three noncollinear points determine a plane.** This statement means that if you have three points not on one line, then only one specific plane can go through those points. The plane is *determined* by the three points because the points show you exactly where the plane is.

  To see how this works, hold your thumb, forefinger, and middle finger so that your three fingertips make a triangle. Then take something flat like a hardcover book and place it so that it touches your three fingertips. There's only one way you can tilt the book so that it touches all three fingers. Your three noncollinear fingertips determine the plane of the book.

- **A line and a point not on the line determine a plane.** Hold a pencil in your left hand so that it's pointing away from you, and hold your right forefinger (pointing upward) off to the side of the pencil. There's only one place something flat could be placed so that it lies along the pencil and touches your fingertip.

- **Two intersecting lines determine a plane.** If you hold two pencils so that they cross each other, there's only one place a flat plane could be placed so that it would rest on both pencils.

- **Two parallel lines determine a plane.** Hold two pencils so that they're parallel. There's only one position in which a plane could rest on both pencils.

Now onto the multiple-plane principles and problems.

Line and plane interactions

Take a look at the following properties about perpendicularity and parallelism of lines and planes. You use some of these properties in 3-D proofs that involve 2-D concepts from previous chapters, such as proving that you've got a particular quadrilateral (see Chapter 11) or proving that two triangles are similar (see Chapter 13).

- **Three parallel planes:** If two planes are parallel to the same plane, then they're parallel to each other.
- **Two parallel lines and a plane:**
  - If two lines are perpendicular to the same plane, then they're parallel to each other.

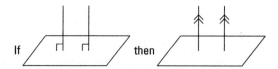

  - If a plane is perpendicular to one of two parallel lines, then it's perpendicular to the other.

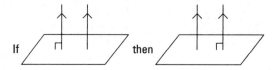

- **Two parallel planes and a line:**
  - If two planes are perpendicular to the same line, then they're parallel to each other.

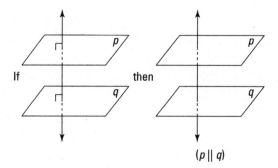

$(p \parallel q)$

Chapter 16: 3-D Space: Proofs in a Higher Plane of Existence

- If a line is perpendicular to one of two parallel planes, then it's perpendicular to the other.

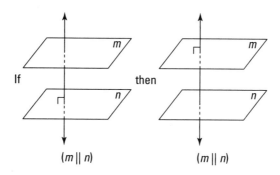

And here's a theorem you need for the example problem that follows.

A plane that intersects two parallel planes: If a plane intersects two parallel planes, then the lines of intersection are parallel. *Note:* Before you use this theorem in a proof, you usually have to use one of the four ways of determining a plane (see the preceding section) to show that the plane that cuts the parallel planes is, in fact, a plane. Steps 6 and 7 in this section's example proof show you how this works.

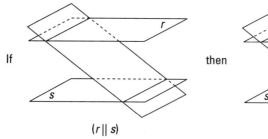

Here's the final proof:

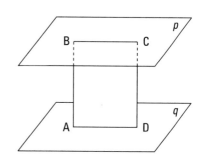

Given: $\overline{AB} \parallel \overline{DC}$
$\overline{AB} \perp p$
$\overline{DC} \perp q$

Prove: $ABCD$ is a rectangle

Statements	Reasons
1) $\overline{AB} \parallel \overline{DC}$	1) Given.
2) $\overline{AB} \perp p$	2) Given.
3) $\overline{DC} \perp p$	3) If a plane is perpendicular to one of two parallel lines, then it's perpendicular to the other.
4) $\overline{DC} \perp q$	4) Given.
5) $p \parallel q$	5) Two planes perpendicular to the same line are parallel to each other.
6) $\overline{AB}$ and $\overline{DC}$ determine plane ABCD	6) Two parallel lines determine a plane.
7) $\overline{BC} \parallel \overline{AD}$	7) If a plane intersects two parallel planes, then the lines of intersection are parallel. (*Note*: Make sure that whenever you use this theorem, you've already *determined* the "cutting" plane, like I did in line 6. You have to do this unless you've been told that it's a plane.)
8) ABCD is a parallelogram	8) A quadrilateral with two pairs of parallel sides is a parallelogram.
9) $\overline{AB} \perp \overline{BC}$	9) If a line is perpendicular to a plane, then it's perpendicular to every line in the plane that passes through its foot.
10) $\angle ABC$ is a right angle	10) Definition of perpendicular.
11) ABCD is a rectangle	11) A parallelogram with a right angle is a rectangle.

Chapter 17

Getting a Grip on Solid Geometry

In This Chapter
- Flat-top solids: The prism and the cylinder
- Pointy-top solids: The pyramid and the cone
- Top-less solids: Spheres

Unlike Chapter 16, which is all about 2-D (and even 1-D) things interacting in three dimensions, this chapter covers 3-D figures you can really sink your teeth into: *solids*. You study cones, spheres, prisms, and other solids of varying shapes, focusing on their two most fundamental characteristics, namely *volume* and *surface area*. To give you an everyday example, the volume of an aquarium (which is technically a prism) is the amount of water it holds, and its surface area is the total area of its glass sides plus its base and top.

Flat-Top Figures: They're on the Level

Flat-top figures (that's what I call them, anyway) are solids with two congruent, parallel bases (the top and bottom). A *prism* — your standard cereal box is one example — has polygon-shaped bases, and a *cylinder* — like your standard soup can — has round bases. But despite the different shape of their bases, the same volume and surface area formulas work for both of them because they share the flat-top structure.

Here are the technical definitions of *prism* and *cylinder* (see Figure 17-1):

✓ **Prism:** A prism is a solid figure with two congruent, parallel, polygonal bases. Its corners are called *vertices,* the segments that connect the vertices are called *edges,* and the flat sides are called *faces.*

A *right prism* is a prism whose faces are perpendicular to the prism's bases. All prisms in this book, and most prisms you find in other geometry books, are right prisms. And when I say *prism,* I mean a right prism.

✔ **Cylinder:** A cylinder is a solid figure with two congruent, parallel bases that have rounded sides (in other words, the bases are not straight-sided polygons); these bases are connected by a rounded surface.

A *right circular cylinder* is a cylinder with circular bases that are directly above and below each other. All cylinders in this book, and almost all cylinders you find in other geometry books, are right circular cylinders. When I say *cylinder,* I mean a right circular cylinder.

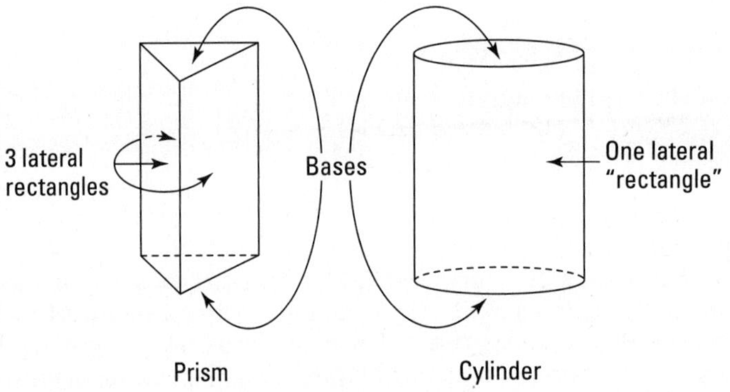

Figure 17-1: A prism and a cylinder with their bases and lateral rectangles.

Now that you know what these things are, here are their volume and surface area formulas.

Volume of flat-top figures: The volume of a prism or cylinder is given by the following formula.

$$Vol_{Flat\text{-}top} = area_{base} \cdot height$$

An ordinary box is a special case of a prism, so you can use the flat-top volume formula for a box, but you probably already know the easier way to compute a box's volume: Vol_{Box} = length · width · height. (Because the length times the width gives you the area of the base, these two methods really amount to the same thing.) To get the volume of a cube, the simplest type of box, you just take the length of one of its edges (sides) and raise it to the third power ($Vol_{Cube} = side^3$).

Surface area of flat-top figures: To find the surface area of a prism or cylinder, use the following formula.

$$SA_{Flat\text{-}top} = 2 \cdot area_{base} + lateral\ area_{rectangle(s)}$$

The shortest distance between two points is... a crooked line?

Check out the box in the figure below. It's 2" tall, 5" wide, and 4" deep. If an ant wants to walk from *A* to *Z* along the outside of the box, what's the shortest possible route, and how long is it?

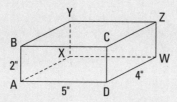

This is a great think-outside-the-box problem. (Get it? The ant must walk on the *outside of the box*. Har-de-har-har.) The key insight you need to solve this problem is that the shortest distance between two points is a straight line. Using that principle, however, requires that you flatten the box so that the path from *A* to *Z* is a straight line.

The ant could take three different "straight-line" paths. (*Note:* Ants are very small, and they can crawl under boxes.) The next figure shows the box with these three routes and the three different ways you could flatten or unfold the box to create straight paths.

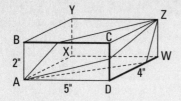
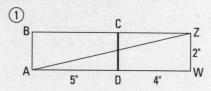
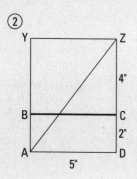
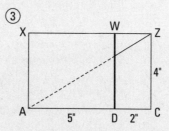

(continued)

> (continued)
>
> Note that the three bolded edges of the box correspond to the three bolded segments of the three "unfolded" rectangles.
>
> You use the Pythagorean Theorem to compute the lengths of the three routes. For route 1, you get $\sqrt{9^2 + 2^2} = \sqrt{85}$; for route 2, $\sqrt{5^2 + 6^2} = \sqrt{61}$; and for route 3, $\sqrt{7^2 + 4^2} = \sqrt{65}$. So the shortest possible route is $\sqrt{61}$, or approximately 7.8 inches. Pretty cool, eh?
>
> If the ant's really smart, he'll know how to pick the shortest route for any box. All he has to do is to make sure he crosses over the longest edge of the box. For this problem, that's the 5" edge.
>
> By the way, if you were to take a string in your hands, hold one end at *A* and the other at *Z*, and then pull it taut, it would end up precisely along one of the three routes shown on the box. Each of the three routes is where a taut string would go, depending on which edge the string crosses over.

Because prisms and cylinders have two congruent bases, you simply find the area of one base and double that value; then you add the figure's lateral area. The *lateral area* of a prism or cylinder is the area of the sides of the figure — namely, the area of everything but the figure's bases (see Figure 17-1). Here's how the two figures compare:

- **The lateral area of a prism is made up of rectangles.** The bases of a prism can be any shape, but the lateral area is always made up of rectangles. So to get the lateral area, all you need to do is find the area of the rectangles using the standard rectangle area formula and then add them up.

   A box, just like any other prism, has a lateral area made up of rectangles (four of them) — but its two bases are also rectangles. So to get the surface area of a box, you simply need to add up the areas of the six rectangular faces — you don't have to bother with using the standard flat-top surface area formula.

   To get the surface area of a cube, a box with six congruent, square faces, you just compute the area of one face and then multiply that by six.

- **The lateral area of a cylinder is basically one rectangle rolled into a tube shape.** Think of the lateral area of a cylinder as one rectangular paper towel that rolls exactly once around a paper towel roll. The base of this rectangle (you know, the part of the towel that wraps around the bottom of the roll) is the same as the circumference of the cylinder's base (for more on circumference, see Chapter 15). And the height of the paper towel is the same as the height of the cylinder.

Time to take a look at these formulas in action.

Given: Prism as shown

ABCD is a square with a diagonal of 8

∠EAD and ∠EDA are 45° angles

Find: 1. The volume of the prism

2. The surface area of the prism

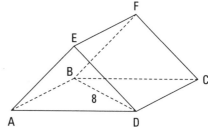

1. Find the volume of the prism.

To use the volume formula, you need the prism's height ($\overline{CD}$) and the area of its base ($\triangle AED$). You've probably noticed that this prism is lying on its side. That's why its height isn't vertical and its base isn't on the bottom.

Get the height first. ABCD is a square, so △BCD (half of the square) is a 45°- 45°- 90° triangle with a hypotenuse of 8. To get the leg of a 45°- 45°- 90° triangle, you divide the hypotenuse by $\sqrt{2}$ (or use the Pythagorean Theorem, noting that $a = b$ in this case; see Chapter 8 for both methods). So that gives you $\frac{8}{\sqrt{2}} = 4\sqrt{2}$ for the length of $\overline{CD}$, which, again, is the height of the prism.

And here's how you get the area of △AED: First, note that AD, like CD, is $4\sqrt{2}$ (because ABCD is a square). Next, because ∠EAD and ∠EDA are given 45° angles, ∠AED must be 90°; thus, △AED is another 45°- 45°- 90° triangle. Its hypotenuse, $\overline{AD}$, has a length of $4\sqrt{2}$, so its legs ($\overline{AE}$ and $\overline{DE}$) are $\frac{4\sqrt{2}}{\sqrt{2}}$, or 4 units long. The area of a right triangle is given by half the product of its legs (because you can use one leg for the triangle's base and the other for its height), so Area$_{\triangle AED} = \frac{1}{2} \cdot 4 \cdot 4 = 8$. You're all set to finish with the volume formula:

$$\text{Vol}_{\text{Prism}} = \text{area}_{\text{base}} \cdot \text{height}$$
$$= 8 \cdot 4\sqrt{2}$$
$$= 32\sqrt{2}$$
$$\approx 45.3 \text{ units}^3$$

2. Find the surface area of the prism.

Having completed part 1, you have everything you need to compute the surface area. Just plug in the numbers:

$$SA_{Prism} = 2 \cdot area_{base} + lateral\ area_{rectangles}$$
$$SA = 2 \cdot 8 + area_{ABCD} + area_{AEFB} + area_{DEFC}$$
$$= 16 + (4\sqrt{2})(4\sqrt{2}) + (4\sqrt{2})(4) + (4\sqrt{2})(4)$$
$$= 16 + 32 + 16\sqrt{2} + 16\sqrt{2}$$
$$= 48 + 32\sqrt{2}$$
$$\approx 93.3\ units^2$$

Now for a cylinder problem: Given a cylinder as shown with unknown radius, a height of 7, and a surface area of 120π units^2, find the cylinder's volume.

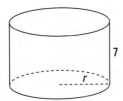

To use the volume formula, you need the cylinder's height (which you know) and the area of its base. To get the area of the base, you need its radius. And to get the radius, you use the surface area formula and solve for r:

$$SA_{Cylinder} = 2 \cdot area_{base} + lateral\ area_{\text{"rectangle"}}$$
$$120\pi = 2\pi r^2 + (\text{"rectangle" base}) \cdot (\text{"rectangle" height})$$

Remember that this "rectangle" is rolled around the cylinder and that the "rectangle's" base is the circumference of the cylinder's circular base. You fill in the equation as follows:

$$120\pi = 2\pi r^2 + (2\pi r)(7)$$
$$120\pi = 2\pi r^2 + 14\pi r$$
$$120\pi = 2\pi(r^2 + 7r) \qquad \text{(Divide both sides by } 2\pi.\text{)}$$
$$60 = r^2 + 7r$$

Now set the equation equal to zero and factor:

$r^2 + 7r - 60 = 0$

$(r + 12)(r - 5) = 0$

$r = -12$ or 5

The radius can't be negative, so it's 5. Now you can finish with the volume formula:

$\text{Vol}_{\text{Cylinder}} = \text{area}_{\text{base}} \cdot \text{height}$

$= \pi r^2 \cdot h$

$= \pi \cdot 5^2 \cdot 7$

$= 175\pi$

$\approx 549.8 \text{ units}^3$

That does it.

Getting to the Point of Pointy-Top Figures

What I call *pointy-top figures* are solids with one flat base and . . . a pointy top! The pointy-top solids are the *pyramid* and the *cone*. Even though the pyramid has a polygon-shaped base and the cone has a rounded base, the same volume and surface area formulas work for both. More details about pyramids and cones follow:

- **Pyramid:** A pyramid is a solid figure with a polygonal base and edges that extend up from the base to meet at a single point. As with a prism, the corners of a pyramid are called *vertices,* the segments that connect the vertices are called *edges,* and the flat sides are called *faces.*

  A *regular pyramid* is a pyramid with a regular-polygon base, whose peak is directly above the center of its base. The lateral faces of a regular pyramid are all congruent. All pyramids in this book, and most pyramids in other geometry books, are regular pyramids. When I use the term *pyramid,* I mean a regular pyramid.

- **Cone:** A cone is a solid figure with a rounded base and a rounded lateral surface that connects the base to a single point.

  A *right circular cone* is a cone with a circular base, whose peak lies directly above the center of the base. All cones in this book, and most cones in other books, are right circular cones. When I refer to a *cone,* I mean a right circular cone.

Now that you have a better idea of what these figures are, check out their volume and surface area formulas.

Volume of pointy-top figures: Here's how to find the volume of a pyramid or cone.

$$\text{Vol}_{\text{Pointy-top}} = \frac{1}{3} \text{area}_{\text{base}} \cdot \text{height}$$

Surface area of pointy-top figures: The following formula gives you the surface area of a pyramid or cone.

$$\text{SA}_{\text{Pointy-top}} = \text{area}_{\text{base}} + \text{lateral area}_{\text{triangle(s)}}$$

The *lateral area* of a pointy-top figure is the area of the surface that connects the base to the peak (it's the area of everything but the base). Here's what this means for pyramids and cones:

- **The lateral area of a pyramid is made up of triangles.** Each lateral triangle has an area of $\frac{1}{2}(\text{base})(\text{height})$. But you can't use the height of the pyramid for the height of its triangular faces, because the height of the pyramid doesn't run along its faces. So instead, you use the pyramid's *slant height*, which is just the ordinary altitude of the triangular faces. (The cursive letter ℓ indicates the slant height.) Figure 17-2 shows how height and slant height differ.

- **The lateral area of a cone is basically one triangle rolled into a cone shape.** The lateral area of a cone is one "triangle" that's been rolled into a cone shape like a snow-cone cup (it's only kind of a triangle because when flattened out, it's actually a sector of a circle with a curved bottom side — see Chapter 15 for details on sectors). Its area is $\frac{1}{2}(\text{base})(\text{slant height})$, just like one of the lateral triangles in a pyramid. The base of this "triangle" equals the circumference of the cone (this works just like the base of the lateral "rectangle" in a cylinder).

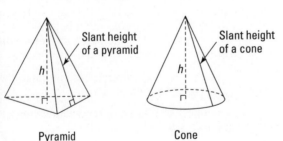

Figure 17-2: A pyramid and a cone with their heights and slant heights.

Chapter 17: Getting a Grip on Solid Geometry

Ready for a pyramid problem?

Given: A regular pyramid as shown

Diagonal $\overline{PT}$ has a length of 12

Find: 1. The pyramid's volume

2. The pyramid's surface area

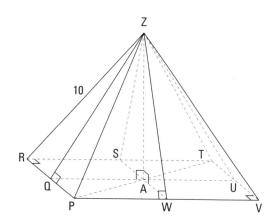

The key to pyramid problems about volume and surface area (and to a lesser degree, prism, cylinder, and cone problems) is *right triangles*. Find them and then solve them with the Pythagorean Theorem or by using your knowledge of special right triangles (see Chapter 8).

Congruent right triangles are all over the place in pyramids. Would you believe that in a pyramid like the one in this problem, there are 28 different right triangles that could be used to solve different parts of the problem? (Many of these triangles aren't shown in this figure.) Here they are:

- Eight right congruent triangles, like △PZW, are on the faces (these have their right angles at W, Q, S, or U).

- Eight right triangles are standing straight up with their right angles at A and one vertex at Z. The four of these triangles with a vertex at Q, S, U, or W (such as △QZA) are congruent, and the other four with a vertex at P, R, T, or V (such as △PZA) are congruent.

- You have four half-squares like △PTV. These are congruent 45°- 45°- 90° triangles.

- Within the base are eight little congruent 45°- 45°- 90° triangles like △PAW.

(In case you're curious, there are 48 other right triangles — depending how you count them — that you're very unlikely to use, which brings the grand total to 76 right triangles!)

Okay, so back to the pyramid problem.

1. **Find the pyramid's volume.**

   To compute the volume of a pyramid, you need its height ($\overline{AZ}$) and the area of its square base, *PRTV*. You can get the height by solving right triangle $\triangle PZA$. The lateral edges of a regular pyramid are congruent; thus, the hypotenuse of $\triangle PZA$, $\overline{PZ}$, is congruent to $\overline{RZ}$, and so its length is also 10. $\overline{PA}$ is half of the diagonal of the base, so it's 6. Triangle *PZA* is thus a 3-4-5 triangle blown up to twice its size, namely a 6-8-10 triangle, so the height, $\overline{AZ}$, is 8 (or you can use the Pythagorean Theorem to get $\overline{AZ}$ — see Chapter 8 for more on this theorem and Pythagorean triples).

   To get the area of square *PRTV*, you can, of course, first figure the length of its sides; but don't forget that a square is a kite, so you can use the kite area formula instead — that's the quickest way to get the area of a square if you know the length of a diagonal (see Chapter 10 for more about quadrilaterals). Because the diagonals of a square are equal, both of them are 12, and you have what you need to use the kite area formula:

   $$\text{Area}_{PRTV} = \frac{1}{2} d_1 d_2$$
   $$= \frac{1}{2}(12)(12)$$
   $$= 72 \text{ units}^2$$

   Now use the pointy-top volume formula:

   $$\text{Vol}_{\text{Pyramid}} = \frac{1}{3} \text{area}_{\text{base}} \cdot \text{height}$$
   $$= \frac{1}{3}(72)(8)$$
   $$= 192 \text{ units}^3$$

   Note: This is *much* less than the volume of the Great Pyramid at Giza (see Chapter 22).

2. **Find the pyramid's surface area.**

   To use the pyramid surface area formula, you need the area of the base (which you got in part 1 of the problem) and the area of the triangular faces. To get the faces, you need the slant height, $\overline{ZW}$.

   First, solve $\triangle PAW$. It's a 45°- 45°- 90° triangle with a hypotenuse ($\overline{PA}$) that's 6 units long; to get the legs, you divide the hypotenuse by $\sqrt{2}$ (or use the Pythagorean Theorem — see Chapter 8). $\frac{6}{\sqrt{2}} = 3\sqrt{2}$, so $\overline{PW}$ and $\overline{AW}$ both have a length of $3\sqrt{2}$. Now you can get $\overline{ZW}$ by using the Pythagorean Theorem with either of two right triangles, $\triangle PZW$ or $\triangle AZW$. Take your pick. How about $\triangle AZW$?

$$(ZW)^2 = (AZ)^2 + (AW)^2$$
$$= 8^2 + \left(3\sqrt{2}\right)^2$$
$$= 64 + 18$$
$$= 82$$
$$ZW = \sqrt{82}$$

Now you're all set to finish with the surface area formula. (One last fact you need is that $\overline{PV}$ is $6\sqrt{2}$ because, of course, it's twice as long as $\overline{PW}$.)

$$SA_{Pyramid} = area_{base} + lateral\ area_{four\ triangles}$$
$$= 72 + 4\left(\tfrac{1}{2} base \cdot slant\ height\right)$$
$$= 72 + 4\left(\tfrac{1}{2} \cdot 6\sqrt{2} \cdot \sqrt{82}\right)$$
$$= 72 + 12\sqrt{164}$$
$$= 72 + 24\sqrt{41}$$
$$\approx 225.7\ units^2$$

That's a wrap.

The water-into-wine volume problem

Here's a great volume brain teaser. Say you have a gallon container of water and a gallon container of wine. You take a ladle, pour a ladle of wine into the water container, and then stir it up. Next, you take a ladle of the water/wine mixture and pour it back into the wine container. Here's the question: After doing both of these transfers, is there more wine in the water or more water in the wine? Come up with your answer before reading on.

Here's how most people's reasoning goes: A ladle of 100 percent wine was poured into the water, but then a ladle of water mixed with a little wine was poured back into the wine container. So with that second pouring, a little wine was going back into the wine container. And thus it seems that there would be less water in the wine container than wine in the water container.

Well, the surprising answer is that the amount of water in the wine is exactly the same as the amount of wine in the water. When you think about it, you'll see that this must be the case. Both containers started with a gallon of liquid, right? Next, a ladle of liquid was poured into the water container, and then a ladle of liquid was taken out of that container and poured back into the wine container. The final result, of course, is that both containers end up, as they started, with a gallon of liquid in them.

Now consider the amount of wine that's in the water at the end of the process. That, of course, is the amount of wine that's missing from the wine container. And because you know that the wine container ends up with a gallon of liquid, *the amount of water in the wine that will fill that container back up to a gallon must be the same amount of wine that's missing*. The amounts must be the same if you end up with a gallon in both containers. Hard to believe but true.

Now for a cone problem:

Given: Cone with base diameter of $4\sqrt{3}$

The angle between the cone's height and slant height is 30°

Find: 1. The cone's volume

2. Its surface area

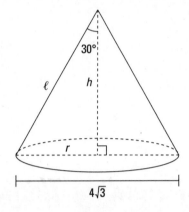

1. **Find the cone's volume.**

    To compute the cone's volume, you need its height and the radius of its base. The radius is, of course, half the diameter, so it's $2\sqrt{3}$. Then, because the height is perpendicular to the base, the triangle formed by the radius, the height, and the slant height is a 30°- 60°- 90° triangle. You can see that h is the long leg and r the short leg, so to get h, you multiply r by $\sqrt{3}$ (see Chapter 8):

    $$h = \sqrt{3} \cdot 2\sqrt{3} = 6$$

    You're ready to use the cone volume formula:

    $$\begin{aligned}\text{Vol}_{\text{Cone}} &= \tfrac{1}{3}\,\text{area}_{\text{base}} \cdot \text{height} \\ &= \tfrac{1}{3}\pi r^2 \cdot h \\ &= \tfrac{1}{3}\pi\left(2\sqrt{3}\right)^2 \cdot 6 \\ &= 24\pi \\ &\approx 75.4 \text{ units}^3\end{aligned}$$

2. **Find the cone's surface area.**

    For the surface area, the only other thing you need is the slant height, ℓ. The slant height is the hypotenuse of the 30°- 60°- 90° triangle, so it's just twice the radius, which makes it $4\sqrt{3}$. Now plug everything into the cone surface area formula:

$$SA_{Cone} = area_{base} + lateral\ area_{"triangle"}$$
$$= \pi r^2 + \frac{1}{2}(base)(slant\ height)$$
$$= \pi r^2 + \frac{1}{2}(2\pi r)(\ell)$$
$$= \pi(2\sqrt{3})^2 + \frac{1}{2}(2\pi \cdot 2\sqrt{3})(4\sqrt{3})$$
$$= 12\pi + 24\pi$$
$$= 36\pi$$
$$\approx 113.1\ units^2$$

Rounding Things Out with Spheres

The sphere clearly doesn't fit into the flat-top or pointy-top categories because it doesn't really have a top at all. Therefore, it has its own unique volume and surface area formulas. First, here's the definition of a sphere.

A *sphere* is the set of all points in 3-D space equidistant from a given point, the sphere's center. (Plain English definition: A sphere is, you know, a ball — duh.) The *radius* of a sphere goes from its center to its surface.

Volume and surface area of a sphere: Use the following formulas for the volume and surface area of a sphere.

- $Vol_{Sphere} = \frac{4}{3}\pi r^3$
- $SA_{Sphere} = 4\pi r^2$

Have a ball with the following sphere problem: What's the volume and surface area of a basketball in a box (a cube, of course) if the box has a surface area of 486 inches^2?

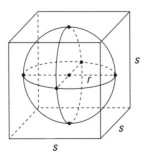

A cube (or any other ordinary box shape) is a special case of a prism, but you don't need to use the fancy-schmancy prism formula, because the surface area of a cube is simply made up of six congruent squares. Call the length of an edge of the cube s. The area of each side is therefore s^2. The cube has six faces, so its surface area is $6s^2$. Set this equal to the given surface area of 486 inches^2 and solve for s:

$$6s^2 = 486$$
$$s^2 = 81$$
$$s = 9 \text{ inches}$$

Thus, the edges of the cube are 9 inches, and because the basketball has the same width as the box it comes in, the diameter of the ball is also 9 inches; its radius is half of that, or 4.5 inches. Now you can finish by plugging 4.5 into the two sphere formulas:

$$\text{Vol}_{\text{Sphere}} = \frac{4}{3}\pi r^3$$
$$= \frac{4}{3}\pi \cdot 4.5^3$$
$$= 121.5\pi$$
$$\approx 381.7 \text{ inches}^3$$

(By the way, this is slightly more than half the volume of the box, which is 9^3, or 729 inches^3.)

Now here's the surface area solution:

$$\text{SA}_{\text{Sphere}} = 4\pi r^2$$
$$= 4\pi \cdot 4.5^2$$
$$= 81\pi$$
$$\approx 254.5 \text{ inches}^2$$

This sphere, in case you're curious, is the actual size of an official NBA basketball. To end this chapter, here's a quick trivia question for you (come up with your guess before reading the answer). Now that you know that the diameter of a basketball is 9 inches, what do you think the diameter of a basketball hoop is? The surprising answer is that the hoop is a full two times as wide — 18 inches!

Part VII
Placement, Points, and Pictures: Alternative Geometry Topics

In this part . . .

In these chapters, you move away from the main course of theorems, properties, and two-column proofs so you can check out some new and different ways of thinking about shapes. In Chapter 18, you analyze shapes as objects in the coordinate plane (you know, that grid with the *x*- and *y*-axes). In Chapter 19, you discover how to move shapes around in the coordinate plane. And in Chapter 20, you think about shapes as collections of points that satisfy given conditions. You also find out how to draw certain shapes using just a compass and straightedge. Although this part requires you to stretch your brain muscles a little (or *neurons* for you sticklers for physiological accuracy), I hope these new topics come as a welcome change and challenge.

Chapter 18

Coordinate Geometry

In This Chapter
- Finding a line's slope and a segment's midpoint
- Calculating the distance between two points
- Doing geometry proofs with algebra
- Working with equations of lines and circles

*I*n this chapter, you investigate the same sorts of things you see in previous chapters: perpendicular lines, right triangles, circles, perimeter, area, the diagonals of quadrilaterals, and so on. What's new about this chapter on coordinate geometry is that these familiar geometric objects are placed in the *x-y* coordinate system and then analyzed with algebra. You use the *coordinates* of the points of a figure — points like (*x*, *y*) or (10, 2) — to prove or compute something about the figure. You reach the same kind of conclusions as in previous chapters; it's just the methods that are different.

The *x-y,* or *Cartesian,* coordinate system is named after René Des*cartes* (1596–1650). Descartes is often called the father of coordinate geometry, despite the fact that the coordinate system he used had an *x*-axis but no *y*-axis. There's no question, though, that he's the one who got the ball rolling. So if you like coordinate geometry, you know who to thank (and if you don't, you know who to blame).

Getting Coordinated with the Coordinate Plane

I have a feeling that you already know all about how the *x-y* coordinate system works, but if you need a quick refresher, no worries. Figure 18-1 shows you the lay of the land of the coordinate plane.

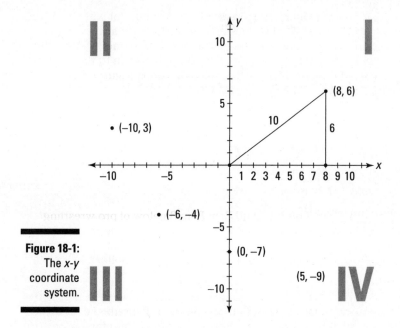

Figure 18-1: The x-y coordinate system.

Here's the lowdown on the coordinate plane you see in Figure 18-1:

- The *horizontal* axis, or *x*-axis, goes from left to right and works exactly like a regular number line. The *vertical* axis, or *y*-axis, goes — ready for a shock? — up and down. The two axes intersect at the *origin*, (0, 0).

- Points are located within the coordinate plane with pairs of coordinates called *ordered pairs* — like (8, 6) or (–10, 3). The first number, the *x-coordinate*, tells you how far you go right or left; the second number, the *y-coordinate*, tells you how far you go up or down. For (–10, 3), for example, you go *left* 10 and then *up* 3.

- Going counterclockwise from the upper-right-hand section of the coordinate plane are *quadrants* I, II, III, IV:

    - All points in quadrant I have two positive coordinates, (+, +).
    - In quadrant II, you go left (negative) and then up (positive), so it's (–, +).
    - In quadrant III, it's (–, –).
    - In quadrant IV, it's (+, –).

Because all coordinates in quadrant I are positive, it's often the easiest quadrant to work in.

- The Pythagorean Theorem (see Chapter 8) comes up a lot when you're using the coordinate system because when you go right and then up to plot a point (or left and then down, and so on), you're tracing along the legs of a right triangle; the segment connecting the origin to the point then becomes the hypotenuse of the right triangle. In Figure 18-1, you can see the 6-8-10 right triangle in quadrant I.

The Slope, Distance, and Midpoint Formulas

Like Shane Douglas, Chris Candido, and Bam Bam Bigelow of pro wrestling fame, the slope, distance, and midpoint formulas are sort of the Triple Threat of coordinate geometry. If you have two points in the coordinate plane, the three most basic questions you can ask about them are the following:

- What's the distance between them?
- What's the location of the point halfway between them (the midpoint)?
- How much is the segment that connects the points tilted (the slope)?

These three questions come up in a plenitudinous and plethoric passel of problems. In a minute, you'll see how to use the three formulas to answer these questions.

But right now, I just want to caution you not to mix up the formulas — which is easy to do because all three formulas involve points with coordinates (x_1, y_1) and (x_2, y_2). My advice is to focus on *why* the formulas work instead of just memorizing them by rote. That'll help you remember the formulas correctly.

The slope dope

The *slope* of a line basically tells you how steep the line is. You may have used the slope formula before this in an Algebra I class. But in case you've forgotten it (or in case you were out with mono the day your teacher detailed its finer points), here's a refresher on the formula and also the straight dope on some common types of lines.

Slope formula: The slope of a line containing two points — (x_1, y_1) and (x_2, y_2) — is given by the following formula (a line's slope is often represented by the letter *m*):

$$\text{Slope} = m = \frac{y_2 - y_1}{x_2 - x_1} = \frac{\text{rise}}{\text{run}}$$

Note: It doesn't matter which points you designate as (x_1, y_1) and (x_2, y_2); the math works out the same either way. Just make sure that you plug your numbers into the right places in the formula.

The *rise* is the "up distance," and the *run* is the "across distance" shown in Figure 18-2. To remember this, note that you *rise up* but you *run across,* and also that "rise" rhymes with "*y*"s.

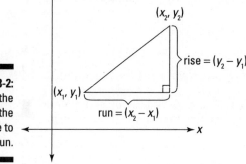

Figure 18-2: Slope is the ratio of the rise to the run.

Take a look at the following list and Figure 18-3, which show you that the slope of a line increases as the line gets steeper and steeper:

- A horizontal line has no steepness at all, so its slope is zero. A good way to remember this is to think about driving on a horizontal, flat road — the road has zero steepness or slope.
- A slightly inclined line might have a slope of, say, $\frac{1}{5}$.
- A line at a 45° angle has a slope of 1.
- A steeper line could have a slope of 5.
- A vertical line (the steepest line of all) sort of has an infinite slope, but math people say that its slope is *undefined.* (It's undefined because with a vertical line, you don't go across at all, and thus the *run* in $\frac{\text{rise}}{\text{run}}$ would be zero, and you can't divide by zero). Think about driving up a vertical road: You can't do it — it's impossible. And it's impossible to compute the slope of a vertical line.

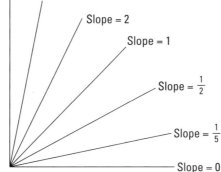

Figure 18-3: The slope tells you how steep a line is.

The lines you see in Figure 18-3 have *positive* slopes (except for the horizontal and vertical lines). Now I introduce you to lines with negative slopes, and I give you a couple of ways to distinguish the two types of slopes:

- Lines that go up to the right have a positive slope. Going from left to right, lines with positive slopes go uphill.
- Lines that go down to the right have a negative slope. Going from left to right, lines with negative slopes go downhill.

Tip: A line with a *N*egative slope goes in the direction of the middle part of the capital letter *N*. See Figure 18-4.

Figure 18-4: A negative slope goes up to the left and down to the right.

Just like lines with positive slopes, as lines with negative slopes get steeper and steeper, their slopes keep "increasing"; but here, *increasing* means becoming a larger and larger negative number (which is technically *decreasing*).

Here are some pairs of lines with special slopes:

- **Slopes of parallel lines:** The slopes of parallel lines are equal.

  For two vertical lines, however, there's a minor technicality: If both lines are vertical, you can't say that their slopes are equal. Their slopes are both undefined, so they have the same slope, but because *undefined* doesn't equal anything, you can't say that *undefined = undefined*.

- **Slopes of perpendicular lines:** The slopes of perpendicular lines are opposite reciprocals of each other, such as $\frac{7}{3}$ and $-\frac{3}{7}$ or -6 and $\frac{1}{6}$.

  This rule works unless one of the perpendicular lines is horizontal (slope = 0) and the other line is vertical (slope is undefined).

Going the distance with the distance formula

If two points in the *x-y* coordinate system are straight across from each other or directly above and below each other, finding the distance between them is a snap. It works just like finding the distance between two points on a number line: You just subtract the smaller number from the larger. Here are the formulas:

- Horizontal distance = right$_{x\text{-coordinate}}$ − left$_{x\text{-coordinate}}$
- Vertical distance = top$_{y\text{-coordinate}}$ − bottom$_{y\text{-coordinate}}$

Distance formula: Finding diagonal distances is a bit trickier than computing horizontal and vertical distances. For this, mathematicians whipped up the distance formula, which gives the distance between two points (x_1, y_1) and (x_2, y_2):

$$\text{Distance} = \sqrt{(x_2 - x_1)^2 + (y_2 - y_1)^2}$$

Note: Like with the slope formula, it doesn't matter which point you call (x_1, y_1) and which you call (x_2, y_2).

Figure 18-5 illustrates the distance formula.

The distance formula is simply the Pythagorean Theorem ($a^2 + b^2 = c^2$) solved for the hypotenuse: $c = \sqrt{a^2 + b^2}$. See Figure 18-5 again. The legs of the right triangle (*a* and *b* under the square root symbol) have lengths equal to $(x_2 - x_1)$ and $(y_2 - y_1)$. Remember this connection, and if you forget the distance formula, you'll be able to solve a distance problem with the Pythagorean Theorem instead.

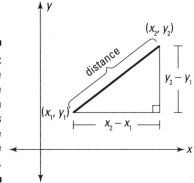

Figure 18-5:
The distance between two points is also the length of the hypotenuse.

Don't mix up the slope formula with the distance formula. You may have noticed that both formulas involve the expressions $(x_2 - x_1)$ and $(y_2 - y_1)$. That's because the lengths of the legs of the right triangle in the distance formula are the same as the *rise* and the *run* from the slope formula. To keep the formulas straight, just focus on the fact that slope is a *ratio* and distance is a *hypotenuse*.

Meeting each other halfway with the midpoint formula

The midpoint formula gives you the coordinates of a line segment's midpoint. The way it works is very simple: It takes the *average* of the x-coordinates of the segment's endpoints and the average of the y-coordinates of the endpoints. These averages give you the location of a point that is exactly in the middle of the segment.

Midpoint formula: To find the midpoint of a segment with endpoints at (x_1, y_1) and (x_2, y_2), use the following formula:

$$\text{Midpoint} = \left(\frac{x_1 + x_2}{2}, \frac{y_1 + y_2}{2} \right)$$

Note: It doesn't matter which point is (x_1, y_1) and which is (x_2, y_2).

The whole enchilada: Putting the formulas together in a problem

Here's a problem that shows how to use the slope, distance, and midpoint formulas.

Given: Quadrilateral PQRS as shown

Solve:
1. Show that PQRS is a rectangle
2. Find the perimeter of PQRS
3. Show that the diagonals of PQRS bisect each other, and find the point where they intersect

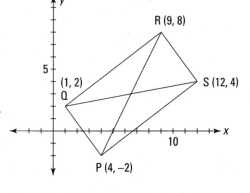

1. Show that PQRS is a rectangle.

The easiest way to show that PQRS is a rectangle is to compute the slopes of its four sides and then use ideas about the slopes of parallel and perpendicular lines (see Chapter 11 for ways to prove that a quadrilateral is a rectangle).

$$\text{Slope} = \frac{y_2 - y_1}{x_2 - x_1}$$

$\text{Slope}_{\overline{QR}} = \frac{8-2}{9-1} = \frac{6}{8} = \frac{3}{4}$ $\text{Slope}_{\overline{QP}} = \frac{2-(-2)}{1-4} = \frac{4}{-3} = -\frac{4}{3}$

$\text{Slope}_{\overline{PS}} = \frac{4-(-2)}{12-4} = \frac{6}{8} = \frac{3}{4}$ $\text{Slope}_{\overline{RS}} = \frac{8-4}{9-12} = \frac{4}{-3} = -\frac{4}{3}$

Seeing these four slopes, you can now conclude that PQRS is a rectangle in two different ways — neither of which requires any further work.

First, because the slopes of $\overline{QR}$ and $\overline{PS}$ are equal, those segments are parallel. Ditto for $\overline{QP}$ and $\overline{RS}$. Quadrilateral PQRS is thus a parallelogram. Then you check any vertex to see whether it's a right angle. Suppose you check vertex Q. Because the slopes of $\overline{QP}$ $\left(-\frac{4}{3}\right)$ and $\overline{QR}$ $\left(\frac{3}{4}\right)$ are opposite reciprocals, those segments are perpendicular, and thus ∠Q is a right angle. That does it because a parallelogram with a right angle is a rectangle (see Chapter 11).

Second, you can see that the four segments have a slope of either $\frac{3}{4}$ or $-\frac{4}{3}$. Thus, you can quickly see that, at each of the four vertices, a pair of perpendicular segments meet. All four vertices are therefore right angles, and a quadrilateral with four right angles is a rectangle (see Chapter 10). That does it.

2. Find the perimeter of PQRS.

Use the distance formula. Because you now know that *PQRS* is a rectangle and that its opposite sides are therefore congruent, you need to compute the lengths of only two sides (the length and the width):

$$\text{Distance} = \sqrt{(x_2 - x_1)^2 + (y_2 - y_1)^2}$$

$$\begin{aligned}\text{Distance}_{P \text{ to } Q} &= \sqrt{(1-4)^2 + (2-(-2))^2} \\ &= \sqrt{(-3)^2 + 4^2} \\ &= \sqrt{25} \\ &= 5\end{aligned}$$

$$\begin{aligned}\text{Distance}_{Q \text{ to } R} &= \sqrt{(9-1)^2 + (8-2)^2} \\ &= \sqrt{8^2 + 6^2} \\ &= \sqrt{100} \\ &= 10\end{aligned}$$

Now that you've got the length and width, you can easily compute the perimeter:

$$\begin{aligned}\text{Perimeter}_{PQRS} &= 2(\text{length}) + 2(\text{width}) \\ &= 2(10) + 2(5) \\ &= 30\end{aligned}$$

3. Show that the diagonals of PQRS bisect each other, and find the point where they intersect.

If you know your rectangle properties (see Chapter 10), you know that the diagonals of *PQRS* must bisect each other. But another way to show this is with coordinate geometry. The term *bisect* in this problem should ring the *midpoint* bell. So use the midpoint formula for each diagonal:

$$\text{Midpoint} = \left(\frac{x_1 + x_2}{2}, \frac{y_1 + y_2}{2}\right)$$

$$\begin{aligned}\text{Midpoint}_{\overline{QS}} &= \left(\frac{1+12}{2}, \frac{2+4}{2}\right) & \text{Midpoint}_{\overline{PR}} &= \left(\frac{4+9}{2}, \frac{-2+8}{2}\right) \\ &= (6.5, 3) & &= (6.5, 3)\end{aligned}$$

The fact that the two midpoints are the same shows that each diagonal goes through the midpoint of the other, and that, therefore, each diagonal bisects the other. Obviously, the diagonals cross at (6.5, 3). That's a wrap.

Proving Properties Analytically

In this section, I show you how to do a proof *analytically,* which means using algebra. You can use analytic proofs to prove some of the properties you see earlier in the book, such as the property that the diagonals of a parallelogram bisect each other or that the diagonals of an isosceles trapezoid are congruent. In previous chapters, you prove this type of thing with ordinary two-column proof methods, using things such as congruent triangles and CPCTC. Here you take a different tack and use the location of shapes in the coordinate system as the basis for your proofs.

Analytic proofs have two basic steps:

1. **Draw your figure in the coordinate system and label its vertices.**
2. **Use algebra to prove something about the figure.**

The following analytic proof walks you through this process: Prove analytically that the midpoint of the hypotenuse of a right triangle is equidistant from the triangle's three vertices. Then show analytically that the median to this midpoint divides the triangle into two triangles of equal area.

Step 1: Drawing a general figure

The first step in an analytic proof is to draw a figure in the *x-y* coordinate system and give its vertices coordinates. You want to put the figure in a convenient position that makes the math work out easily. For example, sometimes putting one of the vertices of your figure at the origin, (0, 0), makes the math easy because adding and subtracting with zeros is so simple. Quadrant I is also a good choice because all coordinates are positive there.

The figure you draw has to represent a general class of shapes, so you make the coordinates letters that can take on any values. You can't label the figure with numbers (except for using zero when you place a vertex at the origin or on the *x*- or *y*-axis) because that'd give the figure an exact size and shape — and then anything you proved would only apply to that particular shape rather than to an entire class of shapes.

Here's how you create your figure for the triangle proof:

- **Choose a convenient position and orientation for the figure in the *x-y* coordinate system.** Because the *x*- and *y*-axes form a right angle at the origin (0, 0), that's the natural choice for the position of the right angle of the right triangle, with the legs of the triangle lying on the two axes.

Then you have to decide which quadrant the triangle should go in. Unless you have some reason to pick a different quadrant, quadrant I is the best way to go.

- **Choose suitable coordinates for the two vertices on the *x*- and *y*-axes.**
  You'd often go with something like $(0, b)$ and $(a, 0)$, but here, because you're going to end up dividing these coordinates by 2 when you use the midpoint formula, the math will be easier if you use $(0, 2b)$ and $(2a, 0)$. Otherwise, you have to deal with fractions. Egad! Figure 18-6 shows the final diagram.

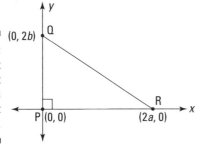

Figure 18-6: A right triangle that represents *all* right triangles.

In an analytic proof, when you decide how to position and label your figure, you must do so in a way such that there is, as mathematicians say, *no loss of generality*. In this proof, for example, you shouldn't give the vertices on the *x*- and *y*-axes coordinates like $(a, 0)$ and $(0, a)$ or $(2a, 0)$ and $(0, 2a)$ because that would mean that the two legs of the right triangle would have the same length — and that would mean that your triangle would be a 45°- 45°- 90° triangle. If you label the vertices like that, then all the conclusions that you draw from this proof would be valid only for 45°- 45°- 90° right triangles, not all right triangles.

The right triangle in the Figure 18-6 *has* been drawn with no loss of generality: With vertices at $(0, 0)$, $(2a, 0)$, and $(0, 2b)$, it can represent every possible right triangle. Here's why this works: Imagine any right triangle of any size or shape, located anywhere in the coordinate system. Without changing its size or shape, you could slide it so that its right angle was at the origin and then rotate it so that its legs would lie on the *x*- and *y*-axes. You could then pick values for *a* and *b* so that $2a$ and $2b$ would work out to equal the lengths of your hypothetical triangle's legs.

Because this proof includes a general right triangle, as soon as this proof is done, you'll have proved a result that's true for every possible right triangle in the universe. All infinitely many of them! Pretty cool, right?

Step 2: Solving the problem algebraically

Okay, so after you complete your drawing (see Figure 18-6 in the preceding section), you're ready to do the algebraic part of the proof. The first part of the problem asks you to prove that the midpoint of the hypotenuse is equidistant from the triangle's vertices. To do that, start by determining the midpoint of the hypotenuse:

$$\text{Midpoint} = \left(\frac{x_1 + x_2}{2}, \frac{y_1 + y_2}{2}\right)$$

$$\text{Midpoint}_{\overline{QR}} = \left(\frac{0 + 2a}{2}, \frac{2b + 0}{2}\right)$$

$$= (a, b)$$

Figure 18-7 shows the midpoint, M, and median $\overline{PM}$.

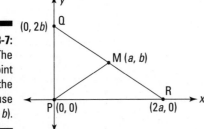

Figure 18-7: The midpoint of the hypotenuse is at (a, b).

To prove the equidistance of M to P, Q, and R, you use the distance formula:

$$\text{Distance} = \sqrt{(x_2 - x_1)^2 + (y_2 - y_1)^2}$$

$$\text{Distance}_{M \text{ to } P} = \sqrt{(a - 0)^2 + (b - 0)^2}$$
$$= \sqrt{a^2 + b^2}$$

$$\text{Distance}_{M \text{ to } Q} = \sqrt{(a - 0)^2 + (b - 2b)^2}$$
$$= \sqrt{a^2 + (-b)^2}$$
$$= \sqrt{a^2 + b^2}$$

These distances are equal, and that completes the equidistance portion of the proof. (Because M is the midpoint of $\overline{QR}$, $\overline{MQ}$ must be congruent to $\overline{MR}$, and thus there's no need to show that the distance from M to R is also $\sqrt{a^2 + b^2}$, though you may want to do so as an exercise.)

For the second part of the proof, you must show that the segment that goes from the right angle to the hypotenuse's midpoint divides the triangle into two triangles with equal areas — in other words, you have to show that Area$_{\triangle PQM}$ = Area$_{\triangle PMR}$. To compute these areas, you need to know the lengths of the base and altitude of both triangles. Figure 18-8 shows the triangles' altitudes.

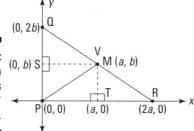

Figure 18-8: Drawing in the altitudes of △PQM and △PMR.

Note that because base $\overline{PR}$ of △PMR is horizontal, the altitude drawn to that base ($\overline{TM}$) is vertical, and thus you know that T is directly below (a, b) at $(a, 0)$. With △PQM (using vertical base $\overline{PQ}$), you create horizontal altitude $\overline{SM}$ and locate point S directly to the left of (a, b) at $(0, b)$.

Now you're ready to use the two bases and two altitudes to show that the triangles have equal areas. To get the lengths of the bases and altitudes, you could use the distance formula, but you don't need to because you can use the nifty shortcut for horizontal and vertical distances from "Going the distance with the distance formula":

Horizontal distance = right$_{x\text{-coordinate}}$ − left$_{x\text{-coordinate}}$

Vertical distance = top$_{y\text{-coordinate}}$ − bottom$_{y\text{-coordinate}}$

For △PQM: Vertical distance$_{\text{Base } \overline{PQ}}$ = $2b - 0 = 2b$
             Horizontal distance$_{\text{Altitude } \overline{SM}}$ = $a - 0 = a$

For △PMR: Horizontal distance$_{\text{Base } \overline{PR}}$ = $2a - 0 = 2a$
             Vertical distance$_{\text{Altitude } \overline{TM}}$ = $b - 0 = b$

Time to wrap this up using the triangle area formula:

$$\text{Area}_{\triangle PQM} = \frac{1}{2}bh \qquad \text{Area}_{\triangle PMR} = \frac{1}{2}(PR)(TM)$$
$$= \frac{1}{2}(PQ)(SM) \qquad\qquad = \frac{1}{2}(2a)(b)$$
$$= \frac{1}{2}(2b)(a) \qquad\qquad = ab$$
$$= ab$$

The areas are equal. That does it.

Deciphering Equations for Lines and Circles

If you've already taken Algebra I, you've probably dealt with graphing lines in the coordinate system. Graphing circles may be something new for you, but you'll soon see that there's nothing to it. Lines and circles are, of course, very different. One is straight, the other curved. One is endless, the other limited. But what they have in common is that neither has a beginning or an end, and you could travel along either one till the end of time. Hmm, I feel a quote from an old TV show coming on . . . "You are about to enter another dimension, a dimension not only of sight and sound but of mind . . . It is a dimension as vast as space and as timeless as infinity . . . Next stop, the Twilight Zone."

Line equations

Talk about the straight and narrow! Lines are infinitely long, perfectly straight, and though it's hard to imagine, infinitely narrower than a strand of hair.

Here are the basic forms for equations of lines:

- **Slope-intercept form:** Use this form when you know (or can easily find) a line's slope and its *y-intercept* (the point where the line crosses the *y*-axis). See the earlier section titled "The slope dope" for details on slope.

    $y = mx + b$,

    where *m* is the slope and *b* is the *y*-intercept $(0, b)$

- **Point-slope form:** This is the easiest form to use when you don't know a line's y-intercept but you do know the coordinates of a point on the line; you also need to know the line's slope.

    $y - y_1 = m(x - x_1)$,

    where m is the slope and (x_1, y_1) is a point on the line

- **Horizontal line:** This form is used for lines with a slope of zero.

    $y = b$,

    where b is the y-intercept

    The b (or the number that's plugged into b) tells you how far up or down the line is along the y-axis. Note that every point along a horizontal line has the same y-coordinate, namely b. In case you're curious, this equation form is a special case of $y = mx + b$ where $m = 0$.

- **Vertical line:** And here's the equation for a line with an undefined slope.

    $x = a$,

    where a is the x-intercept

    The a (or the number that's plugged into a) tells you how far to the right or left the line is along the x-axis. Every point along a vertical line has the same x-coordinate, namely a.

Don't mix up the equations for horizontal and vertical lines. This mistake is extremely common. Because a horizontal line is parallel to the x-axis, you might think that the equation of a horizontal line would be $x = a$. And you might figure that the equation for a vertical line would be $y = b$ because a vertical line is parallel to the y-axis. But as you see in the preceding equations, it's the other way around.

The standard circle equation

In Part V, you see all sorts of interesting circle properties, formulas, and theorems that have nothing to do with a circle's position or location. In this section, courtesy of Descartes, you investigate circles that do have a location; you analyze circles positioned in the x-y coordinate system using analytic methods — that is, with equations and algebra. For example, there's a nice analytic connection between the circle equation and the distance formula because every point on a circle is the same *distance* from its center (see the example problem for more details).

Here are the circle equations:

- **Circle centered at the origin, (0, 0):**

  $$x^2 + y^2 = r^2$$

- **Circle centered at any point (h, k):**

  $$(x - h)^2 + (y - k)^2 = r^2,$$

  where (h, k) is the center of the circle and r is its radius

(As you may recall from an algebra course, it seems backward, but subtracting h from x actually moves the circle to the *right*, and subtracting k from y moves the circle *up*; adding a number to x moves the circle *left*, and adding a number to y moves the circle *down*.)

Ready for a circle problem? Here you go:

Given: Circle C has its center at (4, 6) and is tangent to a line at (1, 2)

Find:
1. The equation of the circle
2. The circle's x- and y-intercepts
3. The equation of the tangent line

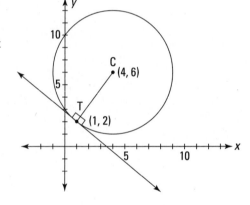

1. Find the equation of the circle.

All you need for the equation of a circle is its center (you know it) and its radius. The radius of the circle is just the distance from its center to any point on the circle, but since the point of tangency is given, that's the easiest point to use. To wit —

$$\begin{aligned}\text{Distance}_{C \text{ to } T} &= \sqrt{(4-1)^2 + (6-2)^2} \\ &= \sqrt{3^2 + 4^2} \\ &= 5\end{aligned}$$

Now you finish by plugging the center coordinates and the radius into the general circle equation:

$$(x - h)^2 + (y - k)^2 = r^2$$
$$(x - 4)^2 + (y - 6)^2 = 5^2$$

2. Find the circle's x- and y-intercepts.

To find the x-intercepts for any equation, you just plug in 0 for y and solve for x:

$$(x-4)^2 + (y-6)^2 = 5^2$$

$$(x-4)^2 + (0-6)^2 = 5^2$$
$$(x-4)^2 + 36 = 25$$
$$(x-4)^2 = -11$$

You can't square something and get a negative number, so this equation has no solution; therefore, the circle has no x-intercepts. (I realize that you can just look at the figure and see that the circle doesn't intersect the x-axis, but I wanted to show you how the math confirms this.)

To find the y-intercepts, plug in 0 for x and solve for y:

$$(0-4)^2 + (y-6)^2 = 5^2$$
$$16 + (y-6)^2 = 25$$
$$(y-6)^2 = 9$$
$$y-6 = \pm\sqrt{9}$$
$$y = \pm 3 + 6$$
$$= 3 \text{ or } 9$$

Thus, the circle's y-intercepts are (0, 3) and (0, 9).

3. Find the equation of the tangent line.

For the equation of a line, you need a point (you have it) and the line's slope. In Chapter 14, you find out that a tangent line is perpendicular to a radius drawn to the point of tangency. So just compute the slope of the radius, and then the opposite reciprocal of that is the slope of the tangent line (for more on slope, see the earlier "The slope dope" section):

$$\text{Slope} = \frac{y_2 - y_1}{x_2 - x_1}$$

$$\text{Slope}_{\text{Radius } \overline{CT}} = \frac{6-2}{4-1} = \frac{4}{3}$$

Therefore,

$$\text{Slope}_{\text{Tangent line}} = -\frac{3}{4}$$

Now you plug this slope and the coordinates of the point of tangency into the point-slope form for the equation of a line:

$$y - y_1 = m(x - x_1)$$
$$y - 2 = -\frac{3}{4}(x - 1)$$

Now clean this up a bit:

$$4(y - 2) = -3(x - 1)$$
$$4y - 8 = -3x + 3$$
$$3x + 4y = 11$$

Of course, if you instead choose to put this in slope-intercept form, you get $y = -\frac{3}{4}x + \frac{11}{4}$. Over and out.

Chapter 19

Changing the Scene with Geometric Transformations

In This Chapter
▶ Reflections: The components of all other transformations
▶ Translations: Slides made of two reflections
▶ Rotations: Revolutions made of two reflections
▶ Glide reflections: A translation plus a reflection (three total reflections)

A transformation takes a "before" figure in the *x-y* coordinate system — say, a triangle, parallelogram, polygon, anything — and turns it into a related "after" figure. The original figure is called the *pre-image,* and the new figure is called the *image.* The transformation may expand or shrink the original figure, warp it into a funny-looking version of itself (like the way those curving amusement park mirrors warp your image), spin the figure around, or slide it to a new position — or the transformation may change the figure in some combination of those ways.

In this chapter, you work with a special subset of transformations called *isometries.* These are the transformations in which the "before" and "after" figures are *congruent,* which, as you know, means that the figures are exactly the same shape and size. I explain the four types of isometries: reflections, translations, rotations, and glide reflections. The discussion starts with reflections, the building blocks of the other three types of isometries.

Note: This chapter asks you to find slopes, midpoints, distances, perpendicular bisectors, and line equations in the coordinate plane. If you need some background on these things, please refer to Chapter 18.

Some Reflections on Reflections

A geometric reflection, like it sounds, works like a reflection in a mirror. Figure 19-1 shows someone in front of a mirror looking at the reflection of a triangle that's on the floor in front of the mirror. *Note:* The image of △*ABC* in the mirror is labeled with the same letters except a *prime* symbol (′) is added to each letter (△*A′B′C′*.) Most transformation diagrams are handled this way.

As you can see, the image of the triangle in the mirror is flipped over compared with the real triangle. Mirrors (and mathematically speaking, reflections) always produce this kind of flipping. Flipping a figure switches its *orientation*, a topic I discuss in the next section.

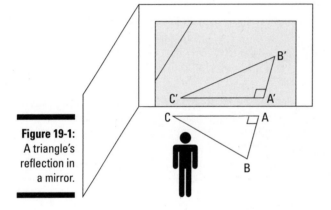

Figure 19-1: A triangle's reflection in a mirror.

Figure 19-2 shows that a reflection can also be thought of as a folding. On the left, you see a folded card with a half-heart shape drawn on it; in the center, you see the folded half-heart that's been cut out; and on the right, you see the heart unfolded. The left and right sides of the heart are obviously the same shape. Each side is the *reflection* of the other side. The crease or fold-line running down the center of the heart is called the *reflecting line,* which I discuss later in this chapter. (I bet you didn't realize that when you were making valentines in first grade, you were dealing with mathematical isometries!)

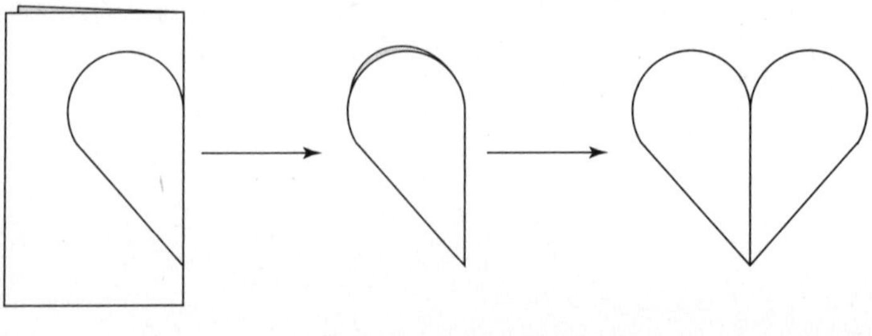

Figure 19-2: Reflections from the heart — will you be my valentine?

Reflections are the building blocks of the other isometries because you can perform the other isometries with a series of reflections:

- Translations are the equivalent of two reflections.
- Rotations are the equivalent of two reflections.
- Glide reflections are the equivalent of three reflections.

I discuss translations, rotations, and glide reflections later in this chapter. But before moving on to them, I briefly discuss orientation and then show you how to do reflection problems.

Getting oriented with orientation

In Figure 19-3, △PQR has been reflected across line *l* to produce △P'Q'R'. Triangles PQR and P'Q'R' are congruent, but their *orientations* are different:

- One way to see that they have different orientations is that you can't get △PQR and △P'Q'R' to stack on top of each other — no matter how you rotate or slide them — without flipping one of them over.
- A second characteristic of figures with different orientations is the clockwise/counterclockwise switch. Notice that in △PQR, you go counterclockwise from P to Q to R, but in the reflected triangle, △P'Q'R', you go clockwise from P' to Q' to R'.

Note that — like with the heart in Figure 19-2 — the reflection shown in Figure 19-3 can be thought of as a folding. If you were to fold this page along line *l*, △PQR would end up stacked perfectly on △P'Q'R', with P on P', Q on Q', and R on R'.

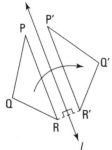

Figure 19-3: Reflecting △*PQR* over line *l* switches the figure's orientation.

310 Part VII: Placement, Points, and Pictures: Alternative Geometry Topics

Reflections and orientation: Reflecting a figure once switches its orientation. When you reflect a figure more than once, the following rules apply:

- If you reflect a figure and then reflect it again over the same line or a different line, the figure returns to its original orientation. More generally, if you reflect a figure an *even* number of times, the final result is a figure with the *same orientation* as the original figure.
- Reflecting a figure an *odd* number of times produces a figure with the *opposite orientation*.

Finding a reflecting line

In a reflection, the *reflecting line,* as you'd probably guess, is the line over which the pre-image is reflected. Figure 19-3 illustrates an important property of reflecting lines: If you form $\overline{RR'}$ by connecting pre-image point R with its image point R' (or P with P' or Q with Q'), the reflecting line, l, is the perpendicular bisector of $\overline{RR'}$.

A reflecting line is a perpendicular bisector: When a figure is reflected, the *reflecting line* is the perpendicular bisector of all segments that connect preimage points to their corresponding image points.

Here's a problem that uses this idea: In the following figure, $\triangle J'K'L'$ is the reflection of $\triangle JKL$ over a reflecting line. Find the equation of the reflecting line using points J and J'. Then confirm that this reflecting line sends K to K' and L to L'.

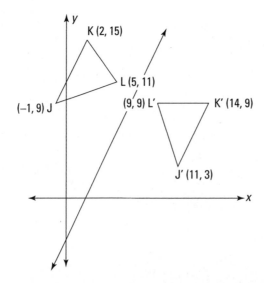

The reflecting line is the perpendicular bisector of segments connecting pre-image points to their image points. Because the perpendicular bisector of a segment goes through the segment's midpoint, the first thing you need to do to find the equation of the reflecting line is to find the midpoint of $\overline{JJ'}$:

$$\text{Midpoint} = \left(\frac{x_1 + x_2}{2}, \frac{y_1 + y_2}{2}\right)$$

$$\text{Midpoint}_{\overline{JJ'}} = \left(\frac{-1 + 11}{2}, \frac{9 + 3}{2}\right)$$

$$= (5, 6)$$

Next, you need the slope of $\overline{JJ'}$:

$$\text{Slope} = \frac{y_2 - y_1}{x_2 - x_1}$$

$$\text{Slope}_{\overline{JJ'}} = \frac{3 - 9}{11 - (-1)} = \frac{-6}{12} = -\frac{1}{2}$$

The slope of the perpendicular bisector of $\overline{JJ'}$ is the opposite reciprocal of the slope of $\overline{JJ'}$ (as I explain in Chapter 18). $\overline{JJ'}$ has a slope of $-\frac{1}{2}$, so the slope of the perpendicular bisector, and therefore of the reflecting line, is 2. Now you can finish the first part of the problem by plugging the slope of 2 and the point (5, 6) into the point-slope form for the equation of a line:

$$y - y_1 = m(x - x_1)$$
$$y - 6 = 2(x - 5)$$
$$y = 2x - 10 + 6$$
$$y = 2x - 4$$

That's the equation of the reflecting line, in slope-intercept form.

To confirm that this reflecting line sends K to K' and L to L', you have to show that this line is the perpendicular bisector of $\overline{KK'}$ and $\overline{LL'}$. To do that, you must show that the midpoints of $\overline{KK'}$ and $\overline{LL'}$ lie on the line and that the slopes of $\overline{KK'}$ and $\overline{LL'}$ are both $-\frac{1}{2}$ (the opposite reciprocal of the slope of the reflecting line $y = 2x - 4$). First, here's the midpoint of $\overline{KK'}$:

$$\text{Midpoint}_{\overline{KK'}} = \left(\frac{2 + 14}{2}, \frac{15 + 9}{2}\right)$$

$$= (8, 12)$$

Plug these coordinates into the equation $y = 2x - 4$ to see whether they work. Because $12 = 2(8) - 4$, the midpoint of $\overline{KK'}$ lies on the reflecting line. Now get the slope of $\overline{KK'}$:

$$\text{Slope}_{\overline{KK'}} = \frac{9 - 15}{14 - 2} = \frac{-6}{12} = -\frac{1}{2}$$

This is the desired slope, so everything's copasetic for K and K'. Now compute the midpoint of $\overline{LL'}$:

$$\text{Midpoint}_{\overline{LL'}} = \left(\frac{5+9}{2}, \frac{11+9}{2}\right)$$
$$= (7, 10)$$

Check that these coordinates work when you plug them into the equation of the reflecting line, $y = 2x - 4$. Because $10 = 2(7) - 4$, the midpoint of $\overline{LL'}$ is on the line. Finally, find the slope of $\overline{LL'}$:

$$\text{Slope}_{\overline{LL'}} = \frac{9-11}{9-5} = \frac{-2}{4} = -\frac{1}{2}$$

This checks. You're done.

Not Getting Lost in Translations

A *translation* — probably the simplest type of transformation — is a transformation in which a figure just slides straight to a new location without any tilting or turning. It shouldn't be hard to see that a translation doesn't change a figure's orientation. See Figure 19-4.

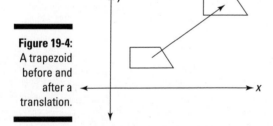

Figure 19-4: A trapezoid before and after a translation.

A translation equals two reflections

It might seem a bit surprising, but instead of sliding a figure to a new location, you can achieve the same end result by reflecting the figure over one line and then over a second line.

You can see how this works by doing the following: Take a blank piece of paper and tear off a little piece of its lower, right-hand corner. Place the sheet of paper in front of you on a desk or table. Now flip the paper to the right over its right edge — you know, so that its right edge doesn't move. You then see the back side of the paper, and the torn-off corner is at the lower, left-hand corner. Finally, flip the paper over to the right again. Now, after the two flips, or two reflections, you see the paper just as it looked originally, except that now it's been slid, or translated, to the right.

Translation line and translation distance: In a translation, the *translation line* is any line that connects a pre-image point of a figure to its corresponding image point; the translation line shows you the direction of the translation. The *translation distance* is the distance from any pre-image point to its corresponding image point.

A translation equals two reflections: A translation of a given distance along a translation line is equivalent to two reflections over parallel lines that

- Are perpendicular to the translation line
- Are separated by a distance equal to half the translation distance

Note: As long as the parallel reflecting lines are separated by half the translation distance, they can be located anywhere along the translation line. And note that if you call the first reflecting line l_1 and call the second reflecting line l_2, then l_1 and l_2 must be placed so that the direction from l_1 to l_2 is the same as the direction from the pre-image to the image.

Is that theorem a mouthful or what? Instead of puzzling over the theorem, take a look at Figure 19-5 to see how reflecting lines work in a translation.

Here are a few points about Figure 19-5. You can see that the translation distance (the distance from Z to Z') is 20; that the distance between the reflecting lines, l_1 and l_2, is half of that, or 10; and that reflecting lines l_1 and l_2 are perpendicular to the translation line, $\overleftrightarrow{ZZ'}$.

I chose to put the two reflecting lines between the pre-image and the image because that's the easiest way to see how they work. But the reflecting lines don't have to be placed there. To give you an idea of another possible placement for the two reflecting lines, imagine grabbing l_1 and l_2 as a single unit and moving them to the right in the direction of the translation line, $\overleftrightarrow{ZZ'}$, till they were both out past $\triangle X'Y'Z'$. With this new placement, you'd flip the pre-image, $\triangle XYZ$, first over l_1 (flipping it up and to the right, beyond l_2), and then second, you'd reflect it over l_2, back down and to the left. The final result would be $\triangle X'Y'Z'$ in the very same place that you see it in Figure 19-5.

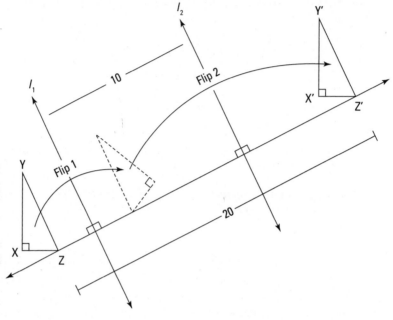

Figure 19-5: After flipping over two reflecting lines, △XYZ moves to △X'Y'Z'.

Finding the elements of a translation

The best way to understand the translation theorem is by looking at an example problem. The next problem shows you how to find a translation line, the translation distance, and a pair of reflecting lines.

In the following figure, pre-image triangle △PQR has been slid down and to the right to image triangle △P'Q'R'.

Given: The coordinates of P, P', Q, and R' as shown

Find:
1. The coordinates of Q' and R
2. The translation distance
3. The equation of a translation line
4. The equations of two different pairs of reflecting lines

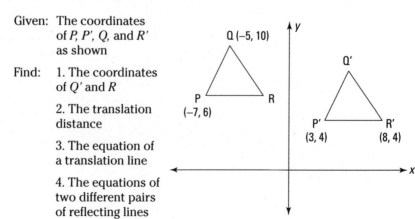

1. **Find the coordinates of Q' and R.**

   From $P(-7, 6)$ to $P'(3, 4)$, you go 10 to the right and 2 down. In a translation, every pre-image point moves the same way to its image point, so to find Q', just begin at Q, which is at $(-5, 10)$, and go 10 to the right and 2 down. That brings you to $(5, 8)$, the coordinates of Q'.

   To get the coordinates of R, you start at R' and go backward (10 left and 2 up). That gives you $(-2, 6)$ for the coordinates of R.

2. **Find the translation distance.**

   The translation distance is the distance between any pre-image point and its image point, such as P and P'. Use the distance formula:

   $$\text{Distance} = \sqrt{(x_2 - x_1)^2 + (y_2 - y_1)^2}$$
   $$\text{Distance}_{P \text{ to } P'} = \sqrt{(3 - (-7))^2 + (4 - 6)^2}$$
   $$= \sqrt{10^2 + (-2)^2}$$
   $$= \sqrt{104} = 2\sqrt{26} \approx 10.2 \text{ units}$$

   This answer tells you that each pre-image point goes a distance of 10.2 units to its image point.

3. **Find the equation of a translation line.**

   For a translation line, you can use any line that connects a point on $\triangle PQR$ with its image point on $\triangle P'Q'R'$. The line connecting P and P' works as well as any other translation line, so work out the equation of $\overleftrightarrow{PP'}$.

   To use the point-slope form, you need a point (you have two, P and P', so take your pick) and the slope of the line. The slope formula gives you — what else? — the slope:

   $$\text{Slope}_{\overleftrightarrow{PP'}} = \frac{4 - 6}{3 - (-7)} = \frac{-2}{10} = -\frac{1}{5}$$

   Now plug this slope and the coordinates of P' into the point-slope form (P would work just as well, but I like to avoid using negative numbers):

   $$y - y_1 = m(x - x_1)$$
   $$y - 4 = -\frac{1}{5}(x - 3)$$

   If you feel like it, you can put this into slope-intercept form with some very simple algebra:

   $$y = -\frac{1}{5}x + \frac{23}{5}$$

   That's the equation of a translation line. Triangle PQR can slide down along this line to $\triangle P'Q'R'$.

4. Find the equations of two different pairs of reflecting lines.

The translation theorem tells you that two reflecting lines that achieve a translation must be perpendicular to the translation line and separated by half the translation distance. There are an infinite number of such pairs of lines. Here's an easy way to come up with one such pair.

Perpendicular lines have slopes that are opposite reciprocals of each other. In part 3 of this problem, you find that $\overleftrightarrow{PP'}$ has a slope of $-\frac{1}{5}$; thus, because the reflecting lines are perpendicular to $\overleftrightarrow{PP'}$, their slopes must be the opposite reciprocal of $-\frac{1}{5}$, which is 5.

For the first reflecting line, you can use the line with a slope of 5 that goes through P at $(-7, 6)$. Use the point-slope form and simplify:

$$y - 6 = 5(x - (-7))$$
$$y = 5x + 41$$

Then, because the translation distance equals the length of $\overline{PP'}$, the distance from P to the midpoint of $\overline{PP'}$ is half the translation distance — the desired distance between reflecting lines. So run your second reflecting line through the midpoint of $\overline{PP'}$. First find the midpoint:

$$\text{Midpoint}_{\overline{PP'}} = \left(\frac{-7+3}{2}, \frac{6+4}{2}\right) = (-2, 5)$$

The second reflecting line, which is parallel to the first, also has a slope of 5. Plug your numbers into the point-slope form and simplify:

$$y - 5 = 5(x - (-2))$$
$$y - 5 = 5x + 10$$
$$y = 5x + 15$$

So you've got your two reflecting lines. If you reflect $\triangle PQR$ over the line $y = 5x + 41$ and then reflect it over $y = 5x + 15$ (it must be in that order), $\triangle PQR$ will land — point for point — on top of $\triangle P'Q'R'$.

TIP

After you know one pair of reflecting lines, you can effortlessly produce as many of these pairs as you want. All reflecting lines will have the *same slope,* and in each pair of lines, *the y-intercepts will be the same distance apart.*

In this problem, all reflecting lines have a slope of 5, and each pair must have y-intercepts that — like $y = 5x + 41$ and $y = 5x + 15$ — are 26 units apart. For example, the following pairs of reflecting lines would also achieve the desired translation:

$$y = 5x + 27 \quad \text{and} \quad y = 5x + 1$$
$$y = 5x + 1{,}000{,}026 \quad \text{and} \quad y = 5x + 1{,}000{,}000$$

Turning the Tables with Rotations

A *rotation* is what you'd expect — it's a transformation in which the pre-image figure rotates or spins to the location of the image figure. With all rotations, there's a single fixed point — called the *center of rotation* — around which everything else rotates. This point can be inside the figure, in which case the figure stays where it is and just spins. Or the point can be outside the figure, in which case the figure moves along a circular arc (like an orbit) around the center of rotation. The amount of turning is called the *rotation angle*.

In this section, you see that a rotation, just like a translation, is the equivalent of two reflections. Then I show you how to find the center of rotation.

A rotation equals two reflections

You can achieve a rotation with two reflections. The way this works is a bit tricky to explain (and the mumbo-jumbo in the following theorem might not help much), so check out Figure 19-6 to get a better handle on this idea.

A rotation equals two reflections: A rotation is equivalent to two reflections over lines that

- Pass through the center of rotation
- Form an angle half the measure of the rotation angle

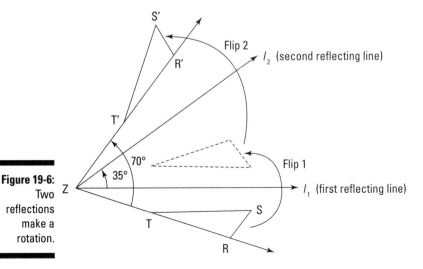

Figure 19-6: Two reflections make a rotation.

In Figure 19-6, you can see that pre-image △RST has been rotated counter-clockwise 70° to image △R'S'T'. This rotation can be produced by first reflecting △RST over line l_1 and then reflecting it again over l_2. The angle formed by l_1 and l_2, 35°, is half of the angle of rotation.

Finding the center of rotation and the equations of two reflecting lines

Just like in the previous section on translations, the easiest way to understand the rotation theorem is by doing a problem: In the following figure, pre-image triangle △ABC has been rotated to create image triangle △A'B'C'.

Find:
1. The center of rotation
2. Two reflecting lines that would achieve the same result as the rotation

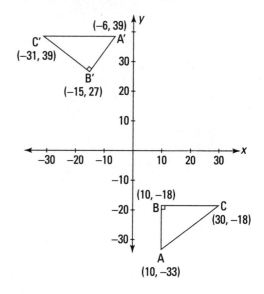

1. Find the center of rotation.

I've got a nifty method for locating the center of rotation. Here's how it works. Take the three segments that connect pre-image points to their image points (in this case, $\overline{AA'}$, $\overline{BB'}$, and $\overline{CC'}$). In all rotations, the center of rotation lies at the intersection of the perpendicular bisectors of such segments (it'd get too involved to explain why, so just take my word for it). Because the three perpendicular bisectors meet at the same point, you need only two of them to find the point of intersection. Any two will work, so find the perpendicular bisectors of $\overline{AA'}$ and $\overline{BB'}$; then you can set their equations equal to each other to find where they intersect.

First get the midpoint of $\overline{AA'}$:

$$\text{Midpoint}_{\overline{AA'}} = \left(\frac{10+(-6)}{2}, \frac{-33+39}{2}\right) = (2, 3)$$

Then find the slope of $\overline{AA'}$:

$$\text{Slope}_{\overline{AA'}} = \frac{39-(-33)}{-6-10} = \frac{72}{-16} = -\frac{9}{2}$$

The slope of the perpendicular bisector of $\overline{AA'}$ is the opposite reciprocal of $-\frac{9}{2}$, namely $\frac{2}{9}$. The point-slope form for the perpendicular bisector is thus

$$y - 3 = \frac{2}{9}(x - 2), \text{ or}$$
$$y = \frac{2}{9}x + \frac{23}{9}$$

Go through the same process to get the perpendicular bisector of $\overline{BB'}$:

$$\text{Midpoint}_{\overline{BB'}} = \left(\frac{10+(-15)}{2}, \frac{-18+27}{2}\right) = \left(-\frac{5}{2}, \frac{9}{2}\right)$$

$$\text{Slope}_{\overline{BB'}} = \frac{27-(-18)}{-15-10} = \frac{45}{-25} = -\frac{9}{5}$$

The slope of the perpendicular bisector of $\overline{BB'}$ is the opposite reciprocal of $-\frac{9}{5}$, which is $\frac{5}{9}$. The equation of the perpendicular bisector is thus

$$y - \frac{9}{2} = \frac{5}{9}\left(x - \left(-\frac{5}{2}\right)\right), \text{ or}$$
$$y = \frac{5}{9}x + \frac{53}{9}$$

Now, to find where the two perpendicular bisectors intersect, set the right sides of their equations equal to each other and solve for x:

$$\frac{2}{9}x + \frac{23}{9} = \frac{5}{9}x + \frac{53}{9}$$
$$-\frac{3}{9}x = \frac{30}{9}$$

Multiply both sides by 9 to get rid of the fractions; then divide:

$-3x = 30$

$x = -10$

Plug –10 back into either equation to get y:

$$y = \frac{2}{9}x + \frac{23}{9}$$
$$= \frac{2}{9}(-10) + \frac{23}{9}$$
$$= \frac{1}{3}$$

You've done it. The center of rotation is $\left(-10, \frac{1}{3}\right)$. Give this point a name — how about point Z?

The following figure shows point Z, ∠AZA', and a little counterclockwise arrow that indicates the rotational motion that would move △ABC to △A'B'C'. If you hold Z where it is and rotate this book counterclockwise, △ABC will spin to where △A'B'C' is now.

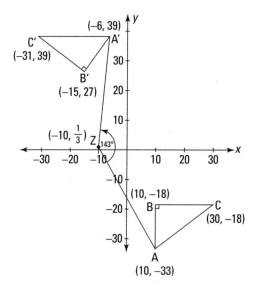

2. Find two reflecting lines that achieve the same result as the rotation.

The rotation theorem tells you that two reflecting lines will achieve this rotation if they go through the center of rotation and form an angle that's half the measure of the rotation angle (as shown in Figure 19-6). An infinite number of pairs of reflecting lines satisfy these conditions, but the following is an easy way to find one such pair.

In this problem, $\triangle ABC$ has been rotated counterclockwise; the amount of rotation is 143°, the measure of $\angle AZA'$. You want an angle half this big for the angle between the two reflecting lines. One way to do this is to cut $\angle AZA'$ in half with its angle bisector. Then you can use the half-angle that goes from side $\overleftrightarrow{ZA}$ to the angle bisector.

So designate $\overleftrightarrow{ZA}$ as the first reflecting line and find its equation by determining its slope and plugging the slope and the coordinates of Z or A into the point-slope form for the equation of a line. If you do the math then clean things up, you should get $y = -\frac{5}{3}x - \frac{49}{3}$.

Again, with $\overleftrightarrow{ZA}$ as the first reflecting line, the second reflecting line will be the angle bisector of $\angle AZA'$. But guess what — you already know this angle bisector because it's one and the same as the perpendicular bisector of $\overline{AA'}$, which you figured out in part 1: $y = \frac{2}{9}x + \frac{23}{9}$. (By the way, if you'd used, say, $\angle BZB'$ instead of $\angle AZA'$, you would've used the perpendicular bisector of $\overline{BB'}$ as the angle bisector of $\angle BZB'$.)

So if you reflect $\triangle ABC$ over $\overleftrightarrow{ZA}$, $y = -\frac{5}{3}x - \frac{49}{3}$, and then over $y = \frac{2}{9}x + \frac{23}{9}$, it'll land precisely where $\triangle A'B'C'$ is. And thus, these two reflections achieve the same result as the counterclockwise rotation about point Z.

Third Time's the Charm: Stepping Out with Glide Reflections

A *glide reflection* is just what it sounds like: You glide a figure (that's just another way of saying *slide* or *translate*) and then reflect it over a reflecting line. Or you can reflect the figure first and then slide it; the result is the same either way. A glide reflection is also called a *walk* because it looks like the motion of two feet. See Figure 19-7.

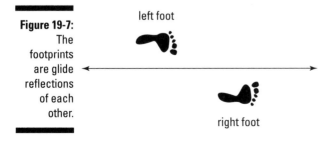

Figure 19-7: The footprints are glide reflections of each other.

A glide reflection is, in a sense, the most complicated of the four types of isometries because it's the composition of two other isometries: a reflection and a translation. If you have a pre-image and an image like the two feet in Figure 19-7, it's impossible to move the pre-image to the image with one simple reflection, one translation, or one rotation (try it with Figure 19-7). The only way to get from the pre-image to the image is with a combination of one reflection and one translation.

A glide reflection equals three reflections

A glide reflection is the combination of a reflection and a translation. And because you can produce the translation part with two reflections (see the earlier "Not Getting Lost in Translations" section), you can achieve a glide reflection with *three* reflections.

You can see in the previous sections that some images are just one reflection away from their pre-images; other images (in translation and rotation problems) are two reflections away. And now you see that glide reflection images are three reflections away from their pre-images. I find it interesting that this covers all possibilities. In other words, every image — no matter where it is in the coordinate system and no matter how it's spun around or flipped over — is either one, two, or three reflections away from its pre-image. Pretty cool, eh?

Finding the main reflecting line

The following theorem tells you about the location of the main reflecting line in a glide reflection, and the subsequent problem shows you, step by step, how to find the main reflecting line's equation.

The main reflecting line of a glide reflection: In a glide reflection, the midpoints of all segments that connect pre-image points with their image points lie on the main reflecting line.

Ready for a glide reflection problem? I do the reflection first and then the translation, but you can do them in either order.

The following figure shows a pre-image parallelogram *ABCD* and the image parallelogram *A'B'C'D'* that resulted from a glide reflection. Find the main reflecting line.

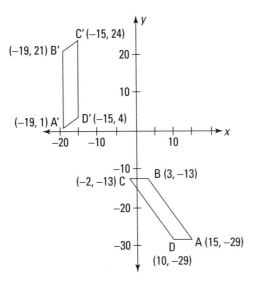

The main reflecting line in a glide reflection contains the midpoints of all segments that join pre-image points with their image points (such as $\overline{CC'}$). You need only two such midpoints to find the equation of the main reflecting line (because you need just two points to determine a line). The midpoints of $\overline{AA'}$ and $\overline{BB'}$ will do the trick:

$$\text{Midpoint}_{\overline{AA'}} = \left(\frac{15+(-19)}{2}, \frac{-29+1}{2}\right) = (-2, -14)$$

$$\text{Midpoint}_{\overline{BB'}} = \left(\frac{3+(-19)}{2}, \frac{-13+21}{2}\right) = (-8, 4)$$

Now simply find the equation of the line determined by these two points:

$$\text{Slope}_{\text{Main reflecting line}} = \frac{-14-4}{-2-(-8)} = \frac{-18}{6} = -3$$

Use this slope and one of the midpoints in the point-slope form and simplify:

$$y - 4 = -3(x - (-8))$$
$$y - 4 = -3x - 24$$
$$y = -3x - 20$$

That's the main reflecting line. If you reflect parallelogram ABCD over this line, it'll then be in the same orientation as parallelogram A'B'C'D' (A to B to C to D will be in the clockwise direction), and ABCD will be perfectly vertical like A'B'C'D'. Then a simple translation in the direction of the main reflecting line will bring ABCD to A'B'C'D' (see Figure 19-8).

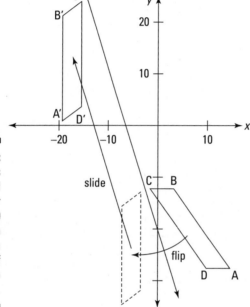

Figure 19-8:
ABCD is reflected over $y = -3x - 20$ and then slid in the direction of the line to A'B'C'D'.

You can achieve the translation to finish this glide reflection with two more reflections. But because I explain how to do such problems in the earlier "Not Getting Lost in Translations" section, I skip this so you can move on to the thrilling material in the next chapter.

Chapter 20

Locating Loci and Constructing Constructions

In This Chapter
- Using the four-step process for finding loci
- Looking at 2-D and 3-D loci
- Copying segments, angles, and triangles with a compass and straightedge
- Using constructions to divide segments and angles

*L*ocus is basically just a fancy word for *set*. In a locus problem, your task is to figure out (and then draw) the geometric object that satisfies certain conditions. Here's a simple example: What's the locus or set of all points 5 units from a given point? The answer is a circle because if you begin with one given point and then go 5 units away from that point in every direction, you get a circle with a radius of 5.

Constructions may be more familiar to you. Your task in construction problems is to use a compass and straightedge either to copy an existing figure, such as an angle or triangle, or to create something like a segment's perpendicular bisector, an angle's bisector, or a triangle's altitude.

What these topics have in common is that both involve drawing sets of points that make up some geometric figure. With locus problems, the challenge isn't drawing the shape; it's figuring out what shape the problem calls for. With construction problems, it's the other way around: You know exactly what shape you want, and the challenge is figuring out how to construct it.

Meeting the Conditions with Loci

Locus: A *locus* (plural: *loci*) is a set of points (usually some sort of geometric object like a line or a circle) consisting of all the points, and only the points, that satisfy certain given conditions.

The process of solving a locus problem can be difficult if you don't go about it methodically. So in this section, I give you a four-step locus-finding method that should keep you from making some common mistakes (such as including too many or too few points in your solution). Next, I take you through several 2-D locus problems using this process, and then I show you how to use 2-D locus problems to solve related 3-D problems.

The four-step process for locus problems

Following is the handy-dandy procedure for solving locus problems I promised you. Don't worry about understanding it immediately. It'll become clear to you as soon as you do some problems in the subsequent sections. (***Warning:*** Even though you'll often come up empty when working through steps 2 and 3, don't neglect to check them!):

1. **Identify a pattern.**

   Sometimes the key pattern will just sort of jump out at you. If it does, you're done with step 1. If it doesn't, find a single point that satisfies the given condition or conditions of the locus problem; then find a second such point; then, a third; and so on until you recognize a pattern.

2. **Look outside the pattern for points to add.**

   Look outside the pattern you identified in step 1 for additional points that satisfy the given condition(s).

3. **Look inside the pattern for points to exclude.**

   Look inside the pattern you found in step 1 (and possibly, though much less likely, any pattern you may have found in step 2) for points that fail to satisfy the given condition(s) despite the fact that they belong to the pattern.

4. **Draw a diagram and write a description of the locus solution.**

Two-dimensional locus problems

In 2-D locus problems, all the points in the locus solution lie in a plane. This is usually but not always the same plane as the given geometric object. Take a look at how the four-step solution method works in a few 2-D problems.

Chapter 20: Locating Loci and Constructing Constructions

Problem one

What's the locus of all points 3 units from a given circle whose radius is 10 units?

1. **Identify a pattern.**

    This is likely a problem in which you can immediately picture a pattern without going through the one-point-at-a-time routine. When you read that you want all points that are 3 units from a circle, you can see that a bigger circle will do the trick. Figure 20-1 shows the given circle of radius 10 and the circle of radius 13 that you'd draw for your solution.

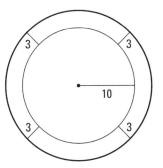

Figure 20-1: Points 3 units away from the original circle that form another circle.

2. **Look outside the pattern for points to add.**

    Do you see what step 1 leaves out? Right — it's a smaller circle with a radius of 7 inside the original circle (see Figure 20-2). I suppose my "missing" this second circle may seem a bit contrived, and granted, many people immediately see that the solution should include both circles. However, people often do focus on one particular pattern (the biggest circle in this problem) to the exclusion of everything else. Their minds sort of get in a rut, and they have trouble seeing anything other than the first pattern or idea that they latch onto. And that's why it's so important to explicitly go through this second step of the four-step method.

3. **Look inside the pattern for points to exclude.**

    All points of the 7-unit-radius and 13-unit-radius circles satisfy the given condition, so no points need to be excluded.

4. **Draw the locus and describe it in words.**

    Figure 20-2 shows the locus, and the caption gives its description.

Figure 20-2:
The locus of points 3 units from the given circle is two circles concentric with the original circle with radii of 7 and 13 units.

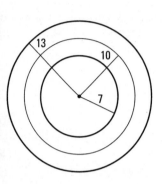

Like a walk in the park, right?

Problem two

What's the locus of all points equidistant from two given points?

   1. **Identify a pattern.**

      Figure 20-3 shows the two given points, *A* and *B,* along with four new points that are each equidistant from the given points.

Figure 20-3: Identifying points that work.

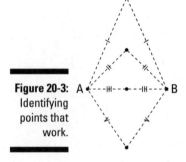

   Do you see the pattern? You got it — it's a vertical line that goes through the midpoint of the segment that connects the two given points. In other words, it's that segment's perpendicular bisector.

Chapter 20: Locating Loci and Constructing Constructions

2. Look outside the pattern.

This time you come up empty in step 2. Check any point *not* on the perpendicular bisector of $\overline{AB}$, and you see that it's *not* equidistant from A and B. Thus, you have no points to add.

3. Look inside the pattern.

Nothing noteworthy here, either. Every point on the perpendicular bisector of $\overline{AB}$ is, in fact, equidistant from A and B. (You may recall that this follows from the second equidistance theorem from Chapter 9.) Thus, no points should be excluded. (**Repeat Warning:** Don't allow yourself to get a bit lazy and skip steps 2 and 3!)

4. Draw the locus and describe it in words.

Figure 20-4 shows the locus, and the caption gives its description.

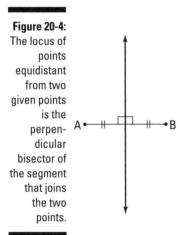

Figure 20-4: The locus of points equidistant from two given points is the perpendicular bisector of the segment that joins the two points.

Now suppose problem two had been worded like this instead: What's the locus of the vertices of isosceles triangles having a given segment for a base?

Look back at Figure 20-4. For the tweaked problem, $\overline{AB}$ is the base of the isosceles triangles. Because the vertex joining the congruent legs of an isosceles triangle is equidistant from the endpoints of its base (points A and B), the solution to this tweaked problem is identical to the solution to problem two — *except,* that is, when you get to steps 2 and 3.

In step 1 of the tweaked problem, you find the same perpendicular bisector pattern, so it's easy to fall into the trap of thinking that the perpendicular bisector is the final solution. But when you get to step 2, you should realize that you have to add the given points *A* and *B* to your solution because, of course, they're vertices of all the triangles.

And when you get to step 3, you should notice that you have to exclude a single point from the locus: The midpoint of $\overline{AB}$ can't be part of the solution because it's on the same line as *A* and *B*, and you can't use three collinear points for the three vertices of a triangle. The locus for this tweaked problem is, therefore, the perpendicular bisector of $\overline{AB}$, *plus* points *A* and *B*, *minus* the midpoint of $\overline{AB}$.

If a point needs to be excluded, there must be something *special* or *unusual* about it. When looking for points that may need to be excluded from a locus solution, check points in special locations such as

- The *given points*
- *Midpoints* and *endpoints* of segments
- *Points of tangency* on a circle

Note how this tip applies to the preceding problem: The point you had to exclude in step 3 is the midpoint of a segment (the second bullet in the list).

Although the points you had to add (the given points) are also listed in the tip, this situation is unusual. Most of the time, points that must be added in step 2 are not the sorts of special, isolated points in this list. Instead, they typically form their own pattern beyond the first pattern you spotted (you see this in problem one, where I "missed" the inner circle).

Problem three

Given points *P* and *R*, what's the locus of points *Q* such that ∠*PQR* is a right angle?

1. **Identify a pattern.**

    This pattern may be a bit tricky to find, but if you start with points *P* and *R* and try to find a few points *Q* that make a right angle with *P* and *R*, you'll probably begin to see a pattern emerging. See Figure 20-5.

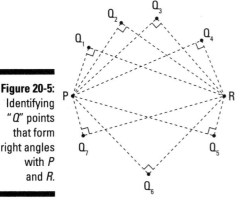

Figure 20-5: Identifying "Q" points that form right angles with P and R.

See the pattern? The Q points are beginning to form a circle with diameter $\overline{PR}$ (see Figure 20-6). This makes sense if you think about the inscribed-angle theorem from Chapter 15: In a circle with $\overline{PR}$ as its diameter, semicircular arc $\overset{\frown}{PR}$ would be 180°, so all inscribed angles PQR would be one-half of that, or 90°.

2. **Look outside the pattern.**

    Nope, nothing to add here. Any point Q inside the circle you identified in step 1 creates an *obtuse* angle with P and R, and any point Q outside the circle creates an *acute* angle with P and R. All the right angles are on the circle. (The location of the three types of angles follows from the angle-circle theorems from Chapter 15.)

3. **Look inside the pattern.**

    Bingo. See what points have to be excluded? It's the given points P and R. If Q is at the location of either given point, all you have left is a segment ($\overline{QR}$ or $\overline{PQ}$), so you no longer have the three distinct points you need to make an angle.

4. **Draw the locus and describe it in words.**

    Figure 20-6 shows the locus, and the caption gives its description. Note the hollow dots at P and R, which indicate that those points aren't part of the solution.

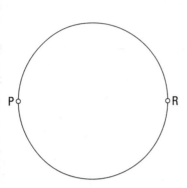

Figure 20-6:
Given points P and R, the locus of points Q such that ∠PQR is a right angle is a circle with diameter $\overline{PR}$, minus points P and R.

Three-dimensional locus problems

With 3-D locus problems, you have to determine the locus of all points in *3-D space* that satisfy the given conditions of the locus. In this short section, instead of doing 3-D problems from scratch, I just want to discuss how 3-D locus problems compare with 2-D problems.

TIP

You can use the four-step locus method to solve 3-D locus problems directly, but if this seems too difficult or if you get stuck, try solving the 2-D version of the problem first. The 2-D solution often points the way to the 3-D solution. Here's the connection:

- ✔ The 3-D solution can often (but not always) be obtained from the 2-D solution by *rotating* the 2-D solution about some line. (Often this line passes through some or all of the given points.)

- ✔ The solution to the 2-D version of a 3-D locus problem is always a *slice* of the solution to the 3-D problem (that's a slice in the sense that a circle is a slice of a sphere or, in other words, that a circle is the intersection of a plane and a sphere).

To get a handle on this 3-D tip, take a look at the 3-D versions of the 2-D locus problems from the previous section (I have a reason for giving them to you out of order).

The 3-D version of problem two

Look back at Figure 20-4, which shows the solution to problem two. Consider the same locus question, but make it a 3-D problem: What's the locus of points in *3-D space* equidistant from two given points?

The answer is a *plane* (instead of a line) that's the perpendicular bisector plane of the segment joining the two points. Note a couple of things about this solution:

- You can obtain the 3-D solution (the perpendicular bisector plane) by rotating the 2-D solution (the perpendicular bisector line) about $\overleftrightarrow{AB}$, the line that passes through the two given points.

- The 2-D solution is a slice of the 3-D solution. It might seem odd to call the 2-D solution a slice because it's only a line, but if you slice or cut the 3-D solution (a plane) with another plane, you get a line.

Now consider the tweaked version of problem two. Its solution, you may recall, is the same as the solution to problem two (a perpendicular bisector), but with two points added and a single point omitted. This 2-D solution can help you visualize the solution to the related 3-D problem. The solution to the 3-D version is the 2-D solution rotated about $\overleftrightarrow{AB}$, namely the perpendicular bisector *plane*, plus points A and B, minus the midpoint of $\overline{AB}$.

The 3-D version of problem three

The 3-D version of problem three (Given points P and R, what's the locus of points Q *in space* such that $\angle PQR$ is a right angle?) is another problem where you can obtain the 3-D solution from the 2-D solution by doing a rotation. Figure 20-6 shows and describes the 2-D solution: It's a circle minus the end points of diameter $\overline{PR}$. If you rotate this 2-D solution about $\overleftrightarrow{PR}$, you obtain the 3-D solution: a sphere with diameter $\overline{PR}$, minus points P and R.

The 3-D version of problem one

Figure 20-2 shows and describes the solution to problem one: two concentric circles. But unlike the other 3-D problems in this section, the solution to the 3-D version of this problem (What's the locus of all points *in space* that are 3 units from a given circle whose radius is 10 units?) cannot be obtained by rotating the 2-D solution. However, the 2-D solution can still help you visualize the 3-D solution because the 2-D solution is a slice of the 3-D solution. Can you picture the 3-D solution? It's a donut shape (a *torus* in mathspeak) that's 26 units wide and that has a 14-unit-wide "donut hole." Imagine slicing a donut in half in a bagel slicer. The flat face of either half donut would have a small circle where the donut hole was and a big circle around the outer edge, right? Look back at the two bold circles in Figure 20-2. They represent the two circles you'd see on the half donut. (Time for a break: How 'bout a cream-filled or a cruller?)

Drawing with the Bare Essentials: Constructions

Geometers since the ancient Greeks have enjoyed the challenge of seeing what geometric objects they could draw using only a compass and a straightedge. A *compass,* of course, is that thing with a sharp point and an attached pencil that you use to draw circles. A *straightedge* is just like a ruler but without marks on it. The whole idea behind these *constructions* is to draw geometric figures from scratch or to copy other figures using these two simplest-possible drawing tools and nothing else. (By the way, you can use a ruler instead of a straightedge when you're doing constructions, but just remember that you're not allowed to measure any lengths with it.)

In this section, I go through methods for doing nine basic constructions. After mastering these nine, you can use the methods on more-advanced problems.

Note: When I want you to draw an arc, I use a special notation. In parentheses, I first name the point where you should place the point of your compass (this is the center of the arc); then I indicate how wide you should open the compass (this is the radius of the arc). The radius can be given as the length of a specific segment or with a single letter. So, for instance, when I want you to draw an arc that has a center at M and a radius of length MN, I write "arc (M, MN)"; or when I'm talking about an arc with a center at T and a radius of r, I write "arc (T, r)."

Three copying methods

In this section, you discover the techniques for copying a segment, an angle, and a triangle.

Copying a segment

The key to copying a given segment is to open your compass to the length of the segment; then, using that amount of opening, you can mark off another segment of the same length.

>Given: $\overline{MN}$
>
>Construct: A segment $\overline{PQ}$ congruent to $\overline{MN}$

Here's the solution (see Figure 20-7):

1. **Using your straightedge, draw a working line, *l*, with a point *P* anywhere on it.**

2. **Put your compass point on point *M* and open it to the length of $\overline{MN}$.**

      The best way to make sure you've opened it to just the right amount is to draw a little arc that passes through *N*. In other words, draw arc (*M, MN*).

   3. **Being careful not to change the amount of the compass's opening from step 2, put the compass point on point *P* and construct arc (*P, MN*) intersecting line *l*.**

      You call this point of intersection point *Q*, and you're done.

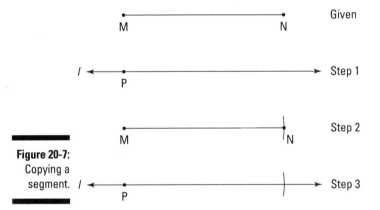

Figure 20-7: Copying a segment.

Copying an angle

The basic idea behind copying a given angle is to use your compass to sort of measure how wide the angle is open; then you create another angle with the same amount of opening.

   Given: ∠*A*

   Construct: An ∠*B* congruent to ∠*A*

Refer to Figure 20-8 as you go through these steps:

   1. **Draw a working line, *l*, with point *B* on it.**
   2. **Open your compass to any radius *r*, and construct arc (*A, r*) intersecting the two sides of ∠*A* at points *S* and *T*.**
   3. **Construct arc (*B, r*) intersecting line *l* at some point *V*.**
   4. **Construct arc (*S, ST*).**
   5. **Construct arc (*V, ST*) intersecting arc (*B, r*) at point *W*.**
   6. **Draw $\overrightarrow{BW}$ and you're done.**

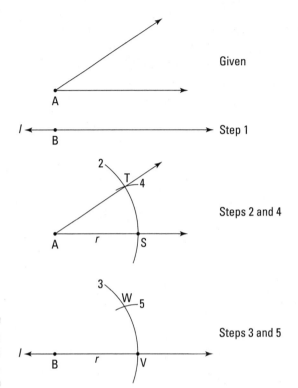

Figure 20-8: Copying an angle.

Copying a triangle

The idea here is to use your compass to "measure" the lengths of the three sides of the given triangle and then make another triangle with sides congruent to the sides of the original triangle. (The fact that this method works is related to the SSS method of proving triangles congruent; see Chapter 9.)

   Given: △DEF

   Construct: △JKL ≅ △DEF

As you work through these steps, refer to Figure 20-9:

1. Draw a working line, *l*, with a point *J* on it.
2. Use the earlier "Copying a segment" method to construct segment $\overline{JK}$ on line *l* that's congruent to $\overline{DE}$.
3. Construct
    a. Arc (D, DF)
    b. Arc (J, DF)

4. Construct

   a. Arc (*E, EF*)

   b. Arc (*K, EF*) intersecting arc (*J, DF*) at point *L*

5. Draw $\overline{JL}$ and $\overline{KL}$ and you're done.

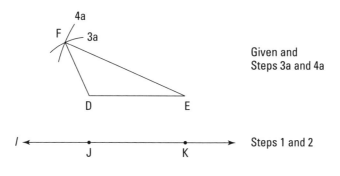

Figure 20-9: Copying a triangle.

Bisecting angles and segments

The next couple of constructions show you how to divide angles and segments exactly in half.

Bisecting an angle

To bisect an angle, you use your compass to locate a point that lies on the angle bisector; then you just use your straightedge to connect that point to the angle's vertex. Let's do it.

   Given: ∠*K*

   Construct: $\overrightarrow{KZ}$, the bisector of ∠*K*

Check out Figure 20-10 as you work through this construction:

1. **Open your compass to any radius *r*, and construct arc (*K*, *r*) intersecting the two sides of ∠*K* at *A* and *B*.**

2. **Use any radius *s* to construct arc (*A*, *s*) and arc (*B*, *s*) that intersect each other at point *Z*.**

   Note that you must choose a radius *s* that's long enough for the two arcs to intersect.

3. **Draw $\overrightarrow{KZ}$ and you're done.**

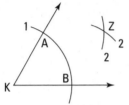

Figure 20-10: Bisecting an angle.

Constructing the perpendicular bisector of a segment

To construct a perpendicular bisector of a segment, you use your compass to locate two points that are each equidistant from the segment's endpoints and then finish with your straightedge. (The method of this construction is very closely related to the first equidistance theorem from Chapter 9.)

Given: $\overline{CD}$

Construct: $\overleftrightarrow{GH}$, the perpendicular bisector of $\overline{CD}$

Up for a challenge? Construct the trisectors of an angle

In the text, you see the relatively easy method for bisecting an angle — cutting an angle into two equal parts. Now, it might not seem that dividing an angle into *three* equal parts would be much harder. But in fact, it's not just difficult — it's *impossible*. For over 2,000 years, mathematicians tried to find a compass-and-straightedge method for trisecting an angle — to no avail. Then, in 1837, Pierre Wantzel, using the very esoteric mathematics of abstract algebra, proved that such a construction is mathematically impossible. Despite this airtight proof, quixotic (foolhardy?) amateur mathematicians continue trying, to this day, to discover a trisection method.

Chapter 20: Locating Loci and Constructing Constructions 339

Figure 20-11 illustrates this construction process:

1. **Open your compass to any radius *r* that's more than half the length of $\overline{CD}$, and construct arc (*C*, *r*).**

2. **Construct arc (*D*, *r*) intersecting arc (*C*, *r*) at points *G* and *H*.**

3. **Draw $\overleftrightarrow{GH}$.**

    You're done — $\overleftrightarrow{GH}$ is the perpendicular bisector of $\overline{CD}$.

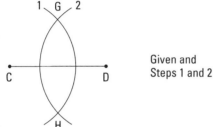

Figure 20-11: Constructing a perpendicular bisector.

Given and Steps 1 and 2

Two perpendicular line constructions

In this section, I give you — at no extra charge — two more methods for constructing perpendicular lines under different given conditions.

Constructing a line perpendicular to a given line through a point on the given line

This perpendicular line construction method is closely related to the method in the preceding section. And like the previous method, this method uses concepts from the first equidistance theorem. The only difference here is that this time, you don't care about bisecting a segment; you care only about drawing a perpendicular line through a point on the given line.

Given: $\overleftrightarrow{EF}$ and point *W* on $\overleftrightarrow{EF}$

Construct: $\overleftrightarrow{WZ}$ such that $\overleftrightarrow{WZ} \perp \overleftrightarrow{EF}$

As you work through this construction, take a look at Figure 20-12:

1. **Using any radius *r*, construct arc (*W*, *r*) that intersects $\overleftrightarrow{EF}$ at *X* and *Y*.**

2. **Using any radius *s* that's greater than *r*, construct arc (*X*, *s*) and arc (*Y*, *s*) intersecting each other at point *Z*.**

3. **Draw $\overleftrightarrow{WZ}$.**

   That's it; $\overleftrightarrow{WZ}$ is perpendicular to $\overleftrightarrow{EF}$ at point W.

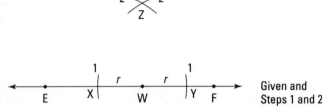

Figure 20-12: Constructing a perpendicular line through a point on a line.

Constructing a line perpendicular to a given line through a point not on the given line

For a challenge, read the following *given* and *construct* and then see whether you can do this construction before reading the solution.

> Given: $\overleftrightarrow{AZ}$ and point J not on $\overleftrightarrow{AZ}$
> Construct: $\overleftrightarrow{JM}$ such that $\overleftrightarrow{JM} \perp \overleftrightarrow{AZ}$

Figure 20-13 can help guide you through this construction:

1. **Open your compass to a radius r (r must be greater than the distance from J to $\overleftrightarrow{AZ}$), and construct arc (J, r) intersecting $\overleftrightarrow{AZ}$ at K and L.**

2. **Leaving your compass open to radius r (other radii would also work), construct arc (K, r) and arc (L, r) — on the side of $\overleftrightarrow{AZ}$ that's opposite point J — intersecting each other at point M.**

3. **Draw $\overleftrightarrow{JM}$, and that's a wrap.**

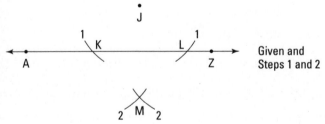

Figure 20-13: Constructing a perpendicular line through a point *not* on a line.

Constructing parallel lines and using them to divide segments

For the final two constructions, you find out how to construct a line parallel to a given line; then you use that technique to divide a segment into any number of equal parts.

Constructing a line parallel to a given line through a point not on the given line

This construction method is based on one of the lines-cut-by-a-transversal theorems from Chapter 10 *(if corresponding angles are congruent, then lines are parallel)*.

Given: $\overleftrightarrow{UW}$ and point X not on $\overleftrightarrow{UW}$
Construct: $\overleftrightarrow{XZ}$ such that $\overleftrightarrow{XZ} \parallel \overleftrightarrow{UW}$

As you try this construction, follow the steps shown in Figure 20-14:

1. **Through X, draw a line l that intersects $\overleftrightarrow{UW}$ at some point V.**

2. **Using the earlier "Copying an angle" method, construct $\angle YXZ \cong \angle XVW$.**

    I've labeled the four arcs you draw in order: 2a, 2b, 2c, and 2d.

3. **Draw $\overleftrightarrow{XZ}$, which is parallel to $\overleftrightarrow{UW}$.**

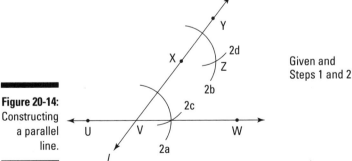

Figure 20-14: Constructing a parallel line.

Dividing a segment into any number of equal subdivisions

The following example shows you how to divide a segment into three equal parts, but the method works for dividing a segment into any number of equal parts. Because this construction technique involves drawing parallel lines, it's related to the same theorem referred to in the preceding construction: *if corresponding angles are congruent, then lines are parallel.* This technique also makes use of the Side-Splitter Theorem from Chapter 13.

Given: $\overline{GH}$

Construct: The two trisection points of $\overline{GH}$

Check out Figure 20-15 for this construction:

1. **Draw any line *l* through point G.**
2. **Open your compass to any radius *r*, and construct arc (G, r) intersecting line *l* at a point you'll call X.**
3. **Construct arc (X, r) intersecting *l* at a point Y.**
4. **Construct arc (Y, r) intersecting *l* at a point Z.**
5. **Draw $\overleftrightarrow{ZH}$.**
6. **Using the preceding parallel-line construction method, construct lines through Y and X parallel to $\overleftrightarrow{ZH}$.**

These two lines will intersect $\overline{GH}$ at its trisection points. That does it for this problem.

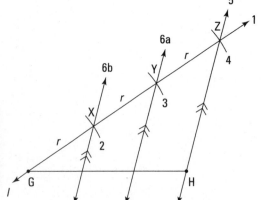

Figure 20-15: Dividing a segment into equal parts.

And as for this book (except for a couple of minor chapters, a glossary, and an appendix), a-thaa-a-thaa-a-thaa-a-thaa-a-that's all, folks!

Part VIII
The Part of Tens

In this part . . .

Chapter 21 should make you sit up and pay attention because this is where you get my top ten things to use as reasons in two-column proofs. This is a good checklist to keep handy when you're working on proofs (in addition to the proof strategies listed on the Cheat Sheet). If you get stuck while doing a proof, do not under any circumstances give up until you've at least run through these checklists. In Chapter 22, you take a break from the actual study of geometry principles, formulas, proofs, and the like as I give you a tour of ten wonders of the geometric world including some of the most famous geometry problems ever solved.

Chapter 21
Ten Things to Use as Reasons in Geometry Proofs

In This Chapter
- Segment and angle postulates and theorems
- Parallel-line theorems
- A circle theorem
- Triangle definitions, postulates, and theorems

Here's the top ten list of definitions, postulates, and theorems that you should absotively, posilutely know how to use in the reason column of geometry proofs. They'll help you tackle any proof you might run across. Whether a particular reason is a definition, postulate, or theorem doesn't matter much because you use them all the same way.

The Reflexive Property

The *Reflexive Property* says that any segment or angle is congruent to itself. You often use the Reflexive Property, which I introduce in Chapter 9, when you're trying to prove triangles congruent or similar. Be careful to notice all shared segments and shared angles in proof diagrams. Shared segments are usually pretty easy to spot, but people sometimes fail to notice shared angles like the one shown in Figure 21-1.

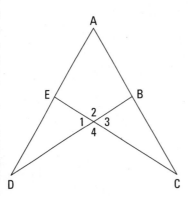

Figure 21-1: Angle *A* is one of the vertex angles of both △ACE and △ADB. Angles 1 and 3 are vertical angles, as are angles 2 and 4.

Vertical Angles Are Congruent

I cover the vertical-angles-are-congruent theorem in Chapter 5. This theorem isn't hard to use, as long as you spot the vertical angles. Remember — everywhere you see two lines that come together to make an X, you have *two* pairs of congruent vertical angles (the ones on the top and bottom of the X, like angles 2 and 4 in Figure 21-1, and the ones on the left and right sides of the X, like angles 1 and 3).

The Parallel-Line Theorems

There are ten parallel-line theorems that involve a pair of parallel lines and a transversal (which intersects the parallel lines). See Figure 21-2. Five of the theorems use parallel lines to show that angles are congruent or supplementary; the other five use congruent or supplementary angles to show that lines are parallel. Here's the first set of theorems:

If lines are parallel, then . . .

- Alternate interior angles, like ∠4 and ∠5, are congruent.
- Alternate exterior angles, like ∠1 and ∠8, are congruent.
- Corresponding angles, like ∠3 and ∠7, are congruent.
- Same-side interior angles, like ∠4 and ∠6, are supplementary.
- Same-side exterior angles, like ∠1 and ∠7, are supplementary.

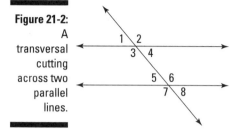

Figure 21-2:
A transversal cutting across two parallel lines.

And here are the ways to prove lines parallel:

- If alternate interior angles are congruent, then lines are parallel.
- If alternate exterior angles are congruent, then lines are parallel.
- If corresponding angles are congruent, then lines are parallel.
- If same-side interior angles are supplementary, then lines are parallel.
- If same-side exterior angles are supplementary, then lines are parallel.

The second five theorems are the reverse of the first five. I discuss parallel lines and transversals more fully in Chapter 10.

Two Points Determine a Line

Not much to be said here — whenever you have two points, you can draw a line through them. Two points *determine* a line because only one particular line can go through both points. You use this postulate in proofs whenever you need to draw an auxiliary line on the diagram (see Chapter 10).

All Radii of a Circle Are Congruent

Whenever you have a circle in your proof diagram, you should think about the all-radii-are-congruent theorem (and then mark all radii congruent) before doing anything else. I bet that just about every circle proof you see will use congruent radii somewhere in the solution. (And you'll often have to use the theorem in the preceding section to draw in more radii.) I discuss this theorem in Chapter 14.

If Sides, Then Angles

Isosceles triangles have two congruent sides and two congruent base angles. The if-sides-then-angles theorem says that if two sides of a triangle are congruent, then the angles opposite those sides are congruent (see Figure 21-3). Do not fail to spot this! When you have a proof diagram with triangles in it, always check to see whether any triangle looks like it has two congruent sides. For more information, flip to Chapter 9.

Figure 21-3: Going from congruent sides to congruent angles.

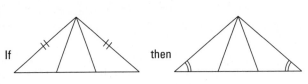

If Angles, Then Sides

The if-angles-then-sides theorem says that if two angles of a triangle are congruent, then the sides opposite those angles are congruent (see Figure 21-4). Yes, this theorem is the converse of the if-sides-then-angles theorem, so you may be wondering why I didn't put this theorem in the preceding section. Well, these two isosceles triangle theorems are so important that each deserves its own section (besides, they already have to share space in Chapter 9).

Figure 21-4: Going from congruent angles to congruent sides.

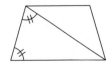

The Triangle Congruence Postulates and Theorems

Here are the five ways to prove triangles congruent (see Chapter 9 for details):

- **SSS (side-side-side):** If the three sides of one triangle are congruent to the three sides of another triangle, then the triangles are congruent.
- **SAS (side-angle-side):** If two sides and the included angle of one triangle are congruent to two sides and the included angle of another triangle, then the triangles are congruent.
- **ASA (angle-side-angle):** If two angles and the included side of one triangle are congruent to two angles and the included side of another triangle, then the triangles are congruent.
- **AAS (angle-angle-side):** If two angles and a non-included side of one triangle are congruent to two angles and a non-included side of another triangle, then the triangles are congruent.
- **HLR (hypotenuse-leg-right angle):** If the hypotenuse and a leg of one right triangle are congruent to the hypotenuse and a leg of another right triangle, then the triangles are congruent.

CPCTC

CPCTC stands for *corresponding parts of congruent triangles are congruent*. It has the feel of a theorem, but it's really just the definition of congruent triangles. When doing a proof, after proving triangles congruent, you use CPCTC on the next line to show that some parts of those triangles are congruent. CPCTC makes its debut in Chapter 9.

The Triangle Similarity Postulates and Theorems

Here are the three ways to prove triangles similar — that is, to show they have the same shape (Chapter 13 can fill you in on the details):

- **AA (angle-angle):** If two angles of one triangle are congruent to two angles of another triangle, then the triangles are similar.
- **SSS~ (side-side-side similar):** If the ratios of the three pairs of corresponding sides of two triangles are equal, then the triangles are similar.
- **SAS~ (side-angle-side similar):** If the ratios of two pairs of corresponding sides of two triangles are equal and the included angles are congruent, then the triangles are similar.

Chapter 22
Ten Cool Geometry Problems

In This Chapter
- Getting in touch with Greek mathematics
- Looking at wonders of the ancient and modern worlds
- Appreciating Earthly calculations
- Checking out other real-world applications

This chapter is sort of a geometry version of *Ripley's Believe It or Not.* I give you ten geometry problems involving some famous and not-so-famous historical figures (Archimedes, Tsu Chung-Chin, Christopher Columbus, Eratosthenes, Galileo Galilei, Buckminster Fuller, and Walter Bauersfeld), some everyday objects (soccer balls, crowns, and bathtubs), some great architectural achievements (the Golden Gate Bridge, the Parthenon, the geodesic dome, and the Great Pyramid), some science problems (figuring out the circumference of the Earth and the motion of a projectile), some geometric objects (parabolas, catenary curves, and truncated icosahedrons), and, lastly, the most famous number in mathematics, pi. So here you go — ten wonders of the geometric world.

Eureka! Archimedes's Bathtub Revelation

Archimedes (Syracuse, Sicily; 287–212 B.C.) is widely recognized as one of the four or five greatest mathematicians of all time (Carl Friedrich Gauss and Isaac Newton are some other all-stars). He made important discoveries in mathematics, physics, engineering, military tactics, and . . . headwear?

The king of Syracuse, a colony of ancient Greece, was worried that a goldsmith had cheated him. The king had given the goldsmith some gold to make a crown, but he thought that the goldsmith had kept some of the gold for himself, replaced it with less-expensive silver, and made the crown out of the mixture. The king couldn't prove it, though — at least not until Archimedes came along.

Archimedes talked to the king about the problem, but he was stumped until he sat in a bathtub one day. As he sat down, the water overflowed out of the tub. "Eureka!" shouted Archimedes (which means "I've found it!" in Greek, in case you were wondering). At that instant, he realized that the volume of water he displaced was equal to the volume of his body, and that gave him the key to solving the problem. He got so excited that he leapt from the bathtub and ran out into the street half naked.

What Archimedes figured out was that if the king's crown were pure gold, it would displace the same volume of water as a lump of pure gold with the same weight as the crown. But when Archimedes and the king tested the crown, it displaced more water than the lump of gold. This meant that the crown was made of more material than the lump of gold, and it was therefore less dense. The goldsmith had cheated by mixing in some silver, a metal lighter than gold. Case solved. Archimedes was rewarded handsomely, and the goldsmith lost his head — kerplunk!

Determining Pi

Pi (π) — the ratio of a circle's circumference to its diameter — begins with 3.14159265... and goes on forever from there. (There's a story about courting mathematicians who would go for long walks and recite hundreds of digits of pi to each other, but I wouldn't recommend this approach unless you're in love with a math buff.)

Archimedes, of bathtub fame (see the preceding section), was the first one to make a mathematical estimate of π. His method was to use two regular 96-sided polygons: one inscribed inside a circle (which was, of course, slightly smaller than the circle) and the other circumscribed around the circle (which was slightly larger than the circle). The measure of the circle's circumference was thus somewhere between the perimeters of the small and large 96-gons. With this technique, Archimedes managed to figure out that π was between 3.140 and 3.142. Not too shabby.

Although Archimedes's calculation was pretty accurate, the Chinese overtook him not too long after. By the fifth century A.D., Tsu Chung-Chin discovered a much more accurate approximation of π: the fraction $\frac{355}{113}$, which equals about 3.1415929. This approximation is within 0.00001 percent of π!

The Golden Ratio

Here's another famous geometry problem with a connection to ancient Greece. (When it came to mathematics, physics, astronomy, philosophy, drama, and the like, those ancient Greeks sure did kick some serious butt.) The Greeks used a number called the *golden ratio,* or *phi* (ϕ), which equals $\frac{\sqrt{5}+1}{2}$ or approximately 1.618, in many of their architectural designs. The Parthenon on the Acropolis in Athens is an example. The ratio of its width to its height is ϕ : 1. See Figure 22-1.

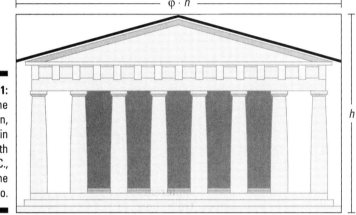

Figure 22-1: The Parthenon, built in the fifth century B.C., features the golden ratio.

The *golden rectangle* is a rectangle with sides in the ratio of ϕ : 1. This rectangle is special because when you divide it into a square and a rectangle, the new, smaller rectangle also has sides in the ratio of ϕ : 1, so it's *similar* to the original rectangle (which means that they're the same shape; see Chapter 14). Then you can divide the smaller rectangle into a square and a rectangle, and then you can divide the next rectangle, and so on. See Figure 22-2. When you connect the corresponding corners of each similar rectangle, you get a spiral that happens to be the same shape as the spiraling shell of the nautilus — amazing!

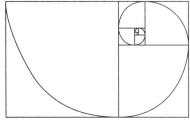 

Figure 22-2: The spirals of the golden rectangle and the nautilus.

The Circumference of the Earth

Contrary to popular belief, Christopher Columbus didn't discover that the Earth is round. Eratosthenes (276–194 B.C.) made that discovery about 1,700 years before Columbus. Eratosthenes was the head librarian in Alexandria, Egypt, the center of learning in the ancient world. He estimated the circumference of the Earth with the following method: He knew that on the summer solstice, the longest day of the year, the angle of the sun above Syene, Egypt, would be 0°, in other words, the sun would be directly overhead. So on the summer solstice, he measured the angle of the sun above Alexandria by measuring the shadow cast by a pole and got a 7.2° angle. Figure 22-3 shows how it worked.

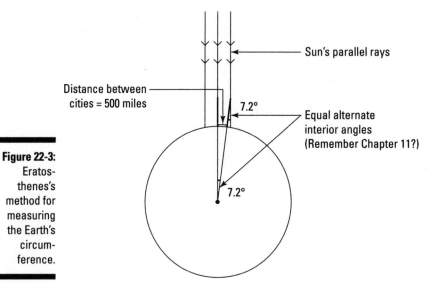

Figure 22-3: Eratosthenes's method for measuring the Earth's circumference.

Eratosthenes divided 360° by 7.2° and got 50, which told him that the distance between Alexandria and Syene (500 miles) was $\frac{1}{50}$ of the total distance around the Earth. So he multiplied 500 by 50 to arrive at his estimate of the Earth's circumference: 25,000 miles. This estimate was only 100 miles off the actual circumference of 24,900 miles.

The Great Pyramid of Khufu

Just 150 miles from Alexandria is the Pyramid of Khufu in Giza, Egypt. Also known as the Great Pyramid, it's the largest pyramid in the world. But how big is it really? Well, the sides of the pyramid's square base are each 745 feet long, and the height of the pyramid is 449 feet. To use the pointy-top volume formula, Vol = $\frac{1}{3}$ bh (see Chapter 18), you first need the area of the pyramid's base: 745 · 745, or 555,025 ft². The volume of the pyramid, then, is $\frac{1}{3}$(555,025)(449), or about 83,000,000 ft³. That's about 6.5 million tons of rock, and the pyramid used to be even bigger before the elements eroded some of it away.

Distance to the Horizon

Here's still more evidence that Columbus didn't discover that the Earth is round. Although many of the people who lived inland in the fifteenth century may have thought that the Earth was flat, no sensible person living on the coast could possibly have held this opinion. Why? Because people on the coast could see ships gradually drop below the horizon as the ships sailed away.

You can use a very simple formula to figure out how far the horizon is from you (in miles): Distance to horizon = $\sqrt{1.5 \cdot height}$, where *height* is your height (in feet) plus the height of whatever you happen to be standing on (a ladder, a mountain, anything). If you're standing on the shore, then you can also estimate the distance to the horizon by simply dividing your height in half. So if you're 5'6" (5.5 feet) tall, the distance to the horizon is only about 2.75 miles!

The Earth curves faster than most people think. On a small lake — say, 2.5 miles across — there's a 1-foot-tall bulge in the middle of the lake due to the curvature of the Earth. On some larger bodies of water, if conditions are right, you can actually perceive the curvature of the Earth when this sort of bulge blocks your view of the opposing shore.

Projectile Motion

Projectile motion is the motion of a "thrown" object (baseball, bullet, or whatever) as it travels upward and outward and then is pulled back down by gravity. The study of projectile motion has been important throughout history, but it really got going in the Middle Ages, once people developed cannons, catapults, and similar war machines. Soldiers needed to know how to point their cannons so their cannonballs would hit their intended targets.

Galileo Galilei (A.D. 1564–1642), who's famous for demonstrating that the Earth revolves around the sun, was the first to unravel the riddle of projectile motion. He discovered that projectiles move in a parabolic path (like the parabola $y = -\frac{1}{4}x^2 + x$, for example). Figure 22-4 shows how a cannonball (if aimed at a certain angle and fired at a certain velocity) would travel along this parabola.

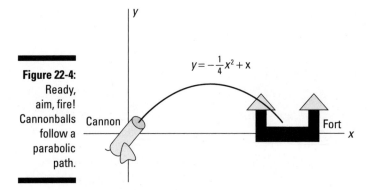

Figure 22-4: Ready, aim, fire! Cannonballs follow a parabolic path.

Without air resistance, a projectile fired at a 45° angle (exactly half of a right angle) will travel the farthest. When you factor in air resistance, however, maximum distance is achieved with a shallower firing angle of 30° to 40°, depending on several technical factors.

Golden Gate Bridge

The Golden Gate Bridge was the largest suspension bridge in the world for nearly 30 years after it was finished in 1937. Today it's number eight, but it's still an internationally recognized symbol of San Francisco.

The first step in building a suspension bridge is to hang very strong cables between a series of towers. When these cables are first hung, they hang in the shape of a *catenary curve;* this is the same kind of curve you get if you take a piece of string by its ends and hold it up. To finish the bridge, though, the hanging cables obviously have to be attached to the road part of the bridge. Well, when evenly-spaced vertical cables are used to attach the road to the main, curving cables, the shape of the main cables changes from a catenary curve to the slightly pointier parabola (see Figure 22-5). The extra weight of the road changes the shape. Pretty cool, eh?

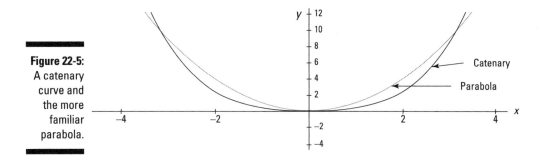

Figure 22-5: A catenary curve and the more familiar parabola.

The Geodesic Dome

A *geodesic dome* looks a lot like a sphere, but it's actually formed from a very large number of triangular faces that are arranged in a spherical pattern. Geodesic domes are extremely strong structures because the interlocking triangle pattern distributes force evenly across the surface — actually, they're the sturdiest type of structure in the world. Geodesic dome principles have also been used to create *buckyballs,* which are tiny, microscopic structures made out of carbon atoms that are extremely strong (some of them are harder than diamonds).

If you've ever heard of the geodesic dome, you've probably heard of Buckminster Fuller. Fuller patented the geodesic dome in the U.S. and went on to build many high-profile buildings based on the concept. However, although he seems to have come up with the idea on his own, Fuller wasn't actually the first one to build a geodesic dome; an engineer named Walter Bauersfeld had already come up with the idea and built a dome in Germany.

A Soccer Ball

"Hey, you want to go outside and kick around the truncated icosahedron?" That's geekspeak for a soccer ball. Seriously, though, a soccer ball is a fascinating geometric shape. It begins with an *icosahedron* — that's a regular polyhedron with 20 equilateral-triangle faces. Take a look at Figure 22-6.

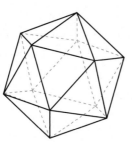

Figure 22-6: An icosahedron: Cut off all the points, and you get a soccer ball.

On the surface of an icosahedron, each vertex (the pointy tips that stick out) has a group of five triangles around it. To get a truncated icosahedron, you cut the pointy tips off, and you then get a regular pentagon where each tip was. Each of the equilateral triangles, meanwhile, becomes a regular hexagon, because when you cut off the three corners of a triangle, the triangle gets three new sides.

If you don't believe me, go get a soccer ball and count up the pentagons and hexagons. You should count 12 regular pentagons and 20 regular hexagons. Play ball!

Part IX
Appendixes

In this part . . .

Here, I organize the most important geometry formulas and definitions — just for you. If you forget a formula that you need to do a problem, or if you're skipping around in the book and you don't have a definition you need, then check here first.

Appendix A
Formulas and Other Important Stuff You Should Know

This appendix contains over three dozen of this book's most important geometry formulas, theorems, properties, and so on that you use for calculations. If you get stumped while working on a problem and can't come up with a formula, this is the place to look.

Triangle Stuff

- **Sum of the interior angles of a triangle:** $180°$
- **Area:** $\text{Area}_\triangle = \frac{1}{2} \text{base} \cdot \text{height}$
- **Hero's area formula:** $\text{Area}_\triangle = \sqrt{S(S-a)(S-b)(S-c)}$, where a, b, and c are the lengths of the triangle's sides and $S = \frac{a+b+c}{2}$ (S is the semiperimeter, half the perimeter)
- **Area of an equilateral triangle:** $\text{Area}_{\text{Equilateral}\triangle} = \frac{s^2\sqrt{3}}{4}$, where s is a side of the triangle
- **The Pythagorean Theorem:** $a^2 + b^2 = c^2$, where a and b are the legs of a right triangle and c is the hypotenuse
- **Common Pythagorean triples (side lengths in right triangles):**
  - 3-4-5
  - 5-12-13
  - 7-24-25
  - 8-15-17

- **Ratios of the sides in special right triangles:**
    - The sides opposite the angles in a 45°-45°-90° triangle are in the ratio of $1 : 1 : \sqrt{2}$.
    - The sides opposite the angles in a 30°-60°-90° triangle are in the ratio of $1 : \sqrt{3} : 2$.
- **Altitude-on-Hypotenuse Theorem:** If an altitude is drawn to the hypotenuse of a right triangle as shown in the following figure, then
    - The two triangles formed are similar to the given triangle and to each other:

    $\triangle ACB \sim \triangle ADC \sim \triangle CDB$
    - $h^2 = xy$
    - $a^2 = yc$ and $b^2 = xc$

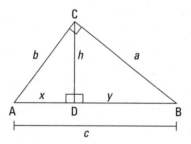

Polygon Stuff

- **Area formulas:**
    - **Parallelogram:** Area = base · height
    - **Rectangle:** Area = base · height
    - **Kite or rhombus:** Area = $\frac{1}{2}$ diagonal$_1$ · diagonal$_2$
    - **Square:** Area = side^2 or $\frac{1}{2} d_1 d_2$
    - **Trapezoid:** Area = $\frac{\text{base}_1 + \text{base}_2}{2}$ · height
    - **Regular polygon:** Area = $\frac{1}{2}$ perimeter · apothem

Appendix A: Formulas and Other Important Stuff You Should Know 363

- **Sum of the interior angles in an *n*-sided polygon:**
  $$\text{Sum}_{\text{Interior Angles}} = (n-2)180°$$

- **Measure of each interior angle of a regular (or other equiangular) *n*-sided polygon:**
  $$\text{Measure}_{\text{Interior Angle}} = \frac{(n-2)180°}{n} \text{ or } 180 - \frac{360°}{n} \text{ (the supplement of an exterior angle)}$$

- **Sum of the exterior angles (one at each vertex) of any polygon:**
  $$\text{Sum}_{\text{Exterior Angles}} = 360°$$

- **Measure of each exterior angle of a regular (or other equiangular) *n*-sided polygon:**
  $$\text{Measure}_{\text{Exterior Angle}} = \frac{360°}{n}$$

- **Number of diagonals that can be drawn in an *n*-sided polygon:**
  $$\text{Number of diagonals} = \frac{n(n-3)}{2}$$

Circle Stuff

- **Circumference:** $C = 2\pi r$ or πd, where r is the radius of the circle and d is its diameter

- **Area:** $\text{Area}_{\text{Circle}} = \pi r^2$

- **Arc length:** The length of an arc (part of the circumference) is equal to the circumference of the circle ($2\pi r$) times the fraction of the circle represented by the arc.
  $$\text{Arc length} = \left(\frac{\text{degree measure of arc}}{360°}\right) \cdot 2\pi r$$

- **Sector area:** The area of a sector (a pizza-slice shape cut out of a circle) is equal to the area of the circle (πr^2) times the fraction of the circle represented by the sector.
  $$\text{Area}_{\text{Sector}} = \left(\frac{\text{degree measure of sector's arc}}{360°}\right) \cdot \pi r^2$$

- **Measure of an angle . . .**
  - **On a circle:** $\text{Measure}_{\angle \text{ On a Circle}} = \frac{1}{2}(\text{measure of arc})$

    In the following figure, $\angle A = \frac{1}{2}\left(m\widehat{PQ}\right)$

  - **Inside a circle:** $\text{Measure}_{\angle \text{ Inside a Circle}} = \frac{1}{2}(\text{measure of arc} + \text{measure of vertical angle's arc})$

    In the figure, $\angle B = \frac{1}{2}\left(m\widehat{TU} + m\widehat{RS}\right)$

- **Outside a circle:** $\text{Measure}_{\angle \text{Outside a Circle}} = \frac{1}{2}(\text{measure of big arc} - \text{measure of small arc})$

  In the figure, $\angle C = \frac{1}{2}\left(m\widehat{YZ} - m\widehat{WX}\right)$

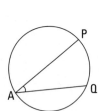 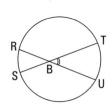 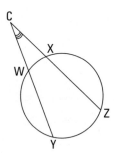

Angle *on* a circle   Angle *inside* a circle   Angle *outside* a circle

- ✓ **Chord-Chord Power Theorem:** When two chords of a circle intersect, the product of the parts of one chord is equal to the product of the parts of the other chord.

  In the figure, for instance, $5 \cdot 4 = 2 \cdot 10$.

- ✓ **Tangent-Secant Power Theorem:** When a tangent and a secant of a circle meet at an external point, the measure of the tangent squared is equal to the product of the secant's external part and its total length.

  The figure shows that $8^2 = 4(4 + 12)$.

- ✓ **Secant-Secant Power Theorem:** When two secants of a circle meet at an external point, the product of one secant's external part and its total length is equal to the product of the other secant's external part and its total length.

  In the figure, $4(4 + 2) = 3(3 + 5)$.

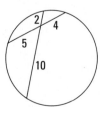 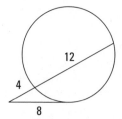 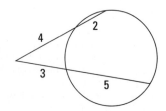

Chord-Chord
Power Theorem:
$5 \cdot 4 = 2 \cdot 10$

Tangent-Secant
Power Theorem:
$8^2 = 4 \cdot 16$

Secant-Secant
Power Theorem:
$4 \cdot 6 = 3 \cdot 8$

3-D Geometry Stuff

- **Flat-top objects (prisms and cylinders):**
  - **Volume:** $\text{Vol}_{\text{Flat-top}} = \text{area}_{\text{base}} \cdot \text{height}$
  - **Surface area:** $\text{SA}_{\text{Flat-top}} = 2 \cdot \text{area}_{\text{base}} + \text{area}_{\text{lateral rectangles}}$

  For a cylinder, the single lateral rectangle that wraps around the cylinder has a length equal to the circumference of the cylinder's base; its width is the cylinder's height.

- **Pointy-top objects (pyramids and cones):**
  - **Volume:** $\text{Vol}_{\text{Pointy-top}} = \frac{1}{3} \text{area}_{\text{base}} \cdot \text{height}$
  - **Surface area:** $\text{SA}_{\text{Pointy-top}} = \text{area}_{\text{base}} + \text{area}_{\text{lateral triangles}}$

  For a pyramid, the base of each lateral triangle is a side of the pyramid's base, and the height of the triangle is the slant height of the pyramid. For a cone, one "triangle" wraps around the cone; its base is equal to the circumference of the cone's base, and its height is equal to the slant height of the cone. (See Chapter 17 for more on slant height.)

- **Sphere:**
  - **Volume:** $\text{Vol}_{\text{Sphere}} = \frac{4}{3} \pi r^3$
  - **Surface area:** $\text{SA}_{\text{Sphere}} = 4\pi r^2$

Coordinate Geometry Stuff

- **Slope formula:** Given two points (x_1, y_1) and (x_2, y_2), the slope of the line that goes through the points is

  $$m = \frac{y_2 - y_1}{x_2 - x_1} = \frac{\text{rise}}{\text{run}}$$

  It doesn't matter which point is designated as (x_1, y_1) and which is designated as (x_2, y_2).

  - The slopes of parallel lines are equal.
  - The slopes of perpendicular lines are opposite reciprocals of each other.

- **Midpoint formula:** Given a segment with endpoints (x_1, y_1) and (x_2, y_2), the coordinates of its midpoint are

  $$(x, y) = \left(\frac{x_1 + x_2}{2}, \frac{y_1 + y_2}{2} \right)$$

  It doesn't matter which point is (x_1, y_1) and which is (x_2, y_2).

- **Distance formula:** Given two points (x_1, y_1) and (x_2, y_2), the distance between the points is

  $$d = \sqrt{(x_2 - x_1)^2 + (y_2 - y_1)^2}$$

  It doesn't matter which point is (x_1, y_1) and which is (x_2, y_2).

- **Equations of a line:**
  - **Slope-intercept form:** $y = mx + b$, where m is the slope and b is the y-intercept
  - **Point-slope form:** $y - y_1 = m(x - x_1)$, where m is the slope and (x_1, y_1) is a point on the line
  - **Horizontal line:** $y = b$, where b is the y-intercept
  - **Vertical line:** $x = a$, where a is the x-intercept

- **Equation of a circle:** $(x - h)^2 + (y - k)^2 = r^2$, where (h, k) is the center of the circle and r is its radius

Appendix B
Glossary

acute angle: An angle that measures less than 90°

acute triangle: A triangle with three acute angles

adjacent angles: Angles that share a vertex (corner) and that have a side in common

adjacent sides: Sides of a polygon that share a vertex

alternate exterior angles: When a transversal cuts across two parallel lines, angles on the outside of the parallel lines and on opposite sides of the transversal are called alternate exterior angles; alternate exterior angles are congruent

alternate interior angles: When a transversal cuts across two parallel lines, angles between the parallel lines and on opposite sides of the transversal are called alternate interior angles; alternate interior angles are congruent

altitude: A segment drawn from a vertex of a triangle to its opposite side such that it forms right angles with the opposite side; commonly called the *height* of a triangle

angle: A figure formed by two rays that have a common endpoint

angle bisector: A ray that cuts an angle in half

apothem: A line segment that connects the center of a regular polygon to the midpoint of one of the polygon's sides; it's the perpendicular bisector of the side

arc: A portion of a circle's edge, or circumference

area: The number of square units of space within the boundary of a closed region, such as a polygon

base (of an isosceles triangle): In a non-equilateral isosceles triangle, the side that isn't congruent to either of the other sides

base (of a 3-D figure): In a pyramid or cone (the pointy-top solids), the face opposite the pointy top; in a prism or cylinder (the flat-top solids), both the bottom and the top are called bases

base angle (of an isosceles triangle): Either of the angles formed by the base and a leg; base angles are congruent

base angles (of an isosceles trapezoid): Either of the two pairs of angles that share a base; base angles are congruent

bases (of a trapezoid): Both of the trapezoid's parallel sides

bisect: To divide a segment or an angle into two congruent (equal) parts

central angle: An angle whose vertex is the center of a circle

centroid (of a triangle): The point where the three medians of a triangle meet

chord: Any segment that joins two points on a circle

circle: A set of points in a plane that are all equidistant from a single point (the circle's center)

circumcenter: The point where a triangle's three perpendicular bisectors meet

circumference: The distance around a circle (or the circle's "perimeter")

circumscribed: Drawn around; a polygon is circumscribed about a circle if the polygon is drawn outside the circle such that each of the polygon's sides is tangent to (just touches) the circle; a circle is circumscribed about a polygon if it's drawn outside of the polygon such that each of the polygon's vertices lies on the circle

The triangle is circumscribed about the circle.

The circle is circumscribed about the square.

collinear points: Points that lie on the same line

common tangent: A line that's tangent to two circles

complementary angles: Two angles whose measures add up to 90°

concentric circles: Circles that share the same center but aren't congruent

cone: A solid 3-D figure that has a circle for its base and comes to a single point at its tip

congruent: Identical in shape and size (sort of a fancy way of saying *equal*)

congruent angles: Angles that have the same measure

congruent circles: Circles that have congruent radii

congruent segments: Line segments that have the same length

construction: A drawing made using a compass and straightedge

coordinate (Cartesian) plane: A plane in which the location of every point can be given by two values (usually *x* and *y*)

coplanar: Lying in the same plane (said of points or shapes)

corresponding angles: When a transversal cuts across two parallel lines, non-adjacent angles on the same side of the transversal — one of which is between the parallel lines and one of which is outside the parallel lines — are called corresponding angles; corresponding angles are congruent

cylinder: A 3-D figure that has two congruent circular bases that are parallel

degree: A common unit of measurement for an angle; there are 360° in a circle

diagonal: A segment that goes from one vertex (corner) of a polygon to a non-adjacent vertex

diameter: A chord that passes through a circle's center; it is two radii in length

edge: A segment where two faces of a 3-D shape meet

equiangular: A polygon is equiangular if all of its angles are congruent

equilateral: A polygon is equilateral if all of its sides are congruent

equilateral triangle: A triangle with three congruent sides and three congruent angles

exterior angle: An angle that's adjacent to and supplementary to an interior angle of a polygon

face: A flat side of a 3-D shape; faces are polygons

foot: The point where a line intersects a plane

glide reflection (walk): An isometry composed of a translation (the "glide" part) and a reflection; glide reflections reverse orientation

height (of a 3-D shape): In a pyramid or cone, the length of the segment from the pointy top straight down to the base (it's perpendicular to the base); in a prism or cylinder, the perpendicular distance between the two parallel bases

hypotenuse: In a right triangle, the side opposite the right angle; it's always the longest side of the triangle

image: When a pre-image object (the "before") goes through a transformation, the result (the "after") is called the image

incenter: The point where a triangle's three angle bisectors meet

inscribed: Drawn inside; a polygon is inscribed in a circle if it's drawn inside the circle such that each of its vertices lies on the circle; a circle is inscribed in a polygon if it's drawn inside the polygon such that each of the polygon's sides is tangent to (just touches) the circle

interior angle: An angle inside a polygon between two of its sides

intersection (of two geometric objects): Where the two objects touch or overlap

isometry: A transformation that doesn't change the size or shape of the original object

isosceles trapezoid: A trapezoid in which the nonparallel sides (the legs) are congruent

isosceles triangle: A triangle with at least two congruent sides (and two congruent angles)

kite: A quadrilateral in which two disjoint pairs of consecutive sides are congruent

legs (of an isosceles triangle): In a non-equilateral isosceles triangle, the two congruent sides

legs (of a trapezoid): The nonparallel sides of a trapezoid

line: A straight, infinitely thin "line" that continues forever in both directions (technically an undefined term)

line segment: A portion of a line that has a finite length; the two ends are called endpoints

locus: A set of points (which might form a line or curve or other object) that satisfy certain given conditions; the plural is *loci*

median (of a trapezoid): The segment joining the midpoints of a trapezoid's nonparallel sides

median (of a triangle): A segment that connects a vertex of a triangle with the midpoint of the opposite side

midpoint: A point that divides a segment into two congruent segments

minute: In angle measurement, $\frac{1}{60}$ of a degree

non-collinear points: Points that don't lie on the same line

non-coplanar: Not lying in the same plane

obtuse angle: An angle that measures more than 90° and less than 180°

obtuse triangle: A triangle that has one obtuse angle and two acute angles

orientation: In transformations, the relationship between an image and its pre-image in terms of the order of their vertices. Two figures with the same orientation can be moved so that they stack perfectly on top of each other without either one being flipped over; figures with opposite orientations can't stack unless one of them is flipped. When two figures have opposite orientations, the vertices of one figure read in the clockwise direction correspond to the vertices of the other figure read in the counterclockwise direction. Reflections reverse orientation.

orthocenter: The point where a triangle's three altitudes meet

parallel lines: Coplanar lines that don't cross, no matter how far they're extended

parallelogram: A four-sided figure that has two pairs of parallel sides

pentagon: A five-sided polygon

perimeter: The sum of the lengths of a polygon's sides

perpendicular: Forming right angles (said of lines, rays, or segments)

perpendicular bisector: The line that both is perpendicular to a segment and cuts the segment in half

plane: A flat, infinitely thin surface that goes on forever in every direction (technically an undefined term)

point: An infinitely small dot (technically an undefined term)

polygon: A closed shape (no gaps or openings) with straight sides

postulate: A geometrical statement that's assumed to be true without proof; technically, postulates are a bit different from theorems, but in this book, you use them in exactly the same way

pre-image: A geometrical object (the "before" shape) that gets transformed into its image (the "after" shape)

prism: A 3-D figure that has two parallel, congruent bases; all of its faces are polygons

pyramid: A pointy-top, 3-D figure that has a polygon for its base and triangles for its lateral faces

quadrilateral: A four-sided polygon

radius (of a circle): A segment that goes from the center of a circle to a point on the circle; the plural is *radii*

radius (of a regular polygon): A segment joining the polygon's center to one of its vertices (corners)

ray: A portion of a line that originates at a point and goes on forever in only one direction

rectangle: A quadrilateral with four right angles

reflection: An isometry in which a pre-image flips across a line to an image of opposite orientation

regular polygon: A polygon that's both equilateral (with equal sides) and equiangular (with equal angles)

rhombus: A quadrilateral with four congruent sides

right angle: A 90° angle

right triangle: A triangle that contains one right angle and two acute angles

rotation: An isometry that spins an object about a fixed point; rotations do not change orientation

same-side exterior angles: When a transversal cuts across two parallel lines, angles on the outside of the parallel lines and on the same side of the transversal are called same-side exterior angles; same-side exterior angles are supplementary (add up to 180°)

same-side interior angles: When a transversal cuts across two parallel lines, angles between the parallel lines and on the same side of the transversal are called same-side interior angles; same-side interior angles are supplementary (add up to 180°)

scalene triangle: A triangle that has no congruent sides

secant: A line that intersects a circle at two points

second: In angle measurement, $\frac{1}{60}$ of a minute or $\frac{1}{3,600}$ of a degree

sector: A region of a circle bounded by two radii and the arc between those radii; it looks like a pizza slice

segment (of a circle): A region of a circle bounded by a chord and the arc intersected by that chord

similar: Having the same shape; similar figures may be congruent, but they're usually different sizes

slant height: In a pyramid, a segment drawn from its vertex perpendicular to an edge of its base; in a cone, a segment drawn from its vertex to the outside of its base; in short, the slant height of either shape is the slanting height from its top down along its side

slope: An indicator of a line's steepness (see the formula for slope in Appendix A)

sphere: A 3-D shape in which every point on its surface is equidistant from its center

square: A quadrilateral with four congruent sides and four right angles

straight angle: A 180° angle; basically, a line with a point on it

supplementary angles: Two angles whose measures add up to 180°

tangent: A line that touches a circle at a single point

theorem: A mathematical statement that can be proved

transformation: A repositioning of a pre-image object to its image in the x-y coordinate system; if the transformation is an isometry, the size and shape of the pre-image don't change (otherwise, the pre-image may be enlarged, shrunk, or warped out of shape)

translation: An isometry that slides an object without changing its orientation or rotating it

transversal: A line that crosses two parallel lines

trapezoid: A quadrilateral with exactly one pair of parallel sides

triangle: A three-sided polygon

trisect: To divide a segment or an angle into three congruent parts

union (of two geometric objects): Everything that's part of either or both objects

vertex (of an angle): The common endpoint of the two rays that form an angle; the plural is *vertices*

vertex (of a polygon): A common endpoint of two sides of a polygon (a corner)

vertical angles: Congruent angles formed by two intersecting lines that make an X; the vertical angles are on opposite sides of the X

volume: The number of cubic units of space contained within a 3-D object

Index

Numerics

1-D (one-dimensional) geometry, 12, 14
2-D (two-dimensional) geometry, 12–14, 326–331
3 : 4 : 5 family of Pythagorean triples, 115–117
3-4-5 Pythagorean triple, 114
3-D (three-dimensional) geometry
  base, 368
  cone. *See* cone
  cylinder. *See* cylinder
  determining a plane, 269
  flat-top figures, 273–279
  formulas and properties, 365
  height, 354–355, 370
  intersecting line and plane, 270–272
  line perpendicular to plane, 265–268
  locus (loci), 332–333
  overview, 12, 13–14
  pointy-top figures, 279–285
  problems, 332–333
  pyramid. *See* pyramid
  spheres. *See* sphere
  surface area. *See* surface area
  volume. *See* volume
5-12-13 Pythagorean triple, 114
7-24-25 Pythagorean triple, 114
8 : 15 : 17 family of Pythagorean triples, 116
8-15-17 Pythagorean triple, 114
30°- 60°- 90° (right) triangle, 94, 120–122, 190–191, 196–197
45°- 45° -90° (isosceles right) triangle, 118–120, 197–198

• A •

AA (Angle-Angle), 209, 210, 350
AAS (Angle-Angle-Side), 137–139, 349
acute angle, 30, 31, 96, 367
acute triangle, 96, 367
adding
  angles, 40
  arcs, 235
  segments, 39
addition theorem, 39, 62–67, 70
adjacent angles, 31, 32, 367
adjacent sides, 367
algebraic proof, coordinate geometry, 300–302
all-radii-are-congruent theorem, 229, 347
alternate exterior angles, 152, 367
alternate interior angles, 152, 367
altitude. *See also* height
  defined, 96, 367
  drawing, 190, 232
  of triangle, 96–98, 104, 106, 111
Altitude-on-Hypotenuse Theorem, 216–218, 362
analytical proof, coordinate geometry, 14, 298–299
angle
  adding and subtracting, 40
  addition theorem, 63–64, 70
  base. *See* base angle
  bisecting, 42–43, 337–338
  of circle. *See* angle, of circle
  copying, 335–336
  defined, 23, 367
  diagrams, 23, 44–45
  equilateral triangle, 93–94
  if-angles-then-sides theorem, 135, 348
  if-sides-then-angles theorem, 135, 348
  measurement, 30, 36–39, 42–44, 154, 199–201
  pairs of, 31–33, 156
  Parallel-Lines Theorems, 346–347
  parallelogram, 163, 164
  in polygon, 199–201, 363
  rotation, 317
  subtraction theorem, 67–68
  transversal. *See* transversal
  triangle, 96. *See also* triangle
  trisecting, 42–43, 338
  vertex of, 23, 374
  writing and reading, 23, 31
angle, of circle
  angle-arc theorems and formulas, 250–257
  central angles, 233–235
  chord-chord angles, 252–253, 257–259, 364
  defined, 250–252, 363
  inside circle, 252–254, 363, 364
  measurements, 250–256, 363, 364
  on circle, 250–252, 363, 364
  outside circle, 254–256, 363, 364
  secant-secant, 254, 260–262, 364
  secant-tangent, 254
  size of, and distance from circle's center, 256–257
  tangent-chord, 250–251
  tangent-secant, 254, 259–260, 364
  tangent-tangent, 254
angle, type of
  acute, 30, 31, 96, 367
  adjacent, 31, 32, 367
  alternate exterior, 152, 367

angle, type of *(continued)*
  alternate interior, 152, 367
  base. *See* base angle
  central, 233–235, 368
  complementary, 32, 59–62, 369
  congruent, 38, 60, 73–75, 251, 369
  corresponding, 152, 204–205, 369
  exterior. *See* exterior angle
  included, 126
  inscribed, 250, 251–252
  interior. *See* interior angle
  obtuse, 30, 31, 96, 371
  reflex, 30, 31
  right, 30, 31, 32, 373
  rotation angle, 317
  straight, 30, 31, 374
  supplementary, 32, 59–62, 374
  vertex, 93
  vertical, 33, 73–74, 346, 374
  Z-angle, 161–162
angle bisector, 42, 104, 105, 367
Angle-Angle (AA), 209, 210, 350
Angle-Angle-Side (AAS), 137–139, 349
Angle-Bisector Theorem, 223–224
Angle-Side-Angle (ASA), 128–130, 349
apothem, 195–197, 245–246, 367
arc. *See also* circle
  and central angle theorems, 233–234
  and chords theorems, 234
  defined, 233, 367
  formulas, 363
  length of, 244–246, 248–249, 363
  measurement, 244
Archimedes (mathematician), 13, 351–352
area
  circle, 176–177, 244, 247–248, 363
  defined, 96, 367

formulas, 98–100, 188–202, 361–362
hexagon, 110–111, 196–197
kite, 188, 189, 192–193, 362
lateral, 276, 280
octagon, 197–199
parallelogram, 188–191, 362
  with Pythagorean Theorem, 109–113
quadrilateral, 187–195
rectangle, 188, 189
regular polygon, 195–199, 245, 362
rhombus, 188, 191–192
sector of circle, 247–249, 363
segment of circle, 248
square, 188
surface. *See* surface area
trapezoid, 188, 189–190, 194–195
triangle, 96–101, 109–113, 196, 300–302, 361
ASA (Angle-Side-Angle), 128–130, 349
auxiliary lines, in geometry proof, 159–161

• *B* •

base
  3-D diagram, 368
  area formula, 98–100
  cylinder, 274
  isosceles triangle, 93, 368
  prism, 273
  trapezoid, 157, 368
  triangle, 96
base angle
  defined, 93
  trapezoid and isosceles trapezoid, 171–172, 368
  triangle, 93, 348, 368
bisect(ing)
  angle, 42–43, 337–338
  defined, 368
  Like Divisions Theorem, 70–73
  segment, 40–41, 338–339

bisector
  angle, 42, 104, 105, 367
  perpendicular. *See* perpendicular bisector
  reflecting line, 310
  segment, 338–339
bubble logic in proofs, 55–57, 88
buckyballs, 357

• *C* •

careers using geometry, 18
Cartesian coordinate system. *See* coordinate system
Cartesian plane, 289–291, 369
CASTC (corresponding angles of similar triangles are congruent), 213–214
catenary curve, 356–357
center of rotation, 317, 318–320
center of triangle, 102–106
central angle, 233–235, 368
centroid of triangle, 102–104, 368
China, geometry use in, 13
chord. *See also* circle
  arcs theorems, 234
  bisected theorem, 234
  central angle theorems, 234
  chord-chord angles, 252–253, 257–259, 364
  congruency and equidistance theorem, 229
  defined, 228–229, 368
  inscribed angle, 250, 251–252
  tangent-chord angle, 250
Chord-Chord Power Theorem, 257–259, 262, 364
Chung-Chin Tsu (mathematician), 352
circle
  all-radii-are-congruent theorem, 347
  angle measuring. *See* angle, of circle

Index

arc. *See* arc
area, 176–177, 244, 247–248, 363
chord. *See* chord
circumference, 244, 246, 363, 368
circumscribed, 104, 105, 368
as common shape, 227
common-tangent problem proof, 238–241
concentric, 369
congruent, 229, 233–234, 347, 369
defined, 228, 368
equations of, 303–305, 366
formulas, 243–262, 363–364
inscribed, 105
pi (π). *See* pi (π)
as polygon, 245–246
proof, 229–230, 235–236
properties, 363–364
radius, 231–232, 347, 372
secant-secant, 254, 260–262, 364
sectors, 363
segments inside, 228–229
tangent to, 237–241, 259–260, 374
tangent-secant, 254, 259–260, 364
theorems, 229, 233–235, 257–262, 347
walk-around problem, 241–242
circumcenter, 104–105, 368
circumcircles, 105
circumference of circle, 244, 363, 368
circumscribed circle, 104, 105, 368
collinear points, 25, 26, 369
Columbus, Christopher, 354
common tangent, 238–241, 369
compass, 334–342, 369
complementary angle, 32, 59–62, 369
complementary angle theorems, 59–62
concentric circles, 333, 369

cone
circular, right, 279
defined, 279, 369
lateral area, 280
pointy-top objects, 279
surface area, 284–285, 365
volume, 280, 284, 365
congruence. *See also* isometry
AAS (Angle-Angle-Side), 137–139, 349
all-radii-are-congruent theorem, 229, 347
angle, 38, 60, 73–75, 251, 369
angle on circle, 251–252
ASA (Angle-Side-Angle), 128–130, 349
circles, 229, 233–234, 347, 369
complementary angle theorems, 59–62
complementary triangle, 60–61
CPCTC (corresponding parts of congruent triangles are congruent), 131–134
defined, 369
HLR (hypotenuse-leg-right angle) theorem, 139–141, 349
if-sides-then-angles theorem, 135, 348
proving that angles are congruent (theorem), 152, 153, 156
radius, 347
segments, 36, 369
and similarity, 203
SSS (Side-Side-Side), 124–126, 130, 349
Substitution Property, 75–78
supplementary angle theorems, 59–62, 153, 156
Transitive Property, 75–78
triangle. *See* triangle congruence
of vertex angle, 346

vertical angle, 346
vertical-angles-as-congruent theorem, 73–75, 346
congruent, defined, 369
congruent angle, 38, 60, 73–75, 152, 251, 369
congruent circles, 229, 233–234, 347, 369
congruent segments, 36, 369
construction
bisecting an angle, 337–338
bisecting a segment, 338–339
copying an angle, 335–336
copying a segment, 334–335
copying a triangle, 336–337
defined, 369
parallel line, 341–342
perpendicular bisector, 338–339
perpendicular line, 339–342
trisecting an angle, 338
coordinate (Cartesian) plane, 289–291, 369
coordinate system
algebraic proof, 300–302
analytical proof, 298–299
coordinate plane, 289–291
distance formula, 294–295, 366
drawing general figure, 298–299
equations of circle, 303–306, 366
equations of line, 302–303, 366
formulas, 291–295, 365–366
midpoint formula, 295, 365
problem, 295–297
properties, 365–366
slope and slope formula, 291–294, 365
coordinates, 289
coplanar. *See also* intersection; parallel line
defined, 369
line, 27–28
points, 25, 26

copying. *See also* locus (loci), locating and constructing
  an angle, 335–336
  a segment, 334–335
  a triangle, 336–337
corresponding angle, 152, 204–205, 213–214, 369
corresponding angles of similar triangles are congruent (CASTC), 213–214
corresponding parts of congruent triangles are congruent (CPCTC), 131–134, 349
corresponding sides, 204–205, 213–216
corresponding sides of similar triangles are proportional (CSSTP), 213–216
CPCTC (corresponding parts of congruent triangles are congruent), 131–134, 349
cross diagonal, 170
CSSTP (corresponding sides of similar triangles are proportional), 213–216
cube, volume of, 274, 285–286
cylinder
  circular, right, 274
  defined, 273, 274, 369
  lateral area, 276
  surface area, 274, 278–279, 365
  volume, 274, 279, 365

• *D* •

da Vinci, Leonardo (artist), 91
deductive reasoning, 2, 15, 43, 54, 56
definition
  as concept, 22
  use in proofs, 53–54
degree, 37, 244, 369
Descartes, René (mathematician), 13, 65, 289, 303
determining a line, 347
determining a plane, 269
diagonal
  for area of kite, 192–193
  defined, 369
  kite, 170
  midpoint formula and bisection, 297
  n-gon, 201–202
  number that can be drawn, 363
  parallelogram, 163, 164–165
  polygon, 201–202, 363
  Pythagorean Theorem, 108–109
  quadrilateral, 162
diagram. *See also specific topics*
  angle, 23, 44–45
  construction of. *See* construction
  flat-top object, 273–279
  general drawing, 298–299
  line, 23
  parallel line, 27
  plane, 24
  point, 23
  pointy-top object, 279–285
  proof, 43–46, 50, 51
  Pythagorean theorem, 108
  ray, 23
  right angle, 31
  segment, 23, 44–45
  straight line, 44
diameter of circle, 228, 369
disjoint, 170
displacement, 351–352
distance
  chord congruency and equidistance theorem, 229
  circle equation, 304–306
  formula, 294–295, 297, 366
  to horizon, 260, 355
  problem solving with, 297, 300–301
  translation distance, 313, 315
dividing a segment, 41, 342
dome, geodesic, 357
drawing constructions. *See* construction
drawing general figure, 298–299
Dunce Cap Theorem, 241–242

• *E* •

Earth, 354, 355
edge, 273, 279, 369
Egyptian pyramids at Giza, 13
Einstein, Albert (physicist), 13
endpoint
  equidistance theorem, 143–144
  midpoint formula, 295, 297, 300, 365
  point excluded from problem solving, 330
equation of circle, 303–306, 366
equation of line. *See* line, equation of
equiangular, 93, 369
equiangular polygons, 363
equidistance theorems, 141–144
equilateral, defined, 93, 369
equilateral triangle
  30°- 60°- 90° (right) triangle, 94, 120–122, 190–191, 196–197
  altitude, 98
  area, 100, 196, 361
  defined, 91, 92, 93–94, 370
  icosahedron, 357–358
Eratosthenes (mathematician), 354
Euclid (Greek mathematician), 17
"Eureka" moment, Archimedes, 351–352
exterior angle
  alternate, 152, 367
  defined, 370
  diagram, 23
  parallel-line theorems, 152–153, 178–179, 346–347
  polygon angles, 199–201, 363
  same-side, 153, 373
  sum measures, 363
external tangent, 238

• F •

face, 273, 279, 370
families of Pythagorean triple triangle, 115–118
finished geometry proof, 88
flat-top object
  cylinder, 274
  defined, 273, 365
  diagram, 273–279
  lateral area, 276
  prism, 273
  shortest distance between two points, 275–276
  surface area, 274, 278–279
  volume, 274, 277
foot, 265, 370
formulas. *See also* properties
  3-D geometry, 365
  angle-arc, 250–257
  arcs, 250–257, 363
  area, 98–100, 188–202, 361–362
  circle, 243–262, 363–364
  circumference, 363
  coordinate system, 291–295, 365–366
  distance formula, 294–295, 297, 366
  equilateral triangle area, 100, 196, 361
  Hero's area, 100, 113, 361
  horizontal distance, 294
  kite area, 188, 189, 192–193
  midpoint, 295, 297, 300, 365
  parallelogram area, 188–191, 362
  polygons, 362–363
  quadrilateral area, 188–190
  rectangle area, 188, 189, 362
  rhombus area, 188, 191–192, 362
  right triangle, 361
  slope, 291–294, 365
  trapezoid area, 188, 189–190, 194–195, 362
  triangles, 361–362
  vertical distance formula, 294

45°-45°-90° (isosceles right) triangle, 118–120, 197–198
four-step problem solving process, 326
Franklin, Ben (inventor), 13
Fuller, Buckminster (scientist), 357

• G •

Galileo Galilei (mathematician), 13, 355
game plan for geometry proofs, 2, 61, 80, 145–146
Gauss, Carl Friedrich (mathematician), 351
generality, loss of, 299
geodesic dome, 357
geometry, overview. *See also* specific topics
  definitions in, 22
  dimensions, 12–14
  historical view, 13
  proof importance, 11, 14–20
  ten interesting problems, 351–358
  uses for, 18–19
geometry proof
  analytical, 14, 298–302
  auxiliary lines in, 159–161
  bubble logic in, 55–57, 88
  circle, 229–230, 235–236
  common-tangent problem, 238–241
  components of, 49–51
  congruent lines, 16–17
  congruent triangles. *See* triangle congruence
  defined, 14
  definitions for, 53–54
  diagram, 43–46, 50, 51
  Euclid and, 17
  everyday approach (song lyrics), 15–16
  example, 56–57
  finished proof, 88
  format, 88
  game plan for, 2, 61, 80, 145–146

  gaps, bridging, 86–87
  getting stuck, 83–85
  given statement, 16–17, 50, 51, 65, 81
  if-then statement, 51–53, 56, 81–83
  importance of, 11, 14–20
  indirect, 146–148
  introduction, 14–17
  jumping ahead, 85–86
  overview, 11, 14–20
  parallel-line theorems, 152–153, 178–179, 346–347
  parallelogram, 162–186
  plane, intersecting, 266–272
  postulate use in, 54–55, 173–175
  property-proof connection, 173–186
  prove statement, 50, 51, 66, 141
  quadrilaterals, 162–186
  reason column, 50, 51, 53–54
  Reflexive Property, 345–346
  similarity, 207–224
  statement column, 50, 51
  Substitution Property, 75–78
  "therefore," 16–17
  Transitive Property, 75–78
  transversal theorem, 153–155
  triangle congruence, 123–141, 145–148
  two-column format, 17, 55–57
  uses for, 18–19
  working backward, 85–87, 145–146
given points, 330
given statement, geometry proof, 16–17, 50, 51, 65, 81
glide reflection
  defined, 321, 370
  equals three reflections (theorem), 309, 322–323
  main reflecting line of (theorem), 322–324

Golden Gate Bridge, 356–357
golden ratio, 353
golden rectangle, 353
"Goldilocks rule" for scalene triangles, 92
Great Pyramid of Khufu, 354–355
Greek mathematicians, 1, 108, 334, 351–353

• H •

half properties, 170
height. *See also* altitude; *specific geometric shapes*
  3-D geometry, 354–355, 370
  area formula, 98–100
  defined, 370
  distance to horizon, 355
  golden ratio (phi), 353
  Great Pyramid of Khufu, 354–355
  plane, 269
  quadrilaterals, 190, 232
  slant, 280, 373
  triangle, 96–98, 104, 106, 111
Hero's area formula, 100, 113, 361
hexagon, area of, 110–111, 196–197
historical view of geometry, 1, 13, 108, 334, 351–353
HLR (hypotenuse-leg-right angle) theorem, 139–141, 349
horizon, distance to, 355
horizontal, 26
horizontal (x) axis, 290. *See also* coordinate system
horizontal distance formula, 294
horizontal line
  equation of, 303, 366
  slope, 292, 293, 303
hypotenuse. *See also* Pythagorean Theorem
  30°- 60°- 90° (right) triangle, 94, 120–122, 190–191, 196–197
  45°- 45°- 90° (isosceles right) triangle, 118–120, 197–198

Altitude-on-Hypotenuse Theorem, 216–218, 362
  defined, 108, 370
  distance formula, 294–295, 366
  right triangle, 96
hypotenuse-leg-right angle (HLR) theorem, 139–141, 349

• I •

icosahedron, 357–358
if-angles-then-sides theorem, 135, 348
if-sides-then-angles theorem, 135, 348
if-then statement, geometry proof, 51–53, 56, 81–83
image, 307, 370
incenter of triangle, 104–105, 370
incircles, 105
incline and slope, 292–293
indirect geometry proof, 146–148
infinite Pythagorean triple triangle, 115
infinite series of triangles, 103
inscribed angle, circle, 250, 251–252
inscribed circle, 352
inscribed shapes, 370
interior angle
  alternate, 152, 367
  defined, 370
  diagram, 23
  parallel-line theorems, 152–153, 178–179, 346–347
  polygon angles, 199–201, 363
  same-side, 153, 373
  sum measures, 363
internal tangent, 238
intersection
  defined, 370
  determining a plane, 269–271
  of diagonals at midpoint, 297
  lines, 28, 270–272

origin, 290
perpedicular. *See* perpendicular line
plane, 29, 266–272
point, 28
transversal. *See* transversal
vertex. *See* vertex (vertices)
irreducible Pythagorean triple triangle, 114–115
isometry. *See also* reflection
  defined, 307, 370
  glide reflection, 309, 321–324, 370
  rotation, 317–321
  translation, 312–316
isosceles trapezoid
  defined, 157, 370
  properties, 172
  quadrilateral relationships, 158–159
isosceles triangle
  45°- 45°- 90° (isosceles right) triangle, 118–120, 197–198
  altitude, 98
  base, 93, 368
  base angle, 368
  defined, 91, 92, 93, 370
  if-angles-then-sides theorem, 135, 348
  if-sides-then-angles theorem, 135, 348
  leg, 371
  theorems, 134–137, 348

• J •

jumping ahead, geometry proof, 85–86
justification in *given* statement, 16–17, 50, 51, 65, 81

• K •

Kepler, Johannes (mathematician), 13
Khufu, Great Pyramid of, 354–355

kite
  area formula, 188, 189, 192–193, 362
  defined, 157
  proofs, 185–186
  properties, 169–171
  quadrilateral relationships, 158–159

• L •

lateral area
  cone, 280
  cylinder, 276
  defined, 276
  flat-top objects, 276
  pointy-top objects, 280–281
  prism, 276
  pyramid, 280
leg. See also side
  isosceles triangle, 93, 371
  Pythagorean theorem, 108
  right triangle, 96
  trapezoid, 157, 371
  of triangle, 108
length
  arc, 244–246, 248–249, 363
  of segment, 36
Like Divisions Theorem, 70–73
Like Multiples Theorem, 70–73
line
  auxiliary lines in proof, 159–161
  defined, 22, 371
  determining a plane, 269
  diagram, 23
  equation of. See line, equation of
  horizontal, 26, 27
  intersecting plane and, 270–272
  Line-Plane Perpendicularity Theorem, 265–268
  pairs, 27–29
  and planes, 270–272
  point-slope form, 303, 306, 366
  points, determining, 347
  slope, 291–294
  slope-intercept form, 302, 306, 366
  translation, 313
  two points determine a line postulate, 347
  type of. See line, type of
  writing and reading, 22
line, equation of
  coordinating system, 302–303, 366
  horizontal, 303, 366
  point-slope, 303, 306, 366
  reflecting line, 316, 318–321
  slope, 302–303
  slope-intercept, 302, 306, 366
  tangent line, 305–306
  translation line, 313, 315
  vertical, 303, 366
line, type of
  coplanar, 27–28
  intersecting, 269
  main reflecting line, 322–324
  non-coplanar, 28–29
  parallel. See parallel line
  perpendicular. See perpendicular line
  ray. See ray
  reflecting. See reflecting line
  segment. See segment
  tangent, 237–238
  translation line, 313, 315
  transverse. See transversal
  vertical, 26, 27
linear pair of angles, 32
Line-Plane Perpendicularity Theorem, 265–268
locus (loci), locating and constructing. See also point
  2-D geometry, 326–331
  3-D geometry, 332–333
  defined, 326, 371
  four-step problem solving process, 326
  problem solving, 326–333
  three-dimensional problems, 332–333
  two-dimensional problems, 326–331
logic
  bubble, in proof, 55–57, 88
  everyday approach (song lyrics), 15–16
  if-then, 51–53, 56, 81–83

• M •

main diagonal, 170
mathematikoi, 108
measurement. See also pi (π); specific topics
  angle, 30, 36–39, 42–44, 154, 199–201
  angle of circle, 250–256, 363, 364
  arc, 233, 244
  segment, 35–36, 154
median, 102, 188, 371
Mesopotamia, geometry in, 13
Midline Theorem, 211–212
midpoint
  center of rotation, 319–320
  centroid of triangle, 102–104, 368
  defined, 41, 371
  formula, 295, 297, 300, 365
  Like Divisions Theorem, 70–73
  main reflecting line of glide reflection (theorem), 322–324
  point excluded from problem solving, 330
  problem solving with, 297, 300
  reflecting line, 311–312, 316
minute, 371
mirror. See reflection
moon, size of, 229
motion of projectile, 355–356

• N •

nautilus, golden ratio, 353
negative slope, 293
Newton, Isaac
　(mathematician), 13, 351
n-gon, diagonals in, 201–202,
　363. *See also* polygon
non-collinear points,
　25, 269, 371
non-coplanar, 28–29, 371
non-coplanar points, 26
numerics
　1-D (one-dimensional)
　　shapes, 12, 14
　2-D (two-dimensional)
　　geometry, 12–14, 326–331
　3 : 4 : 5 family of
　　Pythagorean triple,
　　115–117
　3-4-5 Pythagorean triple, 114
　3-D (three-dimensional)
　　geometry. *See* 3-D (three-
　　dimensional) geometry
　5-12-13 Pythagorean
　　triple, 114
　7-24-25 Pythagorean
　　triple, 114
　8 : 15 : 17 family of
　　Pythagorean triple, 116
　8-15-17 Pythagorean
　　triple, 114
　30°- 60°- 90° (right) triangle,
　　94, 120–122, 190–191,
　　196–197
　45°- 45°- 90° (isosceles
　　right) triangle, 118–120,
　　197–198

• O •

oblique, 28
obtuse angle, 30, 31, 96, 371
obtuse triangle, 96, 98, 371
octagon, 197–199
1-D (one-dimensional)
　geometry, 12, 14
ordered pair, 290
orientation, 308, 309–310, 371

origin, 290, 304–306
orthocenter of triangle,
　104, 106, 371

• P •

pairs
　angles, 31–33, 156
　ordered, on coordinate
　　plane, 290
　triangle congruence,
　　123–134
parabola, 355–356, 357
parallel line
　constructing, 341–342
　construction, 341–342
　coplanar, 27
　defined, 27, 371
　determining a plane,
　　269–271
　diagrams, 27
　line parallel to a given line
　　through point not on
　　given line, 341
　naming, 27
　properties, 151–157
　proving that angles are
　　parallel (theorem), 153
　slopes, 293, 294
　transversal. *See* transversal
parallel plane, determining,
　269–271
parallel-line theorems,
　152–153, 178–179,
　346–347
parallelogram
　angles, 163, 164
　area, 188–191, 362
　defined, 157, 372
　diagonals, 163, 164–165
　main reflecting line of glide
　　reflection, 322–324
　proofs, 162–186
　properties, 162–166, 173–186
　property-proof connection,
　　173–186
　quadrilateral relationships,
　　158–159
　reversible properties, 174
　sides, 163, 164
　special, 166–169

Parthenon, 353
pentagon, 372
perimeter
　circumference of circle,
　　244, 246, 363, 368
　defined, 372
　distance formula, 297
　similar polygon,
　　205, 208–209
perpendicular, defined, 372
perpendicular bisector
　circumcenter of triangle,
　　104, 105
　constructions, 338–339
　defined, 372
　equidistance theorems,
　　141–144
　reflecting line as, 310–312
　of a segment, 338–339
　triangle congruence,
　　141–144
perpendicular line
　bisected chords
　　theorem, 229
　construction, 339–342
　line perpendicular to plane,
　　265–268
　Line-Plane Perpendicularity
　　Theorem, 265–268
　perpendicular to plane,
　　265–268
　radius-tangent
　　perpendicularity,
　　237–238
　right angle, 30
　slopes, 293, 294
　through a point not on a
　　given line, 340
　through a point on a given
　　line, 339–340
pi (π)
　area of circle, 176–177, 244,
　　247–248
　calculating, 352
　circumference of circle,
　　244, 246, 363, 368
　defined, 5
Pi Day (March 14), 123
pizza slice. *See* sector of
　circle

plane
- 2-D (two-dimensional) geometry, 12–14, 326–331
- 3-D locus problems, 333
- coordinate (Cartesian), 289–291, 369
- defined, 24, 265, 372
- determining, 269
- diagram, 24
- intersecting line, 270–272
- intersecting plane, 29, 266–272
- Line-Plane Perpendicularity Theorem, 265–268
- lines, 265–268, 270–272
- naming, 24
- parallel, 29, 270
- types, 29

point. *See also* locus (loci), locating and constructing
- circle centered at, and circle equation, 304–306
- defined, 22, 25, 372
- determining a plane, 269
- diagram, 23
- equidistance theorem, 143–144
- excluded from problem solving, 330
- intersection, 28
- line, determining, 347
- point-slope form, 303, 306, 366
- shortest distance between two points, 275–276
- two points determine a line postulate, 160, 347
- types of. *See* point, type of

point, type of
- collinear, 25, 26, 369
- coplanar, 25, 26
- endpoint. *See* endpoint
- equidistant, 141, 143
- midpoint. *See* midpoint
- non-collinear, 25, 269, 371
- non-coplanar, 26
- of tangency, 237, 330
- point-slope form, 303, 306, 366

pointy-top object
- cone, 279
- defined, 365
- diagram, 279–285
- lateral area, 280–281
- pyramid, 279
- surface area, 281–285, 365
- volume, 280, 281–282, 284, 365
- water into wine problem, 283

polygon
- align similar, 205–207
- angles of, 199–201, 363
- area, 195–199, 245, 362
- defined, 12, 372
- formulas, 362–363
- pentagon, 372
- perimeter, 205, 208–209
- properties, 362–363
- radius, 372
- regular, 195–199, 362, 373
- similar, 204–209
- vertex, 374

polygon, type of
- 3-sided (triangle). *See* triangle
- 4-sided (quadrilateral). *See* quadrilateral
- 5-sided (pentagon), 372
- 6-sided (hexagon), 110–111, 196–197
- 8-sided (octagon), 197–199
- *n*-sided, 201–202, 363
- regular, 195–199, 245, 279, 362, 373

positive slope, 293

postulates
- AA (Angle-Angle), 209, 210, 350
- ASA (Angle-Side-Angle), 128–130, 349
- defined, 17, 372
- Euclid's, 17
- SAS. *See* Side-Angle-Side (SAS)
- SSS. *See* Side-Side-Side (SSS)
- two points determine a line, 347

use in proofs, 54–55, 173–175
practical uses of geometry, 18–19
pre-image, 307, 372
prime mark, for reflections, 308
Principia Mathematica (Newton), 13

prism
- defined, 273, 372
- flat-top objects, 273
- lateral area, 276
- right, 273
- surface area, 274, 278, 365
- volume, 274, 277, 365

problem solving, 326–333
projectile motion, 355–356
proof. *See* geometry proof

properties. *See also* formulas
- circle, 363–364
- coordinate system, 365–366
- defined, 162
- half, 170
- isosceles trapezoid, 172
- kite, 169–171
- parallel line, 151–157
- parallelogram, 162–166, 173–186
- polygon, 362–363
- property-proof connection, 173–186
- quadrilateral, 151–157, 162–172
- rectangle, 167–169, 174
- Reflexive Property, 132–133, 345–346
- reversible, 174–175
- rhombus, 167–169, 174
- square, 167–169
- Substitution, 75–78
- 3-D (three-dimensional) geometry, 365
- Transitive Property, 75–78
- transversal, 152–157
- trapezoid, 171–172
- triangle, 361–362
property-proof connection, quadrilaterals, 173–186

proportions
　Angle-Bisector Theorem, 223–224
　CSSTP (corresponding sides of similar triangles are proportional), 213–216
　Side-Splitter Theorem, 219–222
　similar figures, 204–205
prove statement, geometry proof, 50, 51, 66, 141
pyramid
　defined, 279, 372
　Egyptian, 13, 354–355
　lateral area, 280
　pointy-top objects, 279
　regular, 279
　surface area, 281, 282–283, 365
　volume, 280, 281, 282, 365
Pythagoras (mathematician), 13, 108
Pythagorean Theorem
　coordinate system, 291, 294–295
　defined, 108, 361
　distance formula, 294
　historical view, 107–108
　problem solving with, 108–113
Pythagorean triple triangles, 113–118, 122

• Q •

quadrant, coordinate system, 290
quadrilateral. *See also specific quadrilaterals*
　area, 187–195, 362
　area formula, 188–190
　child and parent, 174
　coordinate geometry proof, 296–297
　defined, 12, 151, 157, 372
　parallel-line properties, 151–157
　proofs, 162–186
　properties of, 151–157, 162–172
　property-proof connection, 173–186
　transversal theorems, 152–157
　type of, 157–159

• R •

radius of circle
　all-radii-are-congruent theorem, 229, 347
　and bisected chords theorem, 229
　defined, 228, 229, 372
　extra, for problem solving, 231–232
radius of regular polygon, 372
radius of sphere, 285
radius-tangent perpendicularity, 237–238
ratio. *See also* slope
　3 : 4 : 5 family of Pythagorean triples, 115–117
　8 : 15 : 17 family of Pythagorean triples, 116
　for area of rhombus, 191–192
　families of Pythagorean triple triangle, 115–118
　scalene triangle, 93
　special right triangles, 122, 362
ray
　angle bisector as, 42, 104, 105, 367
　as angle side, 39
　defined, 23, 372
　diagram, 23
　horizontal, 26, 27
　naming, 23
　parallel, 27
　type of, 26–27
　vertical, 26, 27
reason column, geometry proof, 50, 51, 53–54
reasoning
　deductive, 2, 15, 43, 54, 56
　indirect proofs, 148

rectangle
　area formula, 188, 189, 362
　coordinate geometry proof, 296–297
　defined, 157, 372
　golden, 353
　lateral, and flat-top figures, 274
　proof that quadrilateral is, 180–182
　properties, 167–169, 174
　Pythagorean Theorem, 109
　quadrilateral relationships, 158–159, 166–169
　reversible properties, 174
reflecting line
　defined, 310
　equation of, 316, 318–321
　as perpendicular bisector, 310–312
reflection. *See also* glide reflection
　defined, 308, 372
　lines, 310–312, 318–324
　main reflecting line, 322–324
　number of, 309
　orientation, 309–310
reflex angle, 30, 31
Reflexive Property, 132–133, 345–346
regular polygon
　area, 195–199, 245, 362
　defined, 195, 373
　pointy-top figures, 279
reversible properties, 174–175
rhombus
　area, 188, 191–192, 362
　defined, 157, 373
　proof that quadrilateral is, 182–184
　properties, 167–169, 174
　quadrilateral relationships, 158–159, 166–169
　reversible properties, 174
right angle, 30, 31, 32, 373
right circular cone, 279

Index 385

right circular cylinder, 274
right prism, 273
right triangle
   30°-60°-90° (right) triangle, 94, 120–122, 190–191, 196–197
   45°-45°-90° (isosceles right) triangle, 118–120, 197–198
   altitude, 98
   Altitude-on-Hypotenuse Theorem, 216–218, 362
   area, 110–111, 361
   for area of parallelogram, 188–191, 362
   defined, 96, 373
   formulas, 361
   hypotenuse, 108
   leg (side), 108
   pyramid, 281
   Pythagorean triple triangle, 113–118, 122
   ratio of sides, 362
   special, 122, 362
   and trapezoids, 194–195
rise, 292, 295
rotation
   center of, 317, 318–320
   defined, 317, 373
   equals two reflections (theorem), 309, 317–318
   equations of two reflecting lines, 318–321
rotation angle, 317
round-robin tennis, 202
run, 292, 295

• S •

same-side exterior angle, 153, 373
same-side interior angle, 153, 373
SAS (Side-Angle-Side)
   triangle congruence, 126–128, 130, 349
   triangle similarity, 209, 212–213, 350
scalene triangle, 91, 92–93, 98, 373
secant, 254, 373

secant-secant angles, 254, 260–262, 364
Secant-Secant Power Theorem, 260–262, 364
secant-tangent angle, of circle, 254
second, 373
sector of circle
   area of, 247–249, 363
   defined, 247, 373
segment
   addition theorem, 39, 63–64, 70
   auxiliary lines in proof, 159–161
   bisecting, 40, 338–339
   bisector of, 338–339
   of circle, 243–244, 247, 248, 249, 373
   congruent, 36, 369
   copying, 334–335
   defined, 22, 243, 371
   diagram, 23, 44–45
   dividing into equal parts, 342
   equidistance theorem, 143–144
   horizontal, 26, 27
   length of, 36
   measurement, 35–36, 154
   midpoint, 41
   naming, 22, 36
   parallel, 27
   perpendicular bisector of a segment, 338–339
   subtraction theorem, 39, 67–68
   transversal theorems, 152–157, 341
   trisecting, 41
   vertical, 26, 27
semiperimeter, 100
set (locus), 325
shapes, 12–14
side. *See also* leg; *specific geometric shapes*
   adjacent, 367
   corresponding, 204–205
   defined, 23
   equilateral triangle, 93–91

if-angles-then-sides theorem, 135, 348
if-sides-then-angles theorem, 135, 348
included, 128
isosceles triangle, 93
parallelogram, 163, 164
polygon. *See* polygon
right triangle, 362
Side-Splitter Theorem, 219–222
triangle, 91–94, 362
Side-Angle-Side (SAS)
   triangle congruence, 126–128, 130, 349
   triangle similarity, 212–213, 350
Side-Side-Side (SSS)
   triangle congruence, 124–126, 130, 349
   triangle similarity, 209, 211–212, 350
Side-Splitter Theorem, 219–222
similar, defined, 373
similarity. *See also* isometry
   aligning polygon, 205–207
   Altitude-on-Hypotenuse Theorem, 216–218
   Angle-Bisector Theorem, 223–224
   and congruence, 203
   defined, 203, 373
   polygons, 204–209
   Side-Splitter Theorem, 219–222
   triangle. *See* triangle similarity
skew, 28
slant height, 280, 373
slide. *See* glide reflection
slope
   center of rotation, 319–320
   defined, 291, 373
   formula, 292, 365
   incline, 292–293
   line equations, 302–303
   negative, 293

slope *(continued)*
  point-slope form, 303, 306, 366
  positive, 293
  problem solving with, 296
  reflecting line as perpendicular bisector, 311–312, 316
  undefined, 292, 293, 303
  slope-intercept form, 302, 306, 366
slope-intercept form, 302, 306, 366
soccer balls, 357–358
solid geometry. *See also* 3-D (three-dimensional) geometry
  flat-top figures, 273–279
  pointy-top objects, 279–285
space, 3-D, 24
special figures
  parallelogram, 166–169
  right triangle, 122, 362
sphere
  defined, 285, 373
  surface area, 285–286, 365
  volume, 285, 365
spiral, golden ratio, 353
square
  area, 188, 362
  area of octagon, 198–199
  defined, 157, 374
  proof that quadrilateral is, 184
  properties, 167–169
  Pythagorean triple triangle, 114
  quadrilateral relationships, 158–159, 166–169
square roots, Pythagorean Theorem, 112, 113, 118, 122
SSS (Side-Side-Side)
  triangle congruence, 124–126, 130, 349
  triangle similarity, 209, 211–212, 350
statement column, geometry proof, 50, 51
straight angle, 30, 31, 374
straight line, 44
straightedge, 334–342, 369

Substitution Property, 75–78
subtracting
  angles, 40
  arcs, 235
  segments, 39, 67–68
supplementary angle, 32, 59–62, 374
supplementary angle theorems, 59–62, 153, 156
surface area. *See also* area
  3-D geometry, 365
  cone, 284–285, 365
  cylinder, 274, 278–279, 365
  flat-top objects, 274, 278–279, 365
  pointy-top objects, 281–285, 365
  prism, 274, 278, 365
  pyramid, 282–283, 365
  sphere, 285–286, 365

• *T* •

tangent
  to circle, 237–241, 259–260, 369, 374
  common-tangent, 238–241, 369
  Dunce Cap Theorem, 241–242
  equation of, 305–306
  walk-around problem, 241–242
tangent line, 237–238
tangent-chord angle, 250–251
tangent-secant angle, 254, 259–260, 364
Tangent-Secant Power Theorem, 259–260, 262, 364
tangent-tangent angle, of circle, 254
theorem
  AAS (Angle-Angle-Side), 137–139, 349
  addition, 39, 62–67, 70
  all-radii-are-congruent, 229, 347
  Altitude-on-Hypotenuse, 216–218, 362

angle-arc, 250–257
Angle-Bisector, 223–224
arc theorems, 233–234
bisected chords, 229
central angle, 233–234
chord theorems, 229, 234
Chord-Chord Power, 257–259, 262, 364
circle, 229, 233–235, 257–262, 347
complementary angle, 59–62
defined, 17, 374
Dunce Cap, 241–242
equidistance, 141–144, 229
Euclid's, 17
glide reflection equals three reflections, 309, 322–323
HLR (hypotenuse-leg-right angle), 139–141, 349
if-angles-then-sides, 135, 267, 348
if-sides-then-angles, 135, 348
isosceles triangle, 134–137
Like Divisions, 70–73
Like Multiples, 70–73
Line-Plane Perpendicularity, 265–266
main reflecting line of glide reflection, 322–324
Midline, 211–212
parallel-line, 152–153, 178–179, 346–347
perpendicularity and bisected chords theorem, 229
proving that angles are congruent, 152, 153, 156
Pythagorean. *See* Pythagorean Theorem
radius, 229
rotation equals two reflections, 309, 317–318
Secant-Secant Power, 260–262, 364
Side-Splitter, 219–222
subtraction, 39, 67–68
supplementary angle, 59–62, 153, 156
Tangent-Secant Power, 259–260, 262, 364
translation equals two reflections, 309, 312–314

Index

transversal theorems, 152–157, 341
triangle, 134–136, 139–141
use in proofs, 54–55, 59–75
vertical-angles-as-congruent, 73–75, 346
30°- 60°- 90° (right) triangle, 94, 120–122, 190–191, 196–197
3-D (three-dimensional) geometry
  base, 368
  cone. *See* cone
  cylinder. *See* cylinder
  determining a plane, 269
  flat-top figures, 273–279
  formulas and properties, 365
  height, 354–355, 370
  intersecting line and plane, 270–272
  line perpendicular to plane, 265–268
  locus (loci), 332–333
  overview, 12, 13–14
  pointy-top figures, 279–285
  problems, 332–333
  pyramid. *See* pyramid
  spheres. *See* sphere
  surface area. *See* surface area
  volume. *See* volume
3-D (three-dimensional) space, 24
torus, 333
transformation, 307, 374
Transitive Property, 75–78
translation
  defined, 309, 312
  distance, 313, 315
  elements of, 314–316
  equals two reflections (theorem), 309, 312–314
  line, 313, 315–316
  proof, 313–316
transversal
  angles are congruent (theorem), 152, 156
  angles are parallel (theorem), 153
  defined, 152, 374

multiple, parallel lines with, 155–157
parallel-line properties, 152–157
supplementary angle theorem, 59–62, 153, 156
theorems, 152–157, 341
trapezoid
  area, 188, 189–190, 194–195, 362
  bases, 157, 368
  defined, 157, 374
  isosceles, 157, 368, 370
  legs, 157, 371
  median, 188, 371
  properties, 171–172
  quadrilateral relationships, 158–159
triangle
  altitude, 96–98, 104, 106, 111
  angles, 96
  area, 96–101, 109–113, 196, 300–302, 361
  for area of rhombus, 191–192
  for area of trapezoid, 194–195
  base, 96
  center of, 102–106
  centroid, 102–104, 368
  congruence. *See* triangle congruence
  copying, 336–337
  defined, 374
  equiangular, 93
  formulas, 361–362
  geodesic domes, 357
  if-angles-then-sides theorem, 348
  if-sides-then-angles theorem, 348
  inequality principle, 94–95
  infinite series of, 103
  lateral, and pointy-top figures, 280
  median, 102, 371
  properties, 361–362
  Pythagorean triple, 113–118
  semiperimeter, 100
  sides of, 91–94

similar. *See* triangle similarity
type of. *See* triangle, type of
triangle, type of
  acute, 96, 98, 367
  equilateral. *See* equilateral triangle
  isosceles. *See* isosceles triangle
  obtuse, 96, 98, 371
  right. *See* right triangle
  scalene, 91, 92–93, 98, 373
triangle congruence
  AAS (Angle-Angle-Side), 137–139, 349
  ASA (Angle-Side-Angle), 128–130, 349
  CPCTC (corresponding parts of congruent triangles are congruent), 131–134, 349
  defined, 124, 131–132, 349
  versus equidistance theorem, 141–144
  geometric proofs, 123–141, 145–148
  HLR (Hypotenuse-leg-right angle) theorem, 139–141
  isosceles triangle theorems, 134–137
  for pairs, 123–134
  SAS (Side-Angle-Side), 126–128, 130, 349
  SSS (Side-Side-Side), 124–126, 130, 349
triangle inequality principle, 94–95
triangle similarity
  AA (Angle-Angle), 209, 210, 350
  CASTC (corresponding angles of similar triangles are congruent), 213–214
  CSSTP (corresponding sides of similar triangles are proportional), 213–216
  defined, 350
  proof, 209–216

triangle similarity *(continued)*
  Pythagorean triple triangle, 116
  SAS (Side-Angle-Side), 209, 212–213, 350
  SSS (Side-Side-Side), 209, 211–212, 350
trigonometry, 13
trisect(ing)
  angle, 42–43, 338
  defined, 41, 374
  Like Divisions Theorem, 70–73
  segment, 41
trisection point, 41
trisector, angle, 42, 338
Tsu Chung-Chin (mathematician), 352
two equidistant points determine the perpendicular bisector (equidistance theorem), 141–143
two points determine a line postulate, 160
2-D (two-dimensional) geometry, 12–14, 326–331
two-column format, geometry proof, 17, 55–57

• U •

undefined slope, 292, 293, 303
union of two objects, 374
unit segment, 36

• V •

vertex (vertices)
  of angle, 23, 374
  of polygon, 374
  of prism, 273
  of pyramid, 279
vertex angle, 42, 93, 223, 346
vertical, as term, 26
vertical (y) axis, 290. *See also* coordinate system
vertical angle, 73
vertical distance formula, 294
vertical line
  equation of, 303, 366
  slope, 292, 293, 303
vertical-angles-as-congruent theorem, 73–75, 346
volume
  3-D objects, 365
  cone, 280, 284, 365
  crown in Archimedes' bathtub, 351–352
  cube, 274, 285–286
  cylinder, 274, 279, 365
  defined, 374
  flat-top objects, 274, 277, 365
  pointy-top objects, 280, 281–282, 284, 365
  prism, 274, 277, 365
  pyramid, 280, 282, 365
  sphere, 285, 365

• W •

walk (glide) reflection, 309, 321–322, 370
walk-around problem, 241–242
Wantzel, Pierre (mathematician), 338
Washington, George (U.S. President), 13
wine-into-water volume problem, 283

• X •

X angles, 73
x-axis, 290. *See also* coordinate system
x-intercept
  circle equation, 305
  and slope, 303
x-y coordinate system. *See also* coordinate system
  defined, 289–290, 369
  distance formula, 294

• Y •

y-axis, 290. *See also* coordinate system
y-intercept
  circle equation, 305
  and slope, 302–303

• Z •

Z-angle, 161–162
zero slope, 293, 294, 303

BUSINESS, CAREERS & PERSONAL FINANCE

0-7645-9847-3

0-7645-2431-3

Also available:
- Business Plans Kit For Dummies
  0-7645-9794-9
- Economics For Dummies
  0-7645-5726-2
- Grant Writing For Dummies
  0-7645-8416-2
- Home Buying For Dummies
  0-7645-5331-3
- Managing For Dummies
  0-7645-1771-6
- Marketing For Dummies
  0-7645-5600-2
- Personal Finance For Dummies
  0-7645-2590-5*
- Resumes For Dummies
  0-7645-5471-9
- Selling For Dummies
  0-7645-5363-1
- Six Sigma For Dummies
  0-7645-6798-5
- Small Business Kit For Dummies
  0-7645-5984-2
- Starting an eBay Business For Dummies
  0-7645-6924-4
- Your Dream Career For Dummies
  0-7645-9795-7

HOME & BUSINESS COMPUTER BASICS

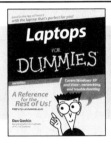
0-470-05432-8

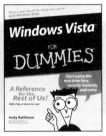
0-471-75421-8

Also available:
- Cleaning Windows Vista For Dummies
  0-471-78293-9
- Excel 2007 For Dummies
  0-470-03737-7
- Mac OS X Tiger For Dummies
  0-7645-7675-5
- MacBook For Dummies
  0-470-04859-X
- Macs For Dummies
  0-470-04849-2
- Office 2007 For Dummies
  0-470-00923-3
- Outlook 2007 For Dummies
  0-470-03830-6
- PCs For Dummies
  0-7645-8958-X
- Salesforce.com For Dummies
  0-470-04893-X
- Upgrading & Fixing Laptops For Dummies
  0-7645-8959-8
- Word 2007 For Dummies
  0-470-03658-3
- Quicken 2007 For Dummies
  0-470-04600-7

FOOD, HOME, GARDEN, HOBBIES, MUSIC & PETS

0-7645-8404-9

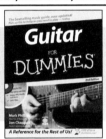
0-7645-9904-6

Also available:
- Candy Making For Dummies
  0-7645-9734-5
- Card Games For Dummies
  0-7645-9910-0
- Crocheting For Dummies
  0-7645-4151-X
- Dog Training For Dummies
  0-7645-8418-9
- Healthy Carb Cookbook For Dummies
  0-7645-8476-6
- Home Maintenance For Dummies
  0-7645-5215-5
- Horses For Dummies
  0-7645-9797-3
- Jewelry Making & Beading For Dummies
  0-7645-2571-9
- Orchids For Dummies
  0-7645-6759-4
- Puppies For Dummies
  0-7645-5255-4
- Rock Guitar For Dummies
  0-7645-5356-9
- Sewing For Dummies
  0-7645-6847-7
- Singing For Dummies
  0-7645-2475-5

INTERNET & DIGITAL MEDIA

0-470-04529-9

0-470-04894-8

Also available:
- Blogging For Dummies
  0-471-77084-1
- Digital Photography For Dummies
  0-7645-9802-3
- Digital Photography All-in-One Desk Reference For Dummies
  0-470-03743-1
- Digital SLR Cameras and Photography For Dummies
  0-7645-9803-1
- eBay Business All-in-One Desk Reference For Dummies
  0-7645-8438-3
- HDTV For Dummies
  0-470-09673-X
- Home Entertainment PCs For Dummies
  0-470-05523-5
- MySpace For Dummies
  0-470-09529-6
- Search Engine Optimization For Dummies
  0-471-97998-8
- Skype For Dummies
  0-470-04891-3
- The Internet For Dummies
  0-7645-8996-2
- Wiring Your Digital Home For Dummies
  0-471-91830-X

* Separate Canadian edition also available
† Separate U.K. edition also available

Available wherever books are sold. For more information or to order direct: U.S. customers visit www.dummies.com or call 1-877-762-2974.
U.K. customers visit www.wileyeurope.com or call 0800 243407. Canadian customers visit www.wiley.ca or call 1-800-567-4797.

SPORTS, FITNESS, PARENTING, RELIGION & SPIRITUALITY

0-471-76871-5

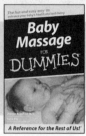
0-7645-7841-3

Also available:
- Catholicism For Dummies
  0-7645-5391-7
- Exercise Balls For Dummies
  0-7645-5623-1
- Fitness For Dummies
  0-7645-7851-0
- Football For Dummies
  0-7645-3936-1
- Judaism For Dummies
  0-7645-5299-6
- Potty Training For Dummies
  0-7645-5417-4
- Buddhism For Dummies
  0-7645-5359-3
- Pregnancy For Dummies
  0-7645-4483-7 †
- Ten Minute Tone-Ups For Dummies
  0-7645-7207-5
- NASCAR For Dummies
  0-7645-7681-X
- Religion For Dummies
  0-7645-5264-3
- Soccer For Dummies
  0-7645-5229-5
- Women in the Bible For Dummies
  0-7645-8475-8

TRAVEL

0-7645-7749-2

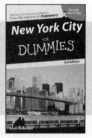
0-7645-6945-7

Also available:
- Alaska For Dummies
  0-7645-7746-8
- Cruise Vacations For Dummies
  0-7645-6941-4
- England For Dummies
  0-7645-4276-1
- Europe For Dummies
  0-7645-7529-5
- Germany For Dummies
  0-7645-7823-5
- Hawaii For Dummies
  0-7645-7402-7
- Italy For Dummies
  0-7645-7386-1
- Las Vegas For Dummies
  0-7645-7382-9
- London For Dummies
  0-7645-4277-X
- Paris For Dummies
  0-7645-7630-5
- RV Vacations For Dummies
  0-7645-4442-X
- Walt Disney World & Orlando For Dummies
  0-7645-9660-8

GRAPHICS, DESIGN & WEB DEVELOPMENT

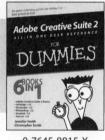
0-7645-8815-X

0-7645-9571-7

Also available:
- 3D Game Animation For Dummies
  0-7645-8789-7
- AutoCAD 2006 For Dummies
  0-7645-8925-3
- Building a Web Site For Dummies
  0-7645-7144-3
- Creating Web Pages For Dummies
  0-470-08030-2
- Creating Web Pages All-in-One Desk Reference For Dummies
  0-7645-4345-8
- Dreamweaver 8 For Dummies
  0-7645-9649-7
- InDesign CS2 For Dummies
  0-7645-9572-5
- Macromedia Flash 8 For Dummies
  0-7645-9691-8
- Photoshop CS2 and Digital Photography For Dummies
  0-7645-9580-6
- Photoshop Elements 4 For Dummies
  0-471-77483-9
- Syndicating Web Sites with RSS Feeds For Dummies
  0-7645-8848-6
- Yahoo! SiteBuilder For Dummies
  0-7645-9800-7

NETWORKING, SECURITY, PROGRAMMING & DATABASES

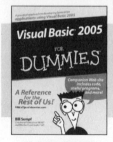
0-7645-7728-X

0-471-74940-0

Also available:
- Access 2007 For Dummies
  0-470-04612-0
- ASP.NET 2 For Dummies
  0-7645-7907-X
- C# 2005 For Dummies
  0-7645-9704-3
- Hacking For Dummies
  0-470-05235-X
- Hacking Wireless Networks For Dummies
  0-7645-9730-2
- Java For Dummies
  0-470-08716-1
- Microsoft SQL Server 2005 For Dummies
  0-7645-7755-7
- Networking All-in-One Desk Reference For Dummies
  0-7645-9939-9
- Preventing Identity Theft For Dummies
  0-7645-7336-5
- Telecom For Dummies
  0-471-77085-X
- Visual Studio 2005 All-in-One Desk Reference For Dummies
  0-7645-9775-2
- XML For Dummies
  0-7645-8845-1

Geometry Workbook FOR DUMMIES®

by Mark Ryan

Wiley Publishing, Inc.

Geometry Workbook For Dummies®

Published by
Wiley Publishing, Inc.
111 River St.
Hoboken, NJ 07030-5774
www.wiley.com

Copyright © 2007 by Wiley Publishing, Inc., Indianapolis, Indiana

Published simultaneously in Canada

No part of this publication may be reproduced, stored in a retrieval system, or transmitted in any form or by any means, electronic, mechanical, photocopying, recording, scanning, or otherwise, except as permitted under Sections 107 or 108 of the 1976 United States Copyright Act, without either the prior written permission of the Publisher, or authorization through payment of the appropriate per-copy fee to the Copyright Clearance Center, 222 Rosewood Drive, Danvers, MA 01923, 978-750-8400, fax 978-646-8600. Requests to the Publisher for permission should be addressed to the Permissions Department, John Wiley & Sons, Inc., 111 River Street, Hoboken, NJ 07030, (201) 748-6011, fax (201) 748-6008, or online at http://www.wiley.com/go/permissions.

Trademarks: Wiley, the Wiley Publishing logo, For Dummies, the Dummies Man logo, A Reference for the Rest of Us!, The Dummies Way, Dummies Daily, The Fun and Easy Way, Dummies.com and related trade dress are trademarks or registered trademarks of John Wiley & Sons, Inc. and/or its affiliates in the United States and other countries, and may not be used without written permission. All other trademarks are the property of their respective owners. Wiley Publishing, Inc., is not associated with any product or vendor mentioned in this book.

LIMIT OF LIABILITY/DISCLAIMER OF WARRANTY: THE PUBLISHER AND THE AUTHOR MAKE NO REPRESENTATIONS OR WARRANTIES WITH RESPECT TO THE ACCURACY OR COMPLETENESS OF THE CONTENTS OF THIS WORK AND SPECIFICALLY DISCLAIM ALL WARRANTIES, INCLUDING WITHOUT LIMITATION WARRANTIES OF FITNESS FOR A PARTICULAR PURPOSE. NO WARRANTY MAY BE CREATED OR EXTENDED BY SALES OR PROMOTIONAL MATERIALS. THE ADVICE AND STRATEGIES CONTAINED HEREIN MAY NOT BE SUITABLE FOR EVERY SITUATION. THIS WORK IS SOLD WITH THE UNDERSTANDING THAT THE PUBLISHER IS NOT ENGAGED IN RENDERING LEGAL, ACCOUNTING, OR OTHER PROFESSIONAL SERVICES. IF PROFESSIONAL ASSISTANCE IS REQUIRED, THE SERVICES OF A COMPETENT PROFESSIONAL PERSON SHOULD BE SOUGHT. NEITHER THE PUBLISHER NOR THE AUTHOR SHALL BE LIABLE FOR DAMAGES ARISING HEREFROM. THE FACT THAT AN ORGANIZATION OR WEBSITE IS REFERRED TO IN THIS WORK AS A CITATION AND/OR A POTENTIAL SOURCE OF FURTHER INFORMATION DOES NOT MEAN THAT THE AUTHOR OR THE PUBLISHER ENDORSES THE INFORMATION THE ORGANIZATION OR WEBSITE MAY PROVIDE OR RECOMMENDATIONS IT MAY MAKE. FURTHER, READERS SHOULD BE AWARE THAT INTERNET WEBSITES LISTED IN THIS WORK MAY HAVE CHANGED OR DISAPPEARED BETWEEN WHEN THIS WORK WAS WRITTEN AND WHEN IT IS READ.

For general information on our other products and services, please contact our Customer Care Department within the U.S. at 877-762-2974, outside the U.S. at 317-572-3993, or fax 317-572-4002.

For technical support, please visit www.wiley.com/techsupport.

Wiley also publishes its books in a variety of electronic formats. Some content that appears in print may not be available in electronic books.

Library of Congress Control Number: 2006932684

ISBN 978-0-471-79940-5

Manufactured in the United States of America

10 9 8 7 6 5

1O/RX/RQ/QW/IN

About the Author

Mark Ryan, a graduate of Brown University and the University of Wisconsin Law School, has been teaching math since 1989. He runs the Math Center in Winnetka, Illinois (www.themathcenter.com), where he teaches high school math courses including an introduction to geometry and a workshop for parents based on a program he developed, *The 10 Habits of Highly Successful Math Students*. In high school, he twice scored a perfect 800 on the math portion of the SAT, and he not only knows mathematics, he has a gift for explaining it in plain English. He practiced law for four years before deciding he should do something he enjoys and use his natural talent for mathematics. Ryan is a member of the Authors Guild and the National Council of Teachers of Mathematics.

Geometry Workbook For Dummies is Ryan's fourth book. *Everyday Math for Everyday Life* was published in 2002, *Calculus For Dummies* (Wiley) in 2003, and *Calculus Workbook For Dummies* (Wiley) in 2005. *Geometry For Dummies 2nd Ed.* (Wiley) unpublished in 2008. His math books have sold over a quarter of a million copies.

A tournament backgammon player and a skier and tennis player, Ryan lives in Chicago.

Author's Acknowledgments

Putting *Geometry Workbook for Dummies* together entailed an enormous amount of work, and I could not have done it alone. I'm grateful to my intelligent, professional, and computer-savvy assistants. Courtney Jones helped a great deal with typing and proofreading the highly technical manuscript. She has a good ear and eye for what's needed in technical writing. Veronica Berns assisted with editing both the book's prose and mathematics. She's a very good writer, and she knows geometry and has a great sense for how to explain it clearly. This is the second book Josh Dillon has helped me with. He also helped edit the book, including the very technical geometrical figures. He has a highly developed aptitude for the many subtleties involved in how to effectively communicate mathematics.

I'm very grateful to my business consultant, Josh Lowitz, Adjunct Associate Professor of Entrepreneurship at the University of Chicago Graduate School of Business. His intelligent and thoughtful contract negotiations on my behalf and his advice on my writing career and on all other aspects of my business have made him invaluable.

The book is a testament to the high standards of everyone at Wiley Publishing. Joyce Pepple, Acquisitions Director, once again handled the contract negotiations with intelligence, honesty, and fairness. Acquisitions Editor Kathy Cox skillfully kept the project on track. Technical Editor Alexis Venter did an excellent and thorough job spotting and correcting the errors that appeared in the book's first draft — some of which would be very difficult or impossible to find without an expert's knowledge of geometry. Copy Editor Danielle Voirol also did a great job correcting mathematical errors, she made many suggestions on how to improve the exposition, and she contributed a great deal to the book's quips and sarcasm. The layout and graphics teams did a fantastic job with the book's thousands of complex equations and mathematical figures. Finally, the book would not be what it is without the contributions of Project Editor Mike Baker. His skillful editing greatly improved the book's exposition, organization, and flow. And in the midst of all the pressing deadlines and the extra work involved in a technical book like this, his calm, positive, and professional attitude made him a pleasure to work with.

Finally, a special thanks to my main assistant, Alex Miller. I don't know what I would have done without her. She helped me with every aspect of the book's production: writing, typing, editing, proofreading, more writing, more editing . . . and still more editing. She also assisted with the creation of a number of geometry problems and proofs for the book. She knows mathematics and how to express it clearly to the novice. She understands the many nuances involved in effective writing. Everything she did was done with great humor, efficiency, and skill.

Publisher's Acknowledgments

We're proud of this book; please send us your comments through our Dummies online registration form located at www.dummies.com/register/.

Some of the people who helped bring this book to market include the following:

Acquisitions, Editorial, and Media Development

Project Editor: Mike Baker

Acquisitions Editor: Kathy Cox

Copy Editor: Danielle Voirol

Editorial Program Coordinator: Hanna K. Scott

Technical Reviewer: Alexsis Venter

Editorial Manager: Christine Meloy Beck

Editorial Assistants: Erin Calligan, David Lutton

Cartoons: Rich Tennant (www.the5thwave.com)

Composition Services

Project Coordinator: Jennifer Theriot

Layout and Graphics: Carrie Foster, Denny Hager, Stephanie D. Jumper, Barbara Moore

Proofreaders: Betty Kish, John Greenough, Jessica Kramer, Christy Pingleton

Indexer: Julie Kawabata

Publishing and Editorial for Consumer Dummies

Diane Graves Steele, Vice President and Publisher, Consumer Dummies

Joyce Pepple, Acquisitions Director, Consumer Dummies

Kristin A. Cocks, Product Development Director, Consumer Dummies

Michael Spring, Vice President and Publisher, Travel

Kelly Regan, Editorial Director, Travel

Publishing for Technology Dummies

Andy Cummings, Vice President and Publisher, Dummies Technology/General User

Composition Services

Gerry Fahey, Vice President of Production Services

Debbie Stailey, Director of Composition Services

Contents at a Glance

Introduction 1

Part I: Getting Started 7
Chapter 1: Introducing Geometry and Geometry Proofs 9
Chapter 2: Points, Segments, Lines, Rays, and Angles 19

Part II: Triangles 49
Chapter 3: Triangle Fundamentals and Other Cool Stuff 51
Chapter 4: Congruent Triangles 83

Part III: Polygons 113
Chapter 5: Quadrilaterals: Your Fine, Four-Sided Friends 115
Chapter 6: Area, Angles, and the Many Sides of Polygon Geometry 151
Chapter 7: Similarity: Size Doesn't Matter 165

Part IV: Circles 193
Chapter 8: Circular Reasoning 195
Chapter 9: Scintillating Circle Formulas 211

Part V: 3-D Geometry and Coordinate Geometry 227
Chapter 10: 2-D Stuff Standing Up 229
Chapter 11: Solid Geometry: Digging into Volume and Surface Area 243
Chapter 12: Coordinate Geometry, Courtesy of Descartes 257
Chapter 13: Transforming the (Geometric) World: Reflections, Rotations, and Translations 271

Part VI: The Part of Tens 285
Chapter 14: Ten (Plus) Incredibly Fantastic Strategies for Doing Proofs 287
Chapter 15: Ten Things You Better Know (for Geometry), or Your Name Is Mudd 291

Index 295

Table of Contents

Introduction 1

    About This Book 1
    Conventions Used in This Book 2
    How to Use This Book 2
    Foolish Assumptions 2
    How This Book Is Organized 3
        Part I: Getting Started 3
        Part II: Triangles 3
        Part III: Polygons 3
        Part IV: Circles 4
        Part V: 3-D Geometry and Coordinate Geometry 4
        Part VI: The Part of Tens 4
    Icons Used in This Book 4
    Where to Go from Here 5

Part I: Getting Started 7

Chapter 1: Introducing Geometry and Geometry Proofs 9

    What Is Geometry? 9
    Making the Right Assumptions 9
    If-Then Logic: If You Bought This Book, Then You Must Love Geometry! 12
    What's a Geometry Proof? 14
    Solutions for Introducing Geometry and Geometry Proofs 17

Chapter 2: Points, Segments, Lines, Rays, and Angles 19

    Basic Definitions 19
    Union and Intersection Problems 20
    Division in the Ranks: Bisection and Trisection 22
    Perfect Hilarity for Perpendicularity 24
    You Complete Me: Complementary and Supplementary Angles 26
    Adding and Subtracting Segments and Angles 30
    Multiplying and Dividing Angles and Segments 33
    X Marks the Spot: Using Vertical Angles 37
    Switching It Up with the Transitive and Substitution Properties 39
    Solutions for Points, Segments, Lines, Rays, and Angles 42

Part II: Triangles 49

Chapter 3: Triangle Fundamentals and Other Cool Stuff 51

    Triangle Types and Triangle Basics 51
    Altitudes, Area, and the Super Hero Formula 55
    Balancing Things Out with Medians and Centroids 58
    Three More "Centers" of a Triangle 59
    The Pythagorean Theorem 64
    Pythagorean Triple Triangles 67
    Unique Degrees: Two Special Right Triangles 70
    Solutions for Triangle Fundamentals and Other Cool Stuff 74

Chapter 4: Congruent Triangles 83
Sizing Up Three Ways to Prove Triangles Congruent 83
Corresponding Parts of Congruent Triangles Are Congruent (CPCTC) 89
Isosceles Rules: If Sides, Then Angles; If Angles, Then Sides 93
Two More Ways to Prove Triangles Congruent 96
The Two Equidistance Theorems 99
Solutions for Congruent Triangles 104

Part III: Polygons 113

Chapter 5: Quadrilaterals: Your Fine, Four-Sided Friends 115
Double-Crossers: Transversals and Their Parallel Lines 115
Quadrilaterals: It's a Family Affair 120
Properties of the Parallelogram and the Kite 123
Properties of Rhombuses, Rectangles, and Squares 127
Properties of Trapezoids and Isosceles Trapezoids 130
Proving That a Quadrilateral Is a Parallelogram or a Kite 132
Proving That a Quadrilateral Is a Rhombus, Rectangle, or Square 136
Solutions for Quadrilaterals: Your Fine, Four-Sided Friends 139

Chapter 6: Area, Angles, and the Many Sides of Polygon Geometry 151
Square Units: Finding the Area of Quadrilaterals 151
A Standard Formula for the Area of Regular Polygons 155
More Fantastically Fun Polygon Formulas 157
Solutions for Area, Angles, and the Many Sides of Polygon Geometry 160

Chapter 7: Similarity: Size Doesn't Matter 165
Defining Similarity 165
Proving Triangles Similar 168
Corresponding Sides and CSSTP — Cats Stalk Silently Then Pounce 172
Similar Rights: The Altitude-on-Hypotenuse Theorem 175
Three More Theorems Involving Proportions 178
Solutions for Similarity: Size Doesn't Matter 183

Part IV: Circles 193

Chapter 8: Circular Reasoning 195
The Segments Within: Radii and Chords 195
Introducing Arcs, and Central Angles 199
Touching on Radii and Tangents 202
Solutions for Circular Reasoning 206

Chapter 9: Scintillating Circle Formulas 211
Pizzas, Slices, and Crusts: Area and "Perimeter" of Circles, Sectors,
  and Segments 211
Angles, Circles, and Their Connections:
  The Angle-Arc Theorems and Formulas 214
The Power Theorems That Be 217
Solutions for Scintillating Circle Formulas 221

Part V: 3-D Geometry and Coordinate Geometry 227

Chapter 10: 2-D Stuff Standing Up 229
Lines Perpendicular to Planes: They're All Right 229
Parallel, Perpendicular, and Intersecting Lines and Planes 233
Solutions for 2-D Stuff Standing Up 238

Chapter 11: Solid Geometry: Digging into Volume and Surface Area 243
Starting with Flat-Top Figures 243
Sharpening Your Skills with Pointy-Top Figures 246
Rounding Out Your Understanding with Spheres 249
Solutions for Solid Geometry 251

Chapter 12: Coordinate Geometry, Courtesy of Descartes 257
Formulas, Schmormulas: Slope, Distance, and Midpoint 257
Mastering Coordinate Proofs with Algebra 260
Using the Equations of Lines and Circles 261
Solutions for Coordinate Geometry, Courtesy of Descartes 264

Chapter 13: Transforming the (Geometric) World: Reflections, Rotations, and Translations 271
Reflections on Mirror Images 271
Lost in Translation 274
So You Say You Want a . . . Rotation? 276
Working with Glide Reflections 278
Solutions for Transforming the World 281

Part VI: The Part of Tens 285

Chapter 14: Ten (Plus) Incredibly Fantastic Strategies for Doing Proofs 287
Look for Congruent Triangles 287
Try to Find Isosceles Triangles 287
Look for Radii, and Draw More Radii 288
Look for Parallel Lines 288
Make a Game Plan 288
Use All the Givens 288
Check Your If-Then Logic 288
Work Backwards 289
Make Up Numbers for Segments and Angles 289
Think Like a Computer 289
Bonus! Number 11 (Like the Amp in This Is Spinal Tap That Goes Up to 11): Do Something! 290

Chapter 15: Ten Things You Better Know (for Geometry), or Your Name Is Mudd 291
The Pythagorean Theorem (the Queen of All Geometry Theorems) 291
Special Right Triangles 291
Area Formulas 292
Sum of Angles 292
Circle Formulas 292

Angle-Arc Theorems 292
Power Theorems 293
Coordinate Geometry Formulas 293
Volume Formulas 293
Surface Area Formulas 294

Index 295

Introduction

If you've already bought this book, then you have my undying respect and admiration (not to mention — cha ching — that with my royalty from the sale of this book, I can now afford, oh, say, half a cup of coffee). And if you're just thinking about buying it, well, what are you waiting for, for crying out loud? Buying this book (and its excellent companion volume, *Geometry For Dummies*), can be an important first step on the road to gaining a solid grasp of a subject — and now I'm being serious — that is full of mathematical richness and beauty. By studying geometry, you take part in a long tradition going back at least as far as Pythagoras (although he was one of the early, well-known mathematicians to study geometry, he was certainly not the first). There is no mathematician, great or otherwise, who has not spent some time studying geometry.

I spend a great deal of time in this book explaining how to do geometry proofs. *Many* students have a lot of difficulty when they attempt their first proofs. I can think of a few reasons for this. First, geometry proofs, as well as the rest of geometry, have a spatial aspect that many students find challenging. Second, proofs lack the cut-and-dried nature of most of the math that students are accustomed to (in other words, with geometry proofs there are *way* more instances where there are many correct ways to proceed, and this takes some getting used to). And third, proofs are, in a sense, only half math. The other half is deductive logic — something new for most students, and something that has a significant *verbal* component. The good news is that if you practice the dozen or so strategies and tips for doing proofs presented in this book, you should have little difficulty getting the hang of it. These strategies and tips work like a charm and make many proofs much easier than they initially seem.

About This Book

Geometry Workbook For Dummies, like *Geometry For Dummies,* is intended for three groups of readers:

- High school students (and possibly junior high students) taking a standard geometry course with the traditional emphasis on geometry proofs
- The parents of geometry students
- Anyone of any age who is curious about this interesting subject, which has fascinated both mathematicians and laymen for well over two thousand years

Whenever possible, I explain geometry concepts and problem solutions with a minimum of technical jargon. I take a common-sense or street-smart approach when explaining mathematics, and I try to avoid the often stiff and formal style used in too many textbooks. You get answer explanations for every practice problem. And with proofs, in addition to giving you the steps of the solutions, I show you the thought process behind the solutions. I supplement the problem explanations with tips, shortcuts, and mnemonic devices. Often, a simple tip or memory trick can make learning and retaining a new, difficult concept much easier. The pages here should contain enough blank space to allow you to write out your solutions right in the book.

Conventions Used in This Book

This book uses certain conventions:

- Variables are in *italics*.
- Important math terms are often in *italics* and are defined when necessary. These terms may be **bolded** when they appear as keywords within a bulleted list. Italics are also used for emphasis.
- As in most geometry books, figures are not necessarily drawn to scale.
- Extra hard problems are marked with an asterisk. Don't try these problems on an empty stomach!
- For proof problems, don't assume that the number of blank lines (where you'll put your solution) corresponds exactly to the number of steps needed for the proof.

How to Use This Book

Like all *For Dummies* books, you can use this book as a reference. You don't need to read it cover to cover or work through all problems in order. You may need more practice in some areas than others, so you may choose to do only half of the practice problems in some sections, or none at all.

However, as you'd expect, the order of the topics in *Geometry Workbook For Dummies* roughly follows the order of the traditional high school geometry course. You can, therefore, go through the book in order, using it to supplement your coursework. If I do say so myself, I expect you'll find that many of the explanations, methods, strategies, and tips in this book will make problems you found difficult or confusing in class seem much easier.

I give hints for many problems, but if you want to challenge yourself, you may want to cover them up and attempt the problem without the hint.

And if you get stuck while doing a proof, you can try reading a little bit of the "game plan" or the solution to the proof. These aids are in the solutions section at the end of every chapter. But don't read too much at first. Read a small amount and see whether it gives you any ideas. Then, if you're still having trouble, read a little more.

Foolish Assumptions

Here's what I'm assuming about you — fool that I am. ("A fool thinks himself to be wise, but a wise man knows himself to be a fool." William Shakespeare).

- You're no slouch — and therefore, you have at least some faint glimmer of curiosity about geometry (or maybe you're totally, stark raving mad with desire to learn the subject?). How could people possibly have no curiosity at all about geometry, assuming they're not in a coma? You are literally surrounded by shapes, and every shape involves geometry.
- You haven't forgotten basic algebra. You need very little algebra for geometry, but you do need some, so you might want to brush up a bit. But even if your algebra is a tad rusty, I doubt you'll have any trouble with the algebra in this book: solving simple equations, using simple formulas, doing square roots, and so on.

✓ You're willing to invest some time and effort in doing these practice problems. With geometry — as with anything — practice makes perfect, and practice sometimes involves struggle. But that's a good thing. Ideally, you should give these problems your best shot before you turn to the solutions. Reading through the solutions can be a good way to learn, but you'll usually remember more if you first push yourself to solve the problems on your own — even if that means going down a few dead ends.

How This Book Is Organized

Like all *For Dummies* books, this one is divided into parts, the parts into chapters, and the chapters into sections. Brilliant! Here's a preview of what you can look forward to.

Part I: Getting Started

Part I gives you an introduction to geometry in general and to geometry proofs in particular. The basic logical structure of proofs is explained, and you go through the first dozen or so theorems you need for proofs. You get several great strategies and tips for doing proofs, and then you do lots of practice problems.

Part II: Triangles

In Part II, you first take a break from doing proofs and work out non-proof problems involving all sorts of interesting things about triangles: area, altitudes, medians, angle bisectors, perpendicular bisectors, the Pythagorean Theorem, families of right triangles, and so on.

Then you get back to proofs. Triangles are the main event in geometry proofs — nothing else comes close. And the triangle proofs in Part II are probably the most important ones in this book. You get lots of practice proving that triangles are congruent and then using CPCTC (Congruent Parts of Congruent Triangles are Congruent). Congruent triangles and their congruent parts form the core or focus of a great number of geometry proofs.

Part III: Polygons

Polygons, schmolygons. In this part, you practice problems involving quadrilaterals (four-sided shapes like parallelograms, rhombuses, rectangles, squares, kites, and trapezoids), pentagons (five sides), hexagons (six sides), and on up. You study the many properties of the different quadrilaterals, and you find out how to prove that a four-sided figure qualifies as a particular type of quadrilateral. After that, you discover how to do cool things like compute the area of a polygon, the number of its diagonals, and the sum of its interior angles. I bet you can hardly wait!

In the last chapter of Part III, you practice problems involving *similar* polygons — that is, polygons of the exact same shape but of different sizes.

Part IV: Circles

No shape is more fundamental than the circle. It is the simplest of all shapes and, at the same time, one rich in mathematical depth and beauty. As you know, it involves the number π (3.14159265...), one of the most important numbers in all of mathematics. The circle has fascinated mathematicians for millennia. Now *you* get a chance to study it. How great is that?

In this part, you work on problems involving radii, diameters, and chords of circles; angles formed inside, outside, and on a circle; and lines tangent to a circle. You also do practice problems involving the area and circumference of circles, the area and arc lengths of sections, and the power theorems. What fun!

Part V: 3-D Geometry and Coordinate Geometry

Pyramids, cones, prisms, cylinders, boxes, cubes, and spheres are the shapes you work on in Chapters 10 and 11. In contrast to all those wispy, unsubstantial two-dimensional shapes you study in Parts I through IV, these full-bodied three-dimensional shapes can actually exist in our real, three-dimensional world. You practice problems involving the volume and surface area of these shapes. You also take a look at parallel planes and 2-D shapes "standing up" in the third dimension.

Then I switch gears on you. Up to this point in the book, all the shapes you work with are sort of floating on the page with no specified location or orientation. You're asked, for example, to compute the area of some triangle, and the problem has nothing to do with where the triangle is or which way it's facing. In Chapters 12 and 13, everything has to do with location. Geometric objects, in this part, have specific locations in the *x-y* coordinate system. You practice coordinate geometry problems involving the distance formula, the midpoint formula, and slope — things you probably studied in algebra. You also study what happens to shapes when you rotate them, slide them, or flip them over.

Part VI: The Part of Tens

Here you get the top ten geometry formulas and the top ten strategies for doing proofs. The formulas come in handy for non-proof problems. The proof strategies can make many otherwise difficult proofs much easier to do. If you learn them well and have them at the ready when you're doing proofs, you won't get stuck very often, and if you do get stuck, you'll know what to do.

Icons Used in This Book

Look for the following icons to quickly spot important information:

Next to this icon are definitions of geometry terms, explanations of geometry principles, and a few things you should know from algebra. You often use geometry definitions in the reason column of two-column proofs.

Introduction

This icon is next to all example problems — like, duh.

This tip icon gives you shortcuts, memory devices, strategies, and so on.

Ignore these icons, and you may end up doing lots of extra work and getting the wrong answer — and then you could fail geometry, become unpopular, and lose any hope of becoming homecoming queen or king. Better safe than sorry, right?

This icon identifies the theorems and postulates that you'll use to form the chain of logic in geometry proofs. You use them in the reason column of two-column proofs. A *theorem* is an if-then statement, like "if angles are supplementary to the same angle, then they are congruent." You use *postulates* basically the same way that you use theorems. The difference between them is sort of a mathematical technicality (which I wouldn't sweat if I were you).

Where to Go from Here

You can go

- To Chapter 1
- To whatever chapter contains the concepts you need to practice
- To *Geometry For Dummies* for more in-depth explanations
- To the movies
- To the beach
- Into your geometry final to kick some @#%$!
- On to bigger and better things

Part I
Getting Started

"This is my old geometry teacher, Mr. Wendt, his wife Doris, and their two children, Obtuse and Acute."

In this Part . . .

One of the main challenges you're faced with in your geometry course is getting a handle on two-column geometry proofs. But proofs really aren't so tough once you get used to them. If you pay careful attention to the explanation in Chapter 1 of the basic structure of proofs and the way their chain of reasoning works, you'll wonder what all the fuss is about. Then in Chapter 2, you can practice your skills on lots of proof problems. And don't forget: Practice makes perfect.

Chapter 1

Introducing Geometry and Geometry Proofs

In This Chapter
- ▶ Defining geometry
- ▶ Examining theorems and if-then logic
- ▶ Geometry proofs — the formal and the not-so-formal

*I*n this chapter, you get started with some basics about geometry and shapes, a couple points about deductive logic, and a few introductory comments about the structure of geometry proofs. Time to get started!

What Is Geometry?

What is geometry?! C'mon, everyone knows what geometry is, right? *Geometry* is the study of shapes: circles, triangles, rectangles, pyramids, and so on. (If you didn't know that, you may want to look into hanging up your protractor.) Shapes are all around you. The desk or table where you're reading this book has a shape. You can probably see a window from where you are, and it's probably a rectangle. The pages of this book are also rectangles. Your pen or pencil is roughly a cylinder (or maybe a right hexagonal prism — see Part V for more on solid figures). Your iPod has a circular dial. Shapes are ubiquitous — in our world, anyway.

For the philosophically inclined, here's an exercise that goes *way* beyond the scope of this book: Try to imagine a world — some sort of different universe — where there aren't various objects with different shapes. (If you're into this sort of thing, check out *Philosophy For Dummies*.)

Making the Right Assumptions

Okay, so geometry is the study of shapes. And how can you tell one shape from another? From the way it looks, of course. But — this may seem a bit bizarre — when you're studying geometry, you're sort of *not* supposed to rely on the way shapes look. The point of this strange treatment of geometric figures is to prohibit you from claiming that something is true about a figure merely because it looks true, and to force you, instead, to *prove* that it's true by airtight, mathematical logic.

When you're working with shapes in any other area of math, or in science, or in, say, architecture or design, paying attention to the way shapes look is very important: their proportions, their angles, their orientation, how steep their sides are, and so on. Only in a geometry course are you supposed to ignore to some degree the appearance of the shapes you study.

10 Part I: Getting Started

When you look at a diagram in this or any geometry book, you *cannot* assume any of the following just from the appearance of the figure:

- **Right angles:** Just because an angle looks like an exact 90° angle, that doesn't necessarily mean it is one.
- **Congruent angles:** Just because two angles look the same size, that doesn't mean they really are. (As you probably know, *congruent* [symbolized by ≅] is a fancy word for "equal" or "same size.")
- **Congruent segments:** Ditto, like for congruent angles. You can't assume segments are the same size.
- **Relative sizes of segments and angles:** Just because, say, one segment is drawn to look longer than another in a diagram, it doesn't follow that the segment really is longer.

Sometimes size relationships are marked on the diagram. For instance, a small L-shaped mark in a corner means that you have a right angle. Tick marks can indicate congruent parts. Basically, if the tick marks match, you know the segments or angles are the same size.

You can assume pretty much anything not on this list; for example, if a line looks straight, it really is straight.

Before doing the following problems, you may want to peek ahead to Chapters 3 and 5 if you've forgotten or don't know the names of various triangles and quadrilaterals.

Q. What can you assume and what can't you assume about *SIMON?*

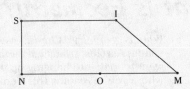

A. You *can* assume that

- $\overline{MN}$ (line segment *MN*) is straight; in other words, there's no bend at point *O*.

    Another way of saying the same thing is that ∠*MON* is a *straight angle* or a 180° angle.

- $\overline{NS}$, $\overline{SI}$, and $\overline{IM}$ are also straight as opposed to curvy.
- Therefore, *SIMON* is a quadrilateral because it has four straight sides.

    (If you couldn't assume that $\overline{MN}$ is straight, there could actually be a bend at point *O* and then *SIMON* would be a pentagon, but that's not possible.)

That's about it for what you can assume. If this figure were anywhere else other than a geometry book, you could safely assume all sorts of other things — including that *SIMON* is a trapezoid. But this *is* a geometry book, so you *can't* assume that.

You also *can't* assume that

- ∠*S* and ∠*N* are right angles.
- ∠*I* is an obtuse angle (an angle greater than 90°).
- ∠*M* is an acute angle (an angle less than 90°).

- ∠*I* is greater than ∠*M* or ∠*S* or ∠*N*, and ditto for the relative sizes of other angles.
- $\overline{NS}$ is shorter than $\overline{SI}$ or $\overline{MN}$, and ditto for the relative lengths of the other segments.
- *O* is the midpoint of $\overline{MN}$.
- $\overline{SI}$ is parallel to $\overline{MN}$.
- *SIMON* is a trapezoid.

The "real" *SIMON* could — weird as it seems — actually look like this:

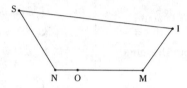

1. What type of quadrilateral is *AMER*? **Note:** See Chapter 5 for types of quadrilaterals.

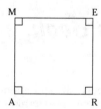

Solve It

2. What type of quadrilateral is *IDOL*?

Solve It

3. Use the figure to answer the following questions (Chapter 3 can fill you in on triangles):

a. Can you assume that the triangles are congruent?

b. Can you conclude that △ABC is acute? Obtuse? Right? Isosceles (with at least two equal sides)? Equilateral (with three equal sides)?

c. Can you conclude that △DEF is acute? Obtuse? Right? Isosceles? Equilateral?

d. What can you conclude about the length of $\overline{EF}$?

e. Might ∠D be a right angle?

f. Might ∠F be a right angle?

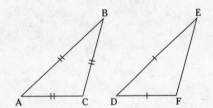

Solve It

4. Can you assume or conclude

a. △ABC ≅ △WXY?

b. △ABD ≅ △CBD?

c. △ABD ≅ △WXZ?

d. △ABC is isosceles?

e. D is the midpoint of $\overline{AC}$?

f. Z is the midpoint of $\overline{WY}$?

g. $\overline{BD}$ is an altitude (height) of △ABC?

h. ∠ADB is supplementary to ∠CDB (∠ADB + ∠CDB = 180°)?

i. △XYZ is a right triangle?

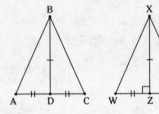

Solve It

If-Then Logic: If You Bought This Book, Then You Must Love Geometry!

Geometry *theorems* (and their first cousins, *postulates*) are basically statements of geometrical truth like "All radii of a circle are congruent." As you can see in this section and in the rest of the book, theorems (and postulates) are the building blocks of proofs. (I may get hauled over by the geometry police for saying this, but if I were you, I'd just glom theorems and postulates together into a single group because, for the purposes of doing proofs, they work the same way. Whenever I refer to theorems, you can safely read it as "theorems and postulates.")

Geometry theorems can all be expressed in the form "*If* blah, blah, blah, *then* blah, blah, blah," like "If two angles are right angles, then they are congruent" (though mathematicians — like you — often write theorems in some shorter way, like "All right angles are congruent"). You may want to flip through the book looking for theorem icons to get a feel for what theorems look like.

An important thing to note here is that the reverse of a theorem is not necessarily true. For example, the statement "If two angles are congruent, then they are right angles" is false. When a theorem does work in both directions, you get two separate theorems, one the reverse of the other.

The fact that theorems are not generally reversible should come as no surprise. Many ordinary statements in *if-then* form are, like theorems, not reversible: "If it's a ship, then it's a boat" is true, but "If it's a boat, then it's a ship" is false, right? (It might be a canoe.)

Geometry definitions (like all definitions), however, are reversible. Consider the definition of *perpendicular:* two lines are perpendicular if they intersect at right angles. Both if-then statements are true: 1) "If lines intersect at right angles, then they are perpendicular," and 2) "If lines are perpendicular, then they intersect at right angles." When doing proofs, you have occasion to use both forms of many definitions.

Q. Read through some theorems.

a. Give an example of a theorem that's not reversible and explain why the reverse is false.

b. Give an example of a theorem whose reverse is another true theorem.

A. A number of responses work, but here's how you could answer:

a. "If two angles are vertical angles, then they are congruent." The reverse of this theorem is obviously false. Just because two angles are the same size, it does not follow that they must be vertical angles (when two lines intersect and form an X, vertical angles are the angles straight across from each other — turn to Chapter 2 for more info).

b. Two of the most important geometry theorems are a reversible pair: "If two sides of a triangle are congruent, then the angles opposite the sides are congruent" and "If two angles of a triangle are congruent, then the sides opposite the angles are congruent." (For more on these isosceles triangle theorems, check out Chapter 4.)

5. Give two examples of theorems that are not reversible and explain why the reverse of each is false. *Hint:* Flip through this book or your geometry textbook looking at various theorems. Try reversing them and ask yourself whether they still work.

Solve It

6. Give two examples of theorems that work in both directions. *Hint:* See the hint for question 5.

Solve It

What's a Geometry Proof?

Many students find two-column geometry proofs difficult, but they're really no big deal once you get the hang of them. Basically, they're just arguments like the following, in which you brilliantly establish that your Labradoodle, Fifi, will not lay any eggs on the Fourth of July:

1. Fifi is a Labradoodle.
2. Therefore, Fifi is a dog, because all Labradoodles are dogs.
3. Therefore, Fifi is a mammal, because all dogs are mammals.
4. Therefore, Fifi will never lay any eggs, because mammals don't lay eggs (okay, okay . . . except for platypuses and spiny anteaters, for you monotreme-loving nitpickers out there).
5. Therefore, Fifi will not lay any eggs on the Fourth of July, because if she will never lay any eggs, she can't lay eggs on the Fourth of July.

In a nutshell: Labradoodle → dog → mammal → no eggs → no eggs on July 4. It's sort of a domino effect. Each statement knocks over the next till you get to your final conclusion.

Check out Figure 1-1 to see what this argument or proof looks like in the standard two-column geometry proof format.

Given: Fifi is a Labradoodle.
Prove: Fifi will not lay eggs on the Fourth of July.

Statements (or Conclusions) These are the specific claims you make.	Reasons (or Justifications) These are the general rules you use to justify your claims. If after each claim you made, I said, "How do you know?" your response to me goes in this column.
I claim...	**How do I know?**
1) Fifi is a Labradoodle.	1) Because it was given as a fact.
2) Fifi is a dog.	2) Because all Labradoodles are dogs.
3) Fifi is a mammal.	3) Because all dogs are mammals.
4) Fifi doesn't lay eggs.	4) Because mammals don't lay eggs.
5) Fifi will not lay eggs on the Fourth of July.	5) Because something that doesn't lay eggs can't lay eggs on the Fourth of July.

Figure 1-1: A standard two-column proof listing statements and reasons.

Chapter 1: Introducing Geometry and Geometry Proofs

Note that the left-hand column contains *specific* facts (about one particular dog, Fifi), while the right-hand column contains *general* principles (about dogs in general or mammals in general). This format is true of all geometry proofs.

Now look at the very same proof in Figure 1-2; this time, the reasons appear in *if-then* form. When reasons are written this way, you can see how the chain of logic flows.

In a two-column proof, the idea or ideas in the *if* part of each reason must come from the statement column somewhere *above* the reason; and the single idea in the *then* part of the reason must match the idea in the statement on *the same line* as the reason. This incredibly important flow-of-logic structure is shown with arrows in the following proof.

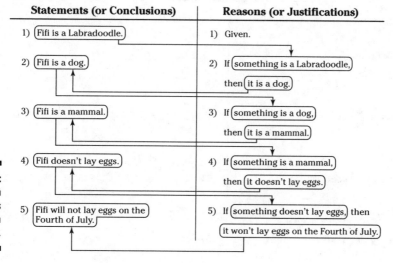

Figure 1-2:
A proof with the reasons written in if-then form.

In the preceding proof, each *if* clause uses only a single idea from the statement column. However, as you can see in the following practice problem, you often have to use more than one idea from the statement column in an *if* clause.

Part I: Getting Started

7. In the following facetious and somewhat fishy proof, fill in the missing reasons in *if-then* form and show the flow of logic as I do in Figure 1-2.

Given: You forgot to set your alarm last night.

You've already been late for school twice this term.

Prove: You will get a detention at school today.

Note: To complete this "proof," you need to know the school's late policy: A student who is late for school three times in one term will be given a detention.

Statements (or Conclusions)	Reasons (or Justifications)
1) I forgot to set my alarm last night.	1) Given.
2) I will wake up late.	2)
3) I will miss the bus.	3)
4) I will be late for school.	4)
5) I've already been late for school twice this term.	5) Given.
6) This will be the third time this term I'll have been late.	6)
7) I'll get a detention at school today.	7)

Solutions for Introducing Geometry and Geometry Proofs

1 *AMER* looks like a square, but you can't conclude that because you can't assume the sides are equal. You do know, however, that the figure is a rectangle because it has four sides and four right angles.

2 *IDOL* also looks like a square, but this time you can't conclude that because you can't assume that the angles are right angles. But because you do know that *IDOL* has four equal sides, you do know that it's a rhombus.

3 Here are the answers (flip to Chapter 3 if you need to go over triangle classification):

  a. No. The triangles look congruent, but you're not allowed to assume that.

  b. The tick marks tell you that △*ABC* is equilateral. It is, therefore, an acute triangle and an isosceles triangle. It is neither a right triangle nor an obtuse triangle.

  c. The tick marks tell you that △*DEF* is isosceles and that, therefore, it is not scalene. That's all you can conclude. It may or may not be any of the other types of triangles.

  d. Nothing. $\overline{EF}$ could be the longest side of the triangle, or the shortest, or equal to the other two sides. And it may or may not have the same length as $\overline{BC}$.

  e. Yes. ∠*D* may be a right angle, though you can't assume that it is.

  f. No. (If you got this question, give yourself a pat on the back.) If ∠*F* were a right angle, △*DEF* would be a right triangle with $\overline{DE}$ its hypotenuse. But $\overline{DE}$ is the same length as $\overline{DF}$, and the hypotenuse of a right triangle has to be the triangle's longest side.

4 Here are the answers:

  a. No. The triangles might not be congruent in any number of ways. For example, you know nothing about the length of $\overline{ZY}$, and if $\overline{ZY}$ were, say, a mile long, the triangles would obviously not be congruent.

  b. No. The triangles would be congruent only if ∠*ADB* and ∠*CDB* were right angles, but you don't know that. Point *B* is free to move left or right, changing the measures of ∠*ADB* and ∠*CDB*.

  c. No. You don't know that ∠*ADB* is a right angle.

  d. No. The figure *looks* isosceles, but you're not allowed to assume that $\overline{AB} \cong \overline{CB}$.

  e. Yes. The tick marks show it.

  f. No. Like with part *a.*, you know nothing about the length of $\overline{ZY}$.

  g. No. You can't assume that $\overline{BD} \perp \overline{AC}$ (the upside-down *T* means "is perpendicular to").

  h. Yes. You *can* assume that $\overline{AC}$ is straight and that ∠*ADC* is 180°; therefore, that ∠*ADB* and ∠*CDB* must add up to 180°.

  i. Yes. ∠*WZY* is 180° and ∠*WZX* is 90°, so ∠*YZX* must also be 90°.

5 Answers vary. One example is "If angles are complementary to the same angle, then they're congruent." The reverse of this is false because many angles, like obtuse angles, do not have complements (obtuse angles are already bigger than 90°, so you can't add another angle to them to get a right angle).

6 Answers vary. Any of the parallel line theorems in Chapter 2 makes a good answer: for example, "If two parallel lines are cut by a transversal, then alternate interior angles are congruent." In short, both of the following are true: "If lines are parallel, then alternate interior angles are congruent," and "If alternate interior angles are congruent, then lines are parallel."

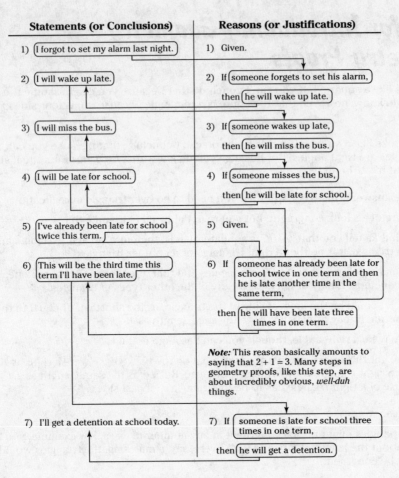

I sure hope it goes without saying that this is *not* an airtight, mathematical proof.

Chapter 2
Points, Segments, Lines, Rays, and Angles

In This Chapter
▶ Walking a fine line: Semi-precise definitions of geometry terms
▶ Working with union and intersection problems
▶ Supplementary and complementary angles (free stuff!)
▶ Turning to right angles
▶ Using angle and segment arithmetic
▶ Spotting vertical angles
▶ Standing in: Substitution and transitivity

In this chapter, you first review the building blocks of geometry: points, segments, lines, rays, and angles. Then you get your first taste of the meat of this course: geometry proofs. Do the practice problems carefully, and make sure you understand their solutions. Everything in the subsequent chapters builds on the important concepts presented here.

For more information on parallel lines, transversals, and the angles they create, see Chapter 5.

Basic Definitions

You've probably known what the following things are for a few years, but here are their definitions and undefinitions anyway. That's right — I said *un*definitions. Technically, *point* and *line* are *un*defined terms, so the first two "definitions" below aren't technically definitions. But if I were you, I wouldn't sweat this technicality.

- **Point:** You know, like a dot except that it actually has no size at all. Or you could say that it's infinitely small (that's pretty small, eh? But even "*infinitely* small" makes a point sound larger than it really is).

- **Line:** A line's like a thin straight wire (actually, it's infinitely thin or, even better, it has *no width at all* — nada). Don't forget that it goes on forever in both directions, which is why you use the little double-headed arrow as in $\overleftrightarrow{AB}$ (read as "line *AB*"; this is the line that goes through points *A* and *B*). Because lines go on forever, no matter how you tilt them or how good your shoehorn is, you can't fit them in the universe.

- **Line segment** or just **segment:** A segment is a section of a line that has two endpoints. If it goes from *C* to *D*, you call it "segment *CD*" and write it like $\overline{CD}$. (You can also call it and write it $\overline{DC}$.)

  Note: CD without the segment bar over it indicates the *length* of the segment as opposed to the segment itself.

20 Part I: Getting Started

- **Ray:** A ray is a section of a line (sort of half a line) that has one endpoint and goes on forever in the other direction. If its endpoint is point *M* and the ray goes through point *N* and then past it forever, you call the half-line "ray *MN*" and write $\overrightarrow{MN}$. The endpoint always comes first.

- **Angle:** Two rays with the same endpoint form an angle. The common endpoint is called the *vertex* of the angle. An *acute angle* is less than 90°; a *right angle* is, of course, a 90° angle; an *obtuse angle* has a measure greater than 90°; and a *straight angle* has a measure of 180° (which is kinda weird, because a 180° angle looks just like a line or a segment like ∠ACE in the example below).

  Note: Technically, angles go on forever, and their sides are rays that go on forever. This is the case even when an angle in a figure has segments for its sides instead of rays. (It's like the rays are really there even though they're not drawn.)

Union and Intersection Problems

And now for something completely different. The *intersection* (∩) of two geometric objects is where they overlap or touch. The *union* (∪) of two objects contains all of each object including the overlapping portion (if any).

Q. For the figure on the right, determine the following and write your answer in as many ways as possible.

  a. $\overrightarrow{AE} \cap \overrightarrow{CA}$
  b. $\overrightarrow{AE} \cup \overrightarrow{CA}$
  c. ∠BDE ∩ $\overleftrightarrow{ED}$
  d. ∠BDE ∪ $\overrightarrow{DE}$

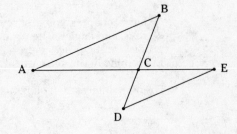

A. Here's how this problem shapes up:

  a. $\overrightarrow{AE} \cap \overrightarrow{CA} = \overline{AC}$ or $\overline{CA}$
  b. $\overrightarrow{AE} \cup \overrightarrow{CA} = \overleftrightarrow{AC}$ or $\overleftrightarrow{CA}$ or $\overleftrightarrow{AE}$ or $\overleftrightarrow{EA}$ or $\overleftrightarrow{CE}$ or $\overleftrightarrow{EC}$

  Tip: If you find some of these union and intersection problems tricky, you're not alone. Here's a great way to do them or to think about them. Imagine that the first object is colored blue and the second, yellow (or you can actually color them). Blue and yellow make green, right? So wherever you see (or imagine) green, that's the intersection. And the union contains anything that's blue or yellow or green. Another way to do these problems is to trace over each object. Wherever you traced twice, that's the intersection. And wherever you did any tracing (once or twice), that's the union.

  Remember: However you do these problems, lines, rays, and angles go on forever even if the diagram makes it look like they end.

  c. ∠BDE ∩ $\overleftrightarrow{ED}$ = $\overrightarrow{DE}$

  If ∠BDE is blue and $\overleftrightarrow{ED}$ is yellow, $\overrightarrow{DE}$ will be green.

  d. ∠BDE ∪ $\overrightarrow{DE}$ = ∠BDE or ∠EDB or ∠CDE or ∠EDC

  Note that sometimes the answer to a union or intersection problem is one of the original objects.

Chapter 2: Points, Segments, Lines, Rays, and Angles

Use the following figure to answer problems 1–6.

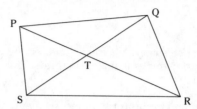

---

1. $\overline{ST} \cap \overline{TQ}$

Solve It

2. $\overline{ST} \cup \overline{PT}$

Solve It

---

3. $\angle RTS \cap \vec{PR}$

Solve It

4. $\vec{RT} \cap \vec{TP}$

Solve It

---

5. $\overline{PS} \cap \overleftrightarrow{QR}$

Solve It

6. $\vec{ST} \cup \vec{QT}$

Solve It

22 Part I: Getting Started

Division in the Ranks: Bisection and Trisection

In this section, you practice something you've understood almost since you first rode a bicycle or tricycle: cutting things in half or in thirds. This geometry is kids' stuff. Check out these definitions:

- **Segment bisection** and **midpoint:** A point, segment, ray, or line that divides a segment into two congruent segments *bisects* the segment. The point of bisection is called the *midpoint* of the segment. The midpoint, obviously, cuts the segment in half.

- **Segment trisection:** Two things (points, segments, rays, lines, or any combination of these) that divide a segment into three congruent segments *trisect* the segment. The points of trisection are called — get this — the *trisection points* of the segment.

- **Angle bisection:** A ray that cuts an angle into two congruent angles *bisects* the angle. It's called the *bisector* of the angle, or the *angle bisector*.

- **Angle trisection:** Two rays that divide an angle into three congruent angles *trisect* the angle. They're called *trisectors* of the angle, or *angle trisectors*.

Q. For the triangle on the right, given that $\vec{CD}$ bisects $\angle ACB$:

   a. Find the measure of $\angle BCD$.

   b. Other than the fact that $\angle ACD \cong \angle BCD$, can you conclude anything else about this figure?

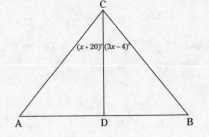

A. Given that $\vec{CD}$ bisects $\angle ACB$:

   a. You can find the measure of $\angle BCD$ in two steps. First, because $\vec{CD}$ bisects $\angle ACB$, $\angle ACD \cong \angle BCD$, so you can set them equal to each other and solve for x:

   $$x + 20 = 3x - 4$$
   $$-2x = -24$$
   $$x = 12$$

   Now plug 12 into the measure of $\angle BCD$ to get your answer:

   $$\angle BCD = 3x - 4$$
   $$= 3 \cdot 12 - 4$$
   $$= 32°$$

   b. Other than the fact that $\angle ACD \cong \angle BCD$, you can conclude nothing else.

   Don't jump to conclusions based on the appearance of figures. For this problem, you know only that $\vec{CD}$ bisects an *angle* ($\angle ACB$). You *cannot* conclude that $\vec{CD}$ bisects the base of the triangle, and therefore you don't know whether D is the midpoint of $\overline{AB}$. You also can't conclude that $\triangle ABC$ has been cut in half. And you can't say that $\overline{AC} \cong \overline{BC}$ or that $\angle A \cong \angle B$.

Chapter 2: Points, Segments, Lines, Rays, and Angles

7. On this number line, Q and R trisect $\overline{PS}$. What are the coordinates of Q and R?

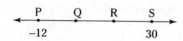

Solve It

8. Given that $\angle 1 = 4x$, $\angle 2 = x + 9$, and $\angle 3 = 5x - 7$, is $\angle STU$ trisected?

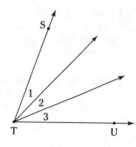

Solve It

9. $\overrightarrow{NP}$ and $\overrightarrow{NQ}$ divide right $\angle MNO$ into $\angle MNP$, $\angle PNQ$, and $\angle QNO$, whose measures are in the ratio 4 : 5 : 6. Determine the measure of $\angle PNO$.

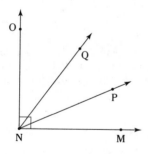

Solve It

*10. Given: $\overrightarrow{BD}$ and $\overrightarrow{BE}$ trisect $\overline{AC}$; $\overline{AD}$ and $\overline{DE}$ have lengths as shown.

    a. Determine DC (the length of the segment).

    b. Can you conclude that $\angle 1 \cong \angle 2$? That $\angle 1 \cong \angle 3$?

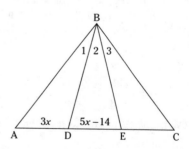

Solve It

Perfect Hilarity for Perpendicularity

You're surrounded by perpendicular things: floors are perpendicular to walls (hopefully), sides of rectangular shapes are perpendicular, the majority of streets that cross are perpendicular, and so on. In this section, you practice problems involving perpendicular lines (and rays and segments). I bet you can hardly wait!

Perpendicular: Lines, rays, or segments that form a right angle are *perpendicular*. The symbol for perpendicularity is ⊥. (Note that you say that lines, rays, or segments are perpendicular and that an angle is a right angle; you do *not* say that an angle is perpendicular.)

All right angles are congruent: If two angles are right angles, then they are *congruent* (they have the same number of degrees).

Many geometry theorems are statements of obvious things. You'll see more of them later in this chapter. But this one about congruent right angles takes the cake in the *well-duh* category. (Put this theorem in your back pocket; you'll use it soon but not in this section.)

Q. In the figure on the right, $\vec{BA} \perp \vec{BC}$, $\angle 1 \cong \angle 3$, and $\angle 2$ is three times as big as $\angle 1$. Find the measure of $\angle 2$.

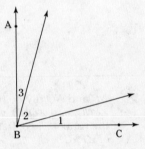

A. Because the rays are perpendicular, $\angle ABC$ is a right angle and thus measures 90°. $\angle 1$ and $\angle 3$ are equal, so you can set them both equal to x. $\angle 2$ is three times as big as $\angle 1$, so its measure is $3x$. Now you have three angles, $\angle 1$, $\angle 2$, and $\angle 3$, whose measures (x, $3x$, and x) must add up to 90. Thus,

$$x + 3x + x = 90$$
$$5x = 90$$
$$x = 18$$

Now, plugging 18 into $3x$ gives you the measure of $\angle 2$:

$$3(18) = 54°$$

Chapter 2: Points, Segments, Lines, Rays, and Angles

11. In the following figure:

a. Find ∠BFC given that ∠DFE measures 25°, that $\overleftrightarrow{AE} \perp \overrightarrow{FC}$, and that $\overrightarrow{FB} \perp \overrightarrow{FD}$.

b. What two objects form the sides of ∠BFC?

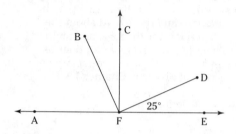

Solve It

12. Given that $\overleftrightarrow{AD} \perp \overleftrightarrow{BE}$, ∠DGC measures 10°, and ∠BGC is four times as large as ∠AGF, find the measure of ∠FGE.

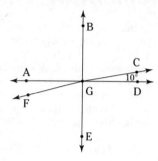

Solve It

13. Given that $\overleftrightarrow{AF} \perp \overleftrightarrow{EH}$, that $\overleftrightarrow{BG}$ bisects ∠FIH, and that $\overrightarrow{IC}$ and $\overrightarrow{ID}$ trisect ∠BIE, find the measure of ∠BID.

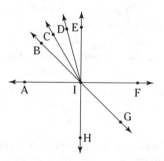

Solve It

14. In the following figure, $\overline{RG} \perp \overline{RY}$, $\overline{RG} \perp \overline{GA}$, and $\overline{RY} \perp \overline{LN}$.

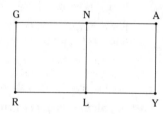

a. Name the angles you know are right angles.

b. Can you conclude that ∠ANL is a right angle?

c. What's ∠GRY ∩ $\overrightarrow{YL}$?

d. What's ∠GRY ∩ $\overrightarrow{LY}$?

Solve It

You Complete Me: Complementary and Supplementary Angles

In this section, you can finally get going with the main focus of this book: proofs. The proofs here are fairly short and straightforward. (Later in this chapter and especially in subsequent chapters, they get harder.) But if these proofs are your first attempts, they may seem difficult, despite how short they are. Work hard and think hard about the logic of these proofs and how they're put together. If you master the logic and method of doing these short proofs, you should be able to handle the longer, harder ones as well. (If you get stuck, you can check out Chapter 14 for some tips; and you may want to review the explanation of if-then logic and the structure of proofs in Chapter 1.)

Before I move on to proofs, however, here are two easy definitions:

- **Complementary angles:** Two angles whose sum is 90° (or a right angle)
- **Supplementary angles:** Two angles whose sum is 180° (or a straight angle)

And here are four easy theorems about pairs of angles that add up to either 90° or 180°:

- **Complements of the same angle are congruent:** If two angles are each complementary to a third angle, then they're congruent to each other (you have three total angles here).

  For example, say you have a 70° angle, ∠C. If ∠A is complementary to ∠C and ∠B is also complementary to ∠C, then ∠A ≅ ∠B (both have to be 20°, right?). Like so many theorems, the idea behind this one is a totally *well-duh* concept.

- **Complements of congruent angles are congruent:** If two angles are complementary to two other congruent angles, then they're congruent (you're working with four total angles).

  For example, if ∠B ≅ ∠C (say they're both 40°), and ∠A is complementary to ∠B and ∠D is complementary to ∠C, then ∠A ≅ ∠D (both have to be 50°).

- **Supplements of the same angle are congruent:** If two angles are each supplementary to a third angle, then they're congruent to each other (three total angles are involved). This theorem works exactly like the first theorem in this list.

- **Supplements of congruent angles are congruent:** If two angles are supplementary to two other congruent angles, then they're congruent (this theorem uses four total angles). This theorem works exactly like the second theorem.

Several theorems (the four preceding and many you'll see later) involve either three or four segments or angles. So when doing a proof, pay attention to whether the proof diagram involves three or four segments or angles. Doing so can help you select the appropriate theorem.

Chapter 2: Points, Segments, Lines, Rays, and Angles

Q. Given: $\overline{KS} \perp \overline{SY}$
$\overline{YU} \perp \overline{UK}$
$\angle RST \cong \angle TUR$

Prove: $\angle KSR \cong \angle YUT$

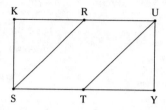

A. *Tip:* Before trying to write down the formal statements and reasons in a two-column proof, it's often a good idea to think through the proof using your own common sense. In other words, try to see why the *prove* statement is true without worrying about how to prove it or worrying about which theorems to use. When you can see why the *prove* statement has to be true, all that remains to be done is to translate your Joe/Jane-six-pack argument into the formal language of a proof.

For example, in this proof, you might say to yourself, "Can I see why angle *KSR* should be congruent to angle *YUT*?" And you could respond, "Sure. Because the segments are perpendicular, angles *KSY* and *YUK* are 90°, and because angle *RST* is congruent to angle *TUR* (say they're both 50°), angle *KSR* has to equal angle *YUT* (they'd both have to equal 40°). Bingo." If you can understand the proof in this common sense way, then all you have to do is put formalwear on this casual line of reasoning.

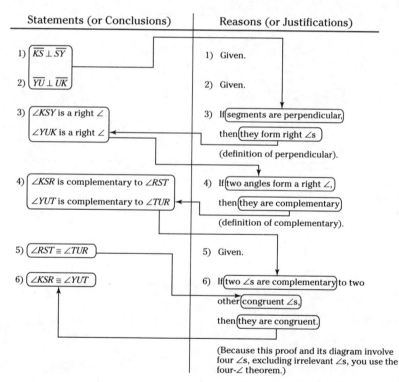

28 Part I: Getting Started

15. Given: ∠1 = 25°
∠2 = 90°
∠4 is complementary to ∠6

Find: The measures of angles 3 through 9

Solve It

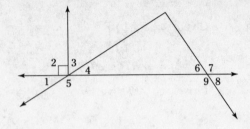

16. The supplement of an angle is 20° greater than twice the angle's complement. Find the angle's measure.

Solve It

Chapter 2: Points, Segments, Lines, Rays, and Angles 29

17. Given: ∠1 ≅ ∠4

Prove: ∠2 ≅ ∠3

Note: For this and all proof problems, you should *not* assume that the number of blank lines is the same as the number of steps needed for the proof.

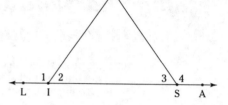

Solve It

Statements	Reasons

18. Given: $\vec{ST} \perp \vec{SA}$
$\vec{SR} \perp \vec{SB}$

Prove: ∠TSR ≅ ∠BSA

Hint: If you get stuck, go to the solution and copy Statement 5 and Reason 3 onto this page and then try the proof again.

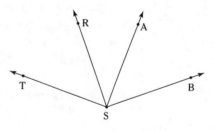

Solve It

Statements	Reasons

Adding and Subtracting Segments and Angles

You get eight more theorems in this section — all of them based on incredibly simple ideas. But despite the fact that the ideas are simple, having to memorize all this mumbo-jumbo lingo may still seem like a pain. If so, I have a tip for you.

Focus on the ideas behind the theorems. Doing so can help you remember how they're worded. And here's another benefit: If you're doing a proof on a quiz or test and you can't remember exactly how to write some theorem, you can just write the idea of the theorem in your own words. If you get the idea right, you may get partial or even full credit, depending on your teacher's grading style. (And if you're just doing the proof for fun — and who wouldn't? — you can get through the proof using some of your own words. After you're done, you can look up the proper wording of the theorem or theorems you couldn't remember.)

For example, one of the following theorems is based on the incredibly simple notion that if you take two sticks of equal length (say 3 inches and 3 inches) and add them end-to-end to two other equal sticks (say 5 inches and 5 inches), you end up with two equal totals (8 inches and 8 inches, of course). If you understand that idea, you've got the theorem in the bag. Adding equal things to equal things produces equal totals.

Without further ado, here are four theorems to use when adding line segments or angles (when writing a proof, students sometimes abbreviate these theorems as "addition"):

- **Segment addition (three total segments):** If a segment is added to two congruent segments, then the sums are congruent.
- **Angle addition (three total angles):** If an angle is added to two congruent angles, then the sums are congruent.
- **Segment addition (four total segments):** If two congruent segments are added to two other congruent segments, then the sums are congruent.
- **Angle addition (four total angles):** If two congruent angles are added to two other congruent angles, then the sums are congruent.

If you're subtracting segments or angles, here are four more theorems to choose from (after you get a handle on these theorems, you may simply write "subtraction"):

- **Segment subtraction (three total segments):** If a segment is subtracted from two congruent segments, then the differences are congruent.
- **Angle subtraction (three total angles):** If an angle is subtracted from two congruent angles, then the differences are congruent.
- **Segment subtraction (four total segments):** If two congruent segments are subtracted from two other congruent segments, then the differences are congruent.
- **Angle subtraction (four total angles):** If two congruent angles are subtracted from two other congruent angles, then the differences are congruent.

Here are a couple huge tips that you can use when working on any proof. You can see them in action in the first example in this section.

- ✔ **Use every given.** You have to do something with every given in a proof. So if you're not sure how to do a proof, don't give up until you've at least asked yourself, "Why did they give me this given? And why did they give me that given?" If you then write down what follows from each given (even if you don't know how that information can help you), you might see how to proceed. You may have a geometry teacher (or mathematician friend) who likes to throw you the occasional curveball, but in every geometry text that I know, the authors don't give you irrelevant givens. And that means that *every given is a built-in hint.*

- ✔ **Work backwards.** Thinking about how a proof will end — what the last and second-to-last lines will look like — is often very helpful. In some proofs, you may be able to work backwards from the final statement to the second-to-last statement and then to the third-to-last statement and maybe even to the fourth-to-last. Doing proofs this way is a little like doing one of those mazes you see in a newspaper or magazine. You can begin by working on a path from the Start point. Then, if you get stuck, you can work on a path from the Finish point, taking that as far as you can. And then you can go back to where you left off and try to connect the ends of the two paths.

Q. Given: *I* is the midpoint of $\overline{RN}$
             R and *N* trisect $\overline{GD}$

      Prove: *I* is the midpoint of $\overline{GD}$

G   R   I   N   D

A. Use every given. In this example proof, pretend that you have no idea how to begin. Just do something with the two givens. Ask yourself why someone would tell you about a midpoint. Well, because that tells you that you have two congruent segments, of course. And why would someone give you the trisection points? Because that given tells you that you have three congruent segments (though you use only two of them).

Statements	Reasons
1) *I* is the midpoint of $\overline{RN}$	1) Given.
2) *R* and *N* trisect $\overline{GD}$	2) Given.
3) $\overline{RI} \cong \overline{IN}$	3) If a point is the midpoint of a segment, then it divides it into two congruent segments.
4) $\overline{GR} \cong \overline{ND}$	4) If two points trisect a segment, then they divide it into three congruent segments.
5) $\overline{GI} \cong \overline{ID}$	5) If two congruent segments are added to two other congruent segments, then the sums are congruent.
6) *I* is the midpoint of $\overline{GD}$	6) If a point divides a segment into two congruent segments, then it is the midpoint of the segment (reverse of definition of midpoint).

Okay, here's where working backwards can help: Say you can figure out lines 3 and 4 in the preceding proof but aren't sure where to go next. No worries. Jump to the end of the proof. You know the final statement has to be the *prove* statement (*I* is the midpoint of $\overline{GD}$). Now ask yourself what you'd need to know to draw that conclusion. Well, to conclude that a point is a midpoint, you need a segment that's been cut into two congruent segments, right? So you don't have to be a mathematical genius to see that the second-to-last statement has to be $\overline{GI} \cong \overline{ID}$.

After you see that point, all you have to do is figure out why *that* would be true. So you then go back to where you left off (line 4), and hopefully you then see that you can add two pairs of congruent segments to get $\overline{GI} \cong \overline{ID}$.

Part I: Getting Started

19. Given: ∠GBU ≅ ∠SBM
Prove: ∠GBM ≅ ∠SBU

Solve It

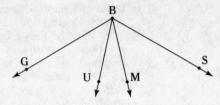

Statements	Reasons

20. Given: R is the midpoint of $\overline{BS}$
U and N trisect $\overline{BS}$

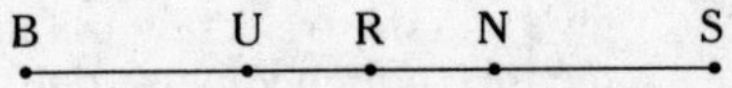

Prove: R is the midpoint of $\overline{UN}$

B • — U • — R • — N • — S

Hint: If you have a hard time with this one, fill in here Statements 4 and 6 and Reason 3 from the solution. But don't do this unless you absolutely have to!

Solve It

Statements	Reasons

21. Given: $\overrightarrow{QY}$ bisects ∠ZQX
∠ZQW ≅ ∠XQJ

Prove: $\overrightarrow{QY}$ bisects ∠WQJ

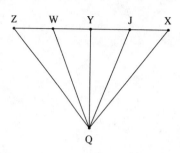

Hint: Don't forget to use all the givens in your proof (you might want to make them your first two steps). If you're really stumped, go to the solution and copy just the *if* part of Reason 3 onto this page.

Solve It

Statement	Reason

Multiplying and Dividing Angles and Segments

The preceding section lets you work on addition and subtraction of segments and angles. Now you graduate to multiplication and division of segments and angles. The two new theorems in this section can be a bit tricky to use correctly, so study these proofs carefully and heed the tips and the warning.

- **Like Multiples:** If two segments (or angles) are congruent, then their like multiples are congruent. This statement just means that if you have, say, two congruent segments, then 3 times one segment equals 3 times the other, or 4 times one equals 4 times the other — another *well-duh* idea.

- **Like Divisions:** If two segments (or angles) are congruent, then their like divisions are congruent. All this statement tells you is that if you have, say, two congruent angles, then ½ of one equals ½ the other, or ⅓ of one equals ⅓ of the other.

If the givens in a proof mention midpoint, bisect, or trisect *twice* (or something else that amounts to the same thing), then there's a pretty good chance that you'll want to use the Like Multiples Theorem or the Like Divisions Theorem in the proof.

Notice that this tip applies to both example problems and the three practice problems.

34 Part I: Getting Started

Q. Given: $\overline{AC} \cong \overline{VX}$
$\overline{AB} \cong \overline{VW}$
$\overleftrightarrow{CX}$ and $\overleftrightarrow{DY}$ trisect both $\overline{BE}$ and $\overline{WZ}$
Prove: $\overline{BE} \cong \overline{WZ}$

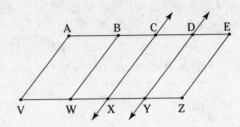

A. Here's the answer:

Statements	Reasons
1) $\overline{AC} \cong \overline{VX}$	1) Given.
2) $\overline{AB} \cong \overline{VW}$	2) Given.
3) $\overline{BC} \cong \overline{WX}$	3) If two congruent segments are subtracted from two other congruent segments, then the differences are congruent (segment subtraction; four-segment version).
4) $\overleftrightarrow{CX}$ and $\overleftrightarrow{DY}$ trisect both $\overline{BC}$ and $\overline{WZ}$	4) Given.
5) $\overline{BE} \cong \overline{WZ}$	5) If segments are congruent, then their like multiples are congruent.

Tip: When, like in the proof here, you go from a statement about small things (like $\overline{BC}$ and $\overline{WX}$) to a statement about big things (like $\overline{BE}$ and $\overline{WZ}$), use the Like *Multiples* Theorem.

Q. Given: $\angle 1 \cong \angle 2$
$\overrightarrow{QU}$ bisects $\angle RQS$
$\overrightarrow{ST}$ bisects $\angle RSQ$
Prove: $\angle RQU \cong \angle RST$

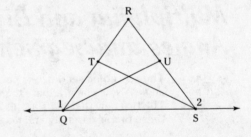

A. Game plan: Okay. You have $\angle 1$ equal to $\angle 2$ (say they're both 120°). Their supplements would then have to be equal (they'd both be 60°). Each of those is bisected, so $\angle RQU$ and $\angle RST$ would both be 30°. Piece of cake.

Chapter 2: Points, Segments, Lines, Rays, and Angles

Statements	Reasons
1) $\angle 1 \cong \angle 2$	1) Given.
2) $\angle RQS$ is supplementary to $\angle 1$ $\angle RSQ$ is supplementary to $\angle 2$	2) If two $\angle$s form a straight $\angle$ (assumed from diagram), then they are supplementary (reverse of definition of supplementary).
3) $\angle RQS \cong \angle RSQ$	3) If two $\angle$s are supplementary to two other congruent $\angle$s, then they are congruent (supplements of congruent angles) (statements 1 and 2).
4) $\overrightarrow{QU}$ bisects $\angle RQS$	4) Given.
5) $\overrightarrow{ST}$ bisects $\angle RSQ$	5) Given.
6) $\angle RQU \cong \angle RST$	6) If $\angle$s are congruent, then their like divisions are congruent (Like Divisions) (statements 3, 4 and 5).

Tip: When, like in this last proof, you go from a statement about big things ($\angle RQS$ and $\angle RSQ$) to a statement about small things (like $\angle RQU$ and $\angle RST$), you use the Like *Divisions* Theorem.

Tip: When you're new to proofs, it's easy to get confused about when to use the definitions of midpoint, bisect, or trisect and when to use the Like Divisions Theorem. So take heed: Use the definitions when you want to show that two or three parts of the *same* segment or *same* angle are equal to each other. Use Like Divisions, in contrast, when you want to show that a part of one segment (or angle) is equal to a part of a *different* segment (or angle).

*22. Given: $\overline{NO} \perp \overline{NI}$
$\overline{NO} \perp \overline{OE}$
$\angle 1 \cong \angle 2$
$\overrightarrow{NI}$ bisects $\angle DNG$
$\overrightarrow{OE}$ bisects $\angle TOG$

Prove: $\angle DNG \cong \angle TOG$

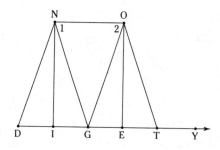

Hint: Want a little help? Check out Statements 1, 2, and 3 on the solution page.

Solve It

Statements	Reasons

Part I: Getting Started

23. Given: $\overline{SD} \cong \overline{UE}$
M is the midpoint of $\overline{SU}$
G is the midpoint of $\overline{DE}$
Prove: $\overline{SM} \cong \overline{GE}$

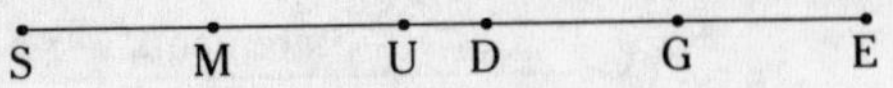

Solve It

Statements	Reasons

24. Given: $\overrightarrow{EA} \perp \overrightarrow{ED}$
$\overrightarrow{VW} \perp \overrightarrow{VZ}$
$\overrightarrow{EB}$ and $\overrightarrow{EC}$ trisect $\angle AED$
$\overrightarrow{VX}$ and $\overrightarrow{VY}$ trisect $\angle WVZ$
Prove: $\angle AEV \cong \angle WVE$

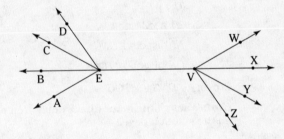

Hint: If this problem seems a bit tough, copy Statement 9 and Reason 9 from the solution and try to work backwards from there.

Solve It

Statements	Reasons

X Marks the Spot: Using Vertical Angles

Don't ask me how they came up with the term *vertical angles*, because these angles have nothing to do with the ordinary meaning of *vertical* (you know, as in vertical and horizontal). Go figure. When two lines cross to make an *X*, the two angles on opposite sides of the *X* are vertical angles. They're automatically equal. As you can see, every *X* has two pairs of vertical angles. If the name had been up to me, I would've called them *x-angles* or *cross angles*.

Vertical angles: Vertical angles are congruent.

Q. Given: $\angle 1 = x^2 + 7$
              $\angle 3 = 3x^2 - 1$
   Find: $\angle 2$

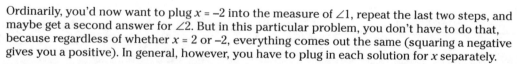

A. $\angle 1$ and $\angle 3$ are vertical angles and are thus congruent, so set them equal to each other and solve for x:

$3x^2 - 1 = x^2 + 7$

$2x^2 = 8$

$x^2 = 4$

$x = \pm 2$

Plug $x = 2$ into the measure of $\angle 1$:

$\angle 1 = x^2 + 7$

$= 2^2 + 7$

$= 11$

Figure the measure of $\angle 2$:

$\angle 2 = 180 - \angle 1$

$= 180 - 11$

$= 169$

Ordinarily, you'd now want to plug $x = -2$ into the measure of $\angle 1$, repeat the last two steps, and maybe get a second answer for $\angle 2$. But in this particular problem, you don't have to do that, because regardless of whether $x = 2$ or -2, everything comes out the same (squaring a negative gives you a positive). In general, however, you have to plug in each solution for x separately.

Remember: Segments and angles must, of course, have *positive* lengths or measures. So if you plug an *x*-value into a segment or angle and your answer is 0 or negative, reject that *x*-value.

Warning: Be careful, however, not to reject *x*-values simply because *they* are 0 or negative. The segments and angles, not *x*, must be positive. There are plenty of problems in which a *negative* solution for *x* gives you a *positive* answer for a segment or angle, and vice versa.

Part I: Getting Started

25. Use the figures to answer the following questions.

   a. Is this possible?

   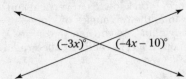

   b. Is this possible?

   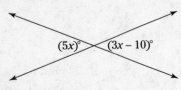

   Solve It

26. Solve for ∠AQB and ∠DQC.

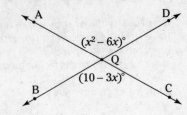

Solve It

27. You can do the following proof in four different ways, using four different sets of theorems. Don't write out four two-column proofs (unless you feel like it). Just write your game plans for the four alternatives. *Hint:* Two of the versions use vertical angles, two use ∠ACT instead, two use angle subtraction, and two use complementary angles (2 + 2 + 2 + 2 = 4 game plans, right?).

   Given: $\vec{CA} \perp \overleftrightarrow{GH}$
   $\vec{CT} \perp \overleftrightarrow{NI}$

   Prove: ∠NCA ≅ ∠HCT (paragraph proof)

   Solve It

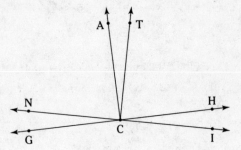

Switching It Up with the Transitive and Substitution Properties

The transitive and substitution properties should be familiar to you from algebra. You may have used the idea of transitivity in this way: If $a = b$ and $b = c$, then $a = c$; or if $a > b$ and $b > c$, then $a > c$. Transitivity works the same in geometry: You use it like above but to show congruence instead of equality (you almost never, however, use the inequality version). And you've certainly used substitution in algebra — like if $x = 2y - 5$ and $4x - 3y = 10$, you can switch the x with the $2y - 5$ (because they're equal, of course) and write $4(2y - 5) - 3y = 10$. This property works the same in geometry: When two objects are congruent, you can switch 'em.

- **Transitive Property (for three segments or angles):** If two segments (or angles) are each congruent to a third segment (or angle), then they're congruent to each other. For example, if $\angle A \cong \angle B$ and $\angle B \cong \angle C$, then $\angle A \cong \angle C$ ($\angle A$ and $\angle C$ are each congruent to $\angle B$, so they're congruent to each other).

- **Transitive Property (for four segments or angles):** If two segments (or angles) are congruent to congruent segments (or angles), then they're congruent to each other. For example, if $\overline{AB} \cong \overline{CD}$, $\overline{CD} \cong \overline{EF}$, and $\overline{EF} \cong \overline{GH}$, then $\overline{AB} \cong \overline{GH}$. ($\overline{AB}$ and $\overline{GH}$ are congruent to the congruent segments $\overline{CD}$ and $\overline{EF}$, so they're congruent to each other.)

- **Substitution Property:** If two geometric objects (segments, angles, triangles, and so on) are congruent and you have a statement involving one of them, you can pull the switcheroo and replace the one with the other. For example, if $\angle A \cong \angle B$ and $\angle B$ is supplementary to $\angle C$, then $\angle A$ is supplementary to $\angle C$.

You use the Transitive Property as the reason when the statement says things are congruent; you use the Substitution Property for the reason when the statement says anything else.

And one more thing: You'll be less likely to mix up substitution with other theorems if you note that like with transitivity, other theorems (addition, subtraction, complements and supplements of congruent angles, and so on) go with statements about congruent things; substitution does not.

Q. Given: $\angle 2 \cong \angle 3$

Prove: $\angle 1 \cong \angle 3$

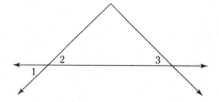

A. Here's how this one unfolds:

Statements	Reasons
1) $\angle 2 \cong \angle 3$	1) Given.
2) $\angle 1 \cong \angle 2$	2) Vertical $\angle$s are congruent.
3) $\angle 1 \cong \angle 3$	3) If two $\angle$s are each congruent to a third $\angle$, then they are congruent to each other (Transitive Property).

Part I: Getting Started

Did it occur to you that you could use substitution instead of transitivity for reason 3? That's correct — sort of. You could use substitution in step 3 because you can essentially put ∠3 where ∠2 is. The switch works this way because transitivity is a special case of substitution. However, you probably want to use the property as I do (rebels excepted), because that's probably what your geometry teacher and mathematician buddies want. (For info on how to keep the properties straight, see the preceding tips in this section.)

Q. Given: ∠2 ≅ ∠3
Prove: ∠1 is supplementary to ∠3

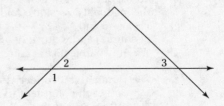

A. The proof, dear reader:

Statements	Reasons
1) ∠1 is supplementary to ∠2	1) If two ∠s form a straight ∠ (assumed from diagram), then they are supplementary.
2) ∠2 ≅ ∠3	2) Given.
3) ∠1 is supplementary to ∠3	3) Substitution (putting ∠3 where ∠2 was).

28. Given: $\overrightarrow{AC}$ bisects ∠BAD
Prove: ∠1 ≅ ∠3

Solve It

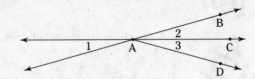

Statements	Reasons

Chapter 2: Points, Segments, Lines, Rays, and Angles 41

29. Given: $\overrightarrow{MB}$ bisects $\angle AMC$
$\overrightarrow{MC}$ bisects $\angle BMD$

Prove: $\angle 4 \cong \angle 6$

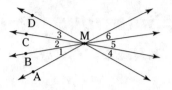

Hint: If you get stuck, copy Statements 1 and 3 and Reason 5 from the solution page and then take it from the top.

Solve It

Statements	Reasons

30. Given: $\overleftrightarrow{TO} \perp \overleftrightarrow{GO}$

Prove: $\angle 1$ is complementary to $\angle 2$

Solve It

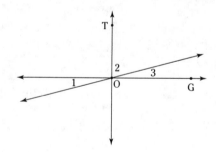

Statements	Reasons

Solutions for Points, Segments, Lines, Rays, and Angles

1. $\overline{ST} \cap \overline{TQ}$ = point T

2. $\overline{ST} \cup \overline{PT}$ = ?

Did you have ∠STP for this one? Nope! Angles go on forever, but you have segments rather than rays. Technically, this union is not an angle. And there is no nice, simple name for this thing. You can't really write it any more simply than you see it in the original problem: $\overline{ST} \cup \overline{PT}$.

3. ∠RTS ∩ $\overrightarrow{PR}$ = $\overrightarrow{TR}$

If you trace over ∠RTS (remembering that it goes out past R forever) and then over $\overrightarrow{PR}$, you trace twice over $\overrightarrow{TR}$, the ray that begins at T and goes out forever past R.

4. $\overrightarrow{RT} \cap \overrightarrow{TP}$ = $\overrightarrow{TP}$

5. $\overline{PS} \cap \overleftrightarrow{QR}$ = ∅ (the empty set)

They don't overlap anywhere.

6. $\overrightarrow{ST} \cup \overrightarrow{QT}$ = $\overleftrightarrow{SQ}$ or $\overleftrightarrow{QS}$ or $\overleftrightarrow{ST}$ or $\overleftrightarrow{TS}$ or $\overleftrightarrow{TQ}$ or $\overleftrightarrow{QT}$

7. You can solve this problem in two steps:

$\overline{PS}$ is trisected, so determine PS and then divide that by 3:

$PS = 30 - (-12) = 42$

$PS \div 3 = 42 \div 3 = 14$

Add 14 to –12 to get Q, and then add 14 more to get R:

$-12 + 14 = 2 \rightarrow Q$

$2 + 14 = 16 \rightarrow R$

8. For ∠STU to be trisected, all three angles must be equal. So first set any two angles equal to each other and solve for x. (Any two work, but I use ∠1 and ∠2.)

$m\angle 1 = m\angle 2$

$4x = x + 9$

$3x = 9$

$x = 3$

Plugging $x = 3$ into the measure of ∠1 or ∠2 determines both of their measures, because you can assume that they're congruent:

$m\angle 1 = 4x$

$= 4 \cdot 3$

$= 12$

Thus, if ∠1 and ∠2 are congruent, they're both 12°. Finally, check whether ∠3 is also 12° when $x = 3$:

$m\angle 3 = 5x - 7$

$= 5 \cdot 3 - 7$

$= 8$

Nope. Thus, ∠STU is not trisected.

Chapter 2: Points, Segments, Lines, Rays, and Angles   43

9 The three angles are in the ratio 4 : 5 : 6, so you first set their measures equal to $4x$, $5x$, and $6x$. Together, the three angles make up a right angle, so set their sum equal to 90° and solve:

$$4x + 5x + 6x = 90$$
$$15x = 90$$
$$x = 6$$

Use $x = 6$ to determine the measure of $\angle PNO$:

$$m\angle PNO = m\angle PNQ + m\angle QNO$$
$$= 5x + 6x$$
$$= 11x$$
$$= 11 \cdot 6$$
$$= 66$$

Of course, you could use $x = 6$ to determine that $\angle PNQ$ is 30° and $\angle QNO$ is 36° and then add them to get 66°.

***10** Check out the answers:

a. Because $\overline{AC}$ is trisected, AD must equal DE. So set them equal to each other, solve for x, and then plug the answer in to get AD and DE:

$$3x = 5x - 14$$
$$-2x = -14$$
$$x = 7$$

Therefore, $AD = 3x = 3 \cdot 7 = 21$. DE is also, of course, 21, and because $\overline{AC}$ is trisected, EC is also 21. $DC = DE + EC$, so $DC = 42$.

b. No, you can't conclude that $\angle 1 \cong \angle 2$ or that $\angle 1 \cong \angle 3$. Despite the fact that we typically think of rays as angle bisectors or trisectors, the given in this problem is that $\overrightarrow{BD}$ and $\overrightarrow{BE}$ trisect a *segment*, $\overline{AC}$. This statement tells you only the location of points D and E; it tells you nothing about how the rays divide up $\angle ABC$. $\angle ABC$ might look trisected, but you can't conclude that it is. As it turns out, it's impossible for $\angle ABC$ to be trisected given that $\overline{AC}$ is trisected. $\angle 1$ would be congruent to $\angle 2$ only if $\triangle ABE$ were isosceles (which you can't conclude). And $\angle 1$ would be congruent to $\angle 3$ only if $\triangle ABC$ were isosceles (which you also can't conclude despite the fact that it looks like it is).

11 And here's another fine solution:

a. Because of the given perpendicularity, you know that $\angle CFE$ and $\angle BFD$ are both 90° angles. Now, $\angle CFD$ and $\angle DFE$ have to add up to 90°, right? So because $\angle DFE$ is 25°, $\angle CFD$ must be 90° – 25°, or 65°. Then, using the same logic, $\angle BFC$ must be 90° – 65°, which is 25°.

b. The sides of $\angle BFC$ are *rays FB* and *FC*. If you said *segment FB*, you're close. Angles go on forever, and their sides are rays that go on forever. Whether or not you can actually see the rays in the figure is irrelevant.

12 Because $\overleftrightarrow{AD} \perp \overleftrightarrow{BE}$, you know that the measures of both $\angle BGD$ and $\angle AGE$ are 90°. You see that the measure of $\angle BGD = \angle BGC + \angle DGC$ (which is 10°). Thus, $\angle BGC = 90° - 10° = 80°$. Then, because $\angle BGC$ (80°) is four times as big as $\angle AGF$, $\angle AGF = 20°$. Finally, $\angle FGE = 90° - 20°$, which is 70°.

13 The given perpendicularity tells you that the four big angles are each 90°. (This loose, nontechnical use of "big" may get me pulled over by the math police; don't try it with your geometry teacher.) Because $\overrightarrow{IG}$ bisects right $\angle HIF$, $\angle GIF$ must be 45°. $\angle EIF$ measures 90°, so add these two up to get 135° for $\angle EIG$. Straight $\angle BIG$ (another "big" angle — don't you just love these geometry puns?) is, of course, 180°, so $\angle BIE$ must be 180° – 135°, or 45°. Now, trisect that 45° to get 15° for the three small angles. And finally, two of these 15° angles make up $\angle BID$, so $\angle BID$ measures 30°.

Part I: Getting Started

14 Here's how you do this gnarly problem:

  a. The three given pairs of perpendicular segments tell you, by the definition of perpendicular, that the following are right angles: $\angle R$, $\angle G$, $\angle RLN$, and $\angle YLN$. Despite the fact that $\angle Y$ and $\angle A$ look like right angles, you can't conclude that. (You can also conclude that $\angle ANL$ and $\angle GNL$ are right angles — see part *b*.)

  b. Yes, you can conclude that $\angle ANL$ is a right angle, though I haven't covered the necessary concepts yet. What? Is it against the law for me to challenge you with a problem before I've presented the relevant ideas? Well, excuse me! Really, though, you probably could've reached this conclusion if you're familiar with rectangles. Because $\angle R$, $\angle G$, and $\angle RLN$ are right angles, the fourth angle in quadrilateral *RGNL*, $\angle GNL$, must also be a right angle; that's because the angles in a quadrilateral have to add up to 360°. (The official explanation of the sum of angles in a polygon is in Chapter 6.) Because $\angle GNL$ is a right angle, the angle's supplement, $\angle ANL$, must also be a right angle.

  c. $\angle GRY \cap \overrightarrow{YL} = \overline{RY}$

  d. $\angle GRY \cap \overrightarrow{LY} = \overrightarrow{LY}$

15 $\angle 1$ and $\angle 5$ are supplementary; $\angle 1$ is 25°, so $\angle 5$ is 155°. Then $\angle 5$ and $\angle 4$ work the same way, so $\angle 4$ is 25°. Because $\angle 2$ is 90°, $\angle 3$ and $\angle 4$ together have to make up another 90° (because $\angle 2$, $\angle 3$, and $\angle 4$ add up to a straight angle, or 180°). Thus, because $\angle 4$ is 25°, $\angle 3$ is 65°. $\angle 4$ and $\angle 6$ are complementary, so $\angle 6$ is also 65°. Finally, going clockwise around the point to $\angle 7$, $\angle 8$, and $\angle 9$, each adjacent pair of angles is supplementary, so $\angle 7$ is 115°, $\angle 8$ is 65°, and $\angle 9$ is 115°.

16 *Tip:* You can often come up with the correct equation for a word problem like this one by reading through the sentence and translating each word or phrase into its mathematical equivalent.

For this problem, first let *x* equal the measure of the angle you're trying to find. Then, because you obtain any angle's complement by subtracting the angle's measure from 90° and obtain any angle's supplement by subtracting the angle's measure from 180°, the measure of the complement of the unknown angle is $90 - x$, and the measure of its supplement is $180 - x$. Now you can do the translation:

The supplement of an angle	is	20 greater than	twice	the angle's complement.
$180 - x$	=	$20 +$	$2 \cdot$	$90 - x$

Write this problem like an ordinary equation and solve for *x*. (But first note that in the following equation, I move the "20 +" to the end of the equation, where it becomes "+ 20." Adding the 20 at the end is more natural. Say you hear someone say, "That's twenty greater than one hundred forty-five." You think 145 + 20, not 20 + 145, right? Either works, of course, but now consider the expression "twenty less than one hundred forty-five." For that, you *have to* subtract the 20 from the 145, not the other way around. Being consistent and putting the 20 at the end is best, regardless of whether you're adding or subtracting.)

Finish the problem:

  $180 - x = 2(90 - x) + 20$
  $180 - x = 180 - 2x + 20$
  $x = 20$

17

Statements (or Conclusions)	Reasons (or Justifications)
1) $\angle 1 \cong \angle 4$	1) Given.
2) $\angle LIS$ is a straight $\angle$ $\angle ASI$ is a straight $\angle$	2) Assumed from diagram. (The vast majority of reasons you'll use in proofs will come from your handy lists of definitions, theorems, postulates, and properties. This is one of the few odd exceptions. Some geometry teachers let you skip this step and go right to step 3.)

Chapter 2: Points, Segments, Lines, Rays, and Angles

3) ∠2 is supplementary to ∠1 ∠3 is supplementary to ∠4	3) If two ∠s form a straight ∠, then they are supplementary (definition of supplementary).
4) ∠2 ≅ ∠3	4) If two ∠s are supplementary to two other congruent ∠s, then they are congruent.

18

Statements (or Conclusions)	Reasons (or Justifications)
1) $\overrightarrow{ST} \perp \overrightarrow{SA}$	1) Given.
2) $\overrightarrow{SR} \perp \overrightarrow{SB}$	2) Given.
3) ∠TSA is a right ∠	3) If two rays are perpendicular, then they form a right ∠. (If you understand the if-then rule for reasons that I explain in Chapter 1, then this reason just about writes itself. The *if* part of this reason must come from a statement above it, namely statement 1 or 2. The only fact in those statements concerns perpendicularity. So basically, this reason has to begin with "If ⊥." And the only thing that can follow "If ⊥" is "then right ∠.")
4) ∠BSR is a right ∠	4) Same as reason 3.
5) ∠TSR is complementary to ∠RSA ∠BSA is complementary to ∠RSA	5) If two ∠s form a right ∠, then they are complementary (definition of complementary).
6) ∠TSR ≅ ∠BSA	6) If two ∠s are each complementary to a third angle, then they are congruent to each other. (This proof and its diagram involve three ∠s, so you use the three-∠ theorem.)

19

Statements	Reasons
1) ∠GBU ≅ ∠SBM	1) Given.
2) ∠GBM ≅ ∠SBU	2) If an ∠(∠UBM) is added to two congruent ∠s (∠GBU and ∠SBM), then the sums are congruent (addition of ∠s; three-∠ version).

This proof brings me to my next tip:

If the angles (or segments) in the *prove* statement are *larger* than the given angles (or segments), the proof may call for one of the *addition* theorems.

20

Statements	Reasons
1) R is the midpoint of $\overline{BS}$	1) Given.
2) U and N trisect $\overline{BS}$	2) Given.
3) $\overline{BR} \cong \overline{RS}$	3) A midpoint divides a segment into two congruent segments (definition of midpoint).
4) $\overline{BU} \cong \overline{NS}$	4) Trisection points divide a segment into three congruent segments (definition of trisection).
5) $\overline{UR} \cong \overline{RN}$	5) If two congruent segments are subtracted from two other congruent segments, then the differences are congruent (subtraction of segments; four-segment version).
6) R is the midpoint of $\overline{UN}$	6) If a point divides a segment into two congruent segments, then it's the midpoint of the segment (reverse of definition of midpoint).

46 Part I: Getting Started

Note that in contrast to the preceding problem, in this proof, the things you're trying to prove something about ($\overline{UR}$ and $\overline{RN}$) are *smaller* than the things in the given ($\overline{BR}$ and $\overline{RS}$ are sort of in the given).

If the segments (or angles) in the *prove* statement are *smaller* than the ones in the given, one of the *subtraction* theorems may be the ticket.

21

Statements	Reasons
1) $\overrightarrow{QY}$ bisects $\angle ZQX$	1) Given.
2) $\angle ZQW \cong \angle XQJ$	2) Given.
3) $\angle ZQY \cong \angle XQY$	3) If a ray bisects an $\angle$, then it divides it into two congruent $\angle$s (definition of bisect).
4) $\angle WQY \cong \angle JQY$	4) If two congruent $\angle$s are subtracted from two other congruent $\angle$s, then the differences are congruent (subtraction of $\angle$s; four-$\angle$ version).
5) $\overrightarrow{QY}$ bisects $\angle WQJ$	5) If a ray divides an $\angle$ into two congruent $\angle$s, then the ray bisects the $\angle$ (reverse of definition of bisect).

*22 Game Plan: You have the right angles and $\angle 1 \cong \angle 2$ (say they're both 70°). So their complements ($\angle ING$ and $\angle EOG$) would both measure 20°. Then because of the bisections, the angles you're trying to prove equal to each other would both be $2 \cdot 20$, or 40°. That's it.

Statements	Reasons
1) $\overline{NO} \perp \overline{NI}$ $\overline{NO} \perp \overline{OE}$	1) Given.
2) $\angle INO$ is a right $\angle$ $\angle EON$ is a right $\angle$	2) Definition of perpendicular (statement 1).
3) $\angle 1 \cong \angle 2$	3) Given.
4) $\angle ING$ is complementary to $\angle 1$ $\angle EOG$ is complementary to $\angle 2$	4) Definition of complementary $\angle$s (statement 2).
5) $\angle ING \cong \angle EOG$	5) Complements of congruent $\angle$s are congruent. (statements 3 and 4).
6) $\overrightarrow{NI}$ bisects $\angle DNG$ $\overrightarrow{OE}$ bisects $\angle TOG$	6) Given.
7) $\angle DNG \cong \angle TOG$	7) If $\angle$s are congruent ($\angle ING$ and $\angle EOG$), then their like multiples are congruent (statements 5 and 6).

23 Game Plan: *SD* equals *UE* (say they're both 10). If *UD* is 2, then both *SU* and *DE* would be 8. Then the midpoints cut each of those in half, so that makes *SM* and *GE* both 4. Bingo.

Statements	Reasons
1) $\overline{SD} \cong \overline{UE}$	1) Given.
2) $\overline{SU} \cong \overline{DE}$	2) If a segment is subtracted from two congruent segments, then the differences are congruent (segment subtraction; three-segment version) (statement 1 and diagram).
3) *M* is the midpoint of $\overline{SU}$	3) Given.
4) *G* is the midpoint of $\overline{DE}$	4) Given.
5) $\overline{SM} \cong \overline{GE}$	5) If segments are congruent ($\overline{SU}$ and $\overline{DE}$), then their like divisions are congruent (half of one equals half of the other) (statements 2, 3, and 4).

Chapter 2: Points, Segments, Lines, Rays, and Angles   47

24 **Game Plan:** You have the two 90° angles. Each is trisected, so all the small angles measure 30°. Because ∠BEA and ∠XVW measure 30°, ∠AEV and ∠WVE each have to be 150°. Sweet.

Statements	Reasons
1) $\vec{EA} \perp \vec{ED}$	1) Given.
2) $\vec{VW} \perp \vec{VZ}$	2) Given.
3) ∠AED is a right ∠ ∠WVZ is a right ∠	3) Definition of perpendicular (1, 2).
4) ∠AED ≅ ∠WVZ	4) All right ∠s are congruent (3).
5) $\vec{EB}$ and $\vec{EC}$ trisect ∠AED	5) Given.
6) $\vec{VX}$ and $\vec{VY}$ trisect ∠WVZ	6) Given.
7) ∠AEB ≅ ∠WVX	7) If ∠s are congruent (the two right ∠s), then their like divisions are congruent (a third of one equals a third of the other) (4, 5, 6).
8) ∠AEV is supplementary to ∠AEB ∠WVE is supplementary to ∠WVX	8) If two ∠s form a straight ∠ (assumed from diagram), then they are supplementary (definition of supplementary).
9) ∠AEV ≅ ∠WVE	9) Supplements of congruent ∠s are congruent (7, 8).

25 As Sherlock Holmes says in *The Adventure of the Beryl Coronet*, "When you have excluded the impossible, whatever remains, however improbable, must be the truth." So go on and solve this problem just like the great detective would:

a. Yes, it's possible:

$$-3x = -4x - 10$$
$$x = -10$$

Plug $x = -10$ into the angles, and you see that the each angle is 30°.

b. Not possible:

$$5x = 3x - 10$$
$$2x = -10$$
$$x = -5$$

Plug $x = -5$ into the angles, and you get negative measures for each angle, which is impossible.

26 Set the vertical angles equal to each other and solve for x:

$$x^2 - 6x = 10 - 3x$$
$$x^2 - 3x - 10 = 0$$
$$(x - 5)(x + 2) = 0$$
$$x = 5 \text{ or } x = -2$$

Now plug each of these two solutions into the original angles. The solution $x = 5$ gives you negative angles, so you reject $x = 5$. The solution $x = -2$ gives you angles of 16°. Because ∠AQB and ∠DQC are the supplements of these angles, they each equal 164°.

27 All four game plans use, of course, the two right angles.

Game Plan 1: You have the two congruent vertical angles. One is the complement of ∠NCA; the other is the complement of ∠HCT. Therefore, you finish with the complements of congruent angles theorem. (Assuming each statement contains only a single fact, this method takes eight steps. Try it.)

Game Plan 2: This method is the same as Game Plan 1 except that you subtract the congruent vertical angles from the congruent right angles. The final reason is, therefore, the four-angle version of *angle subtraction*. (This strategy takes seven steps. Give it a go.)

Game Plan 3: Use ∠ACT. ∠NCA and ∠HCT are both complements of ∠ACT. You're done, because complements of the same angle are congruent. (This method also takes seven steps. Go for it.)

Game Plan 4: This time, you just subtract ∠ACT from the two congruent right angles. The final reason is the three-angle version of *angle subtraction*. (The winner! — only six steps.)

28

Statements	Reasons
1) $\overrightarrow{AC}$ bisects ∠BAD	1) Given.
2) ∠2 ≅ ∠3	2) Definition of bisect.
3) ∠1 ≅ ∠2	3) Vertical ∠s are congruent.
4) ∠1 ≅ ∠3	4) Transitive Property.

29 **Game Plan:** Think backwards — how can you get ∠4 ≅ ∠6? Well, ∠4 and ∠3 are congruent vertical angles, as are ∠6 and ∠1. Thus, if you can get ∠1 ≅ ∠3, you have it. The two bisectors make ∠1 ≅ ∠2 and ∠2 ≅ ∠3. Thus, ∠1 ≅ ∠3 by the Transitive Property. Bingo.

Statements	Reasons
1) $\overrightarrow{MB}$ bisects ∠AMC	1) Given.
2) ∠1 ≅ ∠2	2) Definition of bisect.
3) $\overrightarrow{MC}$ bisects ∠BMD	3) Given.
4) ∠2 ≅ ∠3	4) Definition of bisect.
5) ∠1 ≅ ∠3	5) Transitive Property (for three ∠s).
6) ∠1 ≅ ∠6	6) Vertical angles are congruent.
7) ∠3 ≅ ∠4	7) Vertical angles are congruent.
8) ∠4 ≅ ∠6	8) Transitive Property (for four ∠s). If ∠s (4 and 6) are congruent to congruent ∠s (1 and 3), then they (4 and 6) are congruent to each other.

30

Statements	Reasons
1) $\overrightarrow{TO} \perp \overrightarrow{GO}$	1) Given.
2) ∠TOG is a right ∠	2) Definition of perpendicular.
3) ∠3 is complementary to ∠2	3) Definition of complementary.
4) ∠1 ≅ ∠3	4) Vertical ∠s are congruent.
5) ∠1 is complementary to ∠2	5) Angle substitution.

Part II
Triangles

In this Part . . .

Start off your love affair with triangles by working out non-proof problems that cover concepts such as area, altitudes, medians, angle bisectors, perpendicular bisectors, the Pythagorean Theorem, families of right triangles, and so on. Then you can get back to proofs. Triangles are the main event in geometry proofs — nothing else comes close. You get lots of practice proving that triangles are congruent and then using CPCTC (Congruent Parts of Congruent Triangles are Congruent). Congruent triangles and their congruent parts form the core or focus of a great number of geometry proofs.

Chapter 3
Triangle Fundamentals and Other Cool Stuff

In This Chapter
- Naming triangles by their sides and angles
- Measuring area and height
- Finding a triangle's center of balance
- The "centers" of attention: Orthocenter, incenter, and circumcenter
- Checking out the Pythagorean Theorem
- Triangles whose sides are whole numbers (and their kin)
- Looking at 45°- 45°- 90° and 30°- 60°- 90° triangles

There's no upper limit to how many sides a polygon can have, but the lower limit is three — and that makes the triangle sort of a special shape. And for some reason, the number three seems to have a certain universal appeal: the Three Stooges, the Three Wise Men, three blind mice, Goldilocks and the three bears, Three Dog Night, three strikes and you're out, and so on. So I give you the triangles: three angles, three sides, three medians, three altitudes, three angle bisectors, three perpendicular bisectors, and three "centers" (plus the centroid).

Triangle Types and Triangle Basics

Six basic terms describe different types of triangles. Here's a great way to remember them: A triangle has three sides and three angles. Well, three of the following terms are about sides, and three are about angles.

Every triangle belongs to one of these three categories about sides:

- **Scalene:** A scalene triangle has no equal sides.
- **Isosceles:** An isosceles triangle has at least two equal sides. The two equal sides are called *legs;* the third side is the *base*. The two angles touching the base, called *base angles,* are equal. The angle between the two legs is the *vertex angle*.
- **Equilateral:** An equilateral triangle has three equal sides (thus, every equilateral triangle is also isosceles). Note that an equilateral triangle is also *equiangular* because it has three equal angles (each is 60°). For polygons with four or more sides, the distinction between *equilateral* and *equiangular* is important. Not so for triangles, because both terms refer to the very same triangle.

Part II: Triangles

Every triangle also belongs to one of these three groups concerning angles:

- **Acute:** An acute triangle has three acute angles (angles less than 90°, of course).
- **Right:** A right triangle has one right angle and two acute angles (the two short sides touching the right angle are the *legs*; the longest side across from the right angle is called the *hypotenuse*).
- **Obtuse:** An obtuse triangle has a single obtuse angle (more than 90°); the other two angles are acute.

And here's one more thing about the angles in a triangle that you may already know:

The sum of the measures of the three angles in a triangle is always 180°.

Whenever possible, don't just memorize math formulas, concepts, and so on as bare-naked facts. Instead, look for some reason why they're true or find a connection between the new idea and something you already know. For instance, to remember the sum of the angles in a triangle, picture the triangle you get when you cut a square in half along its diagonal: You can easily see that the three angles of such a triangle are 45°, 45°, and 90° — which add up to 180°.

Q. Classify this triangle as scalene, isosceles, or equilateral.

Hint: Don't forget that the sides of a triangle can't be negative or zero (or imaginary, like $\sqrt{-10}$) and that each pair of sides of a triangle must add up to more than the third side.

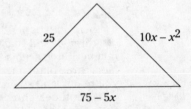

A. This problem isn't simple. Here's how your argument should go: The triangle is scalene unless at least two of the sides are equal. So try the three different pairs of sides and see what happens if you set them equal to each other.

$25 = 75 - 5x$

$5x = 50$

$x = 10$

No good: Plugging $x = 10$ into $10x - x^2$ gives you a side with a length of 0. Next, if

$75 - 5x = 10x - x^2$

$x^2 - 15x + 75 = 0$

Now solve for x with the quadratic formula. "What?" you say. "You expect me to remember the quadratic formula?" Yeah, sure, I know this is a geometry book, but I don't think reviewing some algebra as important as the quadratic formula will kill you. (In fact, I hear that world history and biology injuries outnumber algebra accidents by almost 10 : 1.) Do you have your helmet on?

Chapter 3: Triangle Fundamentals and Other Cool Stuff

$$ax^2 + bx + c = 0$$

$$x = \frac{-b \pm \sqrt{b^2 - 4ac}}{2a}$$

$$= \frac{15 \pm \sqrt{(-15)^2 - 4(1)(75)}}{2}$$

$$= \frac{15 \pm \sqrt{225 - 300}}{2}$$

$$= \frac{15 \pm \sqrt{-75}}{2}$$

No good. A negative under the square root means you have no real solutions. Finally, if

$$25 = 10x - x^2$$
$$x^2 - 10x + 25 = 0$$
$$(x-5)(x-5) = 0$$
$$x = 5$$

Plugging 5 into the third side $(75 - 5x)$ gives you $75 - 5 \cdot 5$, or 50, so when $x = 5$, you might think the three sides could be 25, 25, and 50. "But wait!" you should say. "No triangle can have sides of 25, 25, and 50!"

Remember: Any two sides of a triangle must add up to *more than* the third side. Think about it this way: If you walk, say, from vertex A to vertex B in a triangle, the trip has to be shorter if you walk straight along $\overline{AB}$ than if you go out of your way and walk along the two other sides.

So, back to the problem. Setting the three pairs of sides equal to each other doesn't work, so none of the sides are equal. Therefore, the triangle is scalene. And that's a wrap.

---

1. Classify these triangles as scalene, isosceles, or equilateral.

a)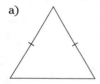

b)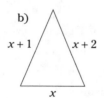

Solve It

2. If △*ISO* is isosceles and its perimeter is more than 10, which side is the base, and how long are the three sides?

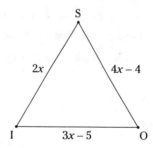

Solve It

Part II: Triangles

3. The angles of a triangle are in the ratio of 4:5:6. Is it an acute, right, or obtuse triangle? Is it scalene, isosceles, or equilateral?

Solve It

4. Classify the following triangles as acute, obtuse, or right.

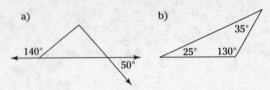

a)   b)

Solve It

5. Are the following statements true *always*, *sometimes*, or *never*?

  a. An equilateral triangle is isosceles.

  b. An isosceles triangle is equilateral.

  c. A right triangle is isosceles.

  d. If two of the angles in a triangle are 70° and 55°, the triangle is isosceles.

  e. The base angles of an obtuse isosceles triangle are each 40°.

  f. The base angles of an acute isosceles triangle are each 40°.

  g. Two of the angles in an obtuse triangle are supplementary (add up to 180°).

  h. Two of the angles in an acute triangle are complementary (add up to 90°).

Solve It

Altitudes, Area, and the Super Hero Formula

In this section, I cover some concepts that you've probably known for a long time, like how to find the area or the height of a triangle. But you'll see some other ideas that I bet you don't know, like why Hero is a real geometry superhero (and that you can find a triangle's area in another way).

First, take a look at the formula for the area of triangle:

$$\text{Area}_\triangle = \tfrac{1}{2}\, \text{base} \cdot \text{height}$$

Area, of course, is usually measured in some kind of units^2, like square feet, square meters, or square centimeters.

The *height*, or *altitude*, of a triangle is just what you'd expect it to be — you know, its height. Think of altitude this way: If you have an actual, physical triangle — say, cut out of cardboard — and you stand it up on a table, its height or altitude is the distance from its peak straight down to the table. Check out the two triangles in Figure 3-1.

Figure 3-1:
An altitude inside a triangle and outside a triangle.

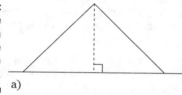
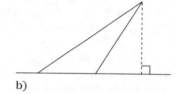

a)   b)

You can stand a triangle up three different ways depending on which side you put flat on the table, so every triangle has three separate altitudes. Depending on which type of triangle you have, the altitudes can have the same or different lengths:

- **Scalene triangles:** The three altitudes have different lengths.
- **Isosceles triangles:** Two of the altitudes have the same length.
- **Equilateral triangles:** All three altitudes have the same length.

Also, as you can see in Figure 3-1b, sometimes an altitude is outside the triangle. This situation occurs when the triangle is obtuse. Two of the three altitudes in every obtuse triangle are outside the triangle; the third altitude is inside the triangle. And for every right triangle, the two legs are also altitudes, and the third altitude is inside the triangle. All three altitudes of an acute triangle are inside the triangle.

The most common way of figuring a triangle's area is by plugging the triangle's base and height into the regular area formula. But if all you know are the triangle's three sides, you can use the following nifty alternate formula attributed to Hero of Alexandria (who lived 10–70 AD — or CE if you prefer).

$$\text{Area}_\triangle = \sqrt{S(S-a)(S-b)(S-c)}$$

In this formula, *a*, *b*, and *c* are the length of the triangle's sides, and *S* is the triangle's *semiperimeter* (half the perimeter: $S = \tfrac{a+b+c}{2}$).

And here's one more triangle area formula for you:

The area of an equilateral triangle with side s is $\dfrac{s^2\sqrt{3}}{4}$.

Q. Given: $AB = 10$

D is the midpoint of $\overline{AB}$

M is the midpoint of $\overline{CD}$

$\overline{AB} \perp \overline{CD}$

$\triangle BDM$ is isosceles

Find: Area of $\triangle ACM$

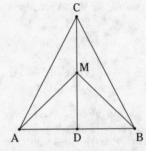

A. To use the ordinary area formula, you need a base and a height of $\triangle ACM$. A base and its corresponding height are always perpendicular, so the 90° angle at D is the place to look. Take this book and rotate it 90° clockwise. Now, picture $\triangle ACM$ standing up on a table, where the table top runs along $\overline{CD}$. Side $\overline{CM}$ is on the table, so that's the base. And the height goes from the peak (A) straight down to the table at D, so the height is $\overline{AD}$. To use the area formula, you need the lengths of $\overline{CM}$ and $\overline{AD}$.

$AB = 10$ and D is the midpoint of $\overline{AB}$, so $AD = 5$. One down, one to go. $\triangle BDM$ is isosceles, so two of its sides are equal. It's also a right triangle with hypotenuse $\overline{MB}$, so the two equal sides have to be $\overline{DM}$ and $\overline{DB}$ (the hypotenuse is always longer than the legs). DB equals AD, which is 5, so DB is 5; thus, so is DM. Because M is the midpoint of $\overline{CD}$, CM is 5 as well. So the base and height are both 5.

Now just use the formula:

$$\text{Area}_{\triangle ACM} = \tfrac{1}{2} \cdot \text{base} \cdot \text{height}$$
$$= \tfrac{1}{2} \cdot 5 \cdot 5$$
$$= 12.5$$

The area is 12.5 units^2.

6. Recalling that some altitudes may be outside the triangle (like in Figure 3-1b earlier in the chapter), draw in the three altitudes of the following triangle.

Solve It

7. Figure the area of the big triangle in four different ways.

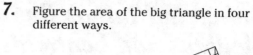

Solve It

8. Compute the area of rectangle *ABDE* and then the areas of △*ACE*, △*AGE*, and △*APE*. What two conclusions can you draw about these areas?

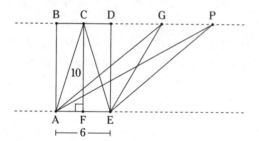

Solve It

9. Given: $MT = 6$
Find: NS

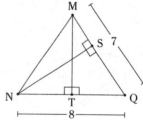

Solve It

Balancing Things Out with Medians and Centroids

Because triangles aren't as symmetrical as, say, circles or rectangles (except for the equilateral triangle), they don't have an obvious center point like circles and rectangles do. In this and the next section, you look at four different "centers" that every triangle has. Of the four, the centroid is probably the best candidate for a triangle's true center.

A triangle's medians point the way to its centroid. Here's everything you've always wanted to know about medians but were afraid to ask:

- **Median:** A *median* of a triangle is a segment joining a *vertex* (corner point) with the midpoint of the opposite side. Every triangle has three medians.

- **Centroid:** The three medians of a triangle intersect at a single point called the *centroid*. (The centroid is the triangle's center of gravity, or balance point.)

- **Position of centroid on median:** Along every median, the distance from the vertex to the centroid is twice as long as the distance from the centroid to the midpoint.

Q. Given △BSF with medians $\overline{BH}$, $\overline{SU}$, and $\overline{FA}$ and centroid L

a. If FL is 12, what's FA?

b. If BH is 12, what's HL?

c. If SL is 12, what's UL?

d. If the area of △BSF is 20 units², what's the area of △BSU?

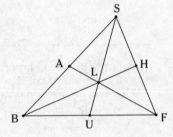

A. The centroid, L, cuts each median into a $\frac{1}{3}$ part and a $\frac{2}{3}$ part. Notice that it's obvious from the figure which is the short part and which is the long part. (Try measuring the parts with your fingers.)

a. FL is $\frac{2}{3}$ of FA, so if FL is 12, FA = 18.

b. HL is $\frac{1}{3}$ of BH, so if BH is 12, HL = 4.

c. A centroid is twice as far from a vertex as it is from the midpoint of the opposite side, so $\overline{SL}$ is twice as long as $\overline{UL}$. SL = 12, so UL = 6.

d. △BSF and △BSU have the same altitude (it goes from point S straight down to $\overline{BF}$, hitting $\overline{BF}$ somewhere between U and F). $\overline{SU}$ is a median, so U is the midpoint of $\overline{BF}$. Thus, $\overline{BU}$, the base of △BSU, is half as long as $\overline{BF}$, the base of △BSF. Therefore, because their altitudes are the same, and because △BSU has a base that's half of the base of △BSF, the area of △BSU must be half of the area of △BSF. The answer is 10 units².

Chapter 3: Triangle Fundamentals and Other Cool Stuff 59

10. Draw in the medians of △ABC. Do they appear to bisect the vertex angles?

Solve It

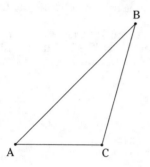

*11. $\overline{NT}$ and $\overline{HO}$ are medians of △NRH. If the area of △NOH is 13, what's the area of △NRT?

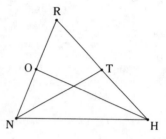

Solve It

Three More "Centers" of a Triangle

The orthocenter, incenter, and circumcenter are three points associated with every triangle. Don't be fooled by the term "center," though. You'll see in a minute why they're called centers, but it's not because these points are near the center of the triangle.

Here's a brief description of each "center":

- **Orthocenter:** Where a triangle's three *altitudes* intersect
- **Incenter:** Where a triangle's three *angle bisectors* intersect; it's the center of a circle *inscribed* in (drawn inside) the triangle
- **Circumcenter:** Where the three *perpendicular bisectors* of the sides intersect; it's the center of a circle *circumscribed* about (drawn around) the triangle

Well, I guess you only get to see why the incenter and the circumcenter are called centers. I don't know why the orthocenter is called a center, but two out of three ain't bad.

If you sketch a few differently shaped triangles, you can see that there isn't always an obvious place where you'd say the center is, like there would be with, for example, a rectangle. I think a triangle's centroid is the best choice for a triangle's center — better than the above so-called "centers." Here's why: The centroid is the triangle's center of gravity, and it always seems to be near what common sense would say is the center.

Of the three "centers" described in this section, two of them (the orthocenter and circumcenter) are sometimes outside of the triangle. The third one (the incenter) is sometimes way at one end of the triangle. Here's the lowdown for the three "centers" plus the centroid:

Part II: Triangles

- For all types of triangles, the centroid and incenter are inside the triangle.
- In an acute triangle, the orthocenter and circumcenter are inside the triangle as well.
- In a right triangle, the orthocenter and circumcenter are on the triangle.
- In an obtuse triangle, the orthocenter and circumcenter are outside the triangle.

Here's a mnemonic device to help you keep the four "centers" straight. It's admittedly not one of my better mnemonics, but it'll probably work just fine, and it's certainly way better than nothing. First, pair up the four "centers" with the lines, rays, or segments that intersect:

- Centroid — Medians
- Circumcenter — Perpendicular bisectors
- Incenter — Angle bisectors
- Orthocenter — Altitudes

Notice that the two terms on the left that begin with consonants pair up with terms on the right that begin with consonants. Ditto for the terms that begin with vowels. The only two terms that contain double vowels *(oi* and *ia)* are paired up. And the two terms with two *t*'s *(orthocenter* and *altitude)* go together. Easy, right?

Q. In the following triangles, identify all marked centroids, orthocenters, incenters, and circumcenters. Try this exercise on your own before reading the solution.

a)

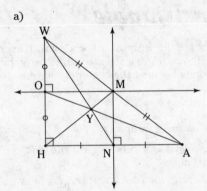

b)

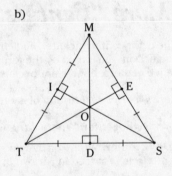

c)

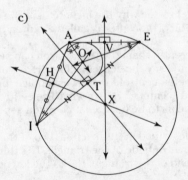

A. In △HWA, the tick marks tell you that M, N, and O are midpoints; therefore, $\overline{HM}$, $\overline{WN}$, and $\overline{AO}$ are medians. Y is thus the centroid. The right-angle marks at the midpoints tell you that $\overleftrightarrow{OM}$ and $\overleftrightarrow{NM}$ are perpendicular bisectors of sides $\overline{HW}$ and $\overline{HA}$. They cross at M, so M is the triangle's circumcenter. (Note that you don't need the third perpendicular bisector; you know that all three intersect at the same point, so any two can show you where the circumcenter is.) Finally, (this one's a bit tricky), you identify point H as the orthocenter. Missing this point is easy because H is part of the triangle. But you can see that $\overline{HW}$ and $\overline{HA}$ are altitudes of △HWA (the two legs of a right triangle are always altitudes), and because $\overline{HW}$ and $\overline{HA}$ intersect at H, H has to be the orthocenter. The incenter of △HWA does not appear on this figure.

Points E, D, and I in △TMS are marked as midpoints, and thus, $\overline{TE}$, $\overline{MD}$, and $\overline{SI}$ are medians. They cross at O, so O is the centroid. The right angle marks on the figure and the tick marks on the angles tell you that $\overline{TE}$, $\overline{MD}$, and $\overline{SI}$ are also altitudes *and* angle bisectors *and* perpendicular bisectors. So, yup, point O is all four points wrapped up into one: the centroid, the orthocenter, the incenter, and the circumcenter. By the way, this overlap happens only in an equilateral triangle. In fact, the four points are *always* four distinct points except when they all come together in an equilateral triangle.

For △IAE, the two circles should make this a no-brainer. Point O, the center of the inscribed circle, is, by definition, the incenter. And point X, the center of the circumscribed circle, is, by definition, the circumcenter. Neither the orthocenter nor the centroid of △IAE appears on this figure.

---

12. Pick and choose: Identify the centroid, the orthocenter, the incenter, and the circumcenter in △XYZ. This figure is drawn to scale.

Solve It

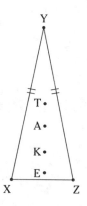

13. Pick and choose: Identify the centroid, the orthocenter, the incenter, and the circumcenter in △ABC. This figure is drawn to scale.

Solve It

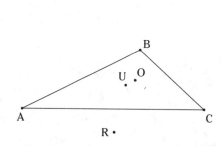

62 Part II: Triangles

14. Pick and choose: Identify the centroid, the orthocenter, the incenter, and the circumcenter in △STU. This figure is drawn to scale.

Solve It

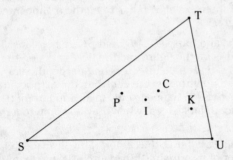

15. What does the fact that Y and R are outside the triangle tell you (this is the same figure as in problem 13)?

Solve It

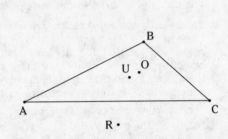

16. For the following triangle, locate (approximately) and draw in its centroid, orthocenter, incenter, and circumcenter. *Hint for problems 16–19:* Just sketch two medians to find the centroid, two perpendicular bisectors to find the circumcenter, and so on.

Solve It

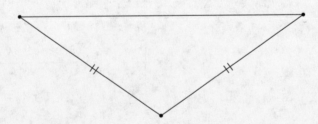

Chapter 3: Triangle Fundamentals and Other Cool Stuff 63

17. For the triangle below, locate (approximately) and draw in its centroid, orthocenter, incenter, and circumcenter.

Solve It

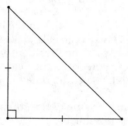

18. For the triangle below, locate (approximately) and draw in its centroid, orthocenter, incenter, and circumcenter.

Solve It

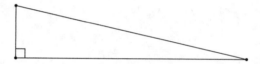

19. For the triangle below, locate (approximately) and draw in its centroid, orthocenter, incenter, and circumcenter.

Solve It

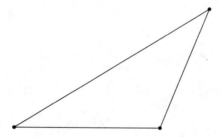

The Pythagorean Theorem

Drum roll, please. Ladieeeees and gentlemen, in the center ring, for your enjoyment and amazement, all the way from Samos, Greece, from over 2,600 years ago, I bring you . . . the Pythagorean Theorem! Pretty thrilling, eh?

The Pythagorean Theorem is certainly one of the most famous theorems in all of mathematics. Mathematicians and lay people alike have studied it for centuries. People have proved it in many different ways. Even President James Garfield was credited with a new, original proof. Well, here you go. As the Scarecrow in *The Wizard of Oz* tried to say after he got his Doctor of Thinkology diploma (a "Th.D.") to prove he had brains:

The Pythagorean Theorem: The sum of the squares of the legs of a right triangle is equal to the square of the hypotenuse.

(Actually, the Scarecrow misstated it as "The sum of the square roots of any two sides of an isosceles triangle is equal to the square root of the remaining side.")

Figure 3-2 contains the well-known 3-4-5 triangle to visually show you the meaning of the Pythagorean Theorem.

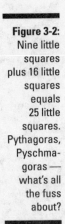

Figure 3-2: Nine little squares plus 16 little squares equals 25 little squares. Pythagoras, Pyschmagoras — what's all the fuss about?

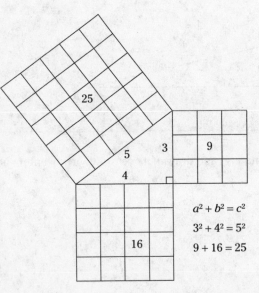

$a^2 + b^2 = c^2$

$3^2 + 4^2 = 5^2$

$9 + 16 = 25$

Chapter 3: Triangle Fundamentals and Other Cool Stuff 65

Q. Calculate the length of the unknown sides in the triangles to the right.

a) b)

A. △ABC: $a^2 + b^2 = c^2$
$7^2 + 8^2 = x^2$
$49 + 64 = x^2$
$113 = x^2$
$x = \sqrt{113}$
≈ 10.6

△XYZ: $a^2 + b^2 = c^2$
$12^2 + y^2 = 13^2$
$144 + y^2 = 169$
$y^2 = 25$
$y = 5$

---

20. Find the length of the unknown side in the following triangle. If the answer is irrational, give your answer in exact, radical (square root) form and in decimal form rounded to two decimal places.

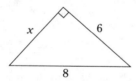

Solve It

21. Find the length of the unknown side in the following triangle. If the answer is irrational, give your answer in exact, radical (square root) form and in decimal form rounded to two decimal places.

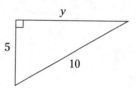

Solve It

66 Part II: Triangles

22. Find the length of the unknown side in the following triangle. If the answer is irrational, give your answer in exact, radical (square root) form and in decimal form rounded to two decimal places.

Solve It

23. Find the length of the unknown side in the following triangle. If the answer is irrational, give your answer in exact, radical (square root) form and in decimal form rounded to two decimal places.

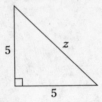

Solve It

24. Find the length of the unknown side in the following triangle. If the answer is irrational, give your answer in exact, radical (square root) form and in decimal form rounded to two decimal places.

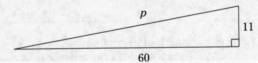

Solve It

25. Find the length of the unknown side in the following triangle. If the answer is irrational, give your answer in exact, radical (square root) form and in decimal form rounded to two decimal places.

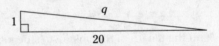

Solve It

26. Find x.

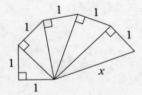

Solve It

27. Find PS, SR, PR, and the area of $\triangle PQR$.

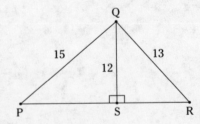

Solve It

Chapter 3: Triangle Fundamentals and Other Cool Stuff

28. Answer the following questions using this figure:

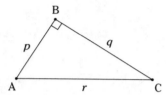

a. Express AC (the length of $\overline{AC}$) in terms of p and q.

b. Express AB in terms of q and r.

c. Express BC in terms of p and r.

Solve It

*29. Find the area of △ MOJ without using Hero's formula.

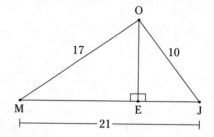

Solve It

Pythagorean Triple Triangles

If you pick any old numbers for two of the sides of a right triangle, the third side usually ends up being irrational — you know, the square root of something. For example, if the legs are 5 and 8, the hypotenuse ends up being $\sqrt{5^2 + 8^2} = \sqrt{89} \approx 9.43398\ldots$ (the decimal goes on forever without repeating). And if you pick whole numbers for the hypotenuse and one of the legs, the other leg usually winds up being the square root of something.

When this doesn't happen — namely, when all three sides are whole numbers — you've got a Pythagorean triple.

Pythagorean Triple: A *Pythagorean triple* (like 3-4-5) is a set of three *whole* numbers that work in the Pythagorean Theorem ($a^2 + b^2 = c^2$) and can thus be used for the three sides of a right triangle.

In this section, you study the four smallest Pythagorean triple triangles: the 3-4-5 triangle; the 5-12-13 triangle; the 7-24-25 triangle; and the 8-15-17 triangle. But infinitely more of them exist. If you're interested, one simple way to find more of them is to take any odd number, say 11, and square it — that's 121. The two consecutive numbers that add up to 121 (60 and 61) give you the two other numbers (to go with 11). So another Pythagorean triple is 11-60-61.

A *family* of right triangles is associated with each Pythagorean triple. For example, the 5:12:13 family consists of the 5-12-13 triangle and all other triangles of the same shape that you'd get by shrinking or blowing up the 5-12-13 triangle. If you shrink it 100 times, you get a $\frac{5}{100}$-$\frac{12}{100}$-$\frac{13}{100}$ triangle. Or you can quadruple each side and get a 20-48-52 triangle or multiply each side by $\sqrt{17}$ to get a 5$\sqrt{17}$-12$\sqrt{17}$-13$\sqrt{17}$ triangle.

68   Part II: Triangles

Q. Find the lengths of the unknown sides in the triangles below by looking for triangle families. (Don't use the Pythagorean Theorem.)

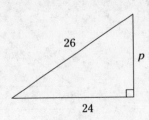
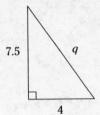

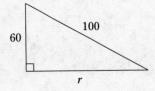
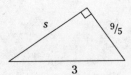

A. For the *p* triangle, you want to first notice that 26 is twice 13. That should ring the $5:12:13$ bell. Then you check that 24 is twice 12, which, of course, it is. Thus, you have a 5-12-13 triangle blown up to twice its size; therefore, *p* is $2 \cdot 5$, or 10.

For the *q* triangle, you recognize the triangle family if you get rid of that pesky decimal. You can do that by multiplying the 7.5 and the 4 by 2, which gives you 15 and 8. Bingo — you have an 8-15-17 triangle shrunk in half. So *q* is half of 17, or 8.5.

For the *r* triangle, first divide the 60 and 100 by 10 — that's 6 and 10. This should ring the $3:4:5$ bell. Doubling 3, 4, and 5 gives you 6, 8, and 10, and then multiplying by 10 gives you 60, 80, and 100, so *r* is 80.

Finally, for the *s* triangle, multiply the 3 and the $9/5$ by the denominator, 5, to get 15 and 9. Then reduce these terms by dividing each by 3: That gives you 5 and 3, and, voilà, you have a triangle in the $3:4:5$ family. One neat way to find *s* is to now take the 4 (because the two given sides became the 5 and 3) and reverse the process: *multiply* by 3 (that's 12) and then *divide* by 5: *s* is $12/5$.

---

30. Without using the Pythagorean Theorem, find the length of the unknown side in the following triangle.

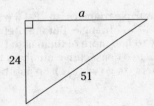

Solve It

31. Without using the Pythagorean Theorem, find the length of the unknown side in the following triangle.

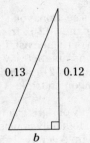

Solve It

Chapter 3: Triangle Fundamentals and Other Cool Stuff 69

32. Without using the Pythagorean Theorem, find the length of the unknown side in the following triangle.

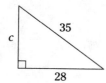

Solve It

33. Without using the Pythagorean Theorem, find the length of the unknown side in the following triangle.

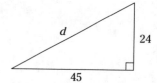

Solve It

*34. Without using the Pythagorean Theorem, find the length of the unknown side in the following triangle.

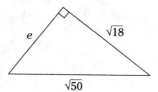

Solve It

*35. Without using the Pythagorean Theorem, find the length of the unknown side in the following triangle.

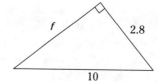

Solve It

*36. Find a, b, c, and d.

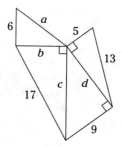

Solve It

37. Find x.

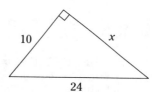

Solve It

Part II: Triangles

38. Find c.

Solve It

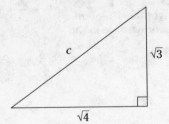

Unique Degrees: Two Special Right Triangles

The Pythagorean triple families of triangles you find in the last section are nice to know because they come up in so many right triangle problems. But mathematically speaking, the two right triangles in this section — the 45°-45°-90° triangle and the 30°-60°-90° triangle — are really more important. The first is exactly half of a square, and the second is exactly half of an equilateral triangle, and this connection to these elemental shapes makes the two triangles ubiquitous in the geometry landscape. Check these triangles out in Figure 3-3:

- The 45°-45°-90° triangle has angles of 45°, 45°, and 90° (duh) and sides in the ratio of $1:1:\sqrt{2}$. This triangle is the shape of half a square, cut along its diagonal.

- The 30°-60°-90° triangle has angles of 30°, 60°, and 90° and sides in the ratio of $1:\sqrt{3}:2$. This triangle is the shape of half an equilateral triangle cut down the middle along its altitude.

Figure 3-3: Two special right triangles.

 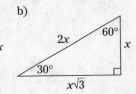

The 45°-45°-90° and 30°-60°-90° triangles are very important in trigonometry and to a lesser extent in calculus. Get to know them forwards, backwards, upside-down, and sideways.

When you use the Pythagorean Theorem, you often end up with a hypotenuse with a square root in it. Because of this, students often mix up the $2x$ and the $x\sqrt{3}$ for the 30°-60°-90° triangle and put the $x\sqrt{3}$ on the hypotenuse. You can avoid this mistake if you remember that $\sqrt{3}$ is less than 2 (think for a few seconds and figure out why it has to be less than 2); because the hypotenuse is always the longest side of a right triangle, the $2x$ has to go on the hypotenuse.

Whenever you sketch a 30°-60°-90° triangle, make sure you make the long leg much longer than the short leg (it doesn't hurt to even exaggerate the relationship a bit). That way, it'll be obvious to you that the short leg touches the 60° angle and that the long leg touches the 30° angle. If you instead get a bit sloppy and draw a 30°-60°-90° triangle so that the legs look about equal, it's easy to get the legs mixed up.

Chapter 3: Triangle Fundamentals and Other Cool Stuff

Q. Find the lengths of the unknown sides in △CBA and △WQX.

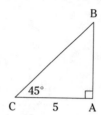 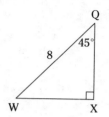

A. You have two ways to solve these problems that really amount to the same thing. First, you can use the ratio of the sides of the 45°- 45°- 90° triangle from Figure 3-3:

leg : leg : hypotenuse

$x : x : x\sqrt{2}$

In △ CBA, one of the legs is 5, so x is 5. Now just plug 5 into $x : x : x\sqrt{2}$ and you have the three sides: 5, 5, and $5\sqrt{2}$.

In △ WQX, the hypotenuse is 8, so you set $x\sqrt{2}$ equal to 8 and solve for x:

$x\sqrt{2} = 8$

$x = \dfrac{8}{\sqrt{2}} = \dfrac{8\sqrt{2}}{2} = 4\sqrt{2}$

So the three sides are $4\sqrt{2}$, $4\sqrt{2}$, and 8.

I prefer the following method: Just think of the 45°- 45°- 90° triangle as the $\sqrt{2}$ triangle (or "root 2 triangle"). Now, if you know the length of a leg and you want the length of the hypotenuse (a *longer* thing), you *multiply* by $\sqrt{2}$. And if you know the hypotenuse and want to figure a leg (a *shorter* thing), you *divide* by $\sqrt{2}$. That's all there is to it.

Q. Find the lengths of the unknown sides in △DEB and △JED.

 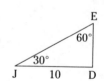

A. You can solve 30°- 60°- 90° triangles with the same two methods you use for 45°- 45°- 90° triangles. The first way — the method of choice of most math teachers because it has that step-by-step, mathematical feel about it — is to use the ratio for the 30°- 60°- 90° triangle from Figure 3-3. The ratio is

short leg : long leg : hypotenuse

$x \ : \ x\sqrt{3} \ : \ 2x$

In △DEB, the hypotenuse is 12; the hypotenuse is represented by $2x$ in the ratio, because you set $2x$ equal to 12 and solve:

$2x = 12$

$x = 6$

So the short side is 6. Now plug $x = 6$ into $x\sqrt{3}$, and you have your three sides: 6, $6\sqrt{3}$, 12.

For △JED, the long leg is 10, and the long leg is represented by $x\sqrt{3}$ in the ratio, so set $x\sqrt{3}$ equal to 10 and solve:

$$x\sqrt{3} = 10$$

$$x = \frac{10}{\sqrt{3}}$$

That's the short side. And plugging $x = \frac{10}{\sqrt{3}}$ into $2x$ gives you $\frac{20}{\sqrt{3}}$ for the hypotenuse.

If your teacher insists on it, go ahead and rationalize the denominator $\left(\frac{20\sqrt{3}}{3}\right)$. (This custom, however, has been pretty much pointless since the advent of calculators that can handle square roots. How long is that? A good 40 years or more. I guess old habits die hard.)

I prefer the quick, easy, street-smart method. First, the relationship between the short leg and the hypotenuse is a total no-brainer. Because the hypotenuse is twice as long as the short leg, as soon as you have one of them, you automatically have the other. So the only real math you have to do is in computing the long leg from the short leg or vice-versa. Think of the 30°- 60°- 90° triangle as the $\sqrt{3}$ triangle (or the "root 3 triangle"). If you know the short leg and you want the long leg (a *longer* thing), you *multiply* by $\sqrt{3}$. And if you have the long leg and want the short leg (a *shorter* thing), you *divide* by $\sqrt{3}$. That's it.

Here's how to use my quick, easy method on △DEB and △JED. For △DEB: *DE* is 12; *EB* is half of that, 6; and *DB* is longer than that, so multiply by $\sqrt{3}$ to get $DB = 6\sqrt{3}$. For △JED: *JD* is 10; *ED* is shorter than that, so divide by $\sqrt{3}$ to get $JD = \frac{10}{\sqrt{3}}$; and *JE* is double that, or $\frac{20}{\sqrt{3}}$. Easy as pie.

39. Find the area of an equilateral triangle whose sides are 10.

Solve It

40. Find the area of a square whose diagonal has a length of 10.

Solve It

*41. Find the perimeter of *QWERT(Y)*.

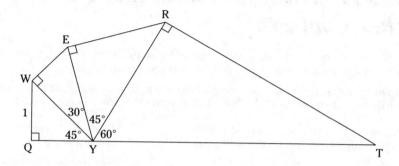

Solve It

*42. Find the area of *KEYBOARD*.

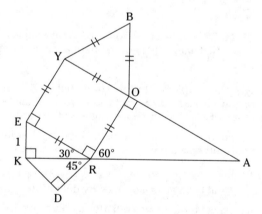

Solve It

Solutions for Triangle Fundamentals and Other Cool Stuff

1 The two tick marks in triangle *a* tell you that those two sides are equal and, thus, the triangle is isosceles. It looks equilateral, but you can't assume that.

Triangle *b* must be scalene, because no matter what *x* is, *x* and *x* + 1 and *x* + 2 will always be three different lengths. Don't be fooled by the fact that the triangle looks isosceles.

2 Because $\triangle ISO$ is isosceles, at least two of the sides must be equal. First try $\overline{IS} \cong \overline{SO}$:

$$2x = 4x - 4$$
$$-2x = -4$$
$$x = 2$$

Plugging $x = 2$ into the three sides gives you sides of 4, 4, and 1; that's not enough 'cause the perimeter is supposed to be more than 10. Try $\overline{SO} \cong \overline{IO}$:

$$4x - 4 = 3x - 5$$
$$x = -1$$

No good. This setup gives you three sides of negative length.

The third pair better work, because that's the only thing left to try:

$$\overline{IO} \cong \overline{IS}$$
$$3x - 5 = 2x$$
$$x = 5$$

Plugging $x = 5$ into the three sides gives you sides of length 10, 10, and 16. Bingo. The base, $\overline{SO}$, is 16 and the legs, $\overline{IO}$ and $\overline{IS}$, are both 10.

3 The angles are in the ratio of 4 : 5 : 6, so set the angles equal to $4x$, $5x$, and $6x$. The angles in a triangle add up to 180°, so

$$4x + 5x + 6x = 180$$
$$15x = 180$$
$$x = 12$$

Plugging $x = 12$ into $4x$, $5x$, and $6x$ gives you three acute angles, 48°, 60°, and 72°, so it's an acute triangle. And because the triangle has three unequal angles, it must have three unequal sides as well. So, it's scalene.

4 For the first triangle, the supplement of the 140° angle is 40°, and the vertical angle across from the 50° angle is, of course, also 50°. So far, you have a 40° angle and a 50° angle. The third angle has to give you a total of 180°, so the third angle is 90°: You have a right triangle.

Did you think the second triangle was obtuse? Good try, but look again. This isn't any type of triangle — not in our universe anyway — because the angles don't add up to 180°.

5 Here are the answers:

a. Always: An equilateral triangle is isosceles by definition.

b. Sometimes: An isosceles triangle is equilateral only when its base is congruent to its legs.

c. Sometimes: A right triangle is isosceles when its legs are congruent (in other words, when it's a 45°- 45°- 90° triangle — see the "Unique Degrees: Two Special Right Triangles" section in this chapter).

d. Always: 70° plus 55° is 125°. The third angle must bring the total to 180°, so it's another 55° angle, and therefore, the triangle is isosceles.

e. Sometimes: If the vertex angle of an obtuse isosceles triangle is 100°, its base angles will both be 40°, so the answer has to be at least *sometimes*. But the answer isn't *always*, because the vertex angle of an obtuse isosceles triangle can have any measure greater than 90° and less than 180°.

f. Never: 40° plus 40° is 80°, so the third angle must be 100°, which makes the triangle obtuse.

g. Never: A triangle can never have two supplementary angles, because they would add up to 180° and there'd be nothing left for the third angle.

Chapter 3: Triangle Fundamentals and Other Cool Stuff

h. Never: If two of the angles in a triangle are complementary, they add up to 90°, and that leaves 90° for the third angle, because all three angles have to total 180°. Thus, the triangle must be a right triangle.

6 Your answer should look roughly like this:

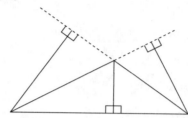

You may want to spin this figure around to make the dotted lines horizontal (like a tabletop) and the altitudes going straight up from the table. That's a good way to picture altitudes and to see where they should go.

7 The area equals 30 units²:

$$\text{Area}_\triangle = \tfrac{1}{2} \, \text{base} \cdot \text{height}$$
$$= \tfrac{1}{2} \cdot 13 \cdot \tfrac{60}{13}$$
$$= 30$$

Now, if you instead use the 12 as the base, the altitude is 5:

$$\text{Area}_\triangle = \tfrac{1}{2} \cdot 12 \cdot 5$$
$$= 30$$

If 5 is the base, the height is 12:

$$\text{Area}_\triangle = \tfrac{1}{2} \cdot 5 \cdot 12$$
$$= 30$$

Finally, use Hero's formula:

$$S = \frac{5 + 12 + 13}{2} = 15$$
$$\text{Area}_\triangle = \sqrt{S(S-a)(S-b)(S-c)}$$
$$= \sqrt{15(15-5)(15-12)(15-13)}$$
$$= \sqrt{15(10)(3)(2)}$$
$$= \sqrt{900}$$
$$= 30$$

8 The area of rectangle *ABDE* is, of course, 6 · 10, or 60 units². The area of △*ACE* is $\tfrac{1}{2} \cdot 6 \cdot 10$, or 30 units². △*AGE* and △*APE* both have the same base as △*ACE* ($\overline{AE}$) and, like △*ACE*, they both have a height of 10 (the vertical distance between the two dotted lines). Thus, the areas of the three triangles are the same.

You can draw these two conclusions:

- The area of a triangle is half of the area of a rectangle with that same base and height. (Do you see why this has to be true? **Hint:** Look at triangles *ACF* and *CAB* and triangles *ECF* and *EDC*.)
- If triangles have the same segment for their bases and their heights are the same, then their areas are equal. Consider this: Imagine that sides $\overline{AP}$ and $\overline{EP}$ are made of elastic, and you grab point *P* and pull it to the right along the dotted line. You could pull it out 1,000 miles or more and the area of △ *APE* would still be only 30 units².

9 For this one, you have to use the area formula twice:

$$\text{Area}_\triangle = \tfrac{1}{2} \cdot b \cdot h$$
$$= \tfrac{1}{2} \cdot 8 \cdot 6$$
$$= 24$$

Now, because $\overline{NS}$ is the altitude drawn to base $\overline{MQ}$, you can figure *NS* by using the area formula backwards:

$$\text{Area}_\triangle = \frac{1}{2} \cdot b \cdot h$$
$$24 = \frac{1}{2} \cdot 7 \cdot h$$
$$48 = 7 \cdot h$$
$$h = \frac{48}{7}$$

So, $NS = \frac{48}{7}$.

10. Here's what your figure should look like:

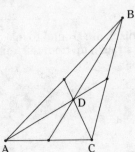

Be careful with this one. Many people take a quick look at the medians and say that it *does* look like they bisect the vertex angles. I hope you didn't jump to that conclusion. If you look carefully, you can see that although ∠BAC and ∠ABC look like they're cut in half, the median from C doesn't even come close to bisecting ∠ACB. ∠ACD in the figure looks like it's somewhere around a 70° angle. But ∠DCB, on the other hand, looks more like a 45° angle.

It turns out that ∠A and ∠B aren't bisected either, though it's pretty close. Only the median to the base of an isosceles triangle bisects the vertex angle (and therefore, all three medians of an equilateral triangle bisect the vertex angles).

*11. At first, you may feel that you've got nothing to go on to solve this problem. You may be thinking, "How can I get the area of △NRT when I don't know anything about it?"

Well, the logic here is quite similar to the reasoning in part *d.* of the example problem. Because $\overline{HO}$ is a median, O is a midpoint, and thus $\overline{NR}$ is twice as long as $\overline{NO}$. Now spin the triangle so that $\overline{NR}$ becomes the base. You can see that △NRH and △NOH have the same height. Because the base of △NRH is twice as long as the base of △NOH, the area of △NRH is twice the area of △NOH — so the area of △NRH is 26 units².

With the same reasoning — this time spinning the triangle so that $\overline{RH}$ is the base — you can conclude that the area of △NRT is half the area of the whole triangle. So, like △NOH, the area of △NRT is 13 units².

12. In △XYZ, if you very roughly sketch the perpendicular bisector of $\overline{XY}$ or $\overline{ZY}$, you can see that it crosses T and doesn't even come close to A, K, or E. So T has to be the circumcenter.

If you sketch the median from Y straight down the middle of the triangle, it passes through all four points. Recalling that the centroid is at the ⅓ point of each median, you can see that point A has to be the centroid. (K and E are nowhere near ⅓ of the way up the median.)

Next, sketch an altitude from, say, angle X perpendicular to $\overline{YZ}$. This segment passes through point E (or close to E, depending on how good your sketching skills are), so E has to be the orthocenter. And if that doesn't convince you, K can't possibly be the orthocenter, because if you draw a line through angle X and point K and that crosses over $\overline{YZ}$, it's easy to see that they're not perpendicular and that, therefore, K is not on an altitude.

Now that you've found the first three points, you have no other choice for K — it has to be the incenter.

Chapter 3: Triangle Fundamentals and Other Cool Stuff 77

13 Here, I skip the lengthy explanation for △ABC (and △STU in the next problem) because the method for identifying the four points is the same as for △XYZ in the preceding problem. You just sketch a median or two, an altitude or two, and so on until, by process of elimination, you've made your picks. Here you go: Y is the orthocenter, O is the incenter, U is the centroid, and R is the circumcenter.

14 Pick 'em: P is the circumcenter, I is the centroid, C is the incenter, and K is the orthocenter.

15 You can conclude that △ABC is an obtuse triangle. Remember that the orthocenter and the circumcenter are *outside* the triangle in obtuse triangles.

16 Your solution should look something like this:

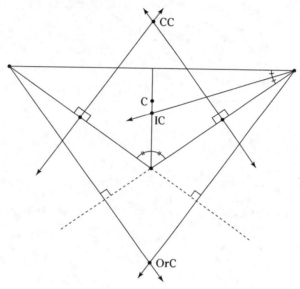

17 Your solution should look something like this:

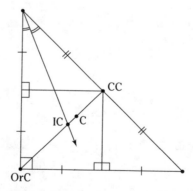

18 Your solution should look something like this:

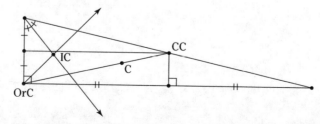

19 Your solution should look something like this:

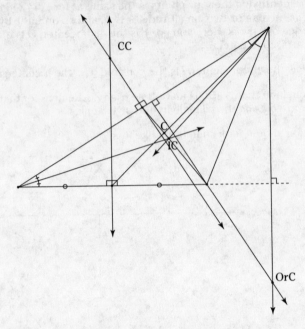

20 You use, of course, $a^2 + b^2 = c^2$ for all six triangles in problems 20 – 25.

$$x^2 + 6^2 = 8^2$$
$$x^2 + 36 = 64$$
$$x^2 = 28$$
$$x = \sqrt{28} = 2\sqrt{7} \approx 5.29$$

21 Here you go:

$$y^2 + 5^2 = 10^2$$
$$y^2 = 75$$
$$y = \sqrt{75} = 5\sqrt{3} \approx 8.66$$

22 Take a look at the answer for the r triangle:

$$10^2 + r^2 = 11^2$$
$$100 + r^2 = 121$$
$$r^2 = 21$$
$$r = \sqrt{21} \approx 4.58$$

23 Check out the answer for this triangle:

$$5^2 + 5^2 = z^2$$
$$50 = z^2$$
$$z = \sqrt{50} = 5\sqrt{2} \approx 7.07$$

24 Here's the solution for the p triangle:

$$11^2 + 60^2 = p^2$$
$$121 + 3600 = p^2$$
$$3721 = p^2$$
$$p = 61$$

25 And here's the last one:

$$1^2 + 20^2 = q^2$$
$$1 + 400 = q^2$$
$$q = \sqrt{401} \approx 20.02$$

26 Number 26 is sort of a domino-effect or chain-reaction problem. You can label the hypotenuses (or is it *hypoteni*, like *hippopotami?*) from left to right as $h_1, h_2, h_3, h_4,$ and x. Now the problem's a walk in the park:

$h_1^2 = 1^2 + 1^2$
$h_1 = \sqrt{2}$

$h_2^2 = 1^2 + \sqrt{2}^2$
$= 1 + 2$
$h_2 = \sqrt{3}$

$h_3^2 = 1^2 + \sqrt{3}^2$
$= 1 + 3$
$h_3 = \sqrt{4} = 2$

$h_4^2 = 1^2 + 2^2$
$= 1 + 4$
$h_4 = \sqrt{5}$

$x^2 = 1^2 + \sqrt{5}^2$
$x = \sqrt{6} \approx 2.45$

27 Here are your answers:

$PS^2 + 12^2 = 15^2$
$PS^2 + 144 = 225$
$PS^2 = 81$
$PS = 9$

$SR^2 + 12^2 = 13^2$
$SR^2 + 144 = 169$
$SR^2 = 25$
$SR = 5$

$PR = PS + SR$
$= 9 + 5$
$= 14$

$\text{Area}_{\triangle PQR} = \frac{1}{2} \text{ base} \cdot \text{height}$
$= \frac{1}{2} \cdot 14 \cdot 12$
$= 84 \text{ units}^2$

28 You know $a^2 + b^2 = c^2$; therefore,

a. $AC^2 = p^2 + q^2$
$AC = \sqrt{p^2 + q^2}$

b. $AB^2 + q^2 = r^2$
$AB^2 = r^2 - q^2$
$AB = \sqrt{r^2 - q^2}$

c. $p^2 + BC^2 = r^2$
$BC^2 = r^2 - p^2$
$BC = \sqrt{r^2 - p^2}$

*29 Label the altitude h and let $\overline{EJ}$ equal x; $\overline{ME}$ is $21 - x$. Then use the Pythagorean Theorem for both triangles and solve the system of two equations with two unknowns.

$\triangle MOE$: $\quad h^2 + (21 - x)^2 = 17^2$
$h^2 + 441 - 42x + x^2 = 289$
$h^2 + x^2 - 42x = -152$

$\triangle JOE$: $\quad h^2 + x^2 = 10^2$
$h^2 + x^2 = 100$

Now subtract the second equation from the first:

$h^2 + x^2 - 42x = -152$
$-(h^2 + x^2 \quad\quad = 100)$
$\overline{\quad\quad\quad -42x = -252}$

$x = \frac{-252}{-42}$
$x = 6$

Next, plug $x = 6$ into the equation for $\triangle JOE$ to get h:

$$h^2 + 6^2 = 100$$
$$h^2 = 64$$
$$h = 8$$

Finally, finish with the area formula:

$$\text{Area}_{\triangle MOJ} = \frac{1}{2}bh$$
$$= \frac{1}{2} \cdot 21 \cdot 8$$
$$= 84 \text{ units}^2$$

30 For the a triangle, reduce the $24:51$ ratio — you know, just like reducing a fraction. Dividing each number by 3, the ratio reduces to $8:17$, so you have a triangle in the $8:15:17$ family. The a is the missing 15 side, and because the triangle is an 8-15-17 triangle blown up 3 times, a is $3 \cdot 15$, or 45.

31 The b triangle should be a no-brainer, because you have a 12 (0.12) and a 13 (0.13) staring you in the face. You know b is thus 0.05, and the triangle is, of course, a 5-12-13 triangle shrunk down 100 times.

32 Do the c triangle just like the a triangle in problem 30. Reducing $28:35$ by 7 gives you $4:5$. Bingo: It's a $3:4:5$ triangle. You find that c is the missing 3 side. Blowing the side back up 7 times gives you 21 for c.

33 The d triangle is also just like the a triangle from problem 30. For the d triangle, the greatest common factor of 24 and 45 is 3. Reducing by 3 gives you 8 and 15, so the triangle's in the $8:15:17$ family. You determine that d is 17 times 3, or 51.

*34 I give you two ways to solve the e triangle. The first method involves simplifiying the radicals:

$$\sqrt{18} = \sqrt{9 \cdot 2} = \sqrt{9}\sqrt{2} = 3\sqrt{2} \text{ and}$$
$$\sqrt{50} = \sqrt{25 \cdot 2} = \sqrt{25}\sqrt{2} = 5\sqrt{2}$$

The $3\sqrt{2}$ and $5\sqrt{2}$ tell you that you have a 3-4-5 triangle blown up $\sqrt{2}$ times. The missing side, e, is thus $4\sqrt{2}$.

The second method is to take your calculator, enter $\frac{\sqrt{18}}{\sqrt{50}}$, and hit *Enter* or =. You get an answer of 0.6. Then "fraction" that, and you get ⅗: Voilà, your triangle is a $3:4:5$ triangle. Now enter $\frac{\sqrt{50}}{5}$ to find out the blow-up multiplier — it's about 1.41. Your approximate answer is 4 times 1.41, or 5.64. (For problems with square roots, this second method can give you only an approximate answer unless you have a super-duper calculator, like the TI-89.)

*35 Just use the calculator trick for the f triangle. Enter $2.8 \div 10$ and hit *Enter*. That gives you 0.28. "Fraction" that and you get ⁷⁄₂₅. Bingo. It's a $7:24:25$ triangle. Dividing 25 by 10 gives you a shrink factor of 2.5. Finally, f equals 24 divided by 2.5, which is 9.6.

*36 I suspect you figured out that you've got to solve for d first, then c, and so on. You know d is, of course, 12. Then that 12 and the 9 are two legs of a 3-4-5 triangle blown up 3 times. Thus, c is 15; b is then 8, of course. Finally, that 8 and the 6 are two legs of another triangle in the $3:4:5$ family, so a is 10.

37 The 10 and the 24 are a 5 and a 12 doubled, so that should ring a bell — the $5:12:13$ bell. But don't answer the bell! In a $5:12:13$ triangle, the 13 represents the hypotenuse, but in this triangle, the 24 (which corresponds to the 12) is the hypotenuse. So this triangle doesn't belong to any of the Pythagorean triple families. You have to solve this triangle with the Pythagorean Theorem:

$$10^2 + x^2 = 24^2$$
$$100 + x^2 = 576$$
$$x^2 = 476$$
$$x = \sqrt{476} \approx 21.82$$

38 Another tricky question. Did you conclude that c equals $\sqrt{5}$? This is *not* a $3:4:5$ triangle, which you can check on your calculator. The ratio of the two legs in a $3:4:5$ triangle is ¾, or 0.75. But if you do $\dfrac{\sqrt{3}}{\sqrt{4}}$ on your calculator, you get something different (0.866), which shows that this triangle is not in the $3:4:5$ family. Solve with the Pythagorean Theorem:

$$c^2 = \sqrt{3}^2 + \sqrt{4}^2$$
$$c^2 = 3 + 4$$
$$c^2 = 7$$
$$c = \sqrt{7}$$

39 Draw your equilateral triangle with its altitude like this:

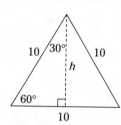

You have the base, so all you need to compute the area is the height. Half of an equilateral triangle is a 30°- 60°- 90° triangle, and you can see that h is the long leg. The short leg is 5; multiply that by $\sqrt{3}$ to get h; h is $5\sqrt{3}$. Finish with the area formula:

$$\text{Area} = \tfrac{1}{2} bh$$
$$= \tfrac{1}{2} \cdot 10 \cdot 5\sqrt{3}$$
$$= 25\sqrt{3} \text{ units}^2$$

You can also, of course, solve this problem with the formula for the area of an equilateral triangle from the section about altitudes and area $\left(A = \dfrac{s^2\sqrt{3}}{4} \right)$. But it's not a bad idea to know the preceding method using the 30°- 60°- 90° triangle because it's useful in its own right. Plus, this method can really come in handy in case you forget the formula.

40 Half a square cut along its diagonal is a 45°- 45°- 90° triangle. The square's diagonal is the hypotenuse of the 45°- 45°- 90° triangle. That's 10, so you divide 10 by $\sqrt{2}$ to get the sides of the square — $\dfrac{10}{\sqrt{2}}$. The area of a square is, of course, s^2, so this square has an area of $\left(\dfrac{10}{\sqrt{2}} \right)^2 = \dfrac{100}{2} = 50 \text{ units}^2$.

*41 $\triangle QWY$ and $\triangle ERY$ are 45°- 45°- 90° triangles, and $\triangle WEY$ and $\triangle RTY$ are 30°- 60°- 90° triangles. You know the drill: In a 45°- 45°- 90° triangle, you multiply or divide by $\sqrt{2}$, and in a 30°- 60°- 90° triangle, you multiply or divide by $\sqrt{3}$. So here you go:

- QW is 1, so since $\triangle QWY$ is a 45°- 45°- 90° $\triangle$, QY is also 1 and WY is $\sqrt{2}$
- WY is $\sqrt{2}$, so WE is $\dfrac{\sqrt{2}}{\sqrt{3}}$
- WE is $\dfrac{\sqrt{2}}{\sqrt{3}}$, so EY is twice that: $\dfrac{2\sqrt{2}}{\sqrt{3}}$
- ER is also $\dfrac{2\sqrt{2}}{\sqrt{3}}$
- ER is $\dfrac{2\sqrt{2}}{\sqrt{3}}$, so RY is $\sqrt{2}\left(\dfrac{2\sqrt{2}}{\sqrt{3}}\right) = \dfrac{4}{\sqrt{3}}$
- RY is $\dfrac{4}{\sqrt{3}}$, so RT is $\sqrt{3}\left(\dfrac{4}{\sqrt{3}}\right) = 4$
- YT is twice RY, so YT is $\dfrac{8}{\sqrt{3}}$

Now just add up all the pieces that make up the perimeter (start at Y and go clockwise):

$$1 + 1 + \frac{\sqrt{2}}{\sqrt{3}} + \frac{2\sqrt{2}}{\sqrt{3}} + 4 + \frac{8}{\sqrt{3}}$$

Simplify:

$$= 6 + \frac{3\sqrt{2}}{\sqrt{3}} + \frac{8}{\sqrt{3}}$$

And rationalize if you feel like it:

$$= 6 + \frac{3\sqrt{6}}{3} + \frac{8\sqrt{3}}{3}$$

$$= 6 + \sqrt{6} + \frac{8\sqrt{3}}{3}$$

$$\approx 6 + 2.45 + 4.62 \approx 13.07$$

42 The total area is made up of three right triangles, an equilateral triangle, and a square. To get the areas of the right triangles, you need their legs. For the equilateral triangle and the square, you need only one side of each (which happen to be equal in this problem).

Here you go:

- KE is 1, so ER is 2 and KR is $\sqrt{3}$
- KR is $\sqrt{3}$, so KD and RD are each $\dfrac{\sqrt{3}}{\sqrt{2}}$
- ER is 2, so YO and OR are also 2
- OR is 2, so OA is $2\sqrt{3}$

Now figure the five areas and then add them up:

$$\text{Area}_{\triangle KER} = \frac{1}{2} \cdot 1 \cdot \sqrt{3} = \frac{\sqrt{3}}{2} \text{ units}^2 \qquad \text{Area}_{\triangle KDR} = \frac{1}{2} \cdot \frac{\sqrt{3}}{\sqrt{2}} \cdot \frac{\sqrt{3}}{\sqrt{2}} = \frac{3}{4} \text{ units}^2$$

$$\text{Area}_{\triangle ROA} = \frac{1}{2} \cdot 2 \cdot 2\sqrt{3} = 2\sqrt{3} \text{ units}^2 \qquad \text{Area}_{\triangle BOY} = \frac{s^2\sqrt{3}}{4} = \frac{2^2\sqrt{3}}{4} = \sqrt{3} \text{ units}^2$$

$$\text{Area}_{EYOR} = 2 \cdot 2 = 4 \text{ units}^2$$

And finally,

$$\text{Area}_{KEYBOARD} = \frac{\sqrt{3}}{2} + \frac{3}{4} + 2\sqrt{3} + \sqrt{3} + 4$$

$$= 3.5\sqrt{3} + 4.75$$

$$\approx 10.81 \text{ units}^2$$

Chapter 4
Congruent Triangles

In This Chapter
▶ Using sides and angles to show triangles congruent
▶ Noting that, naturally, congruent triangles have congruent parts
▶ Trying out the isosceles triangle theorems: If sides, then angles (and vice versa)
▶ Working with the equidistance theorems: Forget CPCTC!

In this chapter, you dive into proofs in a big way. The triangle proofs you do in this and subsequent chapters are real, full-fledged proofs. The proofs in Chapter 2 are basically just warm-up exercises for the longer proofs you do from here on. However, I don't want to diminish the importance of the Chapter 2 material. In fact, Chapter 2 is where you practice using the important theorems and proof techniques that you need for longer proofs, so it's critical that you understand that material before you continue. If you understand Chapter 2, the proofs in this and later chapters probably won't cause you too much trouble (perhaps just the occasional brain hemorrhage).

Sizing Up Three Ways to Prove Triangles Congruent

In a proof, the point at which you prove triangles congruent is sort of like the climax in a novel: Everything builds up to it, and it's the focus or main point or anchor of the proof. Some of the shorter proofs in this chapter end with proving triangles congruent. In longer, more typical proofs, you take things to the next level: You prove triangles congruent and then use that knowledge to prove other things. So proving triangles congruent can either be the final goal of a proof or a stepping stone.

Look for congruent triangles: Proving triangles congruent is critical, and thus you should always check the proof diagram and find *all* pairs of triangles that look the same shape and size. If you find any, you very likely will have to prove one (or more) of the pairs of triangles congruent. And if you can see how to do that, you've probably won at least half the battle.

Okay. So here are the first three of five ways of proving two triangles congruent. (I cover the other two ways in the aptly titled "Two More Ways to Prove Triangles Congruent" section later in the chapter. I don't give you all five at once because I don't want you to blow a geometry fuse from theorem overload.) You're going to use the five triangle theorems all the time.

- **SSS (Side-Side-Side):** If the three sides of one triangle are congruent to the three sides of another triangle, then the triangles are congruent.

- **SAS (Side-Angle-Side):** If two sides and the included angle of one triangle are congruent to two sides and the included angle of another triangle, then the triangles are congruent. (The *included angle* is the mathematician's fancy-pants way of saying "the angle between them.")

- **ASA (Angle-Side-Angle):** If two angles and the included side of one triangle are congruent to two angles and the included side of another triangle, then the triangles are congruent.

And here's one more postulate that comes in handy when trying to prove triangles congruent (this wins first prize in the *well-duh* category):

Reflexive Property: Any segment or angle is congruent to itself. Amazing!

Q. Given: △ABC is isosceles with base $\overline{AC}$ and median $\overline{BM}$

Prove: △ABM ≅ △CBM

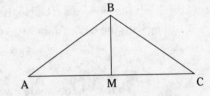

A.

Statements	Reasons
1) △ABC is isosceles with base $\overline{AC}$	1) Given.
2) $\overline{AB} \cong \overline{CB}$	2) Definition of isosceles triangle.
3) $\overline{BM}$ is a median	3) Given.
4) M is the midpoint of $\overline{AC}$	4) Definition of median.
5) $\overline{AM} \cong \overline{CM}$	5) Definition of midpoint.
6) $\overline{BM} \cong \overline{BM}$	6) Reflexive Property.
7) △ABM ≅ △CBM	7) SSS (2, 5, 6).

Note that after SSS in the final step, I indicate the three lines from the *statement* column where the three pairs of sides are shown to be congruent. Doing so is optional, but it's a good idea. It can help you avoid some careless mistakes. **Remember:** The three lines you list must show three congruencies of segments or angles.

Q. Given: B and C trisect $\overline{AD}$

           ∠1 ≅ ∠2

           △BCE is isosceles with base $\overline{BC}$

Prove: △ABE ≅ △DCE

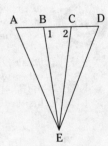

Chapter 4: Congruent Triangles — 85

A.

Statements	Reasons
1) B and C trisect $\overline{AD}$	1) Given.
2) $\overline{AB} \cong \overline{DC}$	2) If a segment is trisected, it is divided into three congruent segments (definition of trisect).
3) $\angle 1 \cong \angle 2$	3) Given.
4) $\angle ABE$ is supplementary to $\angle 1$ $\angle DCE$ is supplementary to $\angle 2$	4) If two angles form a straight angle (assumed from diagram), then they are supplementary.
5) $\angle ABE \cong \angle DCE$	5) If two angles are congruent, then their supplements are congruent.
6) $\triangle BCE$ is isosceles with base $\overline{BC}$	6) Given.
7) $\overline{BE} \cong \overline{CE}$	7) Definition of isosceles triangle.
8) $\triangle ABE \cong \triangle DCE$	8) SAS (2, 5, 7).

Q. Given: $\overline{PS}$ is an altitude of $\triangle QPT$
$\overrightarrow{PS}$ bisects $\angle RPT$
$\overrightarrow{PR}$ bisects $\angle QPS$
$\overrightarrow{PT}$ bisects $\angle UPS$
Prove: $\triangle QPS \cong \triangle UPS$

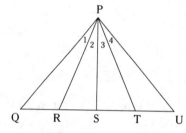

A. **Game plan:** $\overline{PS}$ is an altitude, so it's perpendicular to the base, and the perpendicularity gives you congruent right angles PSQ and PSU — one congruence down, two to go. The two triangles share $\overline{PS}$ — two down, one to go.

So far, you have a pair of congruent angles and a pair of congruent sides, and all the rest of the givens concern the angles near P, so this is almost certainly an ASA problem. All that's left is to show that $\angle QPS$ is congruent to $\angle UPS$. Because of the first bisection, $\angle 2 \cong \angle 3$ (say they're each 20°). Finally, because of the other two bisections, $\angle QPS$ is twice as big as $\angle 2$ and $\angle UPS$ is twice $\angle 3$, so $\angle QPS$ has to be congruent to $\angle UPS$ (they'd both be 40°). Seems simple, huh?

Statements	Reasons
1) $\overline{PS}$ is an altitude of $\triangle QPT$	1) Given.
2) $\overline{PS} \perp \overline{QT}$	2) Definition of altitude.
3) $\angle PSQ$ is a right angle $\angle PSU$ is a right angle	3) Definition of perpendicular.
4) $\angle PSQ \cong \angle PSU$	4) All right angles are congruent. *(continued)*

Four steps probably seems like a lot just to arrive at these two congruent right angles, because as soon as you see an altitude of a triangle, you know you've got two congruent angles. But that's the way proofs work. You have to put down every link in the chain of logic — even incredibly obvious ones. (Like step 2, for example: You can't just jump from step 1 to step 3 even though step 3 is obvious when you know step 1.) Every little step must be spelled out — sort of like if you had to make the logic understandable to a computer. And here's how a computer "thinks":

If *Altitude* then *Perpendicular*

If *Perpendicular* then *Right angles*

If *Right angles* then *Congruent.*

In short, A→P; P→R; R→C. You need this complete chain of logic. Like it or not, those are the rules of the game. And if you're going to play the proof game, you've got to play by the rules.

Statements	Reasons
5) $\overline{PS} \cong \overline{PS}$	5) Reflexive Property. (Well, that was easy.)
6) $\overline{PS}$ bisects ∠RPT	6) Given.
7) ∠RPS ≅ ∠TPS	7) Definition of bisect.
8) $\overline{PR}$ bisects ∠QPS $\overline{PR}$ bisects ∠UPS	8) Given.
9) ∠QPS ≅ ∠UPS	9) If two angles are congruent (∠RPS and ∠TPS), then their like multiples are congruent (∠QPS is double ∠RPS and ∠UPS is double ∠TPS).
10) △QPS ≅ △UPS	10) ASA (4, 5, 9).

1. Given: $\overline{AE}$ is an altitude and a median

Prove: △LEA ≅ △MEA

Solve It

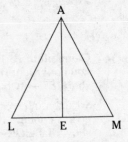

Statements	Reasons

2. Given: $\overline{SQ}$ bisects $\overline{PT}$
∠1 ≅ ∠2

Prove: △PQR ≅ △TSR

Solve It

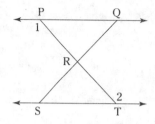

Statements	Reasons

3. Given: △TAG is isosceles with base $\overline{TG}$
$\overline{TH} \cong \overline{GN}$

Prove: △TAN ≅ △GAH

Hint: If this proof has you flummoxed or flabbergasted, try adding here Statements 5 and 6 and Reasons 5 and 6 from the solution.

Solve It

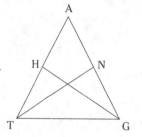

Statements	Reasons

Part II: Triangles

4. Given: ∠AMN is complementary to ∠TAX
∠ATX is complementary to ∠MAN
A is the midpoint of $\overline{TM}$

Prove: △TAX ≅ △MAN

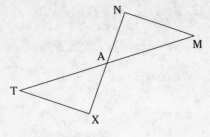

Solve It

Statements	Reasons

*5. Given: △IML and △DRO are isosceles with bases $\overline{IL}$ and $\overline{DO}$
A is the midpoint of $\overline{IM}$
E is the midpoint of $\overline{LM}$
R is the midpoint of $\overline{AO}$ and $\overline{ED}$
$\overline{ID} \cong \overline{LO}$

Prove: △IAO ≅ △LED

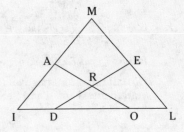

Hint: If you get stuck, copy Statement 4 and Reason 4 from the solution and try again. If you still need a boost, you can copy Statement 8 and Reason 8 as well.

Solve It

Statements	Reasons

Corresponding Parts of Congruent Triangles Are Congruent (CPCTC)

Contrary to popular belief, CPCTC does not stand for *Cows Pull Carts To China*; it's the acronym for *Corresponding Parts of Congruent Triangles are Congruent*.

CPCTC: If two triangles are congruent, then their corresponding parts are congruent.

Here's how you use CPCTC. In a proof, whenever you prove two triangles congruent, you'll use CPCTC on the very next line as the justification for stating that two sides or two angles (of the two triangles) are congruent. Every triangle has six parts: three sides and three angles. You need to use three out of the six parts when you prove two triangles congruent with SSS, SAS, or ASA (see the preceding section for more on these methods of showing congruency). Therefore, there are always three other pairs of sides or angles that you haven't used yet, and you use CPCTC to show that one of those pairs is congruent.

In proofs, you often prove two triangles congruent and then use CPCTC on the following line. This group of two consecutive lines makes up the core or heart of many, many proofs. Here's what the two lines might look like:

Statements	Reasons
...	...
...	...
...	...
7) $\triangle ABC \cong \triangle DEF$	7) SAS.
8) $\overline{BC} \cong \overline{EF}$	8) CPCTC.
...	...
...	...
...	...

In Chapter 2, I tell you to make sure you use every given. Thinking about how to use the givens is essential at the beginning of a proof. Then I give you a tip about working backwards from the end of a proof. Both are great strategies for solving proofs. The preceding tip about using CPCTC right after showing triangles to be congruent is sort of about working at the *middle* of a proof. The key to many proofs is a pair of lines like example lines 7 and 8 in the preceding figure. If you attack proofs like this at their beginning, middle, and end, even the longest, gnarliest proofs won't stand a chance.

For the upcoming CPCTC example proof, the diagram and the givens are identical to those from the first example in the "Sizing Up Three Ways to Prove Triangles Congruent" section. Only the *prove* statement is different.

90 Part II: Triangles

Q. Given: △ABC is isosceles with base $\overline{AC}$ and median $\overline{BM}$
Prove: $\overrightarrow{BM}$ bisects ∠ABC

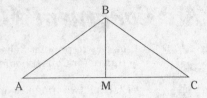

A. This proof is not particularly long or difficult, but — especially if you haven't seen the similar problem in the previous section — it may look just a bit tricky at first glance (you may wonder what the *prove* statement has to do with the givens). Look how easy the proof becomes when you work backwards:

Game plan: Start at the end. You have to prove that $\overrightarrow{BM}$ bisects ∠ABC. The only way to do that is to use the definition of bisect; in other words, you have to show ∠ABM ≅ ∠CBM. Those angles are in triangles that look congruent: △ABM and △CBM. Therefore, if you can show △ABM ≅ △CBM, you can get ∠ABM ≅ ∠CBM with CPCTC. Okay — so if you can prove the triangles congruent, you're home free. At this point, you would go to the beginning of the proof and try to figure out how to show that the triangles are congruent. But here you can cut to the chase, because you already know how to get the triangles congruent from the proof in the preceding section.

Statements	Reasons
1) △ABC is isosceles with base $\overline{AC}$	1) Given.
2) $\overline{AB} ≅ \overline{CB}$	2) Definition of isosceles triangle.
3) $\overline{BM} ≅ \overline{BM}$	3) Reflexive Property.
4) $\overline{BM}$ is a median	4) Given.
5) M is the midpoint of $\overline{AC}$	5) Definition of median.
6) $\overline{AM} ≅ \overline{CM}$	6) Definition of midpoint.
7) △ABM ≅ △CBM	7) SSS (2, 3, 6).
8) ∠ABM ≅ ∠CBM	8) CPCTC.
9) $\overrightarrow{BM}$ bisects ∠ABC	9) Definition of bisect. Or, if you feel like practicing your if-then logic: "If a ray divides an angle into two congruent angles (statement 8), then it bisects the angle (statement 9)."

6. Given: $\overline{AE}$ is an altitude and $\overrightarrow{AE}$ bisects $\angle LAM$

Prove: $\overline{AE}$ is a median

Hint: This proof is quite similar to problem 1 in the preceding section, but try to do it without looking back. If you really need something to go on, however, copy Statement 8 and Reason 9 from the solution instead.

Solve It

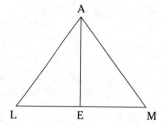

Statements	Reasons

7. Given: △OXE is isosceles with base $\overline{OE}$
$\angle XOE \cong \angle XEO$
$\angle DXE \cong \angle RXO$

Prove: $\overline{DO} \cong \overline{RE}$

There are two different ways to do this proof. *Hint:* The shorter proof goes one step beyond CPCTC; the longer proof uses CPCTC as the final reason. (I realize that idea sounds backwards, but you'll see in a minute that it's correct.)

Solve It

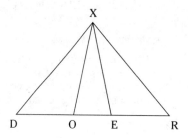

Statements	Reasons

92 Part II: Triangles

8. Given: $\overline{ZA} \cong \overline{XI}$
$\angle MZA \cong \angle FXI$
$\angle FAZ \cong \angle MIX$

Prove: $\overline{MI} \cong \overline{FA}$

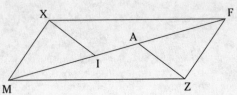

Hint: If this proof freaks you out, write in Statements 5 and 7 and Reason 5 from the solution.

Solve It

Statements	Reasons

***9.** Given: △TAG is isosceles with base $\overline{TG}$
$\overline{TH} \cong \overline{GN}$

Prove: $\triangle THX \cong \triangle GNX$

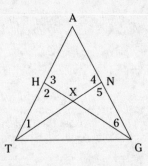

Does this *thang* ring a bell? This proof is the same as problem 3 except that it goes further. But don't look back unless you need a hint. You could also copy Statements 6 and 7 from the solution page with their Reasons; and if that's not enough, copy Statement 10 and its Reason as well.

Solve It

Statements	Reasons

Chapter 4: Congruent Triangles   93

Isosceles Rules: If Sides, Then Angles; If Angles, Then Sides

In this section, you practice problems involving two of the most important and often-used theorems for proofs. Both theorems are about isosceles triangles. But these theorems are really just a single idea that works in both directions:

- **If sides, then angles:** If two sides of a triangle are congruent, then the angles opposite those sides are congruent.
- **If angles, then sides:** If two angles of a triangle are congruent, then the sides opposite those angles are congruent.

These two angle-side theorems come up all the time in proofs. So when you begin a proof, look at the diagram and identify all triangles that look isosceles. Make a mental note that you may have to use one or the other of the theorems for one or more of the isosceles triangles. Because recognizing isosceles triangles is often a cinch — and because it's so easy to use the theorems — be glad when you get these "gimmies" in a proof. On the other hand, if you fail to notice that the theorems should be used, the proof may become impossible. Forewarned is forearmed.

Q. Given: $\overline{OE}$ and $\overline{OT}$ trisect $\overline{NW}$
$\overline{ON} \cong \overline{OW}$

Prove: $\triangle ONE \cong \triangle OWT$

A. **Game plan:** First, you look at the diagram and see two isosceles triangles ($\triangle NOW$ and $\triangle EOT$), so you're rarin' to use one of the angle-side theorems. Sure enough, one of the givens is $\overline{ON} \cong \overline{OW}$, so that gives you $\angle N \cong \angle W$. The trisection gives you $\overline{NE} \cong \overline{WT}$, and, voilà, you have SAS.

Statements	Reasons
1) $\overline{ON} \cong \overline{OW}$	1) Given.
2) $\angle N \cong \angle W$	2) If sides, then angles.
3) $\overline{OE}$ and $\overline{OT}$ trisect $\overline{NW}$	3) Given.
4) $\overline{NE} \cong \overline{WT}$	4) Definition of segment trisection.
5) $\triangle ONE \cong \triangle OWT$	5) SAS (1, 2, 4).

Part II: Triangles

Q. Given: ∠A ≅ ∠E
$\overline{GM}$ is a median

Prove: △MAG ≅ △MEG

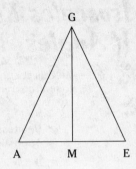

A.

Statements	Reasons
1) ∠A ≅ ∠E	1) Given.
2) $\overline{AG} \cong \overline{EG}$	2) If angles, then sides. (**Repeat warning:** Do not fail to spot this!)
3) $\overline{GM}$ is a median	3) Given.
4) M is the midpoint of $\overline{AE}$	4) Definition of median.
5) $\overline{AM} \cong \overline{EM}$	5) Definition of midpoint.
6) △MAG ≅ △MEG	6) SAS (2, 1, 5). (Note that if you add one more step for $\overline{GM}$ reflexive, you can finish with SSS instead of SAS. This 6-step solution is a fairly unusual proof where you have side-by-side triangles like this but don't use the Reflexive Property.)

10. Given: ∠1 ≅ ∠2
∠3 ≅ ∠4

Prove: △RAY ≅ △BAN

Solve It

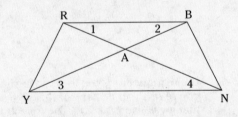

Statements	Reasons

11. Given: △QRS is isosceles with base $\overline{QS}$
∠QPT ≅ ∠STP

Prove: $\overline{PQ} \cong \overline{TS}$

Solve It

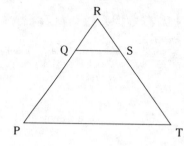

Statements	Reasons

12. Given: $\overline{AB} \cong \overline{ED}$
B is the midpoint of $\overline{AC}$
D is the midpoint of $\overline{EC}$
∠ABF ≅ ∠EDF

Prove: $\overline{BF} \cong \overline{DF}$

Hint: If you have a hard time with this one, copy Statement 3 and Reason 3 from the solution. If you're still stuck, copy Statement 7 and Reason 7 as well.

Solve It

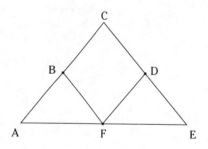

Statements	Reasons

Two More Ways to Prove Triangles Congruent

Mathematicians use five ways of showing triangles congruent. The previous sections in this chapter let you practice problems with SSS, SAS, and ASA. Now you get the final two methods, AAS and HL:

- **AAS (Angle-Angle-Side):** If two angles and a nonincluded side of one triangle are congruent to the corresponding parts of another triangle, then the triangles are congruent.

- **HL (Hypotenuse-Leg):** If the hypotenuse and a leg of one right triangle are congruent to the hypotenuse and a leg of another right triangle, then the triangles are congruent.

AAS works in nearly the same way as SSS, SAS, and ASA. If two triangles have congruent angles, then congruent angles, then congruent sides (in that order, going around the triangles clockwise or counterclockwise), then the triangles are congruent. Like SSS, SAS, and ASA, AAS works with any type of triangle.

You can prove triangles congruent with SSS, SAS, ASA, and AAS, but *not* with ASS or SSA. SAA would work (but you just call it AAS). In short, every three-letter combination of *A*'s and *S*'s works unless it spells *ass* or is *ass* backwards (SSA). (AAA — which you get to in Chapter 7 — also "works," but not to show triangles congruent. You use it to show that triangles are similar.)

HL is a bit different from the other four theorems because it works only with right triangles. For this reason, if I were writing my own book, I'd add the letter *R* and call it HLR (for Hypotenuse, Leg, Right angle). Wait a minute — I *am* writing my own book! Okay, so contrary to other books, I will call it HLR. (Please go to the section intro and the theorem icon, scratch out HL, and replace it with HLR.) HLR is a better name because its three letters make you focus on the fact that when you use HLR — just like with SSS, SAS, ASA, and AAS — you need *three* things to prove two triangles congruent.

Note that, in terms of *A*'s and *S*'s, HLR is *ass* backwards (SSA), because going around the triangle you use a *Side* (the Hypotenuse), a *Side* (the Leg), and an *Angle* (the Right angle) in that order. Thus, HLR is a valid, special case of SSA and an exception to the general invalidity of SSA.

Before going on to problems, I have one more theorem for you. It's yet another in the *well-duh* category. I don't use it in the following example problems, but you do need it for your practice problems.

If two angles are both congruent and supplementary, then they're right angles.

Chapter 4: Congruent Triangles

Q. Given: ∠RTP ≅ ∠RPT
∠PQR ≅ ∠TSR

Prove: △PQR ≅ △TSR

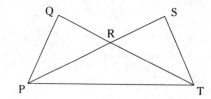

A.

Statements	Reasons
1) ∠RTP ≅ ∠RPT	1) Given.
2) $\overline{PR} \cong \overline{TR}$	2) If angles, then sides.
3) ∠PRQ ≅ ∠TRS	3) Vertical angles are congruent.
4) ∠PQR ≅ ∠TSR	4) Given.
5) △PQR ≅ △TSR	5) AAS (4, 3, 2).

Q. Given: $\overline{RS} \perp \overline{RO}$
$\overline{AE} \perp \overline{AN}$
$\overline{OE} \cong \overline{NS}$
$\overline{RS} \cong \overline{AE}$

Prove: △ROS ≅ △ANE

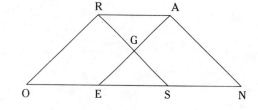

A.

Statements	Reasons
1) $\overline{RS} \perp \overline{RO}$ $\overline{AE} \perp \overline{AN}$	1) Given.
2) ∠ORS is a right ∠ ∠NAE is a right ∠	2) Definition of perpendicular.
3) $\overline{OE} \cong \overline{NS}$	3) Given.
4) $\overline{OS} \cong \overline{NE}$	4) Segment addition. (If a segment ($\overline{ES}$) is added to two congruent segments ($\overline{OE}$ and $\overline{NS}$), then the sums ($\overline{OS}$ and $\overline{NE}$) are congruent.)
5) $\overline{RS} \cong \overline{AE}$	5) Given.
6) △ROS ≅ △ANE	6) HLR (4, 5, 2). (Note that for HLR, you need to state only that you have two right angles, not that they are congruent.)

98 Part II: Triangles

13. Given: $\overline{IN} \cong \overline{IO}$
$\angle OWN \cong \angle NKO$

Prove: $\triangle WOZ \cong \triangle KNZ$

Hint: If you get stuck, copy Statements 5 and 6 with their Reasons onto this page.

Solve It

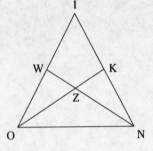

Statements	Reasons

*14. Given: $\angle TIN$ and $\angle EAR$ are right angles
$\overline{AX}$ bisects $\overline{TI}$
$\overline{IZ}$ bisects $\overline{EA}$
$\overline{AX}$ and $\overline{IZ}$ trisect $\overline{TE}$
$\overline{XI} \cong \overline{ZE}$

Prove: $\overline{IN} \cong \overline{AR}$

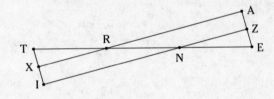

Solve It

Statements	Reasons

Chapter 4: Congruent Triangles   99

*15. Given: $\overline{AB} \cong \overline{CD}$
               $\angle BFA \cong \angle BFE$
               $\angle DEC \cong \angle DEF$
               $\overline{BC} \perp \overline{AB}$
               $\overline{AD} \perp \overline{DC}$
      Prove: $\overline{FB} \cong \overline{ED}$

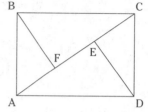

Hint: If you're stumped, write in Statements 5 and 6 and Reasons 5 and 6 from the solution.

Solve It

Statements	Reasons

The Two Equidistance Theorems

Throughout this chapter, I emphasize how important it is to pay attention to the congruent triangles in proof diagrams because the key to so many proofs is showing the triangles congruent and then using CPCTC. Now I muddy the waters a bit by giving you two theorems that you can often use *instead* of proving triangles congruent. You may get proofs in which you see congruent triangles, so it looks like you should try to show that the triangles are congruent, but you don't have to — one of the *equidistance* theorems can give you a shortcut to the final conclusion.

Now you have to be doubly on your toes: looking for congruent triangles and thinking about ways to prove them congruent and, at the same time, being ready to avoid the congruent triangle issue with the equidistance shortcut.

The equidistance theorems:

- If two points are each (one at a time) equidistant from the endpoints of a segment, then those points determine the perpendicular bisector of the segment. (Loose, short form: If *two* pairs of congruent segments, then perpendicular bisector.)

- If a point is on the perpendicular bisector of a segment, then it's equidistant from the endpoints of the segment. (Loose, short form: If perpendicular bisector, then *one* pair of congruent segments.)

These theorems are a royal mouthful. The best way to understand them is visually. For the first theorem, consider Figure 4-1.

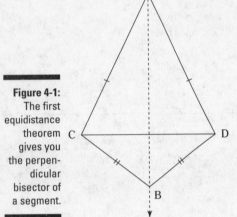

Figure 4-1: The first equidistance theorem gives you the perpendicular bisector of a segment.

Here's how the theorem works. If you have one point (like A) that's equally distant from the endpoints of a segment ($\overline{CD}$) and another point (like B) that's also equally distant from those endpoints, then the two points (A and B) determine (show you where to draw) the perpendicular bisector of that segment. The dashed line in the figure, $\overleftrightarrow{AB}$, is the perpendicular bisector of $\overline{CD}$, which means — as I'm sure you know or can figure out — that it's perpendicular to $\overline{CD}$ and cuts $\overline{CD}$ in half. I didn't mark the perpendicularity or the bisection in the figure because I wanted to mark only the *if* part of the theorem. The figure should also make clear the meaning of the loose, short form of the theorem: If *two* pairs of congruent segments ($\overline{AC} \cong \overline{AD}$ and $\overline{BC} \cong \overline{BD}$), then perpendicular bisector ($\overleftrightarrow{AB}$ is the perpendicular bisector of $\overline{CD}$).

For the second theorem, consider Figure 4-2.

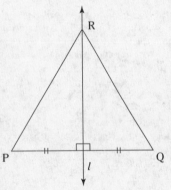

Figure 4-2: The second equidistance theorem lets you know that you have congruent segments.

The second theorem tells you that if you start with a segment (like $\overline{PQ}$) and its perpendicular bisector (like line l) and a point is on the perpendicular bisector (like R), then R is equally distant from the endpoints of the segment. The figure also illustrates the loose, short form of the theorem: If perpendicular bisector (line l is the perpendicular bisector of $\overline{PQ}$), then *one* pair of congruent segments ($\overline{RP} \cong \overline{RQ}$).

Chapter 4: Congruent Triangles   101

Q. Given: $\overline{TA} \cong \overline{TO}$
$\angle PAD \cong \angle POD$

Prove: $\overleftrightarrow{PT}$ is the perpendicular bisector of $\overline{AO}$

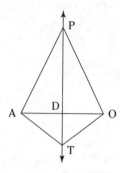

A.

Statements	Reasons
1) $\angle PAD \cong \angle POD$	1) Given.
2) $\overline{PA} \cong \overline{PO}$	2) If angles, then sides.
3) $\overline{TA} \cong \overline{TO}$	3) Given.
4) $\overleftrightarrow{PT}$ is the perpendicular bisector of $\overline{AO}$	4) If two points are each equidistant from the endpoints of a segment, then they determine the perpendicular bisector of that segment.

Q. Given: $\triangle TIC$ and $\triangle TOC$ are isosceles triangles with base $\overline{TC}$

Prove: $\triangle TAC$ is isosceles

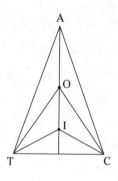

A.

Statements	Reasons
1) $\triangle TIC$ and $\triangle TOC$ are isosceles	1) Given.
2) $\overline{TI} \cong \overline{CI}$ $\overline{TO} \cong \overline{CO}$	2) Definition of isosceles.
3) $\overleftrightarrow{OI}$ is the perpendicular bisector of $\overline{TC}$	3) If two points are each equidistant from the endpoints of a segment, then they determine the perpendicular bisector of that segment.
4) $\overline{TA} \cong \overline{CA}$	4) If a point is on the perpendicular bisector of a segment, then it is equidistant from the endpoints of that segment.
5) $\triangle TAC$ is isosceles	5) Definition of isosceles.

Tip: When you see an unlabeled point in a problem, you don't need to use that point in your proof. Note that in the preceding figure, no letter labels the point where the perpendicular bisector $\overleftrightarrow{OI}$ intersects $\overline{TC}$. This tip doesn't help so much in this particular problem, but for some proofs, this built-in hint can work wonders.

102 Part II: Triangles

16. Given: ∠TIP ≅ ∠TOP
∠HIP ≅ ∠HOP

Prove: $\overline{IP} \cong \overline{OP}$

Solve It

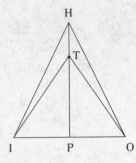

Statements	Reasons

17. Given: ∠SAT and ∠SET are right angles
$\overline{SA} \cong \overline{SE}$

Prove: ∠ANT ≅ ∠ENT

Solve It

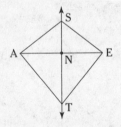

Statements	Reasons

18. Given: $\overline{RS} \cong \overline{CS}$
$\angle ARS \cong \angle ACS$

Prove: $\overline{RY} \cong \overline{CY}$

Hint: If you're stuck, copy Statement 2 and Reason 2 from the solution. And if that doesn't help, copy Statement 5 and Reason 5 as well.

Solve It

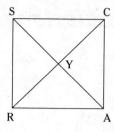

Statements	Reasons

19. Given: $\angle PAN \cong \angle PIN$
$\overline{AN} \cong \overline{IN}$

Prove: a. $\overline{LA} \cong \overline{LI}$

b. (optional) To see what a great shortcut the equidistance theorems are, do this same proof again without using either equidistance theorem.

Solve It

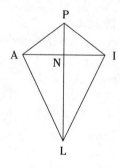

Statements	Reasons

Solutions for Congruent Triangles

1

Statements	Reasons
1) $\overline{AE} \cong \overline{AE}$	1) Reflexive.
2) $\overline{AE}$ is an altitude	2) Given.
3) $\overline{AE} \perp \overline{LM}$	3) Definition of altitude. (Why would they tell you about an altitude? Because it's perpendicular.)
4) $\angle LEA$ is a right angle $\angle MEA$ is a right angle	4) Definition of perpendicular. (What does perpendicularity tell you? Right angles, of course.)
5) $\angle LEA \cong \angle MEA$	5) All right angles are congruent.
6) $\overline{AE}$ is a median	6) Given.
7) E is the midpoint of $\overline{LM}$	7) Definition of median. (Why would they tell you about a median? Because it goes to a midpoint.)
8) $\overline{LE} \cong \overline{ME}$	8) Definition of midpoint. (What do you know about a midpoint? Congruent segments, of course.)
9) $\triangle LEA \cong \triangle MEA$	9) SAS (1, 5, 8).

2

Statements	Reasons
1) $\angle PRQ \cong \angle TRS$	1) Vertical angles are congruent. (This step should be a no-brainer. Always check for vertical angles.)
2) $\overline{SQ}$ bisects $\overline{PT}$	2) Given. (Now, why would they tell you this? Only one possible reason . . .)
3) $\overline{PR} \cong \overline{TR}$	3) Definition of bisect. (So far, you have one pair of congruent angles and one pair of congruent sides, so this problem has to end with either ASA or SAS, right? If it's ASA, you need to show $\angle QPR \cong \angle STR$. If it's SAS, you need to show $\overline{SR} \cong \overline{QR}$. Which seems more promising? It's ASA, because the remaining given concerns $\angle 1$ and $\angle 2$, which are right next to $\angle QPR$ and $\angle STR$. Don't forget: Every given is a built-in hint.)
4) $\angle 1 \cong \angle 2$	4) Given.
5) $\angle QPR \cong \angle STR$	5) Supplements of congruent angles are congruent.
6) $\triangle PQR \cong \triangle TSR$	6) ASA (1, 3, 5).

3

Statements	Reasons
1) △TAG isosceles with base $\overline{TG}$	1) Given.
2) $\overline{TA} \cong \overline{GA}$	2) Definition of isosceles triangle.
3) ∠A ≅ ∠A	3) Reflexive Property. (Did you notice that ∠A is in the same position in both of the *prove* triangles? If not, open your eyes!)
4) $\overline{TH} \cong \overline{GN}$	4) Given.

If you're not sure where to go from here, try this technique. You know △TAG is isosceles, so make up a length for sides TA and GA, say, 10. Then make up a length for congruent segments TH and GN, say, 7. What follows? That HA and NA are both 3, of course, and therefore, that they're congruent. And that's all you need for SAS.

Statements	Reasons
5) $\overline{HA} \cong \overline{NA}$	5) If congruent segments are subtracted from congruent segments, then the differences are congruent.
6) △TAN ≅ △GAH	6) SAS (2, 3, 5).

4

Statements	Reasons
1) ∠TAX ≅ ∠MAN	1) Vertical angles are congruent. (Always look for vertical angles!)
2) ∠AMN is complementary to ∠TAX ∠ATX is complementary to ∠MAN	2) Given.
3) ∠AMN ≅ ∠ATX	3) Complements of congruent angles are congruent. Or, if you prefer the long way, "If two angles are congruent (statement 1), then their complements are congruent (statement 3)."
4) A is the midpoint of $\overline{TM}$	4) Given.
5) $\overline{TA} \cong \overline{MA}$	5) Definition of midpoint.
6) △TAX ≅ △MAN	6) ASA (1, 5, 3).

***5** **Game plan:** The big triangle is isosceles, so you have two equal sides. The midpoints cut those sides in half, so all four halves are equal and IA equals LE. The little triangle is also isosceles, with equal sides $\overline{DR}$ and $\overline{OR}$. Then, because R is a midpoint (for two segments), DE is twice DR and OA is twice OR; thus DE equals OA. The last given is $\overline{ID} \cong \overline{LO}$. Say the segments are each 4 units long; and say DO is 8. That makes IO and LD both 12. That's a good bingo — SSS.

Part II: Triangles

Statements	Reasons
1) △IML is isosceles with base $\overline{IL}$	1) Given.
2) $\overline{IM} \cong \overline{LM}$	2) Definition of isosceles triangle.
3) A is the midpoint of $\overline{IM}$ E is the midpoint of $\overline{LM}$	3) Given.
4) $\overline{IA} \cong \overline{LE}$	4) Like Divisions. Or, to make sure you're following proper if-then logic, "If two segments are congruent (statement 2), then their like divisions are congruent (statement 4)."
5) △DRO is isosceles with base $\overline{DO}$	5) Given.
6) $\overline{OR} \cong \overline{DR}$	6) Definition of isosceles triangle.
7) R is the midpoint of $\overline{AO}$ R is the midpoint of $\overline{ED}$	7) Given.
8) $\overline{OA} \cong \overline{DE}$	8) Like Multiples. Basically, if two segments are congruent ($\overline{OR}$ and $\overline{DR}$), then twice one ($\overline{OA}$) equals twice the other ($\overline{DE}$).
9) $\overline{ID} \cong \overline{LO}$	9) Given.
10) $\overline{IO} \cong \overline{LD}$	10) If a segment is added to two congruent segments, then the sums are congruent ($\overline{DO}$ is added to $\overline{ID}$ and $\overline{LO}$).
11) △IAO ≅ △LED	11) SSS (4, 8, 10).

6 **Game plan:** Work backwards. To prove $\overline{AE}$ is a median, you need to show $\overline{LE} \cong \overline{ME}$. And you can probably get that with CPCTC after showing that the triangles are congruent. 'Nuff said.

Statements	Reasons
1) $\overline{AE}$ is an altitude	1) Given.
2) $\overline{AE} \perp \overline{LM}$	2) Definition of altitude.
3) ∠LEA is a right angle ∠MEA is a right angle	3) Definition of perpendicular.
4) ∠LEA ≅ ∠MEA	4) Right angles are congruent.
5) $\overline{AE} \cong \overline{AE}$	5) Reflexive.
6) $\overline{AE}$ bisects ∠LAM	6) Given.
7) ∠LAE ≅ ∠MAE	7) Definition of bisect.
8) △LEA ≅ △MEA	8) ASA (4, 5, 7).
9) $\overline{LE} \cong \overline{ME}$	9) CPCTC.
10) E is the midpoint of $\overline{LM}$	10) Definition of midpoint.
11) $\overline{AE}$ is a median	11) Definition of median.

Chapter 4: Congruent Triangles

7. Game plan (shorter version): Always take a quick glance at proof diagrams and look for triangles that look congruent. In this diagram, you should see two such pairs: △DXO and △RXE and △DXE and △RXO. The three givens basically hand you that second pair of triangles on a silver platter (with ASA). You get $\overline{DE} \cong \overline{RO}$ with CPCTC; then subtract $\overline{OE}$ and you're done.

Statements	Reasons
1) △OXE is isosceles with base $\overline{OE}$	1) Given.
2) $\overline{OX} \cong \overline{EX}$	2) Definition of isosceles.
3) ∠XOE ≅ ∠XEO	3) Given.
4) ∠DXE ≅ ∠RXO	4) Given.
5) △DXE ≅ △RXO	5) ASA (3, 2, 4).
6) $\overline{DE} \cong \overline{RO}$	6) CPCTC.
7) $\overline{DO} \cong \overline{RE}$	7) Segment subtraction. (If a segment is subtracted from two congruent segments, then the differences are congruent. Say DE and RO are both 8 and OE is 3. Then DO and RE would both be 5).

TIP

The fact that you can do this proof (and many others) in more than one way shows that geometry proofs aren't quite as cut and dried as some other types of math problems. And because of that, you should adopt a flexible approach when doing proofs. Don't assume that there's just one precise way of doing a proof and that you're sunk if you can't find it. Be flexible, use some trial and error, use your imagination. Try anything — don't worry about whether it's "right." Be patient with yourself, and don't expect to always solve a proof on your first try. You have to be willing to try something, find yourself in a dead end, and then go back to the drawing board to try something else. And although it's true that, as a general rule, the fewer steps in your proof, the better, you shouldn't worry too much about that. Most teachers don't mind if your method is a little longer than the shortest possible proof. They may, however, take off a few points if your proof is *way* longer than it has to be.

Game plan (longer version): Say you notice the other pair of congruent triangles, △DXO and △RXE, and realize that the proof can, thus, end with CPCTC. You then have to find a way to show △DXO ≅ △RXE. You already have $\overline{OX} \cong \overline{EX}$. Then you can say ∠DOX ≅ ∠REX, because their supplements are congruent. Finally, you can get ∠DXO ≅ ∠RXE by subtracting the middle angle, ∠OXE, from the overlapping congruent angles, ∠DXE and ∠RXO. Not bad, right?

Statements	Reasons
1) △OXE is isosceles with base $\overline{OE}$	1) Given.
2) $\overline{OX} \cong \overline{EX}$	2) Definition of isosceles.
3) ∠XOE ≅ ∠XEO	3) Given.
4) ∠DOX ≅ ∠REX	4) Supplements of congruent angles are congruent.
5) ∠DXE ≅ ∠RXO	5) Given.
6) ∠DXO ≅ ∠RXE	6) Angle subtraction.
7) △DXO ≅ △RXE	7) ASA (4, 2, 6).
8) $\overline{DO} \cong \overline{RE}$	8) CPCTC.

8 **Game plan:** You have two pairs of congruent triangles that include $\overline{ZA}$ and $\overline{XI}$ — which pair should you shoot for? Stay flexible. Look at the pair including $\angle MXI$ and $\angle FZA$. (If you can prove that pair congruent, you can finish with CPCTC.) You have $\overline{XI}$ and $\angle MIX$. Can you get the third element you need for SAS or ASA? To use SAS, you'd need to know that $\overline{MI} \cong \overline{FA}$, but that's what you're trying to prove, so that won't work. For ASA, you'd need to work your way to $\angle MXI \cong \angle FZA$, but there doesn't seem to be a way to get that. You appear to be at a dead end, so it's time to go back to the drawing board. (By the way, winding up in a dead end like this is par for the course with geometry proofs. Trying to show $\triangle MXI \cong \triangle FZA$ is a perfectly good idea. You have two out of three triangle parts you need for something like ASA, and the triangles are a natural choice because of the possibility of finishing the proof with CPCTC. Don't let dead ends like this frustrate you.)

Time to try the second pair of triangles. Look at $\triangle MAZ$. You have $\angle MZA$ and side $\overline{ZA}$. Can you get $\angle MAZ$ and finish with ASA? Yes. That's it. You have $\angle FAZ \cong \angle MIX$ (say, for instance, they're both 50°), so $\angle MAZ \cong \angle FIX$ (they'd both be 130°). (It's not a bad idea to sometimes actually make up an angle measure like 50° and then write it on the diagram to help you see that the two 130° angles would be congruent.) You now finish with CPCTC and segment subtraction.

Statements	Reasons
1) $\overline{ZA} \cong \overline{XI}$	1) Given.
2) $\angle MZA \cong \angle FXI$	2) Given.
3) $\angle FAZ \cong \angle MIX$	3) Given.
4) $\angle MAZ \cong \angle FIX$	4) Supplements of congruent angles are congruent.
5) $\triangle MAZ \cong \triangle FIX$	5) ASA (2, 1, 4).
6) $\overline{MA} \cong \overline{FI}$	6) CPCTC.
7) $\overline{MI} \cong \overline{FA}$	7) Subtraction.

*__9__ **Game plan (quickie version):** You get $\triangle TAN \cong \triangle GAH$ like you do with problem 3 in the first section. Then you get $\angle 3 \cong \angle 4$ (CPCTC), their supplements, $\angle 2 \cong \angle 5$, and then $\angle 1 \cong \angle 6$ (CPCTC). Bingo! You have ASA.

Statements	Reasons
1) $\triangle TAG$ is isosceles with base $\overline{TG}$	1) Given.
2) $\overline{TA} \cong \overline{GA}$	2) Definition of isosceles.
3) $\overline{TH} \cong \overline{GN}$	3) Given.
4) $\overline{HA} \cong \overline{NA}$	4) Subtraction.
5) $\angle A \cong \angle A$	5) Reflexive.
6) $\triangle TAN \cong \triangle GAH$	6) SAS (2, 5, 4).
7) $\angle 3 \cong \angle 4$	7) CPCTC.
8) $\angle 2 \cong \angle 5$	8) Supplements of congruent angles are congruent.
9) $\angle 1 \cong \angle 6$	9) CPCTC.
10) $\triangle THX \cong \triangle GNX$	10) ASA (8, 3, 9).

Chapter 4: Congruent Triangles 109

10

Statements	Reasons
1) $\angle 1 \cong \angle 2$	1. Given.
2) $\overline{RA} \cong \overline{BA}$	2. If angles, then sides.
3) $\angle 3 \cong \angle 4$	3. Given.
4) $\overline{YA} \cong \overline{NA}$	4. If angles, then sides.
5) $\angle RAY \cong \angle BAN$	5. Vertical angles are congruent.
6) $\triangle RAY \cong \triangle BAN$	6. SAS (2, 5, 4).

11 **Game plan:** As soon as you see two triangles in the diagram that look like they're isosceles and then that $\angle QPT \cong \angle STP$ is in the given, you should immediately realize that you have $\overline{PR} \cong \overline{TR}$. The rest is in the bag.

Statements	Reasons
1) $\angle QPT \cong \angle STP$	1) Given.
2) $\overline{PR} \cong \overline{TR}$	2) If angles, then sides.
3) $\triangle QRS$ is isosceles with base $\overline{QS}$	3) Given.
4) $\overline{QR} \cong \overline{SR}$	4) Definition of isosceles triangle.
5) $\overline{PQ} \cong \overline{TS}$	5) If two congruent segments ($\overline{QR}$ and $\overline{SR}$) are subtracted from two other congruent segments ($\overline{PR}$ and $\overline{TR}$), then the differences ($\overline{PQ}$ and $\overline{TS}$) are congruent. (If PR and TR were both 10, and QR and SR were both 3, then PQ and TS would both be 7, right?)

12 **Game plan:** Maybe I sound like a broken record, but make sure you notice the three triangles that look isosceles and that, therefore, you likely have to use one of the angle-side theorems. You should also, of course, notice that the two small triangles look congruent and that the proof therefore probably ends with CPCTC. So your goal is to show that the two triangles are congruent. You could work backwards — you already have a pair of congruent sides and a pair of congruent angles in those triangles. To finish with SAS, you'd need to use $\overline{BF} \cong \overline{DF}$, but that's what you're trying to prove. Your other option is ASA, and for that to work, you'd need $\angle A \cong \angle E$. Can you get that? This should be the light-bulb-going-on moment of the proof. You should be thinking, "I could get angle A congruent to angle E by *if sides, then angles* if I knew that $\overline{AC}$ is congruent to $\overline{EC}$." Then you can go back to the givens and see that you can, in fact, show that congruency with the Like Multiples Theorem (see Chapter 2).

Another, equally good approach to the proof is to begin by looking at the givens. You see that $\overline{AB}$ and $\overline{ED}$ are congruent and that B and D are midpoints. Now, why would they tell you about those midpoints? Midpoints cut segments in half, of course. So if $\overline{AB} \cong \overline{ED}$, then twice $\overline{AB}$ ($\overline{AC}$) equals twice $\overline{ED}$ ($\overline{EC}$). Then, naturally, you go from $\overline{AC} \cong \overline{EC}$ to $\angle A \cong \angle E$, and the rest is a cake walk.

Part II: Triangles

Statements	Reasons
1) $\overline{AB} \cong \overline{ED}$	1) Given.
2) B is the midpoint of $\overline{AC}$ D is the midpoint of $\overline{EC}$	2) Given.
3) $\overline{AC} \cong \overline{EC}$	3) Like Multiples. (If segments are congruent ($\overline{AB}$ and $\overline{ED}$), then twice one ($\overline{AC}$) is congruent to twice the other ($\overline{EC}$).)
4) $\angle A \cong \angle E$	4) If sides, then angles.
5) $\angle ABF \cong \angle EDF$	5) Given.
6) $\triangle ABF \cong \triangle EDF$	6) ASA (4, 1, 5).
7) $\overline{BF} \cong \overline{DF}$	7) CPCTC.

13

Statements	Reasons
1) $\overline{IN} \cong \overline{IO}$	1) Given.
2) $\angle ION \cong \angle INO$	2) If sides, then angles. (If you missed this, hand over your protractor! This proof is not totally easy, but this step should be.)
3) $\overline{ON} \cong \overline{NO}$	3) Reflexive. (Do you see why I reversed the letters?)
4) $\angle OWN \cong \angle NKO$	4) Given.
5) $\triangle WON \cong \triangle KNO$	5) AAS (4, 2, 3).
6) $\overline{OW} \cong \overline{NK}$	6) CPCTC.
7) $\angle WZO \cong \angle KZN$	7) Vertical angles are congruent.
8) $\triangle WOZ \cong \triangle KNZ$	8) AAS (7, 4, 6).

*__14__

Statements	Reasons
1) $\angle TIN$ and $\angle EAR$ are right angles	1) Given.
2) $\overline{XI} \cong \overline{ZE}$	2) Given.
3) $\overline{AX}$ bisects $\overline{TI}$ $\overline{IZ}$ bisects $\overline{EA}$	3) Given.
4) $\overline{TI} \cong \overline{EA}$	4) Like Multiples. (If segments are congruent ($\overline{XI}$ and $\overline{ZE}$), then twice one ($\overline{TI}$) is congruent to twice the other ($\overline{EA}$).)
5) $\overline{AX}$ and $\overline{IZ}$ trisect $\overline{TE}$	5) Given.
6) $\overline{TR} \cong \overline{RN} \cong \overline{NE}$	6) Definition of trisect.
7) $\overline{TN} \cong \overline{NR}$	7) Segment addition.
8) $\triangle TIN \cong \triangle EAR$	8) HLR (7, 4, 1).
9) $\overline{IN} \cong \overline{AR}$	9) CPCTC.

15

	Statements		Reasons
1)	$\overline{BC} \perp \overline{AB}$ $\overline{AD} \perp \overline{DC}$	1)	Given.
2)	∠ABC is a right angle ∠CDA is a right angle	2)	Definition of perpendicular.
3)	$\overline{AB} \cong \overline{CD}$	3)	Given.
4)	$\overline{AC} \cong \overline{CA}$	4)	Reflexive.
5)	△ABC ≅ △CDA	5)	HLR (4, 3, 2).
6)	∠BAC ≅ ∠DCA	6)	CPCTC.
7)	∠BFA ≅ ∠BFE	7)	Given.
8)	∠BFA and ∠BFE are right angles	8)	If two angles are both congruent and supplementary, then they are right angles.
9)	∠DEC ≅ ∠DEF	9)	Given.
10)	∠DEC and ∠DEF are right angles	10)	If two angles are both congruent and supplementary, then they are right angles.
11)	∠BFA ≅ ∠DEC	11)	All right angles are congruent.
12)	△BFA ≅ △DEC	12)	AAS (11, 6, 3).
13)	$\overline{FB} \cong \overline{ED}$	13)	CPCTC.

16

	Statements		Reasons
1)	∠TIP ≅ ∠TOP	1)	Given.
2)	$\overline{TI} \cong \overline{TO}$	2)	If angles, then sides.
3)	∠HIP ≅ ∠HOP	3)	Given.
4)	$\overline{HI} \cong \overline{HO}$	4)	If angles, then sides.
5)	$\overleftrightarrow{HT}$ is the perpendicular bisector of $\overline{IO}$	5)	If two points are each equidistant from the endpoints of a segment, then they determine the perpendicular bisector of that segment.
6)	$\overline{IP} \cong \overline{OP}$	6)	Definition of bisect.

17

	Statements		Reasons
1)	∠SAT and ∠SET are right angles	1)	Given.
2)	$\overline{SA} \cong \overline{SE}$	2)	Given.
3)	$\overline{ST} \cong \overline{ST}$	3)	Reflexive.
4)	△SAT ≅ △SET	4)	HLR (3, 2, 1).
5)	$\overline{AT} \cong \overline{ET}$	5)	CPCTC.
6)	$\overleftrightarrow{ST}$ is the perpendicular bisector of $\overline{AE}$	6)	If two points are each equidistant from the endpoints of a segment, then they determine the perpendicular bisector of that segment.
7)	∠ANT and ∠ENT are right angles	7)	Definition of perpendicular.
8)	∠ANT ≅ ∠ENT	8)	All right angles are congruent.

Part II: Triangles

18

Statements	Reasons
1) $\overline{RS} \cong \overline{CS}$	1) Given.
2) $\angle SRY \cong \angle SCY$	2) If sides, then angles.
3) $\angle ARS \cong \angle ACS$	3) Given.
4) $\angle ARY \cong \angle ACY$	4) Angle subtraction.
5) $\overline{RA} \cong \overline{CA}$	5) If angles, then sides.
6) $\overleftrightarrow{SA}$ is the perpendicular bisector of $\overline{RC}$	6) If two points are each equidistant from the endpoints of a segment, then they determine the perpendicular bisector of that segment.
7) $\overline{RY} \cong \overline{CY}$	7) Definition of bisect.

19 a.

Statements	Reasons
1) $\angle PAN \cong \angle PIN$	1) Given.
2) $\overline{PA} \cong \overline{PI}$	2) If angles, then sides.
3) $\overline{AN} \cong \overline{IN}$	3) Given.
4) $\overleftrightarrow{PA}$ is the perpendicular bisector of $\overline{AI}$	4) If two points are each equidistant from the endpoints of a segment then they determine the perpendicular bisector of that segment. (Note that when using the equidistance theorem, it doesn't matter that one of the two points (like *N* here) that's equidistant from the endpoints of the segment is actually on the segment.)
5) $\overline{LA} \cong \overline{LI}$	5) If a point is on the perpendicular bisector of a segment, then it's equidistant from the endpoints of that segment.

b.

Statements	Reasons
1) $\angle PAN \cong \angle PIN$	1) Given.
2) $\overline{PA} \cong \overline{PI}$	2) If angles, then sides.
3) $\overline{AN} \cong \overline{IN}$	3) Given.
4) $\triangle PAN \cong \triangle PIN$	4) SAS (2, 1, 3).
5) $\angle APN \cong \angle IPN$	5) CPCTC.
6) $\overline{PL} \cong \overline{PL}$	6) Reflexive.
7) $\triangle APL \cong \triangle IPL$	7) SAS (2, 5, 6).
8) $\overline{LA} \cong \overline{LI}$	8) CPCTC.

Part III
Polygons

In this Part...

You practice problems involving quadrilaterals, pentagons, hexagons, and more. You study the properties of the different quadrilaterals, and you find out how to prove that a four-sided figure qualifies as a particular type of quadrilateral. Next, you discover how to do cool things like compute the area of a polygon, the number of its diagonals, and the sum of its interior angles. I bet you can hardly wait! Finally, you practice problems involving *similar* polygons — that is, polygons of the exact same shape but of different sizes.

Chapter 5

Quadrilaterals: Your Fine, Four-Sided Friends

In This Chapter
- Crossing the tracks safely — parallel lines with transversals
- Classifying quadrilaterals — it's a family affair
- Checking out the properties of parallelograms, kites, and trapezoids
- Working with squares, rhombuses, and rectangles

*I*f you've mastered three-sided figures, you're ready to move up to four-sided figures — *quadrilaterals*. In this chapter, I tell you the defining characteristics of squares, rectangles, and kites, and I give you some pretty tidy definitions of parallelograms, rhombuses, and trapezoids as well. I also explain the properties of these different figures. Finally, I show you how to use the properties to prove that a figure is a certain type of quadrilateral — sort of like "if it walks like a duck and it quacks like a duck. . . ."

Before moving on to quadrilaterals, take a look at some important parallel line concepts that come in handy for parallelogram problems among other things.

Double-Crossers: Transversals and Their Parallel Lines

Take a look at Figure 5-1, which contains two parallel lines, the line that crosses over them (called a *transversal*), and the eight angles. Whenever you have such a situation, the following terminology applies.

Angles formed by parallel lines and a transversal:

- The pair of angles 1 and 5 (also 2 and 6, 3 and 7, and 4 and 8) are called *corresponding angles*.
- The pair of angles 3 and 6 (as well as 4 and 5) are *alternate interior angles*.
- Angles 1 and 8 (and angles 2 and 7) are called *alternate exterior angles*.
- Angles 3 and 5 (and 4 and 6) are *same-side interior angles*.
- Angles 1 and 7 (and 2 and 8) are *same-side exterior angles*.

Now, although knowing all this fancy terminology is nice, and although you need it for the following theorems (not to mention that little matter of your teacher's testing you on these terms), there's a simpler way to summarize everything you need to know about Figure 5-1.

Figure 5-1:
Parallel lines and a transversal: Angles, angles everywhere with lots of them to link.

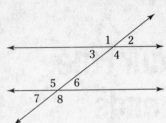

When you have two parallel lines cut by a transversal, you get four acute angles and four obtuse angles (except when you get eight right angles). All the acute angles are congruent, all the obtuse angles are congruent, and every acute angle is supplementary to every obtuse angle.

Parallel-lines-with-transversal theorems: If two parallel lines are cut by a transversal, then

- Corresponding angles are congruent.
- Alternate interior angles are congruent.
- Alternate exterior angles are congruent.
- Same-side interior angles are supplementary.
- Same-side exterior angles are supplementary.

And now for something completely different — kind of. Say that you don't know that the lines are parallel. Well, all the preceding theorems work in reverse, so you can use the following reverse theorems to prove that lines are parallel:

Lines-cut-by-a-transversal theorems: Two lines are parallel if they're cut by a transversal such that

- Two corresponding angles are congruent.
- Two alternate interior angles are congruent.
- Two alternate exterior angles are congruent.
- Two same-side interior angles are supplementary.
- Two same-side exterior angles are supplementary.

Q. Given: $a \parallel b$
Find: $\angle 1$

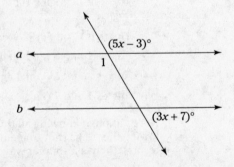

A. The $(5x-3)°$ angle and the $(3x+7)°$ angle are same-side exterior angles, so according to the theorem, they're supplementary. Supplementary angles add up to 180°, so

$$(5x - 3) + (3x + 7) = 180$$
$$8x + 4 = 180$$
$$8x = 176$$
$$x = 22$$

Plugging 22 into $(5x-3)°$ gives you 107° for that angle, and because that angle and ∠1 are vertical angles (see Chapter 2), ∠1 is also 107°.

Q. Given: ∠1 ≅ ∠2
∠G ≅ ∠P
$\overline{GI} \cong \overline{PU}$

Prove: $\overline{BG} \parallel \overline{WP}$

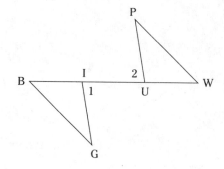

A.

Statements	Reasons
1) ∠1 ≅ ∠2	1) Given.
2) ∠BIG ≅ ∠WUP	2) Supplements of congruent angles are congruent.
3) $\overline{GI} \cong \overline{PU}$	3) Given.
4) ∠G ≅ ∠P	4) Given.
5) △BIG ≅ △WUP	5) ASA (2, 3, 4).
6) ∠B ≅ ∠W	6) CPCTC.
7) $\overline{BG} \parallel \overline{WP}$	7) If alternate interior angles are congruent, then lines are parallel.

Tip: If you have any difficulty seeing that ∠B and ∠W are indeed alternate interior angles, rotate the figure counterclockwise till the parallel segments $\overline{PW}$ and $\overline{BG}$ are horizontal, and then extend $\overline{PW}$, $\overline{BG}$, and $\overline{BW}$ in both directions, turning them into lines (you know, with arrows). After you do this, you'll be looking at the familiar scheme of parallel lines cut by a transversal, like in Figure 5-1.

118 Part III: Polygons

1. List all the pairs of angles such that if you know they're supplementary, you can prove lines *m* and *n* parallel.

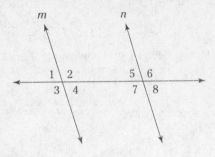

Solve It

2. Are lines *p* and *q* parallel?

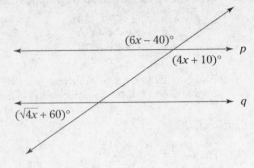

Solve It

***3.** Given that lines *o* and *z* are parallel, solve for *x*. **Hint**: You don't need to use anywhere near all the angles I've numbered. (If I had numbered only the angles you need, I would've given away the solution).

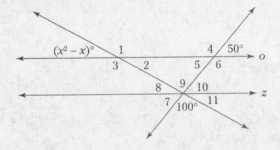

Solve It

***4.** Given: Circle Q
  $\overline{TP} \cong \overline{SR}$
  $m\angle Q = 110$
Prove: $\overline{TS} \parallel \overline{PR}$ (paragraph proof)

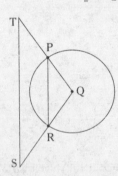

Solve It

Chapter 5: Quadrilaterals: Your Fine, Four-Sided Friends 119

5. Given: $\vec{BL}$ bisects $\angle QBP$
$\overline{BL} \parallel \overline{JP}$

Prove: $\triangle PBJ$ is isosceles

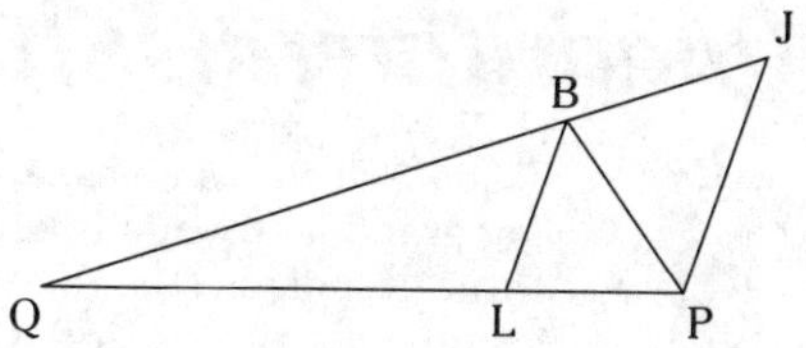

Solve It

Statements	Reasons

6. Given: $\overline{RA} \parallel \overline{TI}$
$\overline{RA} \cong \overline{TI}$
$\overline{DI} \cong \overline{NA}$

Prove: $\overline{DR} \parallel \overline{TN}$

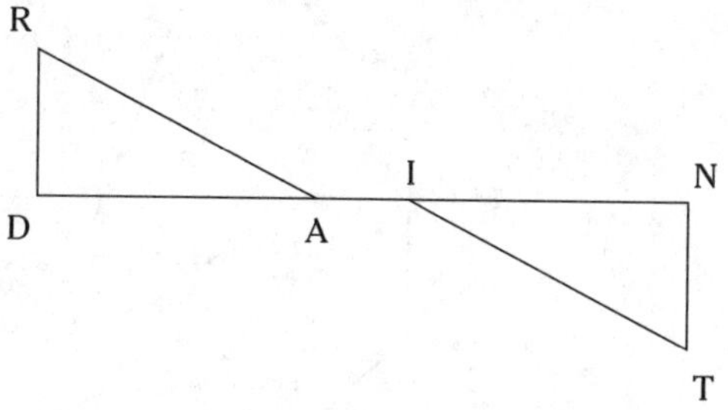

Solve It

Statements	Reasons

Quadrilaterals: It's a Family Affair

A *quadrilateral* is any shape with four straight sides. In the family tree of quadrilaterals, you've got granddaddy quadrilateral, his three kids (the kite, the parallelogram, and the trapezoid), three grandkids (the rhombus, the rectangle, and the isosceles trapezoid), and a single great-grandchild (the square). Check out the family tree in Figure 5-2 and the definitions that follow:

- **Kite:** A quadrilateral in which two disjoint pairs of consecutive sides are congruent (in other words, one side can't be used in both pairs) — it often looks just like the kites you're used to flying
- **Parallelogram:** A quadrilateral that has two pairs of parallel sides
- **Rhombus:** A quadrilateral with four congruent sides; a rhombus is both a kite and a parallelogram
- **Rectangle:** A quadrilateral with four right angles; a rectangle is a parallelogram
- **Square:** A quadrilateral with four congruent sides and four right angles; a square is both a rhombus and a rectangle
- **Trapezoid:** A quadrilateral with exactly one pair of parallel sides; the parallel sides are called the *bases,* and the nonparallel sides are the *legs*
- **Isosceles trapezoid:** A trapezoid with congruent legs

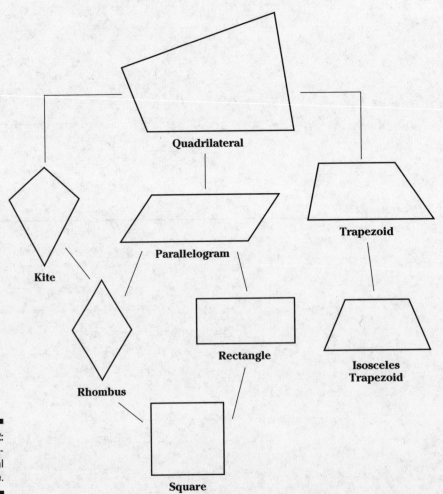

Figure 5-2: The quadrilateral family tree.

Chapter 5: Quadrilaterals: Your Fine, Four-Sided Friends **121**

Q. Given: Rhombus *PQRS*
Prove: Its diagonals are perpendicular bisectors of each other

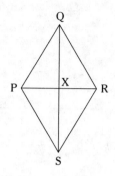

A.

Statements	Reasons
1) *PQRS* is a rhombus	1) Given.
2) $\overline{PQ} \cong \overline{QR} \cong \overline{RS} \cong \overline{SP}$	2) Definition of rhombus.
3) $\overline{QS}$ is the perpendicular bisector of $\overline{PR}$	3) If two points (Q and S) are each equidistant from the endpoints of a segment, then they determine the perpendicular bisector of that segment (one of the equidistance theorems from Chapter 4).
4) $\overline{PR}$ is the perpendicular bisector of $\overline{QS}$	4) If two points (P and R) are each equidistant from the endpoints of a segment, then they determine the perpendicular bisector of that segment.

7. Determine whether the following statements are true *always*, *sometimes*, or *never*:

   a. A parallelogram is a square

   b. A rhombus is a rectangle

   c. A kite is a trapezoid

   d. A rectangle is a kite

   e. A quadrilateral is an isosceles trapezoid

   f. A square is a kite

   g. A kite is a rhombus

   h. A parallelogram is a kite

   i. An isosceles trapezoid is a rectangle

Solve It

122 Part III: Polygons

8. Given: *JOHN* is a parallelogram
Prove: ∠J ≅ ∠H

Solve It

Statements	Reasons

9. Given: *MARY* is a parallelogram
Prove: $\overline{MA} \cong \overline{RY}$
Hint: The figure is incomplete. Use your drawing skills.

Solve It

Statements	Reasons

Chapter 5: Quadrilaterals: Your Fine, Four-Sided Friends

10. Given: *TRAP* is a trapezoid with bases $\overline{TR}$ and $\overline{PA}$
△*TRI* is isosceles with base $\overline{TR}$

Prove: *TRAP* is isosceles

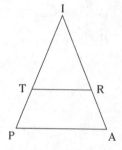

Statements	Reasons

Properties of the Parallelogram and the Kite

You may be wondering what these two quadrilaterals have in common and why they're in this section together. I hate to disappoint you, but they have little in common, and I put them together because I couldn't fit all the quadrilaterals in one section and I had to split them up somehow! So, without further ado, let me present to you the properties of the parallelogram and the kite.

Properties of the parallelogram (see Figure 5-3):

- Opposite sides are parallel by definition.
- Opposite sides are congruent.
- Opposite angles are congruent.
- Consecutive angles are supplementary (for example, ∠*BCD* is supplementary to ∠*CDA*).
- The diagonals bisect each other.

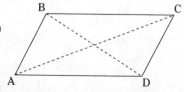

Figure 5-3: Parallelogram.

Part III: Polygons

Properties of the kite (see Figure 5-4):

- Two disjoint pairs of consecutive sides are congruent by definition ($\overline{PQ} \cong \overline{RQ}$ and $\overline{PS} \cong \overline{RS}$).
- The diagonals are perpendicular.
- One diagonal ($\overline{QS}$, the *main diagonal*) is the perpendicular bisector of the other diagonal ($\overline{PR}$, the *cross diagonal*). ("Main diagonal" and "cross diagonal" are good and useful terms, but you won't find them in other geometry books because I made them up.)
- The main diagonal bisects a pair of opposite angles ($\angle Q$ and $\angle S$).
- The opposite angles at the endpoints of the cross diagonal are congruent ($\angle P$ and $\angle R$).

These last three properties are called the *half properties* of the kite.

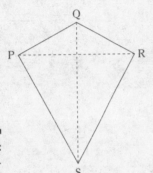

Figure 5-4: Kite.

Q. Given: *MINT* is a parallelogram

           GILT is a square

   Prove: $\triangle MIG \cong \triangle NTL$

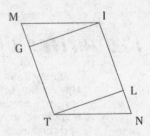

A. I do this proof two different ways (I know of at least four ways to do it). Don't forget that when it comes to proofs, there's often more than one way to skin a cat.

Method 1

Statements	Reasons
1) *MINT* is a parallelogram	1) Given.
2) $\overline{MT} \cong \overline{IN}$	2) Property of parallelogram.
3) *GILT* is a square	3) Given.
4) $\overline{GT} \cong \overline{LI}$	4) Definition of a square.
5) $\overline{MG} \cong \overline{NL}$	5) Segment subtraction.
6) $\angle M \cong \angle N$	6) Property of parallelogram.
7) $\overline{MI} \cong \overline{NT}$	7) Property of parallelogram.
8) $\triangle MIG \cong \triangle NTL$	8) SAS (5, 6, 7).

Chapter 5: Quadrilaterals: Your Fine, Four-Sided Friends 125

Method 2

Statements	Reasons
1) *GILT* is a square	1) Given.
2) ∠*IGT* is a right angle ∠*TLI* is a right angle	2) Definition of a square.
3) $\overline{IG} \perp \overline{MT}$ $\overline{TL} \perp \overline{NI}$	3) Definition of perpendicular.
4) ∠*MGI* is a right angle ∠*NLT* is a right angle	4) Definition of perpendicular.
5) $\overline{IG} \cong \overline{TL}$	5) Definition of a square.
6) *MINT* is a parallelogram	6) Given.
7) $\overline{MI} \cong \overline{NT}$	7) Property of parallelogram.
8) △*MIG* ≅ △*NTL*	8) HLR (7, 5, 4).

11. Given: *PCTR* is a kite
$\overline{PC} \cong \overline{PR}$

Prove: ∠*CAT* ≅ ∠*RAT*

Solve It

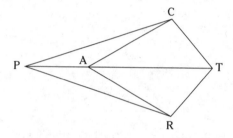

Statements	Reasons

Part III: Polygons

***12.** Given: *KITE* is — guess what — a kite
Find: The lengths of all sides

Solve It

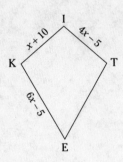

---

13. Given: *DEFG* is a parallelogram
Find: The measures of all angles

Solve It

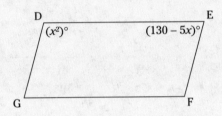

---

14. Given: *YZGH* is a parallelogram
$\overline{ZA} \cong \overline{HO}$

Prove: $\overline{YO} \cong \overline{GA}$

This proof can be done two ways, using two different pairs of congruent triangles.

Hint: With either method, the final reason of the proof is — as I'm sure you can guess — CPCTC.

Solve It

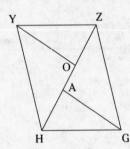

Statements	Reasons

Chapter 5: Quadrilaterals: Your Fine, Four-Sided Friends **127**

15. Given: NQRM is a parallelogram
              $\overline{NO} \cong \overline{RS}$

Prove: $\overline{SL} \cong \overline{OP}$

As with problem 14, you can do this proof two ways, again using two different pairs of ≅ △s.

Solve It

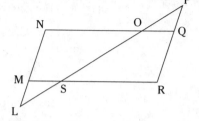

Statements	Reasons

Properties of Rhombuses, Rectangles, and Squares

Keep referring to the quadrilateral family tree in Figure 5-2. Knowing how the different quadrilaterals are related to each other can really help you remember their properties. For example, you can find out from the preceding section that the diagonals of a parallelogram bisect each other. So because rhombuses, rectangles, and squares are all parallelograms, they automatically share that property.

Properties of the rhombus (see Figure 5-5):

- The properties of a parallelogram apply (the ones that matter here are parallel sides, opposite angles congruent, and consecutive angles supplementary).
- All sides are congruent by definition.
- The diagonals bisect the angles.
- The diagonals are perpendicular bisectors of each other.

Properties of the rectangle:

- The properties of a parallelogram apply (the ones that matter here are parallel sides, opposite sides congruent, and diagonals bisect each other).
- All angles are right angles by definition.
- The diagonals are congruent.

128 Part III: Polygons

Properties of the square:

✔ The properties of a rhombus apply (the ones that matter here are parallel sides, diagonals are perpendicular bisectors of each other, and diagonals bisect the angles).

✔ The properties of a rectangle apply (the only one that matters here is diagonals are congruent).

✔ All sides are congruent by definition.

✔ All angles are right angles by definition.

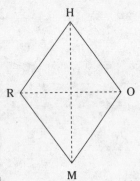

Figure 5-5: Rhombus.

Q. If a rhombus has sides of length 10 and one diagonal measuring 12, what's the length of the other diagonal?

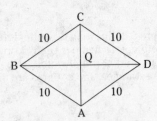

A. The diagonals in a rhombus are perpendicular bisectors of each other, so if AC is 12, AQ must be 6. $\triangle ABQ$ is a right triangle, so you can solve for BQ with the Pythagorean Theorem or by recognizing that $\triangle ABQ$ is in the 3 : 4 : 5 family (see Chapter 3). Either way, BQ is 8, and thus, BD is 16.

Q. Given: $BORE$ is a rectangle
              $OB = DR = 10$

Find: The length of $\overline{BE}$

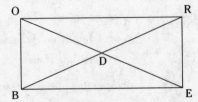

A. In a rectangle, the diagonals are congruent and they bisect each other, so because DR is 10, both diagonals are 20. You have right triangle BOE with a leg of 10 and a hypotenuse of 20. You should recognize this figure as a 30°- 60°- 90° right triangle (see Chapter 3) where OB is x and BE is $x\sqrt{3}$. $\overline{BE}$ thus has a length of $10\sqrt{3}$.

16. Given: RECT is a rectangle
Find: x and y

Solve It

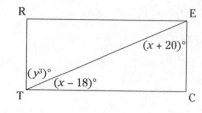

17. Given: MATH is a rhombus
∠QMA measures 60°
HQ is 8
Find: Measures of all sides and angles in △QAT

Solve It

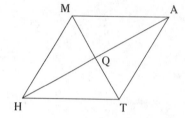

18. Given: HOBU is a rhombus
$\overline{MB} \cong \overline{RH}$
Prove: $\overline{MO} \parallel \overline{RU}$

Note: You can do this proof in a few different ways.

Solve It

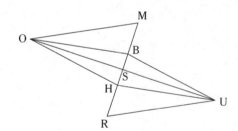

Statements	Reasons

Part III: Polygons

19. Given: ANGL is a rectangle
Find: x, y, z, and AL

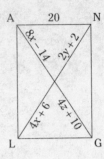

---

20. Given: QRVT is a rhombus
Prove: ∠1 ≅ ∠2

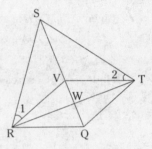

Statements	Reasons

Properties of Trapezoids and Isosceles Trapezoids

I bet you're dying to add to your list of properties, so here you go — last but not least, the trapezoid and the isosceles trapezoid:

Properties of the trapezoid:

- The bases are parallel by definition.
- Each lower base angle is supplementary to the adjacent upper base angle.

Properties of the isosceles trapezoid:

- The properties of a trapezoid apply by definition (parallel bases).
- The legs are congruent by definition.
- The lower base angles are congruent.
- The upper base angles are congruent.
- Any lower base angle is supplementary to any upper base angle.
- The diagonals are congruent.

Q. Given: PQRS is an isosceles trapezoid with bases $\overline{PS}$ and $\overline{QR}$
∠PQX is 85°
∠PSR is 75°

Find: ∠QXR

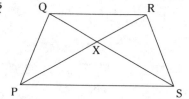

A. An isosceles trapezoid has congruent legs and congruent diagonals. Using those congruent legs and diagonals and $\overline{PS} \cong \overline{PS}$, you get △PQS ≅ △SRP by SSS. ∠SRX is congruent to ∠PQX by CPCTC, so ∠SRX is 85°. The angles in △PRS need to add up to 180°, and you have 85° and 75° (∠PSR) so far, so ∠RPS has to be 20°. ∠QSP is congruent to ∠RPS by CPCTC, so it's 20° as well. The angles in △PXS have to add up to 180°, so ∠PXS is 180 – 20 – 20, or 140°. Finally, ∠QXR and ∠PXS are vertical angles, so ∠QXR is also 140°.

21. Given: QXJW is an isosceles trapezoid with bases $\overline{QW}$ and $\overline{XJ}$
Prove: △QXZ ≅ △WJZ

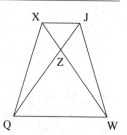

Statements	Reasons

*22. Given: ZOID is a trapezoid with bases $\overline{ZD}$ and $\overline{OI}$
△ZCD is isosceles with base $\overline{ZD}$

Prove: ZOID is isosceles

Hint: Before beginning this problem, you may want to review the ten strategies for proofs in Chapter 14. Three of the strategies are especially helpful for this problem.

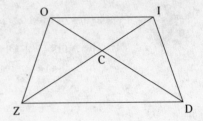

Statements	Reasons

Proving That a Quadrilateral Is a Parallelogram or a Kite

In the last few sections, you can find the definitions of the various quadrilaterals and their properties. In this and the next section, "Proving That a Quadrilateral Is a Rhombus, Rectangle, or Square," you move on to proving that a specific quadrilateral is of a particular type. These three things (definitions, properties, and methods of proof) are related, but there are important differences among them.

You can always use a defining characteristic of a particular quadrilateral to prove that a figure is that particular quadrilateral. For example, a parallelogram is *defined* as a quadrilateral with two pairs of parallel sides, and you can *prove* that a quadrilateral is a parallelogram by showing just that. With other properties, however, the process isn't so simple.

Some properties can be used as a proof method, but others cannot. For example, one of the properties of a parallelogram is that its diagonals bisect each other, and proving that the diagonals of a quadrilateral bisect each other is one of the five ways of proving that a quadrilateral is a parallelogram. On the other hand, one of the properties of a rectangle is that its diagonals are congruent; however, you *can't* prove that a quadrilateral is a rectangle by showing its diagonals congruent because some kites, isosceles trapezoids, and no-name quadrilaterals also have congruent diagonals.

Chapter 5: Quadrilaterals: Your Fine, Four-Sided Friends 133

Ways to prove that a quadrilateral is a parallelogram: You have a parallelogram if

- Both pairs of opposite sides of a quadrilateral are parallel (reverse of definition).
- Both pairs of opposite sides of a quadrilateral are congruent (converse of property).
- Both pairs of opposite angles of a quadrilateral are congruent (converse of property).
- The diagonals of a quadrilateral bisect each other (converse of property).
- One pair of opposite sides of a quadrilateral are both parallel and congruent (note that this is neither the reverse of the definition nor the converse of a property).

Ways to prove that a quadrilateral is a kite: You've got a kite if

- Two disjoint pairs of consecutive sides of a quadrilateral are congruent (reverse of definition).
- One of the diagonals of a quadrilateral is the perpendicular bisector of the other (converse of property).

Q. Given: $\angle NOT \cong \angle BAD$

$\angle DOA \cong \angle TAO$

Prove: DOTA is a parallelogram

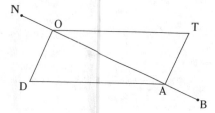

A.

Statements	Reasons
1) $\angle NOT \cong \angle BAD$	1) Given.
2) $\overline{OT} \parallel \overline{DA}$	2) If alternate exterior angles are congruent, then lines are parallel.
3) $\angle DOA \cong \angle TAO$	3) Given.
4) $\overline{DO} \parallel \overline{AT}$	4) If alternate interior angles are congruent, then lines are parallel.
5) DOTA is a parallelogram	5) If both pairs of opposite sides of a quadrilateral are parallel, then it is a parallelogram.

Part III: Polygons

Q. Given: ∠1 ≅ ∠2
$\overrightarrow{LY}$ bisects ∠GLF

Prove: GOFL is a kite

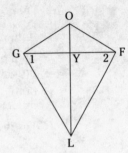

A.

Statements	Reasons
1) ∠1 ≅ ∠2	1) Given.
2) $\overline{GL} \cong \overline{FL}$	2) If angles, then sides.
3) $\overrightarrow{LY}$ bisects ∠GLF	3) Given.
4) ∠GLY ≅ ∠FLY	4) Definition of bisect.
5) $\overline{LO} \cong \overline{LO}$	5) Reflexive.
6) △GLO ≅ △FLO	6) SAS (2, 4, 5).
7) $\overline{GO} \cong \overline{FO}$	7) CPCTC.
8) GOFL is a kite	8) If two disjoint pairs of consecutive sides of a quadrilateral are congruent, then it is a kite (2 and 7).

23. Given: DEAL is a parallelogram
∠DEN ≅ ∠ALO

Prove: NEOL is a parallelogram

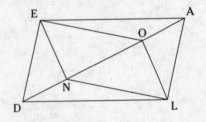

Statements	Reasons

24. Given: *EMNA* is a parallelogram
                $\overline{XE} \cong \overline{RN}$ and $\overline{LE} \cong \overline{IN}$

     Prove: *LXIR* is a parallelogram

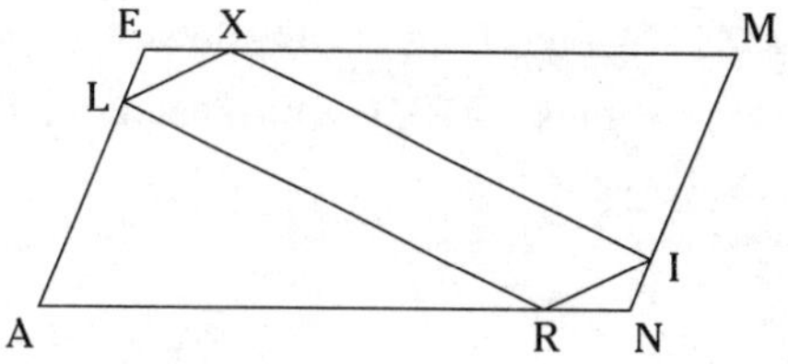

Solve It

Statements	Reasons

25. Do the same kite problem as the *GOFLY* example problem, but this time use the second kite proof method instead of the first.

Solve It

Statements	Reasons

Part III: Polygons

***26.** Given: Diagram as shown

Prove: JKLM is a parallelogram (paragraph proof)

Solve It

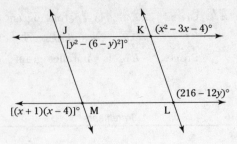

Proving That a Quadrilateral Is a Rhombus, Rectangle, or Square

Note that when proving that a quadrilateral is a rhombus, rectangle, or square, you sometimes go directly from, say, quadrilateral to rhombus or quadrilateral to square — like you do when proving a quadrilateral to be a parallelogram or a kite. But at other times, you first have to prove (or be given) that the quadrilateral is a particular quadrilateral. For example, some methods for proving that a quadrilateral is a rhombus require that you know that the quadrilateral is a parallelogram.

Ways to prove that a quadrilateral is a rhombus: You have a rhombus if

- All sides of a quadrilateral are congruent (reverse of definition).
- The diagonals of a quadrilateral bisect all angles (converse of property).
- The diagonals of a quadrilateral are perpendicular bisectors of each other (converse of property).
- Two consecutive sides of a parallelogram are congruent.
- Either diagonal of a parallelogram bisects two angles.
- The diagonals of a parallelogram are perpendicular.

Ways to prove that a quadrilateral is a rectangle: You're dealing with a rectangle if

- All angles in a quadrilateral are right angles (reverse of definition).
- A parallelogram contains a right angle.
- The diagonals of a parallelogram are congruent.

Ways to prove that a quadrilateral is a square: You have a square if

- A quadrilateral has four congruent sides and four right angles (reverse of definition).
- A quadrilateral is both a rhombus and a rectangle.

Chapter 5: Quadrilaterals: Your Fine, Four-Sided Friends 137

Q. Given: *KNOT* is a parallelogram
             ∠*TOK* ≅ ∠*NOK*

    Prove: *KNOT* is a rhombus

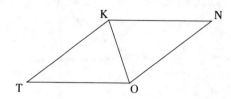

A.

Statements	Reasons
1) ∠*TOK* ≅ ∠*NOK*	1) Given.
2) *KNOT* is a parallelogram	2) Given.
3) $\overline{KN} \parallel \overline{TO}$	3) Property of parallelogram.
4) ∠*TOK* ≅ ∠*NKO*	4) Alternate interior angles are congruent.
5) ∠*NOK* ≅ ∠*NKO*	5) Transitivity.
6) $\overline{NK} \cong \overline{NO}$	6) If angles, then sides.
7) *KNOT* is a rhombus	7) If one pair of consecutive sides of a parallelogram is congruent, then it is a rhombus.

27. Given: *QVST* is a parallelogram
             ∠*UQR* ≅ ∠*USR*

    Prove: *QVST* is a rhombus

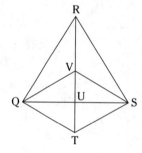

Statements	Reasons

Part III: Polygons

***28.** Given: ∠MAK ≅ ∠AMR
∠KRM ≅ ∠RKA
$\overline{MK} \parallel \overline{AR}$

Prove: MARK is a rectangle

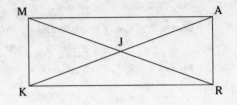

Solve It

Statements	Reasons

***29.** Given: ∠OHR and ∠FKI are right angles
$\overline{AR} \parallel \overline{FE}$
$\overline{GH} \cong \overline{GK}$

Prove: GHJK is a square

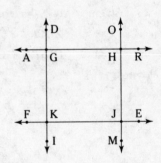

Solve It

Statements	Reasons

Solutions for Quadrilaterals: Your Fine, Four-Sided Friends

1 For the lines to be parallel, you need the same side interior angles or the same side exterior angles to be supplementary, according the interior- and exterior-angle theorems. Therefore, these theorems give you the following pairs: angles 2 and 5 and angles 4 and 7 (the same-side interior angles) and angles 1 and 6 and angles 3 and 8 (the same-side exterior angles).

Congrats if you saw that more answers are correct. If you recognize, for example, that angles 1 and 7 are supplementary, you can prove m parallel to n. Why? Angles 6 and 7 are congruent vertical angles, right? So if angles 1 and 7 are supplementary, angles 1 and 6 have to be supplementary as well. Then the theorem tells you that the lines are parallel. By the same logic, the following supplementary pairs also allow you to prove the lines parallel: angles 2 and 8, angles 3 and 5, and angles 4 and 6.

2 The two angles on the top are vertical angles; thus, because all vertical angles are congruent,

$$6x - 40 = 4x + 10$$
$$2x = 50$$
$$x = 25$$

Plugging $x = 25$ into $(6x - 40)°$ gives you $110°$ for that angle.

Plugging $x = 25$ into $(\sqrt{4x} + 60)°$ gives you $70°$ for that angle.

Because those two angles are same-side exterior angles, they have to be supplementary for the lines to be parallel. $110°$ and $70°$ sum to $180°$ so, yes, the lines are parallel.

*3 *Warning:* When working on a problem that has more than one transversal, make sure you *use only one transversal at a time* when using the theorems to compare various angles. For example, *the theorems tell you nothing* about how angles 4 and 11 compare to each other because those angles use two different transversals. Ditto for angles 2 and 10.

Here goes. The $50°$ angle and $\angle 10$ are corresponding angles, so $\angle 10$ is also $50°$. Next, angles 10 and 11 have to add up to $80°$, because together they form an angle that's supplementary to the $100°$ angle. That makes $\angle 11 = 30°$. Because the $x^2 - x$ angle and $\angle 11$ are alternate exterior angles, they're congruent, so $x^2 - x = 30$. Now you can solve for x:

$$x^2 - x = 30$$
$$x^2 - x - 30 = 0$$
$$(x - 6)(x + 5) = 30$$
$$x = 6 \text{ or } -5$$

Note that both 6 and –5 are valid answers. In geometry problems, you often reject negative answers because segment lengths and angle measures can't be negative. But here, plugging –5 into $(x^2 - x)°$ gives you a *positive* angle (namely $30°$), so –5 is a perfectly good answer.

*4 The two radii are congruent, so $\triangle PQR$ is isosceles (see Chapter 3). That makes $\angle QPR$ congruent to $\angle QRP$. Then, to make the angles in $\triangle PQR$ add up to $180°$, $\angle QPR$ and $\angle QRP$ must each be $35°$.

Because $\overline{TP}$ and $\overline{SR}$ are congruent, you can add them to the congruent radii, making $\overline{TQ}$ and $\overline{SQ}$ congruent. Thus, the big triangle is isosceles with $\angle T \cong \angle S$. By the same reasoning as before, $\angle T$ and $\angle S$ are also $35°$ angles. Finally, $\angle T$ and $\angle QPR$ are corresponding angles. Because they're both $35°$, the lines have to be parallel.

5

Statements	Reasons
1) $\vec{BL}$ bisects ∠QBP	1) Given.
2) ∠QBL ≅ ∠LBP	2) Definition of bisect.
3) $\overline{BL} \parallel \overline{JP}$	3) Given.
4) ∠QBL ≅ ∠J	4) If lines are parallel, corresponding angles are congruent (using transversal $\overline{QJ}$).
5) ∠LBP ≅ ∠BPJ	5) If lines are parallel, alternate interior angles are congruent (using transversal $\overline{BP}$).
6) ∠J ≅ ∠BPJ	6) Transitivity (2, 4, 5).
7) $\overline{BP} \cong \overline{BJ}$	7) If angles, then sides.
8) △PBJ is isosceles	8) Definition of isosceles triangle.

6

Statements	Reasons
1) $\overline{RA} \cong \overline{TI}$	1) Given.
2) $\overline{RA} \parallel \overline{TI}$	2) Given.
3) ∠DAR ≅ ∠NIT	3) If lines are parallel, then alternate exterior angles are congruent.
4) $\overline{DI} \cong \overline{NA}$	4) Given.
5) $\overline{DA} \cong \overline{NI}$	5) Segment subtraction (subtracting $\overline{AI}$ from both segments).
6) △DAR ≅ △NIT	6) SAS (1, 3, 5).
7) ∠D ≅ ∠N	7) CPCTC.
8) $\overline{DR} \parallel \overline{TN}$	8) If alternate interior angles are congruent, then the lines are parallel.

7 One way to do these always-sometimes-never problems is to look at the quadrilateral family tree and follow these guidelines:

- If you go *up* from the first figure to the second, the answer is *always*.
- If you go *down*, the answer is *sometimes*.
- If you can make the connection by going *down and then up* (like from a kite to a rectangle or vice versa), it's *sometimes*.
- And if the only way to make the connection is by going *up and then down*, the answer is *never*.

Here are the correct answers to problem 7:

a. Sometimes

b. Sometimes (when it's a square)

c. Never

d. Sometimes (when it's a square)

e. Sometimes

f. Always

g. Sometimes

h. Sometimes (when it's a rhombus)

i. Never

8

Statements	Reasons
1) $JOHN$ is a parallelogram	1) Given.
2) $\overline{JO} \parallel \overline{NH}$	2) Definition of parallelogram.
3) $\angle J$ is supplementary to $\angle N$	3) If lines are parallel, then same-side interior angles are supplementary (using transversal $\overline{JN}$).
4) $\overline{JN} \parallel \overline{OH}$	4) Definition of parallelogram.
5) $\angle H$ is supplementary to $\angle N$	5) If lines are parallel, then same-side interior angles are supplementary (using transversal $\overline{NH}$).
6) $\angle J \cong \angle H$	6) Supplements of the same angle are congruent.

9

Statements	Reasons
1) $MARY$ is a parallelogram	1) Given.
2) Draw $\overline{AY}$ ($\overline{MR}$ would work as well)	2) Two points determine a line.
3) $\overline{MA} \parallel \overline{YR}$	3) Definition of parallelogram.
4) $\angle MAY \cong \angle RYA$	4) If lines are parallel, then alternate interior angles are congruent (using parallel segments $\overline{MA}$ and $\overline{YR}$ and transversal $\overline{AY}$).
5) $\overline{AR} \parallel \overline{MY}$	5) Definition of parallelogram.
6) $\angle MYA \cong \angle RAY$	6) If lines are parallel, then alternate interior angles are congruent (using parallel segments $\overline{AR}$ and $\overline{MY}$ with transversal $\overline{AY}$).
7) $\overline{AY} \cong \overline{YA}$	7) Reflexive.
8) $\triangle MAY \cong \triangle RYA$	8) ASA (4, 7, 6).
9) $\overline{MA} \cong \overline{RY}$	9) CPCTC.

10

Statements	Reasons
1) △TRI is isosceles with base TR	1) Given.
2) $\overline{TI} \cong \overline{RI}$	2) Definition of isosceles triangle.
3) ∠ITR ≅ ∠IRT	3) If sides, then angles.
4) TRAP is a trapezoid with bases $\overline{TR}$ and $\overline{PA}$	4) Given.
5) $\overline{TR} \parallel \overline{PA}$	5) Definition of trapezoid.
6) ∠P ≅ ∠ITR	6) If lines are parallel, then corresponding angles are congruent (using transversal $\overline{IP}$).
7) ∠A ≅ ∠IRT	7) If lines are parallel, then corresponding angles are congruent (using transversal $\overline{IA}$).
8) ∠P ≅ ∠A	8) Transitivity (3, 6, 7).
9) $\overline{IP} \cong \overline{IA}$	9) If angles, then sides.
10) $\overline{PT} \cong \overline{AR}$	10) Subtraction (2 and 9).
11) TRAP is isosceles	11) Definition of isosceles trapezoid.

11

Statements	Reasons
1) PCTR is a kite	1) Given.
2) $\overline{PC} \cong \overline{PR}$	2) Given.
3) $\overline{TC} \cong \overline{TR}$	3) Property of a kite (because $\overline{PC} \cong \overline{PR}$, the other disjoint pair of sides must also be congruent).
4) $\overline{TP}$ bisects ∠CTR	4) Property of a kite.
5) ∠CTP ≅ ∠RTP	5) Definition of bisect.
6) $\overline{AT} \cong \overline{AT}$	6) Reflexive.
7) △CAT ≅ △RAT	7) SAS (3, 5, 6).
8) ∠CAT ≅ ∠RAT	8) CPCTC.

***12** Set the consecutive sides $\overline{KI}$ and $\overline{TI}$ equal to each other (because, by definition, two pairs of consecutive sides are congruent), and solve:

$$x + 10 = 4x - 5$$
$$-3x = -15$$
$$x = 5$$

Plugging $x = 5$ into the three sides gives you $5 + 10$, or 15 for KI; $4(5) - 5$, or 15 for TI; and $6(5) - 5$, or 25 for KE. $\overline{TE}$ must be congruent to $\overline{KE}$, so it also measures 25.

But wait! There's another possibility. Don't forget — figures don't have to be drawn to scale. $\overline{KI}$ and $\overline{KE}$ could be a congruent pair of sides with $\overline{TI}$ and $\overline{TE}$ the other congruent pair. Thus,

$$x + 10 = 6x - 5$$
$$-5x = -15$$
$$x = 3$$

Doing the math with $x = 3$ gives you the following equally valid set of lengths:

$KI = KE = 13$
$TI = TE = 7$

13 Consecutive angles in a parallelogram are supplementary, so

$$x^2 + (130 - 5x) = 180$$

This is a quadratic equation, so set it equal to 0 and solve by factoring (you can also use the quadratic formula, of course).

$$x^2 - 5x - 50 = 0$$
$$(x - 10)(x + 5) = 0$$
$$x = 10 \text{ or } -5$$

Plugging $x = 10$ into angles D and E gives you $100°$ for $\angle D$ and $80°$ for $\angle E$. $\angle G$ must also be $80°$, and $\angle F$ is $100°$.

But don't reject $x = -5$ just because it's a negative number. Plugging $x = -5$ into angles D and E gives you another valid set of angles (albeit with measures *way* different from what the angles look like in the figure): $\angle D = \angle F = 25°$ and $\angle E = \angle G = 155°$.

14 **Method 1**

Statements	Reasons
1) YZGH is a parallelogram	1) Given.
2) $\overline{ZG} \cong \overline{YH}$	2) Property of parallelogram.
3) $\overline{ZG} \parallel \overline{YH}$	3) Property of parallelogram.
4) $\angle GZA \cong \angle YHO$	4) Alternate interior angles are congruent (using transversal $\overline{ZH}$).
5) $\overline{ZA} \cong \overline{HO}$	5) Given.
6) $\triangle GZA \cong \triangle YHO$	6) SAS (2, 4, 5).
7) $\overline{YO} \cong \overline{GA}$	7) CPCTC.

Method 2

Statements	Reasons
1) YZGH is a parallelogram	1) Given.
2) $\overline{YZ} \cong \overline{HG}$	2) Property of parallelogram.
3) $\overline{YZ} \parallel \overline{HG}$	3) Property of parallelogram.
4) $\angle YZO \cong \angle GHA$	4) Alternate interior angles are congruent (using transversal $\overline{ZH}$).
5) $\overline{ZA} \cong \overline{HO}$	5) Given.
6) $\overline{ZO} \cong \overline{HA}$	6) Subtraction.
7) $\triangle YZO \cong \triangle GHA$	7) SAS (2, 4, 6).
8) $\overline{YO} \cong \overline{GA}$	8) CPCTC.

15 **Method 1 (Using △NLO and △RPS)**

Statements	Reasons
1) NQRM is a parallelogram	1) Given.
2) $\overline{NQ} \parallel \overline{MR}$	2) Property of parallelogram.
3) ∠SON ≅ ∠OSR	3) Alternate interior angles are congruent.
4) $\overline{NO} \cong \overline{RS}$	4) Given.
5) ∠N ≅ ∠R	5) Property of parallelogram.
6) △NLO ≅ △RPS	6) ASA (3, 4, 5).
7) $\overline{LO} \cong \overline{PS}$	7) CPCTC.
8) $\overline{SL} \cong \overline{OP}$	8) Subtraction (subtracting $\overline{OS}$ from $\overline{LO}$ and $\overline{PS}$).

Method 2 (Using △LSM and △POQ)

Statements	Reasons
1) NQRM is a parallelogram	1) Given.
2) $\overline{NM} \parallel \overline{QR}$	2) Property of parallelogram.
3) ∠L ≅ ∠P	3) Alternate interior angles are congruent.
4) $\overline{NQ} \parallel \overline{MR}$	4) Property of a parallelogram.
5) ∠LSM ≅ ∠POQ	5) Alternate exterior angles are congruent.
6) $\overline{NO} \cong \overline{RS}$	6) Given.
7) $\overline{NQ} \cong \overline{MR}$	7) Property of parallelogram.
8) $\overline{QO} \cong \overline{MS}$	8) Subtraction.
9) △LSM ≅ △POQ	9) AAS (3, 5, 8).
10) $\overline{SL} \cong \overline{OP}$	10) CPCTC.

16 You have a rectangle, so △TEC is a right triangle. Its angles have to add up to 180°, so because ∠C is a right angle, the other two angles add up to 90°. Thus,

$$(x - 18) + (x + 20) = 90$$
$$2x + 2 = 90$$
$$2x = 88$$
$$x = 44$$

∠ETC is thus 44 – 18, or 26, and because ∠RTC equals 90°, that leaves 64° for ∠RTE:

$$y^3 = 64$$
$$y = 4$$

17 The diagonals in a rhombus are perpendicular bisectors of each other, so because HQ is 8, AQ is also 8. The sides of a rhombus are equal, so △MAT is isosceles with base $\overline{MT}$. The base angles are congruent, so ∠QTA, like ∠QMA, is 60°. Because ∠AQT is 90°, △QAT is a 30°- 60°- 90° triangle. Its long leg, $\overline{AQ}$, measures 8, so its short leg, $\overline{TQ}$, is $\frac{8}{\sqrt{3}}$ (or $\frac{8\sqrt{3}}{3}$) units long. The hypotenuse, $\overline{AT}$, is twice that, or $\frac{16\sqrt{3}}{3}$ units long. (See Chapter 3 for info on 30°- 60°- 90° triangles.)

18

Statements	Reasons
1) $HOBU$ is a rhombus	1) Given.
2) $\overline{BO} \cong \overline{HU}$	2) Property of a rhombus.
3) $\overline{BO} \parallel \overline{HU}$	3) Property of a rhombus.
4) $\angle OBM \cong \angle UHR$	4) Alternate exterior angles are congruent.
5) $\overline{MB} \cong \overline{RH}$	5) Given.
6) $\triangle OBM \cong \triangle UHR$	6) SAS (2, 4, 5).
7) $\angle M \cong \angle R$	7) CPCTC.
8) $\overline{MO} \parallel \overline{RU}$	8) If alternate interior angles are congruent, then lines are parallel.

19 The diagonals in a rectangle are congruent, and they bisect each other, so all four half-diagonals are equal. You need an equation with a single variable, namely

$$8x - 14 = 4x + 6$$
$$4x = 20$$
$$x = 5$$

Plugging that into $8x - 14$ (or $4x + 6$) gives you 26 for the length of each half-diagonal. Thus,

$$2y + 2 = 26$$
$$2y = 24$$
$$y = 12, \text{ and}$$

$$4z + 10 = 26$$
$$4z = 16$$
$$z = 4$$

Finally, the length of diagonal $\overline{LN}$ is twice 26, or 52. You can then compute AL with the Pythagorean Theorem, or — if you're on your toes — you recognize that $\triangle ANL$ is in the 5 : 12 : 13 family (see Chapter 3). AN is 20 (4 · 5), and LN is 52 (4 · 13), so AL is 4 times 12, or 48.

20

Statements	Reasons
1) $QRVT$ is a rhombus	1) Given.
2) $\overline{VQ}$ is the $\perp$ bisector of $\overline{RT}$	2) Property of a rhombus.
3) $\overline{SR} \cong \overline{ST}$	3) If a point is on the perpendicular bisector of a segment, then it is equidistant from the endpoints of that segment (equidistance theorem).
4) $\overline{RV} \cong \overline{TV}$	4) Property of a rhombus.
5) $\overline{SV} \cong \overline{SV}$	5) Reflexive.
6) $\triangle RSV \cong \triangle TSV$	6) SSS (3, 4, 5).
7) $\angle 1 \cong \angle 2$	7) CPCTC.

21

Statements	Reasons
1) QXJW is an isosceles trapezoid with bases $\overline{QW}$ and $\overline{XJ}$	1) Given.
2) $\overline{QX} \cong \overline{WJ}$	2) The legs of an isosceles trapezoid are congruent.
3) $\overline{WX} \cong \overline{QJ}$	3) The diagonals of an isosceles trapezoid are congruent.
4) $\overline{QW} \cong \overline{WQ}$	4) Reflexive.
5) $\triangle QXW \cong \triangle WJQ$	5) SSS (2, 3, 4).
6) $\angle QXW \cong \angle WJQ$	6) CPCTC.
7) $\angle QZX \cong \angle WZJ$	7) Vertical angles are congruent.
8) $\triangle QXZ \cong \triangle WJZ$	8) AAS (7, 6, 2).

*22

Statements	Reasons
1) $\triangle ZCD$ is isosceles with base $\overline{ZD}$	1) Given.
2) $\overline{ZC} \cong \overline{DC}$	2) Definition of isosceles triangle.
3) $\angle CZD \cong \angle CDZ$	3) If sides, then angles.
4) ZOID is a trapezoid with bases $\overline{ZD}$ and $\overline{OI}$	4) Given.
5) $\overline{OI} \parallel \overline{ZD}$	5) The bases of a trapezoid are parallel.
6) $\angle CZD \cong \angle OIZ$	6) Alternate interior angles are congruent (using transversal $\overline{ZI}$).
7) $\angle CDZ \cong \angle IOD$	7) Alternate interior angles are congruent (using transversal $\overline{DO}$).
8) $\angle OIZ \cong \angle IOD$	8) Transitivity (6, 3, 7).
9) $\overline{OC} \cong \overline{IC}$	9) If angles, then sides.
10) $\angle OCZ \cong \angle ICD$	10) Vertical angles are congruent.
11) $\triangle OCZ \cong \triangle ICD$	11) SAS (2, 10, 9).
12) $\overline{OZ} \cong \overline{ID}$	12) CPCTC.
13) ZOID is isosceles	13) A trapezoid with congruent legs is isosceles.

23)

Statements	Reasons
1) DEAL is a parallelogram	1) Given.
2) $\overline{DE} \parallel \overline{AL}$	2) Property of parallelogram.
3) ∠EDN ≅ ∠LAO	3) If lines are parallel, alternate interior angles are congruent.
4) $\overline{DE} \cong \overline{AL}$	4) Property of parallelogram.
5) ∠DEN ≅ ∠ALO	5) Given.
6) △DEN ≅ △ALO	6) ASA (3, 4, 5).
7) ∠DNE ≅ ∠AOL	7) CPCTC.
8) $\overline{EN} \parallel \overline{LO}$	8) If alternate exterior angles are congruent, lines are parallel.
9) $\overline{EN} \cong \overline{LO}$	9) CPCTC.
10) NEOL is a parallelogram	10) If one pair of opposite sides of a quadrilateral are both parallel and congruent, then it is a parallelogram.

24)

Statements	Reasons
1) EMNA is a parallelogram	1) Given.
2) ∠E ≅ ∠N	2) Opposite angles of a parallelogram are congruent.
3) $\overline{XE} \cong \overline{RN}$ $\overline{LE} \cong \overline{IN}$	3) Given.
4) △LEX ≅ △INR	4) SAS (3, 2, 3).
5) $\overline{XL} \cong \overline{RI}$	5) CPCTC.
6) $\overline{EM} \cong \overline{AN}$	6) Opposite sides of a parallelogram are congruent.
7) $\overline{XM} \cong \overline{RA}$	7) Subtraction.
8) $\overline{EA} \cong \overline{MN}$	8) Opposite sides of a parallelogram are congruent.
9) $\overline{LA} \cong \overline{IM}$	9) Subtraction.
10) ∠A ≅ ∠M	10) Opposite angles of a parallelogram are congruent.
11) △LAR ≅ △IMX	11) SAS (7, 10, 9).
12) $\overline{LR} \cong \overline{IX}$	12) CPCTC.
13) LXIR is a parallelogram	13) If both pairs of opposite sides of a quadrilateral are congruent, then it is a parallelogram.

148 Part III: Polygons

25. The first four steps of this proof are the same as in the example problem, so I pick up with step 5.

Statements	Reasons
5) $\triangle GLY \cong \triangle FLY$	5) ASA (1, 2, 4).
6) $\overline{GY} \cong \overline{FY}$	6) CPCTC.
7) $\overline{OL}$ is the perpendicular bisector of $\overline{GF}$	7) If two points are each equidistant from the endpoints of a segment, then they determine the perpendicular bisector of that segment.
8) GOFL is a kite	8) If one of the diagonals of a quadrilateral is the perpendicular bisector of the other, the quadrilateral is a kite.

*26. I hope this odd problem didn't give you an algebra panic attack. It's not as bad as it looks. The first thing you have to realize is that you don't have to solve for x or y. In fact, solving for the variables is impossible because you don't have any information about the figure that would allow you to write any equations.

First, multiply $(x+1)(x-4)$. That's $x^2 - 3x - 4$. This measure is the same as that of the other x angle, so regardless of the value of x, those angles are congruent. Then, using congruent vertical angles, you can show that $\angle JML$ is congruent to $\angle LKJ$. Now for the y angles. First, simplify $y^2 - (6-y)^2$. That's $y^2 - (36 - 12y + y^2) = 12y - 36$. Then, because $\angle MLK$ is supplementary to the $(216 - 12y)°$ angle, the measure of $\angle MLK = 180 - (216 - 12y) = 12y - 36$. Thus, $\angle MLK$ is congruent to $\angle KJM$. You have two pairs of congruent opposite angles, and so you have a parallelogram.

27.

Statements	Reasons
1) $\angle UQR \cong \angle USR$	1) Given.
2) $\overline{QR} \cong \overline{SR}$	2) If angles, then sides.
3) QVST is a parallelogram	3) Given.
4) $\overline{VT}$ bisects $\overline{QS}$	4) Property of parallelogram.
5) $\overline{QU} \cong \overline{SU}$	5) Definition of bisect.
6) $\overline{VT}$ is the perpendicular bisector of $\overline{QS}$	6) If two points are each equidistant from the endpoints of a segment, then they determine the perpendicular bisector of that segment.
7) QVST is a rhombus	7) If the diagonals of a parallelogram are perpendicular, then it is a rhombus.

Chapter 5: Quadrilaterals: Your Fine, Four-Sided Friends

*28

Statements	Reasons
1) $\angle MAK \cong \angle AMR$	1) Given.
2) $\overline{MJ} \cong \overline{AJ}$	2) If angles, then sides.
3) $\angle KRM \cong \angle RKA$	3) Given.
4) $\overline{KJ} \cong \overline{RJ}$	4) If angles, then sides.
5) $\angle MJK \cong \angle AJR$	5) Vertical angles are congruent.
6) $\triangle MJK \cong \triangle AJR$	6) SAS (2, 5, 4).
7) $\overline{MK} \cong \overline{AR}$	7) CPCTC.
8) $\overline{MK} \parallel \overline{AR}$	8) Given.
9) MARK is a parallelogram	9) If one pair of sides of a quadrilateral are both parallel and congruent, then the quadrilateral is a parallelogram.
10) $\overline{MR} \cong \overline{KA}$	10) Addition (2, 4).
11) MARK is a rectangle	11) If the diagonals of a parallelogram are congruent, then it is a rectangle.

*29

Statements	Reasons
1) $\angle OHR$ and $\angle FKI$ are right angles	1) Given.
2) $\overline{AR} \parallel \overline{FE}$	2) Given.
3) $\angle OHR \cong \angle KJM$	3) If lines are parallel, then alternate exterior angles are congruent.
4) $\angle KJM$ is a right angle	4) Substitution.
5) $\angle FKI \cong \angle KJM$	5) Right angles are congruent.
6) $\overline{DI} \parallel \overline{OM}$	6) If corresponding angles are congruent, then lines are parallel.
7) GHJK is a parallelogram	7) If both pairs of opposite sides of a quadrilateral are parallel, it is a parallelogram.
8) $\overline{GI} \perp \overline{FJ}$	8) If right angle, then perpendicular.
9) $\angle GKJ$ is a right angle	9) If perpendicular, then right angle.
10) GHJK is a rectangle	10) If a parallelogram has a right angle, it is a rectangle.
11) $\overline{GH} \cong \overline{GK}$	11) Given.
12) GHJK is a rhombus	12) If a pair of consecutive sides of a parallelogram are congruent, then it is a rhombus.
13) GHJK is a square	13) If a quadrilateral is both a rhombus and a rectangle, it is a square.

Chapter 6
Area, Angles, and the Many Sides of Polygon Geometry

In This Chapter
▶ Determining the area of quadrilaterals and regular polygons
▶ More fantastically fun polygon formulas
▶ Calculating diagonals and the measures of angles

*I*f you're all proofed-out, you may enjoy this proof-free chapter. Here you work on problems involving formulas for the area of various polygons, the sum of the interior and exterior angles of a polygon, and the number of diagonals of a polygon. If you've always wondered about how many diagonals an octakaidecagon has, you've come to the right place.

Square Units: Finding the Area of Quadrilaterals

Referring to the family tree of quadrilaterals (in Chapter 5) — assuming you don't know it by heart — to remind yourself about which quadrilaterals are special cases of other quadrilaterals is always a good idea. Doing so can help you with area problems because when you know, for example, that a rhombus is a special case of both a parallelogram and a kite, you know that you can use either the parallelogram area formula or the kite area formula when computing the area of a rhombus.

Without further ado, here are the area formulas for quadrilaterals.

Quadrilateral area formulas:

- $\text{Area}_{\text{Parallelogram}} = \text{base} \cdot \text{height}$
- $\text{Area}_{\text{Kite}} = \frac{1}{2} \text{diagonal}_1 \cdot \text{diagonal}_2$
- $\text{Area}_{\text{Square}} = \text{side}^2$, or $\frac{1}{2} \text{diagonal}^2$
- $\text{Area}_{\text{Trapezoid}} = \frac{\text{base}_1 + \text{base}_2}{2} \cdot \text{height}$
  $= \text{median} \cdot \text{height}$

(The *median* of a trapezoid is the segment that connects the midpoints of the legs. Its length equals the average of the lengths of the bases.)

152 Part III: Polygons

Here's a handy guide for the quadrilaterals that don't have an area formula in the preceding list:

- For the area of a rectangle, use the parallelogram formula.
- For the area of a rhombus, use either the parallelogram or the kite formula.
- For the area of an isosceles trapezoid, use, of course, the trapezoid formula.

Q. What's the area of parallelogram *ABCD*?

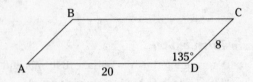

A. *Tip:* For this and many area problems, drawing in altitudes and other perpendicular segments on the diagram can be helpful. And — what often amounts to the same thing — it's a good idea to cut up the figure into right triangles and rectangles.

Draw the altitude from *B* to $\overline{AD}$, and call the length of that segment *h*. $\angle A$ is supplementary to $\angle D$ (a property of parallelograms; see Chapter 5), so $\angle A$ is 45°. Thus, the altitude you drew creates a 45°-45°-90° triangle. The hypotenuse, $\overline{AB}$, is congruent to the opposite side, $\overline{CD}$, and therefore has a length of 8. Using 45°-45°-90° triangle math, *h* equals $\frac{8}{\sqrt{2}}$, or $4\sqrt{2}$ (see Chapter 3 for details).

The area, which equals *base · height*, is thus $20 \cdot 4\sqrt{2}$, or $80\sqrt{2}$ units².

Q. Given: Trapezoid *ABCD* with base of 6 and median, $\overline{PQ}$, 10 units long
$QD = 5$

Find: Area of *ABCD*

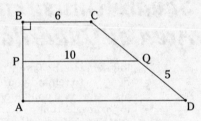

A. For this problem, you can use the trapezoid area formula that uses the median. You know the length of the median, so all you need to compute the area is the trapezoid's height. To get that, first recall that the length of the median, $\overline{PQ}$, is the average of the lengths of the bases, so its length is halfway between them. Since *BC* is 6, or 10 – 4, *AD* must be 10 + 4, or 14. Next, draw an altitude from *C* to $\overline{AD}$, creating a right triangle. The length of its hypotenuse, $\overline{CD}$, is twice 5, or 10, and the triangle's base is 14 – 6, or 8. You have a right triangle in the 3 : 4 : 5 family (namely a 6-8-10 triangle; see Chapter 3), so the altitude is 6 (this has nothing to do, by the way, with the length of $\overline{BC}$, which is coincidentally also 6). Finally, the area, which equals *median · height*, is $10 \cdot 6$, or 60 units².

1. Given: Parallelogram *PQRS* with sides of 7 and 10 and altitudes h_1 and h_2
Find: The ratio $h_1 : h_2$

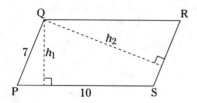

Solve It

2. Given: Parallelogram *GRAM* as shown
Find: *GRAM*'s area and height

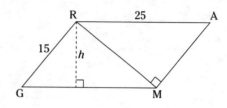

Solve It

3. Given: Trapezoid *WXYZ* with a perimeter of 35 and an area of 55
Find: h

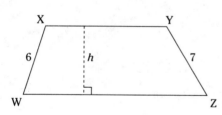

Solve It

*4. The equation of this circle of radius 8 is $x^2 + y^2 = 64$. Estimate its area using the six trapezoids and two triangles.

Hint: Just calculate the areas of the three trapezoids and one triangle in quadrant I; then multiply your result by 4.

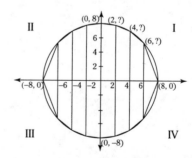

Solve It

154 Part III: Polygons

***5.** Given: Trapezoid *JKLM* with bases 10 and 18 and base angles of 60° and 30°

Find: Area of *JKLM*

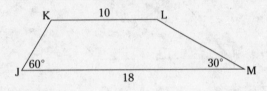

Solve It

6. Given: Kite *ABCD* as shown
△*ABC* is equilateral

Find: Area of *ABCD*

Hint: Use your drawing skills.

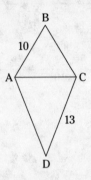

Solve It

7. Find the area of rhombus *RBUS*

Hint: Just connect the dots.

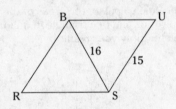

Solve It

8. Find the area of rhombus *QRST*

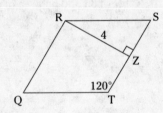

Solve It

A Standard Formula for the Area of Regular Polygons

Why don't we skip the introducing numbers jumbo here and just cut to the chase. Here's the formula for the area of a regular polygon — but first, its definition: A *regular polygon* is a polygon that's both equilateral (with equal sides) and equiangular (equal angles).

Area of a regular polygon:

$$\text{Area}_{\text{Reg. Poly.}} = \tfrac{1}{2} \text{ perimeter} \cdot \text{apothem}$$

An *apothem* of a regular polygon is a segment joining the polygon's center to the midpoint of any side. It's perpendicular to the side.

This area formula, $A = \tfrac{1}{2} pa$, is usually written $A = \tfrac{1}{2} ap$. These formulas are equivalent, of course, so you can use either one, but the way I've written it is a better way to think about what you're actually doing. This polygon formula is based on the formula for the area of a triangle, $A = \tfrac{1}{2} bh$; when you find the area of a regular polygon, you're essentially dividing the polygon into congruent triangles and finding their areas. Because a polygon's perimeter is the counterpart of the triangles' bases and a polygon's apothem is the counterpart of the triangles' heights, $A = \tfrac{1}{2} pa$ is the logical way to write the formula.

As you can see in the following example problem, a regular hexagon can be cut into six equilateral triangles, and an equilateral triangle can be cut into two 30°-60°-90° triangles. For many area problems involving either a hexagon or an equilateral triangle (or both), it's often useful to cut the figure up and make use of one or more 30°-60°-90° triangles. If, instead, the problem involves a square or a regular octagon, adding the right segments to the diagram produces one or more 45°-45°-90° triangles that may be the key to the solution. In other polygon problems, cutting up the polygon into some combination of rectangles and these special triangles can help.

An equilateral triangle is a regular polygon, so to figure its area, you can use the regular polygon formula; however, it also has its own area formula. To wit —

Area of an equilateral triangle with side *s*:

$$\text{Area}_{\text{Equil. }\triangle} = \frac{s^2 \sqrt{3}}{4}$$

156 Part III: Polygons

Q. What's the area of this regular hexagon with a radius of 8? (Yep — that thing is called the *radius*.)

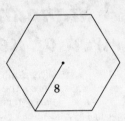

A. You can do this problem two ways, using both of the preceding area formulas.

First, draw in the other five radii, and you can see six congruent isosceles triangles.

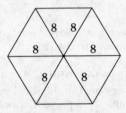

The six angles at the center of the hexagon have to be 60° angles, because all the way around the center is 360°, and 360° ÷ 6 is 60°. An isosceles triangle with a 60° vertex angle is an equilateral triangle, so you have six congruent equilateral triangles.

Method I: Now you can finish using the equilateral triangle formula:

$$\text{Area}_{\text{Equil. }\triangle} = \frac{s^2\sqrt{3}}{4}$$
$$= \frac{8^2\sqrt{3}}{4}$$
$$= 16\sqrt{3}$$

You have six triangles, so the area of the hexagon is $6 \cdot 16\sqrt{3}$, or $96\sqrt{3}$ units².

Method II: Draw in the apothem from the center of the hexagon straight down to the midpoint of the bottom side. That apothem cuts the equilateral triangle into two 30°- 60°- 90° triangles. These triangles have sides with ratios of $1:\sqrt{3}:2$ (see Chapter 3 for more on special right triangles). Each triangle has a hypotenuse of 8, and therefore, a short leg of 4 and a long leg (the apothem) of $4\sqrt{3}$. The perimeter is 6 times 8, or 48, so you're all set to use the polygon formula:

$$\text{Area}_{\text{Reg. Poly.}} = \frac{1}{2}pa$$
$$= \frac{1}{2} \cdot 48 \cdot 4\sqrt{3}$$
$$= 96\sqrt{3} \text{ units}^2$$

9. The span of this regular hexagon is 32. Find its area.

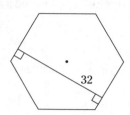

Solve It

10. Find the area of a regular octagon with sides of length 10.

Hint: Cut up the octagon till you create one or more useful 45°- 45°- 90° triangles.

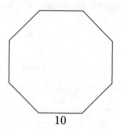

Solve It

More Fantastically Fun Polygon Formulas

The formulas in the preceding section are all about the area of polygons. They're basically meant to show you how much space a polygon takes up. The formulas in this section dive deeper into the building blocks of polygons: the angles and diagonals that give different polygons their unique characteristics.

Definitions of interior and exterior angles:

- An *interior* angle of a polygon is an angle inside the polygon at one of its vertices.
- An *exterior* angle of a polygon is an angle outside the polygon formed by one of its sides and the extension of an adjacent side (see Figure 6-1).

Figure 6-1: ∠EMI is an exterior angle of quadrilateral MILY. Vertical angle ∠OLM is *not* an exterior angle of △AJL.

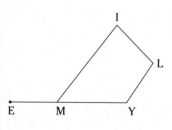
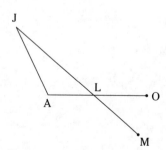

Would you believe me if I told you that regardless of whether a polygon has three sides or a million, the exterior angles of that polygon always add up to 360°? You'd better, because I'm about to tell you formally. And that's just the beginning of the polygon's special properties.

Interior and exterior angle formulas:

- The sum of the measures of the *interior angles* of a polygon with n sides is $(n-2)180$.
- The measure of each *interior angle* of an equiangular n-gon is $\dfrac{(n-2)180}{n}$ or $180 - \dfrac{360}{n}$.
- If you count one exterior angle at each vertex, the sum of the measures of the *exterior angles* of a polygon is always 360°.
- The measure of each *exterior angle* of an equiangular n-gon is $\dfrac{360}{n}$.

Number of diagonals in a polygon: The number of diagonals that you can draw in an n-gon is $\dfrac{n(n-3)}{2}$.

Q. What's the measure of one of the interior angles of a regular 22-gon, and how many diagonals does it have?

A. A regular polygon is equiangular, so you can use either equiangular formula (in the second bullet). The second version is probably easier to use unless you happen to already know the sum of the interior angles (which is the numerator in the first formula). Note that using the second version amounts to finding the supplement of one of the polygon's exterior angles.

$$\text{Interior angle} = 180 - \dfrac{360}{n}$$
$$= 180 - \dfrac{360}{22}$$
$$= 180 - 16\dfrac{4}{11}$$
$$= 163\dfrac{7}{11}°$$

$$\text{Diagonals} = \dfrac{n(n-3)}{2}$$
$$= \dfrac{22(22-3)}{2}$$
$$= 209$$

11. Given: Hexagon TAYLOR with angles as shown
∠ATR ≅ ∠O

Find: ∠1

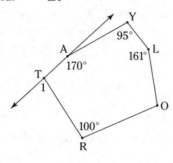

Solve It

12. Find the number of sides in a polygon whose interior angles add up to

a. 1,080°

b. 7,920°

c. $(180x^2 + 180)°$ (for some whole number x)

d. 825°

Solve It

13. Find the sum of all exterior angles and, if you have enough information, the measure of one exterior angle in

a. An equiangular pentagon

b. A regular 18-gon

c. An icosagon (20 sides)

d. An equilateral 80-gon

Solve It

14. How many diagonals can be drawn in a triacontagon (30 sides)?

Solve It

15. How many sides does a polygon have if it has 3 times as many diagonals as sides?

Solve It

*16. What's the measure of one of the interior angles of an equiangular polygon with 54 diagonals?

Solve It

Solutions for Area, Angles, and the Many Sides of Polygon Geometry

1 $\overline{RS}$ is congruent to $\overline{PQ}$ (property of a parallelogram; see Chapter 5), so RS is also 7. The area of a parallelogram equals *base · height,* and obviously, the area must come out the same regardless of which base you use, $\overline{PS}$ or $\overline{RS}$. Thus,

$$\text{base}_1 \cdot \text{height}_1 = \text{base}_2 \cdot \text{height}_2$$
$$10 \cdot h_1 = 7 \cdot h_2$$
$$\frac{h_1}{h_2} = \frac{7}{10}$$

The ratio, $h_1 : h_2$ equals 7 : 10. Note that although you can determine the ratio of the heights, determining h_1 or h_2 or the area of $PQRS$ is impossible.

2 AM is 15, so $\triangle ARM$ is in the 3 : 4 : 5 family of triangles. In fact, it is a 3-4-5 triangle blown up five times. $\overline{RM}$ is the long leg and is thus 4 times 5, or 20 units long.

$\text{Area}_{GRAM} = \text{base} \cdot \text{height}$ $\quad\quad$ GM is 25, so

$\quad\quad = AM \cdot RM$ $\quad\quad\quad\quad\quad$ $\text{Area}_{GRAM} = \text{base} \cdot \text{height}$

$\quad\quad = 15 \cdot 20$ $\quad\quad\quad\quad\quad\quad\quad$ $300 = GM \cdot h$

$\quad\quad = 300 \text{ units}^2$ $\quad\quad\quad\quad\quad\quad$ $300 = 25 \cdot h$

$\quad\quad\quad\quad\quad\quad\quad\quad\quad\quad\quad\quad\quad\quad$ $12 = h$

3 To get h, you need to use the trapezoid area formula, and to use the formula, you need the sum of the lengths of the bases, $\overline{XY}$ and $\overline{WZ}$. (Note that all you need is the sum; you don't need to know the lengths of the individual bases.)

$XY + WZ + 6 + 7 = 35$ $\quad\quad$ $\text{Area}_{WXYZ} = \frac{b_1 + b_2}{2} \cdot h$

$XY + WZ = 22$ $\quad\quad\quad\quad\quad\quad$ $55 = \frac{XY + WZ}{2} \cdot h$

$\quad\quad\quad\quad\quad\quad\quad\quad\quad\quad\quad\quad$ $55 = \frac{22}{2} \cdot h$

$\quad\quad\quad\quad\quad\quad\quad\quad\quad\quad\quad\quad$ $\frac{55}{11} = h$

$\quad\quad\quad\quad\quad\quad\quad\quad\quad\quad\quad\quad$ $5 = h$

*4 The heights of the three sideways trapezoids and the triangle in quadrant I are all equal to 2 (note that the heights run along the *x*-axis). You need the bases of the trapezoids and the triangle to compute their areas, and their bases equal the *y*-coordinates at x = 0, 2, 4, and 6. When x is 0, y is 8, so that left-most base is 8. To find the other bases, plug the *x*-coordinates into the equation of the circle. Thus, you find that

$x^2 + y^2 = 64$

$2^2 + y^2 = 64$ $\quad\quad\quad\quad\quad$ $4^2 + y^2 = 64$ $\quad\quad\quad\quad\quad$ $6^2 + y^2 = 64$

$\quad y^2 = 60$ $\quad\quad\quad\quad\quad\quad\quad$ $y = 4\sqrt{3}$ $\quad\quad\quad\quad\quad\quad\quad$ $y = 2\sqrt{7}$

$\quad y = 2\sqrt{15}$

So the first trapezoid (between x = 0 and x = 2) has bases of 8 and $2\sqrt{15}$, the second trapezoid has bases of $2\sqrt{15}$ and $4\sqrt{3}$, and the third has bases of $4\sqrt{3}$ and $2\sqrt{7}$. The single triangle's base is $2\sqrt{7}$. Compute their areas, add them up, and then multiply that result by 4 (for the four quadrants):

$$\text{Area}_{\text{1st Trap}} = \frac{b_1 + b_2}{2} \cdot h$$
$$= \frac{8 + 2\sqrt{15}}{2} \cdot 2$$
$$= 8 + 2\sqrt{15}$$
$$\text{Area}_{\text{2nd Trap}} = \frac{2\sqrt{15} + 4\sqrt{3}}{2} \cdot 2$$
$$= 2\sqrt{15} + 4\sqrt{3}$$

$$\text{Area}_{\text{3rd Trap}} = \frac{4\sqrt{3} + 2\sqrt{7}}{2} \cdot 2$$
$$= 4\sqrt{3} + 2\sqrt{7}$$
$$\text{Area}_\triangle = \frac{1}{2} \cdot \text{base} \cdot \text{height}$$
$$= \frac{1}{2} \cdot 2\sqrt{7} \cdot 2$$
$$= 2\sqrt{7}$$

$$\text{Total of four areas} = (8 + 2\sqrt{15}) + (2\sqrt{15} + 4\sqrt{3}) + (4\sqrt{3} + 2\sqrt{7}) + 2\sqrt{7}$$
$$= 8 + 4\sqrt{15} + 8\sqrt{3} + 4\sqrt{7}$$
$$\text{Estimate of circle's area} = 4(8 + 4\sqrt{15} + 8\sqrt{3} + 4\sqrt{7})$$
$$= 32 + 16\sqrt{15} + 32\sqrt{3} + 16\sqrt{7}$$
$$\approx 191.7 \text{ units}^2$$

In case you're curious, the area of the circle is

$$\text{Area}_{\text{Circle}} = \pi r^2$$
$$= \pi \cdot 8^2$$
$$= 64\pi$$
$$\approx 201.1 \text{ units}^2$$

The estimate was a little less than 5 percent off.

*5 A good plan of attack here — like with so many polygon problems — is to cut the figure up into right triangles and rectangles. Draw altitudes from K to $\overline{JM}$ and from L to $\overline{JM}$, and call their length h. On the left, you have a 30°- 60°- 90° triangle with h as the long leg, so the short leg (along $\overline{JM}$) is $\frac{h}{\sqrt{3}}$ (because in a 30°- 60°- 90° triangle, the ratio of short leg to long leg is $1:\sqrt{3}$ — see Chapter 3 for details). On the right, you have another 30°- 60°- 90° triangle, but this time h is the short leg. The long leg (along $\overline{JM}$) is thus $h\sqrt{3}$. KL is 10, so the distance between the two altitudes along $\overline{JM}$ is also 10. So you have three pieces along $\overline{JM}$ that add up to 18. Now you can find h:

$$\frac{h}{\sqrt{3}} + 10 + h\sqrt{3} = 18$$
$$\sqrt{3}\left(\frac{h}{\sqrt{3}} + 10 + h\sqrt{3}\right) = 18 \cdot \sqrt{3}$$
$$h + 10\sqrt{3} + 3h = 18\sqrt{3}$$
$$4h = 8\sqrt{3}$$
$$h = 2\sqrt{3}$$

Now that you know h, simply plug it into the trapezoid area formula along with the given bases:

$$\text{Area}_{JKLM} = \frac{b_1 + b_2}{2} \cdot h$$
$$= \frac{10 + 18}{2} \cdot 2\sqrt{3}$$
$$= 28\sqrt{3} \text{ units}^2$$

6 Draw diagonal $\overline{BD}$, which is the perpendicular bisector of $\overline{AC}$, and label the intersection of the diagonals X. Triangle ABC is equilateral, so AC is 10; thus, AX and XC are each 5. Triangle ABX is a 30°- 60°- 90° triangle with a short leg of 5, so its long leg, $\overline{BX}$, has a length of $5\sqrt{3}$. Triangle

XCD is a 5-12-13 right triangle with XD equal to 12 (see Chapter 3 for more on Pythagorean triples). Thus, $\overline{BD}$ has a length of $12 + 5\sqrt{3}$. Finally,

$$\text{Area}_{ABCD} = \tfrac{1}{2} d_1 d_2$$
$$= \tfrac{1}{2} AC \cdot BD$$
$$= \tfrac{1}{2} \cdot 10(12 + 5\sqrt{3})$$
$$= 60 + 25\sqrt{3} \text{ units}^2$$

7. Draw diagonal $\overline{RU}$, which is the perpendicular bisector of $\overline{BS}$, and label the intersection of the diagonals X. That step creates a right triangle with a leg of 8 and a hypotenuse of 15. Careful now — this is *not* an 8-15-17 right triangle (remember, the hypotenuse is the longest side). Use the Pythagorean Theorem to get XU:

$$XU^2 + XS^2 = US^2$$
$$XU^2 + 8^2 = 15^2$$
$$XU^2 = 161$$
$$XU = \sqrt{161}$$

RU is twice XU because the diagonals in a rhombus bisect each other, so RU is $2\sqrt{161}$; using the kite formula,

$$\text{Area}_{RBUS} = \tfrac{1}{2} d_1 d_2$$
$$= \tfrac{1}{2} \cdot 16 \cdot 2\sqrt{161}$$
$$= 16\sqrt{161}$$

8. $\angle S$ and $\angle T$ are supplementary (property of parallelogram — see Chapter 5), so $\angle S$ is 60°, and $\triangle RSZ$ is thus a 30°- 60°- 90° triangle with a long leg of 4. The short leg, $\overline{SZ}$, therefore measures $\tfrac{4}{\sqrt{3}}$, and the hypotenuse, $\overline{RS}$ is twice that, or $\tfrac{8}{\sqrt{3}}$ units long. All sides of a rhombus are equal, so ST is also $\tfrac{8}{\sqrt{3}}$. This time, you use the parallelogram formula to get the desired area:

$$\text{Area}_{QRST} = \text{base} \cdot \text{height}$$
$$= ST \cdot RZ$$
$$= \tfrac{8}{\sqrt{3}} \cdot 4$$
$$= \tfrac{32}{\sqrt{3}}$$
$$\approx 18.5 \text{ units}^2$$

9. If you slide the span up and to the right till it hits the center, you can see that the span is twice the apothem, so the apothem is 16. And as you can see in the example problem, a regular hexagon consists of six equilateral triangles, and its apothem is the altitude of one of these equilateral triangles. See the following figure:

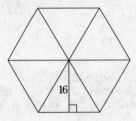

Chapter 6: Area, Angles, and the Many Sides of Polygon Geometry 163

This apothem is the long leg of a 30°-60°-90° triangle, so the short leg is $\frac{16}{\sqrt{3}}$, and the hypotenuse is twice that, or $\frac{32}{\sqrt{3}}$ (the ratio of sides in a 30°-60°-90° triangle is $1:\sqrt{3}:2$; see Chapter 3 for details). And that hypotenuse is a side of the equilateral triangle. Each side of the hexagon is, therefore, $\frac{32}{\sqrt{3}}$, and the perimeter is six times that, or $\frac{192}{\sqrt{3}}$. You're finally ready to use the area formula:

$$\text{Area}_{\text{Reg. Poly.}} = \frac{1}{2} pa$$
$$= \frac{1}{2} \cdot \frac{192}{\sqrt{3}} \cdot 16$$
$$= \frac{1536}{\sqrt{3}}$$
$$= 512\sqrt{3} \text{ units}^2$$

10 Find the right lines to draw? Here they are:

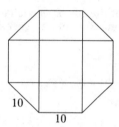

The four triangles are 45°-45°-90° triangles with a hypotenuse of 10. These triangles have sides with lengths in the ratio $1:1:\sqrt{2}$ (see Chapter 3 for more information). The legs of the 45°-45°-90° triangles are thus $\frac{10}{\sqrt{2}}$, or $5\sqrt{2}$. So now you have

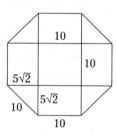

Now add up all the pieces for your total area:

$$\text{Area}_{\text{Octagon}} = 1 \text{ square} + 4 \text{ rectangles} + 4 \text{ triangles}$$
$$= 10^2 + 4(10)(5\sqrt{2}) + 4\left(\frac{1}{2}\right)(5\sqrt{2})(5\sqrt{2})$$
$$= 100 + 4(50\sqrt{2}) + 4(25)$$
$$= 200 + 200\sqrt{2} \text{ units}^2$$

11 *TAYLOR* is a hexagon, so the sum of its interior angles is $(n-2)180 = (6-2)180 = 720°$.

Subtract the four known angles from this value: $720 - (170 + 95 + 161 + 100) = 194°$.

Then, because the two remaining angles, $\angle ATR$ and $\angle O$, are congruent, each must be half of 194°, or 97°. Finally, because $\angle 1$ is the supplement of $\angle ATR$, $\angle 1 = 180 - 97 = 83°$.

12 Here you go:

a. $180(n-2) = 1{,}080$
$n - 2 = 6$
$n = 8$ sides

b. $180(n-2) = 7{,}920$
$n - 2 = 44$
$n = 46$ sides

c. $180(n-2) = 180x^2 + 180$
$n - 2 = x^2 + 1$
$n = x^2 + 3$ sides

d. $180(n-2) = 825$
$n - 2 \approx 4.58$
$n \approx 6.58$ sides

There's no such thing as a polygon with 6.58 sides.

13 Here are the angle measures:

a. The total is 360°. One exterior angle is 360 ÷ 5 = 72°.

b. The total is 360°. One exterior angle is 360 ÷ 18 = 20°.

c. The total is 360°. You can't compute the measure of a single exterior angle because you don't know whether the icosagon is equiangular.

d. The total is 360°. The fact that the 80-gon is equilateral does not tell you whether it's equiangular, so you can't figure the measure of a single exterior angle.

14 Number of diagonals $= \dfrac{n(n-3)}{2}$
$= \dfrac{30(30-3)}{2}$
$= 405$

15 Number of diagonals $= 3 \cdot$ (number of sides)

$\dfrac{n(n-3)}{2} = 3n$
$n^2 - 3n = 6n$
$n^2 - 9n = 0$
$n(n-9) = 0$
$n = 0$ or 9

There's no such thing as a polygon with 0 sides, so the answer is 9.

*16
$\dfrac{n(n-3)}{2} = 54$
$n^2 - 3n = 108$
$n^2 - 3n - 108 = 0$
$(n-12)(n+9) = 0$
$n = 12$ or -9

So you have an equiangular 12-gon; therefore,

One interior angle $= 180° - \dfrac{360°}{n}$
$= 180° - \dfrac{360°}{12}$
$= 150°$

Chapter 7
Similarity: Size Doesn't Matter

In This Chapter
▶ Exploring similar polygons
▶ Proving triangles similar
▶ Looking at the corresponding parts of similar triangles
▶ Breaking it down: Creating similar triangles from a right triangle
▶ Using theorems of proportion

When two triangles, two rectangles, two pentagons, or two of any type of polygon have the same shape (regardless of whether they have the same size), you say that they're *similar* — like if you take a figure and blow it up or shrink it down in a photocopy machine the new image will be the same shape as (and thus *similar* to) the original. Congruent figures are automatically similar, but when you do problems involving two similar figures, you're usually dealing with two things of different sizes that have the same shape. The squiggle symbol, ~, means *is similar to*.

In this chapter, you do problems involving similar triangles and other polygons. Similar polygons have proportional sides, so you also do many problems with proportions. Finally, you practice using theorems — some of which have nothing to do with similarity — that, like similarity theorems, involve proportions.

Defining Similarity

When you're talking about similarity, you have to talk about the two defining characteristics of similar figures.

Similar polygons: In similar polygons, both of the following are true:

✔ **Corresponding angles are congruent.** If objects of different sizes have the same shape, their angles have to be equal. This idea is kinda obvious if you think about it. Imagine you see something like a yield sign on the side of the road. It's a downward-pointing equilateral triangle with three 60° angles. As you get closer, it looks bigger, of course, but regardless of how big or small it looks, the three angles are always 60° angles. If the angles were to change to something other than 60°, the sign would no longer look like a yield sign. It would have morphed into a different shape.

✔ **The ratios of the lengths of corresponding sides are equal.** Say the front door of a house is 7 feet tall and 3 feet wide and that the blueprint for the design of the house contains a door measuring 2.1 inches tall by 0.9 inches wide. Because these two doors are similar rectangles, the ratio of their heights equals the ratio of their widths, and you get the following proportion:

$$\frac{\text{height}_{\text{real door}}}{\text{height}_{\text{blueprint door}}} = \frac{\text{width}_{\text{real door}}}{\text{width}_{\text{blueprint door}}}$$

$$\frac{7 \text{ feet}}{2.1 \text{ inches}} = \frac{3 \text{ feet}}{0.9 \text{ inches}}$$

Both sides of the second equation equal 40, which tells you that the real door is 40 times as tall and 40 times as wide as the door shown in the blueprint (you have to convert all units to inches or feet before calculating this). Such ratios or quotients represent the blow-up or shrink factor, depending on which way you look at it.

Perimeters of similar polygons: The ratio of the perimeters of two similar polygons equals the ratio of any pair of corresponding sides.

Q. Given: Pentagon $ABCDE \sim$ pentagon $VWXYZ$

               Perimeter of $ABCDE$ is 18

Find:

    a. VW

    b. XY

    c. Perimeter of $VWXYZ$

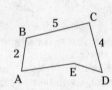
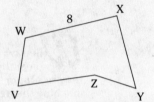

A. In the diagram, the two pentagons have the same *orientation;* in other words, A matches up with V, B matches up with W, and so on. If you were to expand $ABCDE$ a bit and slide it over to the right, it would fit perfectly on top of $VWXYZ$. You wouldn't have to rotate it or flip it upside down to make it fit. But this isn't always the case, so to make sure you're pairing up the correct vertices and sides, pay attention to the way the similarity is written. When someone says $ABCDE \sim VWXYZ$, it means that A pairs up with V, B pairs up with W, and so on, and that $\overline{CD}$ (the third and fourth letters) pairs up with $\overline{XY}$ (also the third and fourth letters). Got it? Fantastic!

    a. $\dfrac{\text{left side}_{VWXYZ}}{\text{left side}_{ABCDE}} = \dfrac{\text{top}_{VWXYZ}}{\text{top}_{ABCDE}}$

$$\frac{VW}{2} = \frac{8}{5}$$

$$VW = \frac{16}{5} = 3.2$$

    b. $\dfrac{\text{right side}_{VWXYZ}}{\text{right side}_{ABCDE}} = \dfrac{\text{top}_{VWXYZ}}{\text{top}_{ABCDE}}$

$$\frac{XY}{4} = \frac{8}{5}$$

$$XY = \frac{32}{5} = 6.4$$

(Or you could just notice that the right side of $ABCDE$ is twice as long as the left side, so you can simply multiply VW by 2 to get XY.)

    c. $\dfrac{\text{perimeter}_{VWXYZ}}{\text{perimeter}_{ABCDE}} = \dfrac{\text{side}_{VWXYZ}}{\text{side}_{ABCDE}}$

$$\frac{\text{perimeter}_{VWXYZ}}{18} = \frac{8}{5}$$

$$\text{perimeter}_{VWXYZ} = \frac{144}{5} = 28.8$$

Chapter 7: Similarity: Size Doesn't Matter 167

1. Given: ABCD ~ EFGH
Find:
  a. All missing angles
  b. All missing sides

Solve It

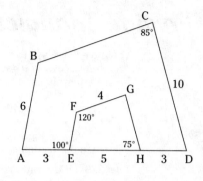

2. Given: △PQR ~ △ZXY
Find:
  a. All missing angles
  b. All missing sides

Solve It

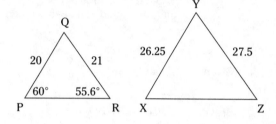

3. Given: ABCDE ~ LMNOP
Perimeter of ABCDE is 30
Find: Perimeter of LMNOP

Solve It

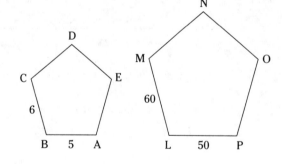

4. Given: △ABC ~ △ACD
Find:
  a. AD
  b. DB

Solve It

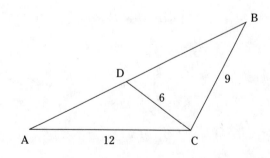

Proving Triangles Similar

You have five ways to prove triangles congruent: SSS, SAS, ASA, AAS, and HLR (see Chapter 4). Now you get three ways to prove triangles similar: SSS~, SAS~, and AA. The most frequently used and by far the easiest to use is AA.

Proving triangles similar:

- **AA:** If two angles of one triangle are congruent to two angles of another triangle, then the triangles are similar.

- **SSS~:** If the ratios of the three pairs of corresponding sides of two triangles are equal, then the triangles are similar.

- **SAS~:** If the ratios of two pairs of corresponding sides of two triangles are equal and the included angles are congruent, then the triangles are similar.

Q. Given: $\angle 1 \cong \angle 3$
$\angle 2 \cong \angle 4$
Prove: $\triangle BLO \sim \triangle WUP$
Find: WU

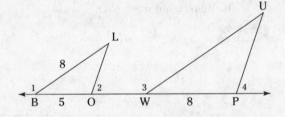

A.

Statements	Reasons
1) $\angle 1 \cong \angle 3$	1) Given.
2) $\angle LBO \cong \angle UWP$	2) Supplements of congruent angles are congruent.
3) $\angle 2 \cong \angle 4$	3) Given.
4) $\angle LOB \cong \angle UPW$	4) Supplements of congruent angles are congruent.
5) $\triangle BLO \sim \triangle WUP$	5) AA (if two angles of one triangle are congruent to two angles of another triangle, then the triangles are similar).

Now find WU. Piece o' cake:

$$\frac{\text{top}_{\triangle WUP}}{\text{top}_{\triangle BLO}} = \frac{\text{base}_{\triangle WUP}}{\text{base}_{\triangle BLO}}$$

$$\frac{WU}{8} = \frac{8}{5}$$

$$WU = \frac{64}{5} = 12.8$$

Chapter 7: Similarity: Size Doesn't Matter **169**

Q. Given: Diagram as shown
Prove: △MAR ~ △BLE (paragraph proof)

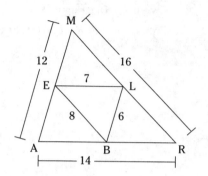

A. No reason to bother with a two-column proof here. All you have to do is to show that all three ratios of corresponding sides are equal, like this: $\frac{12}{6} = \frac{14}{7} = \frac{16}{8}$. Check.

Thus, by SSS~, the triangles are similar. But you still have to show that the right vertices pair up. One way to do this is to pick a vertex, like A, and note that it's across from the longest side of △MAR (16). So it corresponds to L, which is across from the longest side of △BLE (8). Then R and E correspond because both are across from the shortest sides. Lastly, you have no choice, of course, but to pair M with B. Thus, △MAR ~ △BLE.

Q. Given: U is the midpoint of $\overline{RA}$
G is the midpoint of $\overline{RT}$
Prove: △RUG ~ △RAT (paragraph proof)

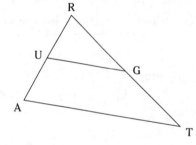

A. Let's skip the two-column mumbo jumbo again. (You can do this proof in two-column format, but that involves all sorts of rigamarole like 1) U is a midpoint; then 2) $\overline{RU} \cong \overline{UA}$; then 3) RU = UA; then 4) RU + UA = RA, then 5) RU + RU = RA; then 6) 2 · RU = RA; then 7) $\frac{RA}{RU} = 2$; and so on, and so on.)

So just use common sense instead. Because U is the midpoint of $\overline{RA}$, you know that $\frac{RU}{RA} = \frac{1}{2}$. G works the same way, so $\frac{RG}{RT} = \frac{1}{2}$. Thus, $\frac{RU}{RA} = \frac{RG}{RT}$. Then, because ∠R ≅ ∠R, △RUG ~ △RAT by SAS~.

5. Given: Diagram as shown
Prove: △PQR ~ △STU (paragraph proof)

Solve It

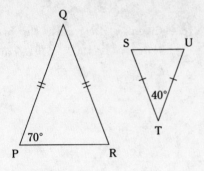

6. Given: △XRT is isosceles with base $\overline{XT}$ and altitude $\overline{RN}$
∠1 ≅ ∠2
Prove: △WGN ~ △TRN

Solve It

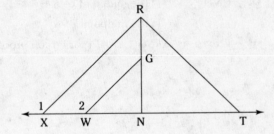

Statements	Reasons

7. Given: Diagram as shown
   a. Prove the triangles similar (paragraph proof)
   b. Write the names of the similar triangles:
      △BCD ~ △_____.

Solve It

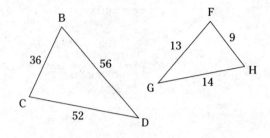

8. Given: $\overline{AT} \parallel \overline{OY}$
   Prove: △BOY is isosceles (paragraph proof)

Solve It

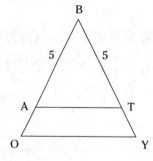

172 Part III: Polygons

9. Given: ∠ELP ≅ ∠IPL
∠EPL ≅ ∠ILP
V is the midpoint of $\overline{LI}$
S is the midpoint of $\overline{PI}$

Prove: △VIS ~ △PEL

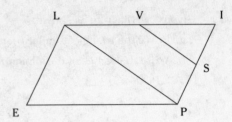

Statements	Reasons

Corresponding Sides and CSSTP — Cats Stalk Silently Then Pounce

Actually, CSSTP stands for *Corresponding Sides of Similar Triangles are Proportional.* You can tell this statement is true from the definition of similar polygons. And if you've done the preceding problems, you've used this concept already when you had to calculate the lengths of the sides of similar triangles and other polygons. What's new here is using CSSTP in formal, two-column proofs.

TIP

You use CSSTP on the line immediately after showing triangles similar, just like you use CPCTC on the line after you show triangles congruent (for more on congruent parts of congruent triangles, see Chapter 4).

Q. Given: ∠T is supplementary to ∠UAC
Prove: TP · AU = AP · TC

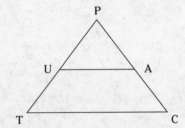

Chapter 7: Similarity: Size Doesn't Matter 173

A. *Tip:* When you're asked to prove that a product equals another product (like $TP \cdot AU = AP \cdot TC$ in this example proof), the proof quite likely involves similar triangles (or perhaps — though less likely — one of the three theorems in the last section of this chapter). So look for similar-looking triangles that contain the four segments in the *prove* statement.

Statements	Reasons
1) $\angle T$ is supplementary to $\angle UAC$	1) Given.
2) $\angle PAU$ is supplementary to $\angle UAC$	2) Two angles that form a straight angle are supplementary.
3) $\angle T \cong \angle PAU$	3) Supplements of the same angle are congruent.
4) $\angle P \cong \angle P$	4) Reflexive Property.
5) $\triangle TPC \sim \triangle APU$	5) AA (3 and 4).
6) $\dfrac{TP}{AP} = \dfrac{TC}{AU}$	6) CSSTP.
7) $TP \cdot AU = AP \cdot TC$	7) Means-Extremes Products Theorem (a fancy name for cross-multiplication).

10. Given: $\angle JKL \cong \angle NML$

             Perimeter of $\triangle JKL$ is 27

       Prove: $\overline{MN} \cong \overline{JL}$ (paragraph proof)

Solve It

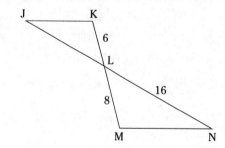

11. Given: $\angle 1 \cong \angle CBD$

Prove: $(AD)^2 = (AB)(AC)$ with a paragraph proof

Solve It

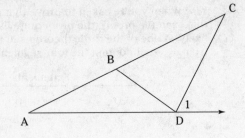

*12. Given: E and O trisect $\overline{AS}$

O bisects $\overline{RM}$

$\angle S \cong \angle MTI$

Prove: $RO \cdot IT = MI \cdot OE$

Solve It

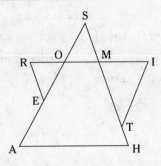

Statements	Reasons

Chapter 7: Similarity: Size Doesn't Matter — 175

13. Given: The triangles are similar
Find: Measures of all angles

Note: Problems 13 and 14 are not CSSTP problems — just more good similar triangle problems.

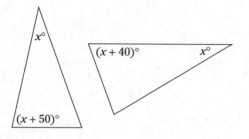

Solve It

14. Indicate whether statements **a.** through **f.** are *always* true, *sometimes* true, or *never* true.

a. If $\triangle ABC \sim \triangle CBA$, then $\overline{AB} \cong \overline{CB}$.

b. If $\angle ABC \cong \angle DEF$, then $\triangle ABC \sim \triangle DEF$.

c. If $\triangle ABC \cong \triangle DEF$, then $\triangle ABC \sim \triangle DEF$.

d. If $\triangle ABC \sim \triangle DEF$, then $\triangle ABC \cong \triangle DEF$.

e. If $\triangle ABC$ is a right triangle and $\triangle DEF$ is an acute triangle, then $\triangle ABC \sim \triangle DEF$.

f. If $\triangle ABC$ and $\triangle DEF$ are both isosceles and $\angle B \cong \angle E$, then $\triangle ABC \sim \triangle DEF$.

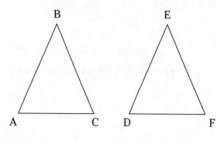

Solve It

Similar Rights: The Altitude-on-Hypotenuse Theorem

If you use the hypotenuse of a right triangle as its base and draw an altitude to it — creating two more, smaller right triangles — all three triangles are similar. Here's the handy-dandy theorem:

Altitude-on-Hypotenuse Theorem: If an altitude is drawn to the hypotenuse of a right triangle as shown in Figure 7-1, then

- The two triangles formed are similar to the given triangle and to each other:

  $\triangle ACB \sim \triangle ADC \sim \triangle CDB$

- $h^2 = xy$

- $a^2 = yc$ and $b^2 = xc$ — note that this is really just one formula or relationship, not two. It works exactly the same on both sides of the big triangle:

  (leg of big $\triangle$)^2 = (part of hypotenuse next to leg) · (whole hypotenuse)

Figure 7-1: Three similar right triangles in one: Triple the pleasure, triple the fun.

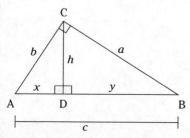

Part III: Polygons

When doing a problem involving an altitude-on-hypotenuse diagram (like Figure 7-1), don't assume that the problem must be solved with the second or third part of the Altitude-on-Hypotenuse Theorem. Sometimes, the easiest way to solve the problem is with the Pythagorean Theorem. And at other times, you can use ordinary similar triangle proportions to solve the problem.

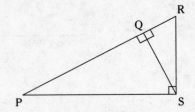

Q. Use the figure to answer the following questions.

   a. If $PQ = 12$ and $QR = 3$, find QS, PS, and RS

   b. If $PR = 13$ and $RS = 5$, find PS, PQ, QR, and QS

A. Here's how this problem plays out:

   a. From the second part of the theorem, $h^2 = xy$, so

$$(QS)^2 = (PQ)(QR)$$
$$= 12 \cdot 3$$
$$= 36$$
$$QS = 6$$

From the third part of the theorem, $a^2 = yc$ and $b^2 = xc$, so

$$(PS)^2 = (PQ)(PR) \quad \text{and} \quad (RS)^2 = (QR)(PR)$$
$$= 12 \cdot 15 \qquad\qquad\qquad = 3 \cdot 15$$
$$= 180 \qquad\qquad\qquad\qquad = 45$$
$$PS = \sqrt{180} \qquad\qquad\qquad RS = \sqrt{45}$$
$$= 6\sqrt{5} \qquad\qquad\qquad\qquad = 3\sqrt{5}$$

Of course, you could also get PS and RS with the Pythagorean Theorem.

   b. PS is 12 (you have a 5-12-13 triangle — see Chapter 3 for info on triangle families). You can get PQ and QR using part three of the theorem:

$$(PS)^2 = (PQ)(PR) \quad \text{and} \quad (RS)^2 = (QR)(PR)$$
$$12^2 = PQ \cdot 13 \qquad\qquad\qquad 5^2 = QR \cdot 13$$
$$PQ = \frac{144}{13} \qquad\qquad\qquad\qquad QR = \frac{25}{13}$$

Of course, you can just calculate one of these lengths and then subtract it from 13 (PR) to get the other.

Finally, you get QS with the second part of the Altitude-on-Hypotenuse Theorem (or the Pythagorean Theorem):

$$(QS)^2 = (PQ)(QR)$$
$$= \left(\frac{144}{13}\right)\left(\frac{25}{13}\right)$$
$$= \frac{12^2 \cdot 5^2}{13^2}$$
$$QS = \frac{12 \cdot 5}{13} = \frac{60}{13}$$

15. Use the figure to calculate these lengths:

a. If $JA = 4$ and $AY = 9$, find JZ and AZ
b. If $JA = 3$ and $JZ = 5$, find AY
c. If $JA = 2$ and $JY = 8$, find YZ
d. If $AZ = 8$ and $AY = 10$, find JY
e. If $JZ = 8$ and $JY = 12$, find AY

Solve It

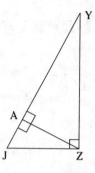

16. If $RQ = 5$ and $RS = 10$, find RT

Hint: You can solve this by using the last two parts of the theorem, but there's an easier way.

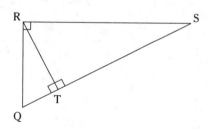

Solve It

*17. Find FL

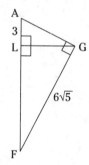

Solve It

Three More Theorems Involving Proportions

In this last section, you practice using three more theorems that involve a proportion. The first two are related to similar triangles.

Side-Splitter Theorem: If a line is parallel to a side of a triangle and it intersects the other two sides, it divides those sides proportionally.

You can use the Side-Splitter Theorem *only* for the four segments on the split sides of the triangle. Do *not* use it for the parallel sides. For the parallel sides, use similar triangle proportions. (Whenever a triangle is divided by a line parallel to one of its sides, the small triangle created is similar to the original, large triangle. This idea follows from the *if two parallel lines are cut by a transversal, then corresponding angles are congruent* theorem [see Chapter 5] and AA.)

The theorem that shall not be named: If three or more parallel lines are intersected by two or more transversals, the parallel lines divide the transversals proportionally. Consider Figure 7-2. Given that the horizontal lines are parallel, the following proportions (among others), follow from the theorem:

$$\frac{AB}{CD} = \frac{PQ}{RS}, \frac{AC}{CD} = \frac{WY}{YZ}$$
$$\frac{PQ}{QS} = \frac{WX}{XZ}, \frac{RS}{QR} = \frac{YZ}{XY}$$

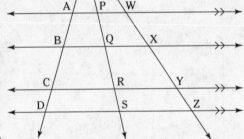

Figure 7-2: Multiple parallel lines with transversals.

The third theorem has nothing to do with similar figures and is in this section just because of the proportion connection:

Angle-Bisector Theorem: If a ray bisects an angle of a triangle, it divides the opposite side into segments that are proportional to the adjacent sides.

When you bisect an angle in a triangle, you *never* get similar triangles (except when you bisect the vertex angle of an isosceles triangle, in which case the resulting triangles are congruent as well as similar). The fact that the Angle-Bisector Theorem is usually in the similar triangle chapter in geometry books despite its having nothing to do with similar triangles may be one reason students often fail to remember the theorem.

Don't forget the Angle-Bisector Theorem. Whenever you see a triangle with one of its angles bisected, you either have an isosceles triangle cut into two congruent triangles or a problem in which you might have to use the Angle-Bisector Theorem.

Chapter 7: Similarity: Size Doesn't Matter   **179**

Q. Given that $\overline{YA} \parallel \overline{DN}$, find LA

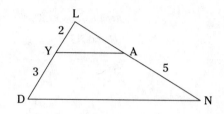

A. By the Side-Splitter Theorem,

$$\frac{LY}{YD} = \frac{LA}{AN}$$

$$\frac{2}{3} = \frac{LA}{5}$$

$$3 \cdot LA = 10$$

$$LA = \frac{10}{3} = 3\,\tfrac{1}{3}$$

Q. Given that $AC = 25$ and $FH = 36$, find BC, AB, IJ, GH, FG, and EF

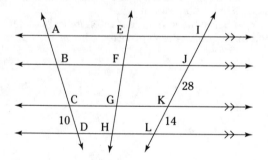

A. First set up a proportion to find BC:

$$\frac{BC}{CD} = \frac{JK}{KL}$$

$$\frac{BC}{10} = \frac{28}{14}$$

$$BC = 20$$

Now, because AC is 25 (given), AB must be 5.

$$\frac{IJ}{JK} = \frac{AB}{BC}$$

$$\frac{IJ}{28} = \frac{5}{20}$$

$$IJ = 7$$

To get FG and GH, note that because the ratio $KL:JK$ is $14:28$ or $1:2$, $GH:FG$ must also equal $1:2$. So let $GH = x$ and $FG = 2x$. Then, because $FH = 36$ (given),

$$x + 2x = 36$$

$$x = 12$$

Therefore, GH is 12 and FG is 24. Finally,

$$\frac{EF}{FG} = \frac{AB}{BC}$$

$$\frac{EF}{24} = \frac{5}{20}$$

$$EF = 6$$

180 Part III: Polygons

Q. Given: $\overrightarrow{AH}$ bisects $\angle CAS$
Find: CH and HS

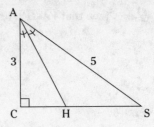

A. By the Angle-Bisector Theorem, $\dfrac{CH}{HS} = \dfrac{AC}{AS}$.

If you set CH equal to x, HS is 4 – x. (You saw that CS is 4, right?) Now substitute:

$$\dfrac{x}{4-x} = \dfrac{3}{5}$$
$$5x = 12 - 3x$$
$$8x = 12$$
$$x = \dfrac{12}{8} = 1.5$$

Thus, CH is 1.5 and HS is 2.5.

Warning: Don't make the mistake of thinking that when an angle in a triangle is bisected, the opposite side will also be cut exactly in half. You can see in this example that side $\overline{CS}$ is *not* bisected. The opposite side often comes very close to being bisected and it often *looks* bisected, but as a matter of fact, the opposite side is divided in half only when you bisect the vertex angle of an isosceles triangle.

18. Given: $\overline{OI} \parallel \overline{GN}$

 a. Prove: $\triangle RIO \sim \triangle RNG$ (paragraph proof)

 b. Find: IN and GN

Solve It

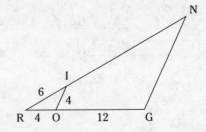

*19. Given: All avenues are parallel to one another.

Along Washington Blvd., it's $\frac{1}{2}$ mile from First Avenue to Fifth and $\frac{1}{10}$ mile from Third to Fourth.

Along Adams, it's $\frac{3}{8}$ mile from First to Third and from Third to Fifth.

Along Jefferson, it's $\frac{4}{5}$ mile from First to Fifth and $\frac{1}{4}$ mile from First to Second.

Find: All unknown distances between the avenues along Washington, Adams, and Jefferson and fill in the distances on the figure below. (I've started it for you on Washington and Jefferson.) You need only the distances from one avenue to the next. In other words, you don't have to calculate the distance from Second to Fourth unless you need it to find one of the smaller distances.

Solve It

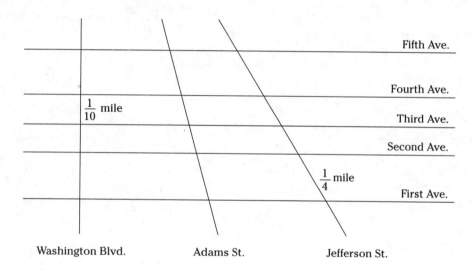

*20. Given: $\vec{AE}$ bisects ∠HAR

Perimeter of △HRA is 27

Find: ER and RA

Note: If you solve both problems 20 and 21, you're a geometry god!

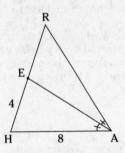

Solve It

**21. Super Challenge problem!

Given: $\vec{ZU}$ bisects ∠EZS

△ZES is isosceles with base $\overline{ZS}$

Dig: Those crazy lengths

Find: Angles 1 through 6; ZE, EU, and ZU; and the areas of △ZEU and △ZUS

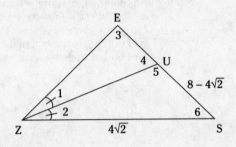

Solve It

Solutions for Similarity: Size Doesn't Matter

1 Here are your answers:

a. Corresponding angles of similar polygons are congruent, so finding the angles should be a cinch:

- $\angle FEH$ is 80° and thus, so is $\angle A$
- $\angle F = 120° = \angle B$
- $\angle C = 85° = \angle G$
- $\angle GHE = 75° = \angle D$
- $\angle GHD = 105°$

b. The ratio of the bases is $\frac{5}{11}$, so all the other ratios must also equal $\frac{5}{11}$:

$$\frac{FE}{6} = \frac{5}{11}$$
$$FE = \frac{30}{11}$$

Then,

$$\frac{4}{BC} = \frac{5}{11}$$
$$BC = \frac{44}{5}$$

Finally,

$$\frac{GH}{10} = \frac{5}{11}$$
$$GH = \frac{50}{11}$$

2 Here are the missing angles and lengths:

a. The angles in $\triangle PQR$ must add up to 180°, so $\angle Q$ is 64.4°. Now just pair up corresponding vertices: P with Z, Q with X, and R with Y. Thus, $\angle Z$ is 60°, $\angle X$ is 64.4°, and $\angle Y$ is 55.6°.

b. $\overline{QR}$ (second and third letters) corresponds to $\overline{XY}$ (second and third letters) and $\overline{PQ}$ corresponds to $\overline{ZX}$, so

$$\frac{QR}{XY} = \frac{PQ}{ZX}$$
$$\frac{21}{26.25} = \frac{20}{ZX}$$
$$ZX = \frac{20 \cdot 26.25}{21} = 25$$

$\overline{PR}$ corresponds to $\overline{ZY}$, so

$$\frac{PR}{27.5} = \frac{21}{26.25}$$
$$PR = \frac{27.5 \cdot 21}{26.25} = 22$$

3 Did you notice the trick? The base $\overline{AB}$ does *not* correspond to the base $\overline{LP}$. $\overline{AB}$ corresponds to $\overline{LM}$. So the expansion factor is $\frac{60}{5} = 12$, not $\frac{50}{5} = 10$. Thus,

$$\frac{\text{perimeter}_{LMNOP}}{\text{perimeter}_{ABCDE}} = \frac{LM}{AB}$$

$$\frac{\text{perimeter}_{LMNOP}}{30} = \frac{60}{5}$$

$$\text{perimeter}_{LMNOP} = 30 \cdot 12 = 360$$

4 And this is how problem 4 plays out:

a. From the way the similarity is written, you can see that $\overline{AC}$ (first and third letters in △ABC) pairs up with $\overline{AD}$ (first and third letters in △ACD) and that $\overline{BC}$ pairs up with $\overline{CD}$. This gives you the desired proportion:

$$\frac{AD}{AC} = \frac{CD}{BC}$$

$$\frac{AD}{12} = \frac{6}{9}$$

$$AD = \frac{72}{9} = 8$$

b. Another way to see how things pair up is to redraw the triangles so they're side by side and in the same orientation. Like this:

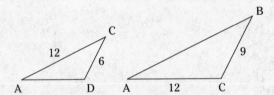

You can see that △ACD had to be flipped over to put it in the same orientation as △ABC.

To get *DB*, you first need *AB*:

$$\frac{\text{top}_{\triangle ABC}}{\text{top}_{\triangle ACD}} = \frac{\text{right side}_{\triangle ABC}}{\text{right side}_{\triangle ACD}}$$

$$\frac{AB}{AC} = \frac{CB}{DC}$$

$$\frac{AB}{12} = \frac{9}{6}$$

$$AB = \frac{108}{6} = 18$$

Finally, you can see in the first figure that $DB = AB - AD$, so $DB = 18 - 8 = 10$.

5 You know that $\angle R \cong \angle P$ by *if sides, then angles* (Chapter 4), so $\angle R$ is 70°. The angles in a triangle add up to 180°, so $\angle Q$ is 40°. You know $\angle S$ and $\angle T$ are also equal by *if sides, then angles*, and together they must sum to 140°, so each is 70°. Because both triangles contain a 40° angle and a 70° angle, △PQR ~ △STU by AA.

After finding that $\angle Q$ is 40°, you could also finish by setting *PQ* and *RQ* equal to *x* and setting *ST* and *UT* equal to *y*. That gives you $\frac{PQ}{ST} = \frac{RQ}{UT}$ (because $\frac{x}{y} = \frac{x}{y}$).

And because $\angle Q \cong \angle T$, △PQR ~ △STU by SAS~.

6

Statements	Reasons
1) △XRT is isosceles with base $\overline{XT}$	1) Given.
2) $\overline{XR} \cong \overline{TR}$	2) Definition of an isosceles triangle.
3) ∠NXR ≅ ∠NTR	3) If sides, then angles.
4) ∠1 ≅ ∠2	4) Given.
5) ∠NXR ≅ ∠NWG	5) Supplements of congruent angles are congruent.
6) ∠NWG ≅ ∠NTR	6) Transitive Property (3 and 5).
7) $\overline{RN}$ is an altitude	7) Given.
8) $\overline{RN} \perp \overline{XT}$	8) Definition of altitude.
9) ∠GNW is a right angle ∠RNT is a right angle	9) Definition of perpendicular.
10) ∠GNW ≅ ∠RNT	10) Right angles are congruent.
11) △WGN ~ △TRN	11) AA (6 and 10).

7 Here's how this problem unfolds:

a. All you have to check is whether

$$\frac{\text{short side}_{\triangle 1}}{\text{short side}_{\triangle 2}} \stackrel{?}{=} \frac{\text{medium side}_{\triangle 1}}{\text{medium side}_{\triangle 2}} \stackrel{?}{=} \frac{\text{long side}_{\triangle 1}}{\text{long side}_{\triangle 2}}$$

$$\frac{36}{9} \stackrel{?}{=} \frac{52}{13} \stackrel{?}{=} \frac{56}{14}$$

$$4 = 4 = 4. \quad \text{Check. Thus, by SSS~, the triangles are similar.}$$

b. *B* and *H* correspond because each is across from a medium-length side. *C* and *F* correspond because each is across from a long side. *D* and *G* are stuck with each other. Therefore, △BCD ~ △HFG.

8 Because $\overline{AT}$ is parallel to $\overline{OY}$, ∠BAT ≅ ∠BOY by *if parallel lines are cut by a transversal, then corresponding angles are congruent* (Chapter 5). Then, because both triangles contain ∠B, △BAT ~ △BOY by AA. Similar triangles have proportional sides, so

$$\frac{BA}{BO} = \frac{BT}{BY}$$

$$\frac{5}{BO} = \frac{5}{BY}$$

$$5 \cdot BY = 5 \cdot BO$$

$$BY = BO$$

$$\overline{BY} \cong \overline{BO}$$

Thus, △BOY is isosceles.

After showing the triangles similar, you can also reason that any triangle similar to an isosceles triangle must also be isosceles, because if you take an isosceles triangle and shrink it or expand it by some factor, the two equal sides remain equal sides.

9.

	Statements		Reasons
1)	V is the midpoint of $\overline{LI}$; S is the midpoint of $\overline{PI}$	1)	Given.
2)	$VI = \frac{1}{2}LI$; $SI = \frac{1}{2}PI$	2)	A midpoint divides a segment into two segments that are each half as long as the original segment. (This reason is true, of course, but I created this "theorem" to avoid having to go through the rigamarole I referred to in the last example in this section.)
3)	$\frac{VI}{LI} = \frac{1}{2}$; $\frac{SI}{PI} = \frac{1}{2}$	3)	Algebra.
4)	$\frac{VI}{LI} = \frac{SI}{PI}$	4)	Substitution.
5)	$\angle I \cong \angle I$	5)	Reflexive Property.
6)	$\triangle VIS \sim \triangle LIP$	6)	SAS~ (4 and 5).
7)	$\angle ELP \cong \angle IPL$	7)	Given.
8)	$\angle EPL \cong \angle ILP$	8)	Given.
9)	$\triangle PEL \sim \triangle LIP$	9)	AA (7 and 8).
10)	$\triangle VIS \sim \triangle PEL$	10)	Transitive Property for similar triangles.

10. The vertical angles are congruent and $\angle JKL \cong \angle NML$, so $\triangle JKL \sim \triangle NML$ by AA. Their sides are proportional; thus,

$$\frac{JL}{NL} = \frac{KL}{ML}$$

$$\frac{JL}{16} = \frac{6}{8}$$

$$JL = \frac{96}{8} = 12$$

Next, JK has to be 9 to make the perimeter of $\triangle JKL$ add up to 27. Finally,

$$\frac{MN}{JK} = \frac{ML}{KL}$$

$$\frac{MN}{9} = \frac{8}{6}$$

$$MN = \frac{72}{6} = 12$$

MN and JL are both 12, so of course, $\overline{MN} \cong \overline{JL}$.

11. You have to prove that a product equals another product — $(AD)^2$ is a product — so the tip in the example problem about looking for similar triangles applies. Thus, you look for triangles that contain AD, AB, and AC — namely $\triangle ABD$ and $\triangle ACD$ — and try to prove that they're similar. You know that $\angle 1 \cong \angle CBD$, so their supplements, $\angle ADC$ and $\angle ABD$, are congruent. Both triangles contain $\angle A$, so $\triangle ABD \sim \triangle ADC$ by AA. (Note the order of the vertices.) Now find a proportion that contains AB and AC and that uses AD twice, and you're done. Here it is:

$$\frac{\text{medium side}_{\triangle ABD}}{\text{medium side}_{\triangle ADC}} = \frac{\text{long side}_{\triangle ABD}}{\text{long side}_{\triangle ADC}}$$

$$\frac{AB}{AD} = \frac{AD}{AC}$$

$$(AD)^2 = (AB)(AC)$$

Chapter 7: Similarity: Size Doesn't Matter

*12

Statements	Reasons
1) E and O trisect $\overline{AS}$	1) Given.
2) $\overline{EO} \cong \overline{SO}$	2) Definition of trisect.
3) O bisects $\overline{RM}$	3) Given.
4) $\overline{RO} \cong \overline{MO}$	4) Definition of bisect.
5) $\angle ROE \cong \angle MOS$	5) Vertical angles are congruent.
6) $\triangle ROE \cong \triangle MOS$	6) SAS (2, 5, 4).
7) $\angle S \cong \angle MTI$	7) Given.
8) $\angle SMO \cong \angle TMI$	8) Vertical angles are congruent.
9) $\triangle MOS \sim \triangle MIT$	9) AA (7 and 8).
10) $\triangle ROE \sim \triangle MIT$	10) Substitution of $\triangle ROE$ for $\triangle MOS$ (lines 6 and 9).
11) $\frac{RO}{MI} = \frac{OE}{IT}$	11) CSSTP.
12) $RO \cdot IT = MI \cdot OE$	12) Means-extremes (cross-multiplication).

13 The triangles are similar, so they must have three pairs of congruent angles according to the definition of similar polygons. Thus, each triangle must contain angles with measures x, $x + 40$, and $x + 50$. These must add up to 180°, so

$$x + (x + 40) + (x + 50) = 180$$
$$3x + 90 = 180$$
$$3x = 90$$
$$x = 30$$

Thus, both triangles contain 30°, 70°, and 80° angles.

14 Here are the answers:

a. Always: You have only one triangle, so if $\triangle ABC \sim \triangle CBA$, $\triangle ABC$ must be congruent to $\triangle CBA$ as well. (Otherwise, $\triangle ABC$ and $\triangle CBA$ would be different sizes, which is impossible.) A (in $\triangle ABC$) corresponds to C (in $\triangle CBA$) and B corresponds to B, so $\overline{AB} \cong \overline{CB}$.

b. Sometimes: With only one pair of congruent angles, the triangles might be similar, but they don't have to be.

c. Always: Congruent triangles are automatically similar as well.

d. Sometimes: Similar triangles can be congruent, but they certainly don't have to be.

e. Never: $\triangle ABC$ contains a right angle. $\triangle DEF$ is acute, so it can't contain a right angle. Thus, the two triangles can't have three pairs of congruent angles, and therefore, they're not similar. (For more info on types of triangles, see Chapter 3.)

f. Sometimes: The answer would be *always* if you were told that $\overline{AC}$ and $\overline{DF}$ are the bases of these isosceles triangles, but the statement is only *sometimes* true because the triangles could look like this:

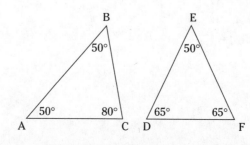

Don't forget: You can't rely on the appearance of the triangles in the figure and conclude, for example, that $\overline{AB} \cong \overline{CB}$ and $\overline{DE} \cong \overline{FE}$.

15 Here are the lengths:

a. Using the third part of the Altitude-on-Hypotenuse Theorem,

$$(JZ)^2 = (JA)(JY)$$
$$= 4 \cdot 13$$
$$= 52$$
$$JZ = \sqrt{52} = 2\sqrt{13} \approx 7.2$$

Then, using the second part of the theorem,

$$(AZ)^2 = (JA)(AY)$$
$$= 4 \cdot 9$$
$$= 36$$
$$AZ = 6$$

b. Using the third part of the theorem,

$$(JZ)^2 = (JA)(JY)$$
$$5^2 = 3 \cdot JY$$
$$JY = \frac{25}{3} = 8\frac{1}{3}$$

Thus,

$$AY = JY - JA$$
$$= \frac{25}{3} - 3$$
$$= \frac{16}{3} = 5\frac{1}{3}$$

Or, alternatively, first note that $AZ = 4$ because $\triangle JAZ$ is a 3-4-5 right triangle. Then finish with the second part of the theorem.

c. Using the third part of the theorem,

$$(YZ)^2 = (AY)(JY)$$
$$= 6 \cdot 8$$
$$= 48$$
$$YZ = \sqrt{48} = 4\sqrt{3} \approx 6.9$$

d. Using the second part of the theorem,

$$(AZ)^2 = (JA)(AY)$$
$$8^2 = JA \cdot 10$$
$$JA = \frac{64}{10} = 6.4$$

$$JY = JA + AY$$
$$= 6.4 + 10$$
$$= 16.4$$

e. Using the third part of the theorem,

$$(JZ)^2 = (JA)(JY)$$
$$8^2 = JA \cdot 12$$
$$JA = \frac{64}{12} = 5\frac{1}{3}$$

$$AY = JY - JA$$
$$= 12 - 5\frac{1}{3}$$
$$= 6\frac{2}{3}$$

16 The Pythagorean Theorem gives you QS:

$$(QS)^2 = (RQ)^2 + (RS)^2$$
$$= 5^2 + 10^2$$
$$= 125$$
$$QS = \sqrt{125} = 5\sqrt{5} \approx 11.2$$

You can now get QT with the third part of the Altitude-on-Hypotenuse Theorem — $(RQ)^2 = (QT)(QS)$ — and then get RT with the Pythagorean Theorem. But you don't have to do all that. You can get RT directly with a proportion from similar triangles:

$$\frac{\text{long leg}_{\triangle QRT}}{\text{long leg}_{\triangle QSR}} = \frac{\text{hypotenuse}_{\triangle QRT}}{\text{hypotenuse}_{\triangle QSR}}$$

$$\frac{RT}{RS} = \frac{RQ}{QS}$$

$$\frac{RT}{10} = \frac{5}{5\sqrt{5}}$$

$$RT = \frac{10}{\sqrt{5}} = 2\sqrt{5} \approx 4.5$$

*17 Set FL equal to x, and then use the last part of the Altitude-on-Hypotenuse Theorem:

$$(FG)^2 = (FL)(FA)$$
$$(6\sqrt{5})^2 = (x)(x+3)$$
$$180 = x^2 + 3x$$
$$x^2 + 3x - 180 = 0$$

Finish by factoring — $(x + 15)(x - 12) = 0$ — or with the quadratic formula. (If you forgot the quadratic formula, I don't want to hear about it. But you can refresh your memory right here.) Here's what the quadratic formula looks like in action:

For an equation in the form $ax^2 + bx + c = 0$,

$$x = \frac{-b \pm \sqrt{b^2 - 4ac}}{2a}$$

$$= \frac{-3 \pm \sqrt{3^2 - 4(1)(-180)}}{2(1)}$$

$$= \frac{-3 \pm \sqrt{729}}{2}$$

$$= \frac{-3 \pm 27}{2}$$

$$= -15 \text{ or } 12$$

FL can't be negative, so FL is 12.

18 Here are the answers:

a. Because $\overline{OI} \| \overline{GN}$, $\angle ROI \cong \angle RGN$ and $\angle RIO \cong \angle RNG$ by *if parallel lines are cut by a transversal, then corresponding angles are congruent*. Thus, $\triangle RIO \sim \triangle RNG$ by AA. (You can also use $\angle R \cong \angle R$ for one of the two pairs of congruent angles.)

TIP Whenever you see parallel lines in a problem involving two or more triangles, the odds are good that some of the triangles are similar (or maybe congruent).

b. The Side-Splitter Theorem gives you *IN*:

$$\frac{RO}{OG} = \frac{RI}{IN}$$

$$\frac{4}{12} = \frac{6}{IN}$$

$$4 \cdot IN = 72$$

$$IN = 18$$

For *GN*, did you fall for my trap? *GN* is not 12, though it sure looks like it should be. The ratio of *OI* : *GN* doesn't equal the 4 : 12 ratio of *RO* : *OG*. Remember, the Side-Splitter Theorem doesn't work for the parallel sides. To get *GN*, you have to use the proportional sides of similar triangles *RIO* and *RNG* (*RG* = *RO* + *OG*, or 4 + 12):

$$\frac{\text{right side}_{\triangle RNG}}{\text{right side}_{\triangle RIO}} = \frac{\text{base}_{\triangle RNG}}{\text{base}_{\triangle RIO}}$$

$$\frac{GN}{OI} = \frac{RG}{RO}$$

$$\frac{GN}{4} = \frac{16}{4}$$

$$GN = 16$$

19 Along Adams, the distance from First to Third equals the distance from Third to Fifth. That's a 1 : 1 ratio. According to the transversals theorem, that 1 : 1 ratio must also hold for Washington and Jefferson.

The whole trip on Washington is half a mile, so you'd have to go 0.25 miles from First to Third and from Third to Fifth. Because it's 0.1 mile from Third to Fourth, Washington runs 0.25 − 0.1, or 0.15 miles, from Fourth to Fifth. The whole trip on Jefferson is 0.8 miles, so each half trip is 0.4 miles. Subtracting the 0.25 mile along First to Second from 0.4 gives you 0.15 miles for the distance from Second to Third.

Now use the distances along Washington from Third to Fourth (0.1 miles) and from Fourth to Fifth (0.15 miles) to get the corresponding distances along Adams and Jefferson. Along Washington, you have a ratio 0.1 : 0.15, which equals 10 : 15, or 2 : 3. Using that ratio on Adams gives you

$$2x + 3x = \frac{3}{8}$$

$$5x = \frac{3}{8}$$

$$x = \frac{3}{40}$$

So the distance along Adams from Third to Fourth ($2x$) is $2\left(\frac{3}{40}\right)$, or $\frac{6}{40}$, or 0.15 miles (that's three times in a row for 0.15 — what a bizarre coincidence! — I didn't plan it that way). From Fourth to Fifth ($3x$) is $3\left(\frac{3}{40}\right)$, or $\frac{9}{40}$, or 0.225 miles.

The calculation works the same along Jefferson, so

$$2x + 3x = 0.4$$

$$5x = 0.4$$

$$x = 0.08$$

Chapter 7: Similarity: Size Doesn't Matter **191**

So the distance along Jefferson is 2(0.08), or 0.16 miles, from Third to Fourth, and 3(0.08), or 0.24 miles, from Fourth to Fifth.

Finally, use the same method with the distances along Jefferson from First to Second (0.25 miles) and from Second to Third (0.15 miles) to get the corresponding distances along Washington and Adams. The ratio along Jefferson is 0.25 : 0.15, which equals 5 : 3. So for Washington, you have

$$5x + 3x = \tfrac{1}{4} \text{ mile}$$
$$8x = \tfrac{1}{4} \text{ mile}$$
$$x = \tfrac{1}{32} \text{ mile}$$

Thus, it's $5\left(\tfrac{1}{32}\right)$ or $\tfrac{5}{32}$ mile from First to Second and $3\left(\tfrac{1}{32}\right)$ or $\tfrac{3}{32}$ mile from Second to Third. For Adams, you have

$$5x + 3x = \tfrac{3}{8} \text{ mile}$$
$$x = \tfrac{3}{64} \text{ mile}$$

So the distance along Adams is $\tfrac{15}{64}$ mile from First to Second and $\tfrac{9}{64}$ mile from Second to Third. That's it. *Finito!*

Street	Washington	Adams	Jefferson
Fourth to Fifth	3/20 (0.15)	9/40 (0.225)	6/25 (0.24)
Third to Fourth	1/10 (0.1)	3/20 (0.15)	4/25 (0.16)
Second to Third	3/32 (0.09375)	9/64 (0.140625)	3/20 (0.15)
First to Second	5/32 (0.15625)	15/64 (0.234375)	1/4 (0.25)
Total (miles)	1/2 (0.5)	3/4 (0.75)	4/5 (0.8)

* **20** Don't forget: When an angle in a triangle is bisected, the opposite side is *not* bisected (unless it's the base of an isosceles triangle). Therefore, you can't conclude that *ER* is 4 or that *RA* is 11.

Set *ER* equal to *x*. Then, because the perimeter of △ *HRA* is 27 and because *AH* and *HE* add up to 12, *ER* and *RA* have to add up to 15. So *RA* is 15 – *x*. Now use the Angle-Bisector Theorem:

$$\tfrac{x}{4} = \tfrac{15-x}{8}$$
$$8x = 60 - 4x$$
$$12x = 60$$
$$x = 5$$

Thus, *ER* is 5 and *RA* is 10.

** **21** △ *ZES* is isosceles with base $\overline{ZS}$, so if you let $\overline{ZE}$ equal *x*, *ES* is also *x*; thus, *EU* is $x - \left(8 - 4\sqrt{2}\right)$, or $x - 8 + 4\sqrt{2}$. Now just use the Angle-Bisector Theorem:

$$\tfrac{EU}{US} = \tfrac{ZE}{ZS}$$
$$\tfrac{x - 8 + 4\sqrt{2}}{8 - 4\sqrt{2}} = \tfrac{x}{4\sqrt{2}}$$

Cross-multiplying gives you

$$4x\sqrt{2} - 32\sqrt{2} + 32 = 8x - 4x\sqrt{2}$$
$$8x\sqrt{2} - 8x = 32\sqrt{2} - 32$$
$$x\sqrt{2} - x = 4\sqrt{2} - 4$$
$$x(\sqrt{2} - 1) = 4(\sqrt{2} - 1)$$
$$x = 4$$

ZE is thus 4, as is *ES*. And then $EU = 4 - (8 - 4\sqrt{2})$, which equals $4\sqrt{2} - 4$.

The three sides of △*ZES* are 4, 4, and $4\sqrt{2}$, so it's a 45°- 45°- 90° triangle. (See Chapter 3 for more on special right triangles.) After you see that, all the angles are easy: ∠3 is 90°, ∠6 is 45°, and ∠1 and ∠2 are each half of 45°, or 22.5°. Because ∠1, ∠3, and ∠4 must sum to 180°, ∠4 has to be 67.5°. ∠5 and ∠4 are supplements, so ∠5 is 112.5°.

Because ∠3 is a right angle, you can get *ZU* with the Pythagorean Theorem:

$$(ZU)^2 = (ZE)^2 + (EU)^2$$
$$= 4^2 + (4\sqrt{2} - 4)^2$$
$$= 16 + (32 - 32\sqrt{2} + 16)$$
$$= 64 - 32\sqrt{2}$$
$$ZU = \sqrt{64 - 32\sqrt{2}}$$
$$= \sqrt{16(4 - 2\sqrt{2})}$$
$$= 4\sqrt{4 - 2\sqrt{2}}$$

Is that a thing of beauty or what?

Finally, the areas are a piece o' cake. △*ZEU* is a right triangle with a base of $4\sqrt{2} - 4$ and a height of 4, so its area ($\frac{1}{2}$base · height) is $8\sqrt{2} - 8$ units². △*ZUS* has the same height of 4 and a base ($\overline{US}$) that measures $8 - 4\sqrt{2}$, so its area is $16 - 8\sqrt{2}$ units².

Part IV
Circles

In this Part . . .

Now you graduate to the ∞-gon (more commonly known as the circle). In all seriousness, the circle really is like a polygon with an infinite number of sides. (Using this idea, the ancient Greeks were able to come up with a very good approximation for π.)

In this part, you practice dozens of problems involving all sorts of nifty circle concepts: the angle-arc theorems, the power theorems, formulas for sectors, arc length, area, circumference, and so on.

Chapter 8
Circular Reasoning

In This Chapter
▶ Circle theorems that use radii and chords
▶ Finding congruent arcs, chords, and central angles
▶ Going off on a tangent

The circle is a paradox of sorts. In one sense, it's the simplest of all shapes, but at the same time, it's rich in complexity and difficult, advanced ideas. Mathematicians have been fascinated by its properties for well over 2,000 years. For example, the ratio of a circle's circumference to its diameter — π (≈ 3.14) — is one of the most important and often-used numbers in all of mathematics.

In this chapter, you study several circle properties by doing proofs. All the proofs here involve circles, but you also see many of the same ideas from earlier chapters like *if angles, then sides* and CPCTC (both in Chapter 4).

The Segments Within: Radii and Chords

In this section, you do proofs involving radii and chords (including diameters).

- **Radius:** Nothing about a circle is more fundamental than its radius. A circle's *radius* — the distance from its center to a point on the circle — tells you its size. In addition to being a measure of distance, a radius is also a segment from a circle's center to a point on the circle.
- **Chord:** A segment that connects two points on a circle is called a *chord*.
- **Diameter:** A chord passing through a circle's center is a *diameter* of the circle. A circle's diameter, as I'm sure you know, is twice as long as its radius.

Here are five circle theorems for your mathematical pleasure (some are the converses of each other):

- **Radii size:** All radii of a circle are congruent (yet another *well-duh* theorem).
- **Perpendicularity and bisected chords:**
  - If a radius is perpendicular to a chord, then it bisects the chord.
  - If a radius bisects a chord (that's not a diameter), then it's perpendicular to the chord.
- **Distance and chord size:**
  - If two chords of a circle are equidistant from the center of the circle, then they're congruent.
  - If two chords of a circle are congruent, then they're equidistant from its center.

Part IV: Circles

If you're looking for tips for completing circle proofs, you've come to the right place. Here are a few that you can put to use in this section:

1. **Draw additional radii on the figure.**

   You should draw radii to points where something else intersects or touches the circle, as opposed to just any old point on the circle.

2. **Open your eyes and notice all the radii — including new ones you've drawn — and mark them all congruent.**

   For some strange reason — despite the fact that *all radii are congruent* is one of the simplest of all theorems — it's very common for people to either fail to notice all the radii in a problem or fail to note that they're all congruent.

3. **Draw in the segment (part of a radius) that goes from the center of the circle to a chord and that's perpendicular to the chord (and which, according to the previous theorem, bisects the chord).**

The purpose of adding radii and partial radii is usually to create right triangles or isosceles triangles that you can then use to solve the problem.

Q. Given: Circle O
$\overline{XY} \cong \overline{ZY}$

Prove: $\overrightarrow{YO}$ bisects $\angle XYZ$

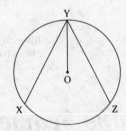

A.

Statements	Reasons
1) Circle O	1) Given.
2) Draw $\overline{OX}$ and $\overline{OZ}$	2) Two points determine a segment.
3) $\overline{OX} \cong \overline{OZ}$	3) All radii are congruent.
4) $\overline{XY} \cong \overline{ZY}$	4) Given.
5) $\overline{YO} \cong \overline{YO}$	5) Reflexive.
6) $\triangle YOX \cong \triangle YOZ$	6) SSS (3, 4, 5).
7) $\angle XYO \cong \angle ZYO$	7) CPCTC.
8) $\overrightarrow{YO}$ bisects $\angle XYZ$	8) Definition of bisect.

Chapter 8: Circular Reasoning

1. Given: Circle S
$\angle P \cong \angle R$
Prove: $\overline{PX} \cong \overline{RX}$

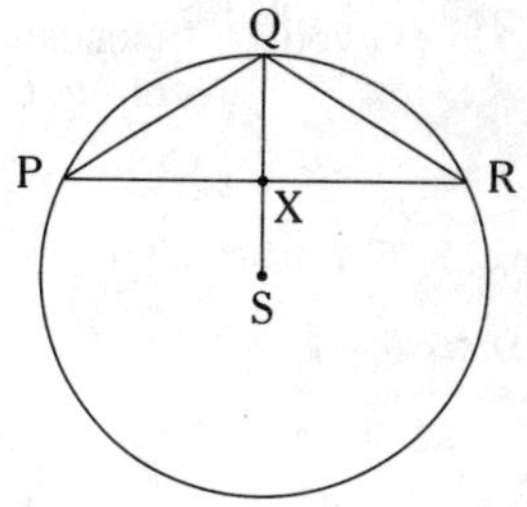

Solve It

Statements	Reasons

2. Use the figure from problem 1, but this time
Given: Circle S
$\overline{QS} \perp \overline{PR}$
Prove: $\overrightarrow{QS}$ bisects $\angle PQR$

Solve It

Statements	Reasons

3. Given: Isosceles trapezoid *ISTR* with bases $\overline{ST}$ and $\overline{IR}$, inscribed in circle *Q*

Circle *Q* has a radius of 5

Find: The area of *ISTR*

Solve It

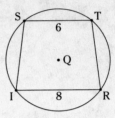

*4. Given: Circle *K*

$\triangle FLT \cong \triangle AYT$

$\overline{KI} \perp \overline{FY}$ and $\overline{KE} \perp \overline{AL}$

Prove: *KITE* is a kite

Hint: Use two of the five theorems from this section.

Solve It

Statements	Reasons

Introducing Arcs and Central Angles

Arcs, chords, and central angles are three peas in a pod.

- **Arc:** An *arc,* as you may know, is simply a curved piece of a circle. Every chord cuts a circle into two arcs: a *minor arc* (the smaller piece) and a *major arc* (the larger), unless the chord is a diameter, in which case both arcs are semicircles.

- **Central angle:** A *central angle* is an angle whose vertex is at the center of a circle. The two sides of a central angle are radii that hit the circle at the opposite ends of an arc or, as mathematicians say, the sides *intercept* an arc. The measure of an arc is the same as the degree measure of the central angle that intercepts it. (For more on arcs, chords, and angles that intercept an arc, please see Chapter 9.)

Congruent circles: Before I get into theorems, here's one more definition: *Congruent circles* are circles with congruent radii.

I know how much you love theorems, so here are six more. But don't sweat it — these six theorems are really just six variations on one simple idea about arcs, chords, and central angles.

- **Central angles and arcs:**
  - If two central angles of a circle (or of congruent circles) are congruent, then their intercepted arcs are congruent. (Short form: If central angles, then arcs.)
  - If two arcs of a circle (or of congruent circles) are congruent, then the corresponding central angles are congruent. (Short form: If arcs, then central angles.)

- **Central angles and chords:**
  - If two central angles of a circle (or of congruent circles) are congruent, then the corresponding chords are congruent. (Short form: If central angles, then chords.)
  - If two chords of a circle (or of congruent circles) are congruent, then the corresponding central angles are congruent. (Short form: If chords, then central angles.)

- **Arcs and chords:**
  - If two arcs of a circle (or of congruent circles) are congruent, then the corresponding chords are congruent. (Short form: If arcs, then chords.)
  - If two chords of a circle (or of congruent circles) are congruent, then the corresponding arcs are congruent. (Short form: If chords, then arcs.)

200 Part IV: Circles

Q. Given: Circle S
∠P ≅ ∠R

Prove: ∠PSQ ≅ ∠RSQ

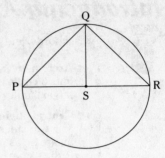

A.

Statements	Reasons
1) Circle S	1) Given.
2) ∠P ≅ ∠R	2) Given.
3) $\overline{PQ} \cong \overline{RQ}$	3) If angles, then sides.
4) ∠PSQ ≅ ∠RSQ	4) If chords, then central angles.

Short and sweet. Try doing this proof with congruent triangles instead. It should take you three extra steps.

---

5. Given: Circle Q
$\overline{AC} \cong \overline{RS}$

Prove: $\overline{AR} \cong \overline{CS}$

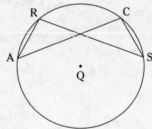

Warning: Note that the first four theorems in this section involve central angles (angles with a vertex at the center of a circle) and that, therefore, they *do not* apply to the angles in the figure. (I cover angles like these in Chapter 9.)

Statements	Reasons

Chapter 8: Circular Reasoning 201

6. Given: Circle Z
$\overline{BA} \cong \overline{CD}$
Prove: $\angle B \cong \angle C$

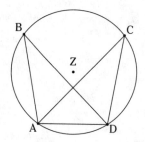

Statements	Reasons

7. Given: Circle $F \cong$ circle U
$\overset{\frown}{MO} \cong \overset{\frown}{IR}$
Prove: MIUF is a parallelogram

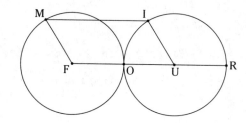

Statements	Reasons

8. Given: Circle $I \cong$ circle L

S is the midpoint of $\overline{CK}$

Prove: $\overline{CR} \parallel \overline{EK}$

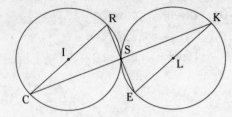

Statements	Reasons

Touching on Radii and Tangents

One of the important ideas in this section is related to something you've seen since you were a little kid. Look at either wheel in the bicycle in Figure 8-1.

Figure 8-1: Two wheels that are tangent to the ground — take a break from geometry and go for a ride.

A line is *tangent* to a circle if it touches the circle at a single point. The important point for this section is that the spoke that goes straight down from the hub in each wheel is *perpendicular* to the ground. Geometrically speaking, the bicycle wheels are, of course, circles, the spokes are radii, the single points where the wheels touch the ground are called *points of tangency,* and the ground is a *tangent* or *tangent line.*

Chapter 8: Circular Reasoning 203

Tangent/radius perpendicularity: A tangent line is perpendicular to the radius drawn to the point of tangency.

Don't forget this important fact! You already know how important it is to notice that all radii in a circle are congruent (and to sometimes draw in more radii). Now you can add this point about radii (and tangents) to your list of critical things to remember. The right angle at the point of tangency often becomes part of a right triangle.

Here's one more fact about tangents before I go through a couple example problems:

Dunce Cap Theorem: Two tangent segments drawn to a circle from the same point are congruent. This is known (by me) as the Dunce Cap Theorem. See Figure 8-2.

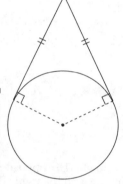

Figure 8-2:
Both sides of a dunce cap are the same length.

The first example and the first practice problem are called *common-tangent problems*, which means a single line is tangent to both circles. The example involves a common *external* tangent (the tangent lies on the same side of both circles). The practice problem involves a common *internal* tangent (the tangent line goes between the two circles). The solution method is the same for both:

1. **First:** Draw both the segment connecting the centers of the two circles and the two radii to the points of tangency (if these segments haven't already been drawn for you).

2. **This is the critical step:** From the *center of the smaller* circle, draw a segment parallel to the common tangent till it hits the radius of the larger circle (or the extension of the radius in a common internal tangent problem).

3. **Finish:** You now have a right triangle and a rectangle and can finish the problem with the Pythagorean Theorem and the simple fact that opposite sides of a rectangle are congruent.

In a common-tangent problem, the segment connecting the centers of the circles is *always* the hypotenuse of a right triangle and the common tangent is *always* the side of a rectangle.

In a common tangent problem, the segment connecting the centers of the circles is *never* one side of a right angle.

204 Part IV: Circles

Q. Given: $\overline{ET}$ is tangent to circle L and circle B with radii as shown

The distance between the centers of the circles is 25

Find: The length of the common tangent, $\overline{ET}$

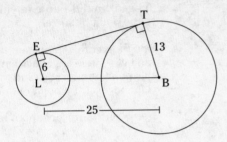

A. The segment connecting the centers of the two circles, as well as the two radii to the points of tangency (step 1 in solving common-tangent problems), are already drawn for you here. Draw the segment described in step 2 of the solution method (from the center of the smaller circle, parallel to the common tangent). See the following:

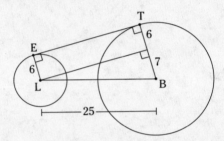

You can see that this new segment creates a rectangle and a right triangle. The opposite sides of a rectangle are congruent, so the radius of 6 on the left gives you the 6 on the right. The larger radius is 13, and 13 – 6 is 7, so you get 7 for the short leg of the right triangle. Its hypotenuse is 25, so you have a 7-24-25 right triangle, and thus the long leg is 24 (see info on Pythagorean triples in Chapter 3). Finally, the long leg of the triangle is also a side of the rectangle, which is congruent to the opposite side, $\overline{ET}$. ET is thus 24 as well.

Q. And now for something completely different, a so-called *walk-around* problem:

Given: Circle T, circle A, and circle N are tangent circles

$TN = 30$, $AN = 17$, and $TA = 23$

Find: The lengths of the radii of the three circles

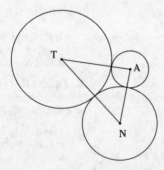

A. Pick a radius, any radius. How about one of the radii of circle T? Call it x. The other radius of circle T is also x, of course. Now you "walk around" $\triangle TAN$, filling in the lengths of the other four radii. In this problem, I walk around the triangle clockwise, though either way would work. TA is 23, so if the radius of circle T is x, the radius of circle A has to be $23 - x$. Then, turning the corner at A, the other radius of circle A is also $23 - x$. Next, because AN is 17, the radius of circle N is $17 - (23 - x)$, or $x - 6$. Turning the corner at N — you're on the home stretch — the other radius of circle N is also $x - 6$. See the following figure:

You can now finish by using the length of $\overline{TN}$:

$$x + (x - 6) = 30$$
$$2x = 36$$
$$x = 18$$

So the radius of circle T is 18, and you get the other radii by plugging in:

Radius$_{Circle\ A}$ = $23 - x = 23 - 18 = 5$

Radius$_{Circle\ N}$ = $x - 6 = 18 - 6 = 12$

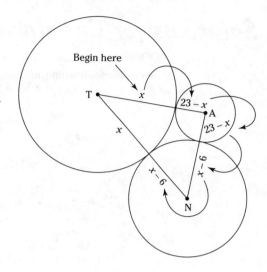

9. Given: $\overline{IT}$ is a common internal tangent of circles S and L with radii as shown

A distance of 5 separates the circles

Find: IT (Get it?)

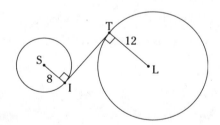

Solve It

10. Given: Diagram as shown

$\overline{WR}$, $\overline{WL}$, $\overline{TL}$, and $\overline{TR}$ are tangent to circle O

Find: TR

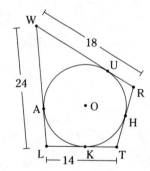

Solve It

Solutions for Circular Reasoning

1 **Game plan:** You see a circle, so think *radii, radii, radii!* Draw in congruent radii $\overline{SP}$ and $\overline{SR}$. You also see an isosceles triangle, and, voilà, you can use *if angles, then sides* to get $\overline{PQ} \cong \overline{RQ}$. Then you can finish with the equidistance theorem (which first appears in Chapter 4).

Statements	Reasons
1) Circle S	1) Given.
2) Draw $\overline{SP}$ and $\overline{SR}$	2) Two points determine a segment.
3) $\overline{SP} \cong \overline{SR}$	3) All radii are congruent.
4) $\angle P \cong \angle R$	4) Given.
5) $\overline{PQ} \cong \overline{RQ}$	5) If angles, then sides.
6) $\overline{QS}$ is the perpendicular bisector of $\overline{PR}$	6) If two points are each equidistant from the endpoints of a segment, then they determine the perpendicular bisector of that segment.
7) $\overline{PX} \cong \overline{RX}$	7) Definition of bisect.

2 **Game plan:** You see a circle, so what should you think? Yep, you got it: *radii, radii, radii!* Well, sorry to disappoint you, but this is one circle problem where you don't need to use extra radii or congruent radii. But don't let up with the radii mantra. It'll serve you well.

You have a radius perpendicular to a chord, so the theorem tells you the chord is bisected. You can use that fact, the right angles, and the Reflexive Property to get the triangles congruent with SAS. Then you finish with — what else? — CPCTC (for more on congruent triangles, see Chapter 4).

Statements	Reasons
1) Circle S with $\overline{QS} \perp \overline{PR}$	1) Given.
2) $\overline{QS}$ bisects $\overline{PR}$	2) If a radius is perpendicular to a chord, then it bisects the chord.
3) $\overline{PX} \cong \overline{RX}$	3) Definition of bisect.
4) $\angle PXQ$ is a right angle $\angle RXQ$ is a right angle	4) Definition of perpendicular.
5) $\angle PXQ \cong \angle RXQ$	5) All right angles are congruent.
6) $\overline{QX} \cong \overline{QX}$	6) Reflexive.
7) $\triangle PQX \cong \triangle RQX$	7) SAS (3, 5, 6).
8) $\angle PQX \cong \angle RQX$	8) CPCTC.
9) $\overrightarrow{QS}$ bisects $\angle PQR$	9) Definition of bisect.

3 **Game plan:** *Radii, radii, radii, radii!* That's right — draw in four of 'em to *I, S, T,* and *R* (actually, you need only two, but it's probably easier to see how the problem works if you draw all four of them; and in any event, when it comes to drawing radii, too many is better than too few). Now you have two isosceles triangles, $\triangle IQR$ and $\triangle SQT$. Next, draw in the altitudes of these triangles. See the following figure:

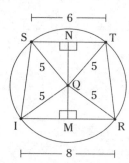

If a radius is perpendicular to a chord, it bisects the chord, so these altitudes bisect the bases of the triangles. That makes *IM* 4 and *SN* 3. And that makes both △*IMQ* and △*SNQ* 3-4-5 triangles (check out Chapter 3 for Pythagorean triples). Pretty sweet, eh? *QM* is thus 3 and *QN*, 4, so *NM* is 7, and that's the height of the trapezoid *ISTR*. You're all set to use the area formula (Chapter 6 provides info on calculating the area of quadrilaterals):

$$\text{Area}_{\text{Trap.}} = \frac{b_1 + b_2}{2} \cdot h$$
$$= \frac{8 + 6}{2} \cdot 7$$
$$= 7 \cdot 7$$
$$= 49 \text{ units}^2$$

Extra credit (well, maybe not exactly *credit*): Show that △*SQI* and △*TQR* are 45°- 45°- 90° triangles. You have two totally different ways of doing this. (This sure is a cool isosceles trapezoid, isn't it: The way it's made up of four congruent 3-4-5 triangles and two congruent 45°- 45°- 90° triangles?)

*★ 4   **Abbreviated game plan:** Why would you be given congruent triangles? It's got to be so you can use CPCTC. You should notice the perpendicular segments drawn to the chords and think about how you could show the chords to be congruent. And you should, as always, also think about what you need at the end of the proof (namely, the two pairs of congruent sides that make a kite) and what you need to do to get there.

Statements	Reasons
1) △*FLT* ≅ △*AYT*	1) Given.
2) $\overline{FT} \cong \overline{AT}$	2) CPCTC.
3) $\overline{LT} \cong \overline{YT}$	3) CPCTC.
4) $\overline{FY} \cong \overline{AL}$	4) Segment addition.
5) $\overline{KI} \perp \overline{FY}$ $\overline{KE} \perp \overline{AL}$	5) Given.
6) $\overline{KI} \cong \overline{KE}$	6) If chords of a circle ($\overline{FY}$ and $\overline{AL}$) are congruent, then they are equidistant from its center.
7) $\overline{KI}$ bisects $\overline{FY}$ $\overline{KE}$ bisects $\overline{AL}$	7) If a radius is perpendicular to a chord, then it bisects the chord.
8) $\overline{IY} \cong \overline{EL}$	8) Like divisions. (Remember that? I knew you would. It's from Chapter 2.)
9) $\overline{IT} \cong \overline{ET}$	9) Segment subtraction (3 and 8).
10) *KITE* is a kite	10) Definition of kite.

5

Statements	Reasons
1) Circle Q $\overline{AC} \cong \overline{RS}$	1) Given.
2) $\widehat{AC} \cong \widehat{RS}$	2) If chords, then arcs.
3) $\widehat{AR} \cong \widehat{CS}$	3) Subtraction (subtracting $\widehat{RC}$ from both $\widehat{AC}$ and $\widehat{RS}$).
4) $\overline{AR} \cong \overline{CS}$	4) If arcs, then chords.

6

Statements	Reasons
1) Circle Z $\overline{BA} \cong \overline{CD}$	1) Given.
2) $\widehat{BA} \cong \widehat{CD}$	2) If chords, then arcs.
3) $\widehat{BD} \cong \widehat{CA}$	3) Addition (of $\widehat{AD}$).
4) $\overline{BD} \cong \overline{CA}$	4) If arcs, then chords.
5) $\overline{AD} \cong \overline{DA}$	5) Reflexive.
6) $\triangle ABD \cong \triangle DCA$	6) SSS (1, 4, 5).
7) $\angle B \cong \angle C$	7) CPCTC.

7

Statements	Reasons
1) Circle $F \cong$ Circle U $\widehat{MO} \cong \widehat{IR}$	1) Given.
2) $\angle MFO \cong \angle IUR$	2) If arcs, then central angles.
3) $\overline{MF} \parallel \overline{IU}$	3) If corresponding angles are congruent, then lines are parallel.
4) $\overline{MF} \cong \overline{IU}$	4) Congruent circles have congruent radii (definition of congruent circles).
5) $MIUF$ is a parallelogram	5) If a quadrilateral has a pair of sides that are both parallel and congruent, then the quadrilateral is a parallelogram.

8

Statements	Reasons
1) Circle $I \cong$ Circle L; S is the midpoint of $\overline{CK}$	1) Given.
2) $\overline{CS} \cong \overline{KS}$	2) Definition of midpoint.
3) $\overset{\frown}{CS} \cong \overset{\frown}{KS}$	3) If chords, then arcs.
4) $\angle CIR \cong \angle KLE$	4) Straight angles are congruent.
5) $\overset{\frown}{CSR} \cong \overset{\frown}{KSE}$	5) If central angles, then arcs.
6) $\overset{\frown}{SR} \cong \overset{\frown}{SE}$	6) Arc subtraction (subtracting congruent arcs from congruent arcs).
7) $\overline{SR} \cong \overline{SE}$	7) If arcs, then chords.
8) $\angle CSR \cong \angle KSE$	8) Vertical angles are congruent.
9) $\triangle CSR \cong \triangle KSE$	9) SAS (2, 7, 6).
10) $\angle C \cong \angle K$	10) CPCTC.
11) $\overline{CR} \parallel \overline{EK}$	11) If alternate interior angles are congruent, then lines are parallel.

Egad! I just saw a much better and easier way of doing this proof. Despite all my years of teaching geometry, I failed to follow my own advice about drawing in more radii. The preceding proof is a good illustration of some of the theorems in this section, but when it comes to proofs, the shorter the better. To wit —

Statements	Reasons
1) Circle $I \cong$ Circle L; S is the midpoint of $\overline{CK}$	1) Given.
2) $\overline{CS} \cong \overline{KS}$	2) Definition of midpoint.
3) Draw $\overline{IS}$ and $\overline{LS}$	3) Two points determine a segment.
4) $\overline{IS} \cong \overline{LS}$	4) Congruent circles have congruent radii.
5) $\overline{IC} \cong \overline{LK}$	5) Congruent circles have congruent radii.
6) $\triangle ICS \cong \triangle LKS$	6) SSS (2, 4, 5).
7) $\angle C \cong \angle K$	7) CPCTC.
8) $\overline{CR} \parallel \overline{EK}$	8) If alternate interior angles are congruent, then lines are parallel.

I guess you're never too old to learn.

9. First, draw $\overline{SL}$. The middle portion of $\overline{SL}$ is the distance between the circles, and you were given that that distance is 5 units; therefore, SL is 8 + 5 + 12, or 25. Next, from the center of the *smaller* circle, S, draw a segment parallel to $\overline{IT}$ till it hits the extension of radius $\overline{TL}$. Your diagram should now look like this:

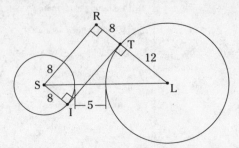

The segment you just drew, $\overline{SR}$, creates a rectangle and a large right triangle. The rectangle has opposite sides of 8. The right triangle has a leg of 8 + 12 = 20 and a hypotenuse of 25. That's in the 3:4:5 family of triangles, so SR is 15, and that makes IT 15. That's it.

10. Set HT equal to x, and walk around clockwise. Both sides of a dunce cap are equal, so KT is also x. KL is then $14 - x$, as is AL. Next, AW is $24 - (14 - x)$, which is $x + 10$. UW is also $x + 10$, and that makes UR equal to $18 - (x + 10)$, or $8 - x$. Finally, HR is also $8 - x$, and because HT is x, TR is $(8 - x) + x$, which is 8. That does it.

Chapter 9
Scintillating Circle Formulas

In This Chapter
- Not quite coming full circle: sectors and segments
- Dealing with angles and their intercepted arcs
- Powering up with power theorems

If you're fully up-to-date on circle proofs, you're ready to move on to some handy circle formulas that help you calculate everything from area to arc length. In this chapter, you find the area and the perimeter of various sections of a circle. You also discover the relationships between angles whose vertices lie on, inside, and outside a circle and the intercepted arcs of these types of angles.

Pizzas, Slices, and Crusts: Area and "Perimeter" of Circles, Sectors, and Segments

In this section, you work on problems involving the area and the circumference/perimeter of circles and parts of circles. (By the way, if the word *segment* in the heading is throwing you, you're not alone. *Segment* is the name of a particular section of a circle, which I show you in a minute. Don't ask me why the bozo who coined this term had to confuse matters by reusing a word with another meaning.)

To start things off, here are two theorems — one about the area of a sector and a related theorem about the length of an arc. Don't worry: Both theorems are based on a very simple idea and are easier than they look. The mathematical examples below each of these theorems correspond to Figure 9-1.

- **Arc length:** The length of an arc (part of the circumference, like $\overarc{AB}$) is equal to the circumference of the circle (πd) times the fraction of the circle represented by the arc. $m\overarc{AB}$ is $30°$, so

$$\text{Length}_{\overarc{AB}} = \left(\frac{m\overarc{AB}}{360}\right)\pi d$$
$$= \frac{30}{360} \cdot \pi \cdot 18$$
$$= \frac{1}{12} \cdot \pi \cdot 18$$
$$= 1.5\pi$$

✔ **Sector area:** The area of a sector (a wedge shape, like sector *AOB* in Figure 9-1) is equal to the area of the circle (πr^2) times the fraction of the circle represented by the sector.

$$\text{Area}_{\text{Sector } AOB} = \left(\frac{m\widehat{AB}}{360}\right)\pi r^2$$

$$= \frac{1}{12} \cdot \pi \cdot 9^2$$

$$= 6.75\pi \text{ units}^2$$

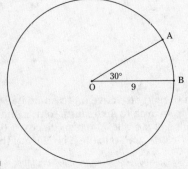

Figure 9-1: A sector and an arc that make up one-twelfth of the circle.

These two theorems are so simple, you may want to do what I do — ignore them. I don't mean ignore the ideas; I mean you don't need to memorize the theorems. Your common sense should tell you that the length of $\widehat{AB}$ above is one-twelfth of the circumference of circle *O* because 30° goes into 360° twelve times. Likewise, the area of sector *AOB* is one-twelfth of the area of circle *O*. When common sense suffices, why clutter your mind with more formulas? Formulae, schmormulae.

Q. What's the area and perimeter of the shaded region? (This shape is the one that the aforementioned bozo decided to call a *segment*.)

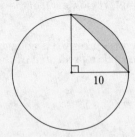

A. The area of the segment, as you can see, equals the area of the sector minus the area of the triangle. The sector measures 90°, so it's one-fourth of the circle. Thus,

$$\text{Area}_{\text{Sector}} = \frac{1}{4}\pi \cdot 10^2$$

$$= 25\pi \text{ units}^2$$

The area of the triangle is a no-brainer because its base and height are both radii:

$$\text{Area}_{\triangle} = \frac{1}{2}bh$$

$$= \frac{1}{2} \cdot 10 \cdot 10$$

$$= 50$$

Thus,

$$\text{Area}_{\text{Segment}} = \text{area}_{\text{sector}} - \text{area}_{\triangle}$$

$$= 25\pi - 50 \text{ units}^2$$

The perimeter equals the hypotenuse of the 45°- 45°- 90° triangle (which you can figure out in your head, right? If not, turn to Chapter 3) plus one-fourth of the circle's circumference. To wit —

$$\text{Perimeter}_{\text{Segment}} = \text{hypotenuse} + \text{arc}$$

$$= 10\sqrt{2} + \frac{1}{4} \cdot 20\pi$$

$$= 10\sqrt{2} + 5\pi$$

Chapter 9: Scintillating Circle Formulas **213**

1. Compute the area and perimeter of these shaded segments made from chords of length 6.

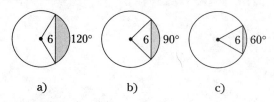

a)   b)   c)

Solve It

2. Compute the shaded areas in the following figures. The inscribed polygons are regular, and each circle has a radius of 10.

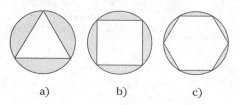

a)   b)   c)

Solve It

3. Compute the shaded areas below. The circumscribed polygons are regular, and each circle has a radius of 10.

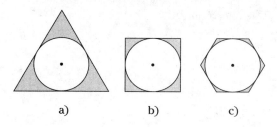

a)   b)   c)

Solve It

*4. Compute the shaded area below. The equilateral triangle is inscribed in a circle, and the three outer arcs are semicircles.

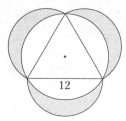

Solve It

Angles, Circles, and Their Connections: The Angle-Arc Theorems and Formulas

Look at the circle in Figure 9-2 with $\overset{\frown}{AC}$ and $\angle ABC$.

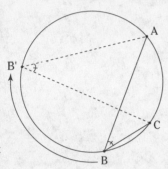

Figure 9-2: For a given arc, no matter where you move the vertex of an inscribed angle, the angle measure doesn't change.

Imagine that $\overline{BA}$ and $\overline{BC}$ are taut, elastic strings and that B is moveable. If you grab B and slide it around the edge of the circle (not crossing over A or C), $\angle B$ always stays the same size even though the distances from B to A and from B to C change. Isn't that cool? This and related ideas are the subjects of this section.

Angle on a circle: The measure of an inscribed angle (Figure 9-3a) or a tangent-chord angle (Figure 9-3b) is *one-half* of the measure of its intercepted arc.

For example, in the circles from Figure 9-3, $\angle Q = \frac{1}{2} m\overset{\frown}{PR}$ and $\angle Y = \frac{1}{2} m\overset{\frown}{XY}$.

a)
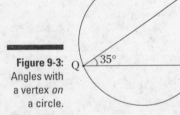

b)
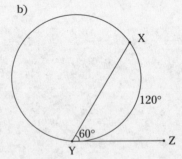

Figure 9-3: Angles with a vertex *on* a circle.

Chapter 9: Scintillating Circle Formulas **215**

Angle inside a circle: The measure of a chord-chord angle is *one-half* the *sum* of the measures of the arcs intercepted by the angle and its vertical angle.

For example, check out Figure 9-4: $\angle CED = \frac{1}{2}\left(m\widehat{AB} + m\widehat{CD}\right)$.

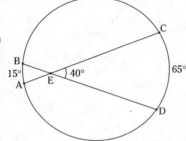

Figure 9-4:
An angle with a vertex *inside* a circle.

Angle outside a circle: The measure of a secant-secant angle (Figure 9-5a), a secant-tangent angle (Figure 9-5b), or a tangent-tangent angle (Figure 9-5c) is *one-half* the *difference* of the measures of the intercepted arcs.

For example, in the circles in Figure 9-5, $\angle C = \frac{1}{2}\left(m\widehat{AE} - m\widehat{BD}\right)$, $\angle R = \frac{1}{2}\left(m\widehat{PS} - m\widehat{QS}\right)$, and $\angle X = \frac{1}{2}\left(m\widehat{WZY} - m\widehat{WY}\right)$.

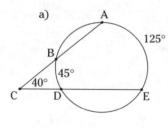
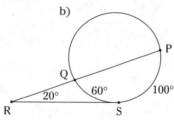
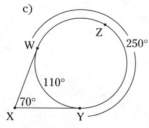

Figure 9-5:
Angles with a vertex *outside* a circle.

Q. Given circle Q and secant-secant ∠ACE as shown, find $m\widehat{AE}$ and $m\widehat{BD}$.

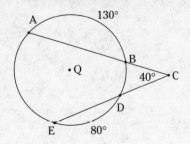

A. Because the measures of the four arcs $\widehat{AE}$, $\widehat{AB}$, $\widehat{BD}$, and $\widehat{ED}$ must add up to 360°, and because $m\widehat{AB}$ and $m\widehat{ED}$ add up to 210°, $m\widehat{AE}$ and $m\widehat{BD}$ must add up to 360° − 210°, or 150°. Now, set $m\widehat{AE}$ equal to x. That makes $m\widehat{BD}$ equal to $150 - x$. ∠C is *outside* the circle, so it equals half the *difference* of the arcs:

$$\angle C = \tfrac{1}{2}\left(m\widehat{AE} - m\widehat{BD}\right)$$
$$40 = \tfrac{1}{2}\left(x - (150 - x)\right)$$
$$80 = 2x - 150$$
$$230 = 2x$$
$$x = 115$$

Thus, $m\widehat{AE}$ is 115° and $m\widehat{BD}$ is 150° − 115°, or 35°.

5. Given the circle and angles as shown, find the measures of angles 1, 2, and 3.

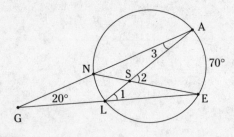

Solve It

6. Given: Diagram as shown, with $\overline{AE}$ tangent to circle Q

Find: ∠B

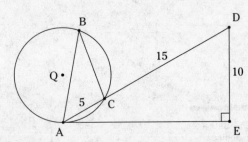

Solve It

*7. Given: Circle *C* with a radius of 5
Find: XZ

Hint: You can use one of the quadrilateral area formulas.

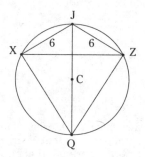

Solve It

8. Given: Circle *Q*
$\angle MTE = 80°$
$\angle AMT = 70°$
Find: $\angle R$

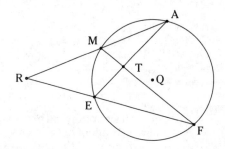

Solve It

The Power Theorems That Be

The figures here look a lot like those from the preceding section because both sections involve angles drawn on the inside and outside of circles. But in this section, you're not investigating the size of the angles; you're looking at the lengths of segments that make up the angles.

Chord-Chord Power Theorem: If two chords of a circle intersect, then the product of the measures of the segments of one chord is equal to the product of the measures of the segments of the other chord.

For example, in Figure 9-6,

$4 \cdot 6 = 3 \cdot 8.$

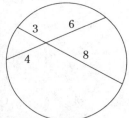

Figure 9-6: The Chord-Chord Power Theorem.

Tangent-Secant Power Theorem: If a tangent and a secant are drawn from an external point to a circle, then the square of the measure of the tangent is equal to the product of the measures of the secant's external part and the entire secant.

In Figure 9-7,

$6^2 = 4 \cdot 9$.

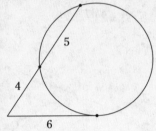

Figure 9-7: The Tangent-Secant Power Theorem.

Secant-Secant Power Theorem: If two secants are drawn from an external point to a circle, then the product of the measures of one secant's external part and that entire secant is equal to the product of the measures of the other secant's external part and that entire secant.

For instance, in Figure 9-8,

$3 \cdot 12 = 4 \cdot 9$.

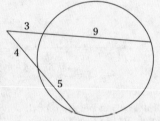

Figure 9-8: The Secant-Secant Power Theorem.

All three of these theorems use the same mathematical idea. On both sides of each equation is a product of two lengths (or one length squared). Each of the four lengths is the distance from the vertex of an angle to the edge of the circle. Therefore, you can think of all three theorems like this:

(vertex to circle) · (vertex to circle) = (vertex to circle) · (vertex to circle)

Pretty nifty, eh? It's because of ideas like this that they pay me the big bucks.

Q. Given: Diagram as shown

Circle Q has a radius of 7

Find: AB and RD

Note: It's difficult to see, but $\overline{DE}$ is a chord of circle Q; $\overline{RE}$ is not tangent to the circle at D or E.

A. First, extend radius $\overline{CQ}$ into a diameter that hits the opposite side of the circle at a point I call X. The diameter has a length of

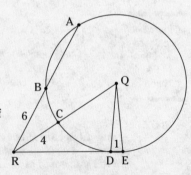

14. Now use the Secant-Secant Power Theorem to get AB:

$$RB \cdot RA = RC \cdot RX$$
$$6 \cdot RA = 4 \cdot 18$$
$$6 \cdot RA = 72$$
$$RA = 12$$

AB is RA minus RB, so AB is 6.

Finding RD is a bit trickier, but it's no big deal. First, set RD equal to y. Then

$$RD \cdot RE = RC \cdot RX$$
$$y(y + 1) = 4 \cdot 18$$
$$y^2 + y = 72$$
$$y^2 + y - 72 = 0$$
$$(y + 9)(y - 8) = 0$$
$$y = -9 \text{ or } 8$$

You can reject –9, so RD has to be 8.

9. Given: Circle O has a radius of $5x + 1$
$OR = 3x - 1$
$NR = 2x + 8$
$RY = 2x + 4$

Find: x

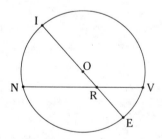

Solve It

10. Given: Diagram as shown
$\overline{EA}$ is tangent to circle Q, which has a radius of 1

Find: EA, EB, and the area of $\triangle EQD$

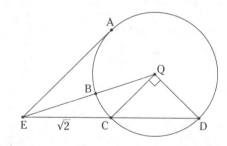

Solve It

11. Given: Circle C has a radius of 4

Find: Area of quadrilateral $QUAD$

Note: Do this problem without using the Pythagorean Theorem.

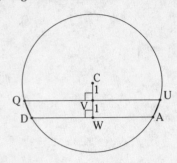

Solve It

***12.** Given: Diagram as shown with circle T

Find: Area of $\triangle TIK$

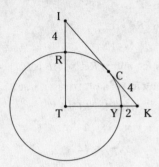

Solve It

Solutions for Scintillating Circle Formulas

1. To tackle these problems, you may need to review some triangle formulas from Chapter 3.

a. You need the radius. Draw the altitude of the triangle to the base of 6. The altitude bisects the 120° vertex angle and thus creates two 30°- 60°- 90° triangles each with a long leg of 3. See the following figure:

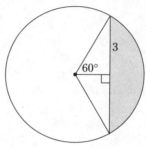

The short leg is $\frac{3}{\sqrt{3}}$, or $\sqrt{3}$, and the hypotenuse, therefore, is $2\sqrt{3}$ — that's the radius. You're all set:

$$\text{Area}_{Segment} = \text{area}_{sector} - \text{area}_\triangle$$
$$= \frac{120°}{360°} \cdot \pi\left(2\sqrt{3}\right)^2 - \frac{1}{2}(6)\left(\sqrt{3}\right)$$
$$= \frac{1}{3}\pi \cdot 12 - 3\sqrt{3}$$
$$= 4\pi - 3\sqrt{3} \text{ units}^2$$

You already have the base of the triangle, so the perimeter is a snap:

$$\text{Perimeter} = 6 + 120° \text{ arc}$$
$$= 6 + \frac{1}{3}\pi d$$
$$= 6 + \frac{1}{3}\pi\left(4\sqrt{3}\right)$$
$$= 6 + \frac{4\pi\sqrt{3}}{3}$$

b. You have a 45°- 45°- 90° triangle with a hypotenuse of 6, so the leg (which is the radius) is $\frac{6}{\sqrt{2}}$, or $3\sqrt{2}$. You're ready to go!

$$\text{Area}_{Segment} = \text{area}_{sector} - \text{area}_\triangle$$
$$= \frac{90°}{360°}\pi\left(3\sqrt{2}\right)^2 - \frac{1}{2} \cdot \text{leg} \cdot \text{leg}$$
$$= \frac{1}{4}\pi \cdot 18 - \frac{1}{2}\left(3\sqrt{2}\right)\left(3\sqrt{2}\right)$$
$$= 4.5\pi - 9 \text{ units}^2$$

$$\text{Perimeter} = 6 + 90° \text{ arc}$$
$$= 6 + \frac{1}{4}\pi d$$
$$= 6 + \frac{1}{4}\pi\left(6\sqrt{2}\right)$$
$$= 6 + \frac{3\pi\sqrt{2}}{2}$$

c. The easiest of the three. You have an equilateral triangle, so the radius is 6 and the sector takes up $\frac{1}{6}$ of the circle:

$$\text{Area}_{\text{Segment}} = \text{area}_{\text{sector}} - \text{area}_{\triangle}$$

$$= \frac{1}{6}(\pi \cdot 6^2) - \frac{6^2\sqrt{3}}{4}$$

$$= 6\pi - 9\sqrt{3} \text{ units}^2$$

$$\text{Perimeter} = 6 + 60° \text{ arc}$$

$$= 6 + \frac{1}{6}(\pi \cdot 12)$$

$$= 6 + 2\pi$$

2 Here's how this problem plays out.

a. Draw the apothem straight down to the base of the triangle (an *apothem* goes from the center of a regular polygon to the midpoint of a side — see Chapter 6). Then draw the circle's radius to one of the triangle's lower vertices. You now have a 30°-60°-90° triangle. The hypotenuse is the radius, so that's 10, and the apothem is the short leg, so that's 5. The long leg is, therefore, $5\sqrt{3}$, and because that's half the base of the triangle, the base is $10\sqrt{3}$ (see Chapter 3 for more on 30°-60°-90° triangles). You know all you need to solve the problem:

$$\text{Shaded area} = \text{circle} - \text{equilateral} \triangle$$

$$= \pi r^2 - \frac{s^2\sqrt{3}}{4}$$

$$= \pi \cdot 10^2 - \frac{(10\sqrt{3})^2\sqrt{3}}{4}$$

$$= 100\pi - 75\sqrt{3} \text{ units}^2$$

b. The radius is 10, and that's half of the square's diagonal, right? A square's a kite, so use the kite formula (Chapter 6):

$$\text{Shaded area} = \text{circle} - \text{square}$$

$$= \pi r^2 - \frac{1}{2}(d_1)(d_2)$$

$$= \pi \cdot 10^2 - \frac{1}{2}(20)(20)$$

$$= 100\pi - 200 \text{ units}^2$$

c. A regular hexagon is made up of six equilateral triangles (which you can read all about in Chapter 6). The radius here is 10, so the legs of the six triangles are also 10.

$$\text{Shaded area} = \text{circle} - \text{hexagon}$$

$$= \pi r^2 - 6\left(\text{area}_{\text{equil. }\triangle}\right)$$

$$= \pi \cdot 10^2 - 6\left(\frac{s^2\sqrt{3}}{4}\right)$$

$$= 100\pi - 6\left(\frac{10^2\sqrt{3}}{4}\right)$$

$$= 100\pi - 150\sqrt{3} \text{ units}^2$$

3 Check out these solutions.

a. Because this problem is so similar to problem 2a, I'll cut to the chase (though note that in this problem, in contrast to 2a, the radius is now the short leg of a 30°-60°-90° triangle).

Shaded area = triangle − circle

$$= \frac{s^2\sqrt{3}}{4} - \pi \cdot 10^2$$

$$= \frac{(20\sqrt{3})^2 \sqrt{3}}{4} - 100\pi$$

$$= 300\sqrt{3} - 100\pi \text{ units}^2$$

b. This one should be a no-brainer:

Shaded area = square − circle

$$= 20^2 - \pi \cdot 10^2$$

$$= 400 - 100\pi \text{ units}^2$$

c. Draw the apothem straight down, and draw a radius of the *hexagon* to one of its lower vertices. You have yet another 30°-60°-90° triangle. The apothem is the circle's radius, so it's 10. That's the long leg of the 30°-60°-90° triangle, so the short leg is $\frac{10}{\sqrt{3}}$. The short leg is half the length of one of the hexagon's sides, so those sides are $\frac{20}{\sqrt{3}}$, and multiplying that by 6 gives you the hexagon's perimeter: $\frac{120}{\sqrt{3}}$, or $40\sqrt{3}$. Use the formula for the area of a regular polygon (Chapter 6) for the hexagon.

Shaded area = hexagon − circle

$$= \tfrac{1}{2}pa - 100\pi$$

$$= \tfrac{1}{2}(40\sqrt{3})(10) - 100\pi$$

$$= 200\sqrt{3} - 100\pi \text{ units}^2$$

*4 The shaded area is everything minus the circle. And *everything* is the equilateral triangle plus the three semicircles. Thus,

Shaded area = triangle + 3 semicircles − circle

The triangle has sides of 12, and the radius (r_1) of each semicircle is 6, so you have everything you need to finish except for the radius (r_2) of the circle. You should be an expert at figuring this out by now. Draw the apothem straight down and draw a radius to one of the triangle's lower vertices. That gives you a 30°-60°-90° triangle. You can take it from there. You should get a radius (hypotenuse) of $4\sqrt{3}$.

$$\text{Shaded area} = \frac{s^2\sqrt{3}}{4} + 3\left(\tfrac{1}{2}\pi r_1\right) - \pi r_2$$

$$= \frac{12^2\sqrt{3}}{4} + 3\left(\tfrac{1}{2}\pi \cdot 6^2\right) - \pi(4\sqrt{3})^2$$

$$= 36\sqrt{3} + 54\pi - 48\pi$$

$$= 36\sqrt{3} + 6\pi \text{ units}^2$$

Piece o' cake.

5 Here's how to find the angle measures.

$$\angle 1 = \tfrac{1}{2} m\widehat{AE}$$

$$= 35°$$

One down, two to go. You have a few ways to finish from this point. Here's one of them: ∠GLA is the supplement of ∠1, so it's 145°. Then, because the angles of △GLA have to add up to 180°, ∠3 is 15°. Two down, one to go.

$m\widehat{NL}$ is twice ∠3, so it's 30°. And so

$$\angle 2 = \tfrac{1}{2}\left(m\widehat{NL} + m\widehat{AE}\right)$$
$$= \tfrac{1}{2}(30 + 70)$$
$$= 50°$$

6 Right △ADE has a hypotenuse of 20 and a leg of 10. That makes it a 30°- 60°- 90° triangle (see Chapter 3 for more on special right triangles). You find that ∠CAE is thus 30°. The measure of tangent-chord ∠CAE is half of its intercepted arc, $\widehat{AC}$, so $m\widehat{AC}$ is 60°. Finally, the measure of inscribed ∠B is also half of $m\widehat{AC}$, so that makes it 30°.

I have a feeling that this problem may have been trickier than this short solution suggests.

*7 ∠JXQ and ∠JZQ both intercept half the circle (a 180° arc), so each angle measures half of 180° — that's 90° of course — giving you right ∠JXQ and right ∠JZQ. The diameter is 10 — that's the hypotenuse — so legs $\overline{XQ}$ and $\overline{ZQ}$ are both 8 (the triangles are in the 3:4:5 family; see Chapter 3). Therefore, you have a 6-6-8-8 kite.

The area of the kite is twice the area of △JXQ, which is $\tfrac{1}{2} \cdot 6 \cdot 8$, or 24. So the area of the kite is 48 units². Now, finish with the kite area formula (from Chapter 6) and solve for the length of diagonal $\overline{XZ}$:

$$\text{Area}_{\text{Kite}} = \tfrac{1}{2} d_1 d_2$$
$$48 = \tfrac{1}{2}(10)(XZ)$$
$$48 = 5(XZ)$$
$$XZ = \tfrac{48}{5}, \text{ or } 9.6$$

Note: After you get the 6-8-10 right triangles, you could also finish with the Altitude-on-Hypotenuse Theorem (see Chapter 7, on similar triangles).

8 ∠AMF is inscribed in the circle, so $m\widehat{AF}$ is twice ∠AMF; thus, $m\widehat{AF}$ is 140°. Then to get ∠R you just need the measure of $\widehat{ME}$:

$$\angle MTE = \tfrac{1}{2}\left(m\widehat{ME} + m\widehat{AF}\right)$$
$$80 = \tfrac{1}{2}\left(m\widehat{ME} + 140\right)$$
$$80 = \tfrac{1}{2} m\widehat{ME} + 70$$
$$m\widehat{ME} = 20°$$

And then

$$\angle R = \tfrac{1}{2}\left(m\widehat{AF} - m\widehat{ME}\right)$$
$$= \tfrac{1}{2}(140 - 20)$$
$$= 60°$$

9 You want to use the Chord-Chord Power Theorem:

(vertex to circle) · (vertex to circle) = (vertex to circle) · (vertex to circle)

For that, you need expressions in *x* for the lengths of the four segments. You have two of them: *NR* and *RV*. To get *IR*, you add $3x - 1$ to the radius of $5x + 1$. That's $8x$. And *RE* is the radius minus $3x - 1$. That's $2x + 2$. Now you have what you need to use the theorem:

$$(2x+8)(2x+4) = (8x)(2x+2)$$
$$4x^2 + 8x + 16x + 32 = 16x^2 + 16x$$
$$-12x^2 + 8x + 32 = 0 \text{ (Now divide both sides by } -4)$$
$$3x^2 - 2x - 8 = 0$$
$$(3x+4)(x-2) = 0$$
$$x = -\tfrac{4}{3} \text{ or } 2$$

You can reject $-\tfrac{4}{3}$, so x is 2.

10. $\triangle CQD$ is an isosceles right triangle, which makes it a 45°- 45°- 90° triangle (see Chapter 3 for more information). The legs are 1, so the hypotenuse, $\overline{CD}$, is $\sqrt{2}$ units long. Now do the last problem first. To get the area of $\triangle EQD$, you need its height. So draw the altitude of $\triangle CQD$ to base $\overline{CD}$. (This altitude of $\triangle CQD$ is also the altitude of $\triangle EQD$.) This altitude cuts $\triangle CQD$ into two smaller 45°- 45°- 90° triangles. Each has a hypotenuse of 1 (the radius) and thus legs (including the altitude) of $\tfrac{1}{\sqrt{2}}$. The rest is child's play:

$$\text{Area}_{\triangle EQD} = \tfrac{1}{2} bh$$
$$= \tfrac{1}{2}(2\sqrt{2})\left(\tfrac{1}{\sqrt{2}}\right)$$
$$= 1 \text{ unit}^2$$

To get EA, use the Tangent-Secant Power Theorem:

$$(\text{vertex to circle}) \cdot (\text{vertex to circle}) = (\text{vertex to circle}) \cdot (\text{vertex to circle})$$
$$(EA)(EA) = (EC)(ED)$$
$$(EA)^2 = (\sqrt{2})(2\sqrt{2})$$
$$(EA)^2 = 4$$
$$EA = 2$$

You use the Tangent-Secant Power Theorem again to get EB. First, set EB equal to x. Don't forget that you always go from vertex to *circle* (not from vertex to center of circle), so you need to use the whole diameter with a length of 2 that goes from B through Q to the other side of the circle:

$$(\text{vertex to circle})(\text{vertex to circle}) = (\text{vertex to circle})(\text{vertex to circle})$$
$$(EB)(EB + \text{diameter}) = (EA)(EA)$$
$$(x)(x+2) = (2)(2)$$
$$x^2 + 2x - 4 = 0$$

Finish up with the quadratic formula (which I first use in Chapter 3):

$$x = \frac{-2 \pm \sqrt{2^2 - 4(1)(-4)}}{2}$$
$$= \frac{-2 \pm \sqrt{20}}{2}$$
$$= \frac{-2 \pm 2\sqrt{5}}{2}$$
$$= -1 \pm \sqrt{5}$$

Reject the negative answer, so EB is $-1 + \sqrt{5}$. Wasn't that fun?

11 First, note that $\overline{CW}$ (which is part of a radius) is perpendicular to chords $\overline{QU}$ and $\overline{DA}$, so it bisects the chords (you can find this circle theorem in Chapter 8). Next, extend $\overline{CW}$ into a diameter (which is 8 units long). Now you can get VQ and VU with the Chord-Chord Power Theorem. The part of the diameter above V is 5 (the radius plus 1), and the part below V is 3. Set the lengths of the congruent segments, $\overline{VQ}$ and $\overline{VU}$, both equal to x, and you're ready to go:

$$(x)(x) = (5)(3)$$
$$x^2 = 15$$
$$x = \sqrt{15}$$

Thus, QU (which is twice VQ) is $2\sqrt{15}$.

Use the same method to get WD and WA. This time, the parts of the diameter are 6 and 2, so

$$(y)(y) = (6)(2)$$
$$y = \sqrt{12}, \text{ or } 2\sqrt{3}$$

Thus, DA (twice WD) is $4\sqrt{3}$.

Finish with the formula for the area of a trapezoid (check out Chapter 6).

$$\text{Area}_{\text{Trap. }QUAD} = \frac{b_1 + b_2}{2} \cdot h$$
$$= \frac{2\sqrt{15} + 4\sqrt{3}}{2} \cdot 1$$
$$= \sqrt{15} + 2\sqrt{3} \text{ units}^2$$

*12 Extend the radius ($\overline{YT}$) into a diameter. Set its length equal to x, and solve with the Tangent-Secant Power Theorem:

$$(KY)(KY + x) = (KC)(KC)$$
$$(2)(2 + x) = (4)(4)$$
$$4 + 2x = 16$$
$$x = 6$$

So the diameter is 6, and the radius is thus 3.

The other radius, $\overline{RT}$, is also, of course, 3 units long. Did you fall for the trap and assume that $\triangle TIK$ is a right triangle? If you did, you would've gotten an incorrect area of $\frac{1}{2}(5)(7)$, or 17.5 units².

However, you *can't* assume $\triangle TIK$ is a right triangle, and, in fact, it's not. Instead, you draw the radius $\overline{TC}$, which is perpendicular to $\overline{IK}$ (because a tangent is always perpendicular to the radius) and is thus the altitude of $\triangle TIK$ to base $\overline{IK}$. All you need to finish is the length of base $\overline{IK}$, and to get that, you need the length of $\overline{IC}$. You use the other diameter (from R straight down):

$$(IC)(IC) = (4)(4 + \text{diameter})$$
$$(IC)^2 = (4)(4 + 6)$$
$$(IC)^2 = 40$$
$$IC = 2\sqrt{10}$$

Finally,

$$\text{Area}_{\triangle TIK} = \frac{1}{2}bh$$
$$= \frac{1}{2}(2\sqrt{10} + 4)(3)$$
$$= 3\sqrt{10} + 6$$
$$\approx 15.5 \text{ units}^2$$

I didn't call this a challenge problem for nothing!

Part V
3-D Geometry and Coordinate Geometry

In this Part . . .

If you've never read Edwin Abbott's book *Flatland*, I highly recommend it. It's about a world that has no third dimension. Everything takes place in a flat plane, and all the characters in the book are two-dimensional geometric shapes. It's an excellent book to read while studying geometry. Needless to say, unlike the fantasy world of *Flatland* and unlike everything in this book so far, our world *does* have three dimensions. And that's what the first two chapters in this Part are all about. In these chapters, you practice problems involving the volume and surface area of spheres, pointy-top figures (cones and pyramids), and flat-top figures (cylinders and prisms). You also get practice working with parallel planes and intersecting planes, and you revisit some of the triangle topics from Chapter 3, but, this time, the triangles are standing up in the third dimension.

After you finish with 3-D objects, you move on to coordinate geometry. As you know, the concepts you've worked with up to this point in this book have nothing to do with location or orientation. There's a good reason for looking at things this way. Ignoring location and orientation is a useful mathematical abstraction that teaches us about things whose truth is independent of location. The volume of a particular pyramid, for example, has nothing to do with where the pyramid is or which way its edges are facing. Location has nothing to do with whether or not two triangles are congruent. But this abstraction isn't exactly realistic. In our world, of course, we *do* care about the location of things So, in this part, you come down to earth and look at geometry from the practical viewpoint of where things are, how far they are from other things, what their orientation is, etc. You also consider what happens to shapes when you spin them around, slide them, or flip them over. You'll see things here which you may remember from algebra like slope, the distance formula, and the midpoint formula. If you like this sort of thing, great. If not, don't blame me, blame Descartes!

Chapter 10

2-D Stuff Standing Up

In This Chapter
▶ Finding your feet: Where lines meet planes
▶ Flying through space in a geometric plane
▶ Parallel pairings: And never the twain shall meet
▶ Speeding through the intersection of lines and planes

Many of the ideas in this chapter should be familiar to you: congruent triangles, CPCTC, parallel lines, quadrilaterals, and so on. (If they're not, take a look at Chapters 1 through 9.) What's new about Chapter 10 is that some of the lines, triangles, and quadrilaterals, instead of lying in a plane, are now standing up in three-dimensional space.

Lines Perpendicular to Planes: They're All Right

This section involves problems about lines that are perpendicular to planes. Lines like this can come in handy, because they create right angles that are just begging for you to use them in a proof. Remember to look for *all* the right angles in the following problems; doing so can make the proofs much easier.

Plane: A *plane* is a flat, two-dimensional shape — you know, like a piece of paper — except that it's infinitely thin and it goes on forever in all directions.

Perpendicularlity of Line to Plane: A *line is perpendicular to a plane* if it's perpendicular to every line in the plane that passes through its foot. (A *foot* is the point where a line intersects a plane.)

Part V: 3-D Geometry and Coordinate Geometry

If a line is perpendicular to two lines that lie in a plane and pass through its foot, then it is perpendicular to the plane.

In short, when writing proofs, you use the definition to say (about a line) *if it's ⊥ to a plane, then it's ⊥ to a line* and the theorem to say *if it's ⊥ to two lines, then it's ⊥ to the plane*.

Q. Given: $\overline{AB} \perp k$

BEDC is a kite with $\overline{BE} \cong \overline{BC}$

Prove: $\triangle AED \cong \triangle ACD$

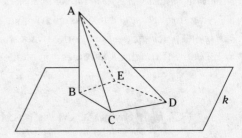

A.

	Statements		Reasons
1)	$\overline{AB} \perp k$	1)	Given.
2)	$\overline{AB} \perp \overline{BE}$ $\overline{AB} \perp \overline{BC}$	2)	If a line is perpendicular to a plane, then it is perpendicular to every line in the plane that passes through its foot (definition of perpendicularity of a line to a plane).
3)	$\angle ABE$ is a right angle $\angle ABC$ is a right angle	3)	Definition of perpendicular.
4)	$\angle ABE \cong \angle ABC$	4)	All right angles are congruent.
5)	$\overline{BE} \cong \overline{BC}$	5)	Given.
6)	$\overline{AB} \cong \overline{AB}$	6)	Reflexive.
7)	$\triangle ABE \cong \triangle ABC$	7)	SAS (5, 4, 6).
8)	$\overline{AE} \cong \overline{AC}$	8)	CPCTC.
9)	$\overline{ED} \cong \overline{CD}$	9)	Property of a kite.
10)	$\overline{AD} \cong \overline{AD}$	10)	Reflexive.
11)	$\triangle AED \cong \triangle ACD$	11)	SSS (8, 9, 10).

Chapter 10: 2-D Stuff Standing Up 231

1. Given: $\overline{BD} \perp p$
$\overrightarrow{AC}$ bisects $\angle BAD$
Prove: C is the midpoint of $\overline{BD}$

Solve It

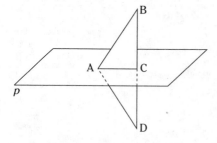

Statements	Reasons

2. Given: $\overline{VX} \perp q$
$\triangle VYZ$ is isosceles with base $\overline{YZ}$
Prove: $\triangle XYZ$ is isosceles

Solve It

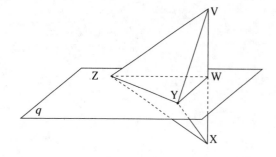

Statements	Reasons

3.
Given: $\overline{ST} \perp r$
$\angle TVU \cong \angle TUV$

Prove: $\angle SVU \cong \angle SUV$

Solve It

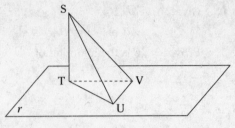

Statements	Reasons

*4.
Given: Circle O in plane p
$\angle AOZ$ and $\angle BOZ$ are right angles
$\overarc{AB} \cong \overarc{BC}$

Prove: $\angle AZB \cong \angle CZB$

Solve It

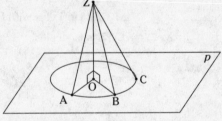

Statements	Reasons

Continued

Statements	Reasons

Parallel, Perpendicular, and Intersecting Lines and Planes

In the preceding section, all the figures contain a single line perpendicular to a single plane. In this section, you move on to figures that involve multiple perpendicularity and/or multiple planes and parallel lines. But first take a look at the four ways to determine a plane.

Determining a plane: Four different sets of geometric objects determine a plane:

- Three noncollinear points

    In plain English, this statement just means that if you have three points not on one line, then only one specific plane contains those points. The plane is *determined* by the three points because they show you exactly where this unique plane is.

- A line and a point not on the line
- Two intersecting lines
- Two parallel lines

Check out the following properties about perpendicularity and parallelism of lines and planes. For the most part, these are *well-duh* properties after you picture what the lines and planes would look like.

- **Three parallel planes:** If two planes are parallel to the same plane, they're parallel to each other.
- **Two parallel lines and a plane:**
    - If two lines are perpendicular to the same plane, they're parallel to each other.
    - If a plane is perpendicular to one of two parallel lines, it's perpendicular to the other.
- **Two parallel planes and a line:**
    - If two planes are perpendicular to the same line, they're parallel to each other.
    - If a line is perpendicular to one of two parallel planes, it's perpendicular to the other.

And here's one more point before getting to the problems:

Intersecting planes: If a plane intersects two parallel planes, the lines of intersection are parallel.

Part V: 3-D Geometry and Coordinate Geometry

Q. Given: $\overline{GN} \perp p$
$\overline{LE} \perp p$
$\overline{GN} \cong \overline{LE}$

Prove: NGLE is a rectangle

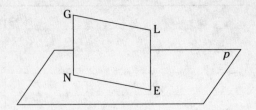

A.

	Statements		Reasons
1)	$\overline{GN} \perp p$ $\overline{LE} \perp p$	1)	Given.
2)	$\overline{GN} \parallel \overline{LE}$	2)	If two lines are perpendicular to the same plane, then they are parallel to each other.
3)	$\overleftrightarrow{GN}$ and $\overleftrightarrow{LE}$ determine a plane, NGLE	3)	Two parallel lines determine a plane. (You need this odd-looking step to ensure that NGLE is a planar quadrilateral. Otherwise it could be a weird, bent, four-sided figure like a "rectangle" bent along one of its diagonals. It could also be a "rectangle" with a curvy surface. Can you picture these shapes?)
4)	$\overline{GN} \cong \overline{LE}$	4)	Given.
5)	NGLE is a parallelogram	5)	If a quadrilateral contains a pair of sides that are both parallel and congruent, then the quadrilateral is a parallelogram.
6)	$\overline{GN} \perp \overline{NE}$	6)	If a line is perpendicular to a plane, then it is perpendicular to every line in the plane that passes through its foot.
7)	$\angle GNE$ is a right angle	7)	Definition of perpendicular.
8)	NGLE is a rectangle	8)	A parallelogram with a right angle is a rectangle.

Q. Given: $p \parallel q$
$\overline{RT} \parallel \overline{SU}$

Prove: $\overline{RS} \cong \overline{TU}$

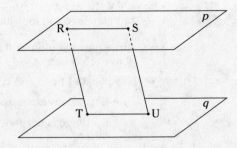

A.

Statements	Reasons
1) $p \parallel q$	1) Given.
2) $\overleftrightarrow{RT} \parallel \overleftrightarrow{SU}$	2) Given.
3) $\overleftrightarrow{RT}$ and $\overleftrightarrow{SU}$ determine a plane, RSUT	3) Two parallel lines determine a plane. (You need this step before you can use the theorem in reason 4.)
4) $\overleftrightarrow{RS} \parallel \overleftrightarrow{TU}$	4) If a plane intersects two parallel planes, then the lines of intersection are parallel.
5) RSUT is a parallelogram	5) If both pairs of opposite sides of a quadrilateral are parallel, the quadrilateral is a parallelogram (definition of parallelogram).
6) $\overline{RS} \cong \overline{TU}$	6) Opposite sides of a parallelogram are congruent.

5. Given: $s \parallel y$

$\triangle EIO$ is isosceles with base $\overline{EO}$

Prove: $\triangle AIU$ is isosceles

Solve It

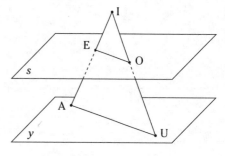

Statements	Reasons

Part V: 3-D Geometry and Coordinate Geometry

6. Give this problem a go:

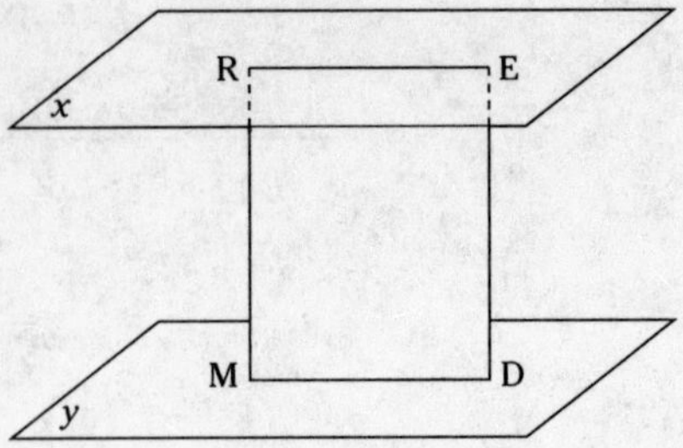

    a. Given: $x \parallel y$
                     $\overline{MR} \parallel \overline{ED}$
                     $\overline{MR} \perp \overline{MD}$

        Prove: MRED is a rectangle
        Answer: Is $\overline{MR} \perp y$?
                     Is $\overline{MR} \perp x$?

    b. Given: $x \parallel y$
                     $\overline{MR} \parallel \overline{ED}$
                     $\overline{MR} \perp y$

        Prove: $\overline{ED} \perp x$

Solve It

a.

Statements	Reasons

b.

Statements	Reasons

**7. Given: $\overline{GN} \perp p$
$\overline{LE} \perp p$
$\overline{GL} \cong \overline{NE}$

Prove: *NGLE* is a rectangle (paragraph proof)

(Note the similarity of this problem to the first example. The one minor difference here makes this problem *much* harder.)

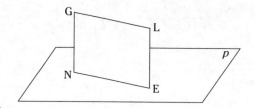

Solve It

---

8. Indicate whether the following statements are true *always*, *sometimes*, or *never*:

 a. Two lines that don't intersect are parallel.
 b. Two lines perpendicular to the same plane are parallel.
 c. Two lines perpendicular to a third line are parallel.
 d. If a line is perpendicular to each of two intersecting lines, then it's perpendicular to the plane determined by the intersecting lines.
 e. Two planes intersect at one point.
 f. Two planes intersect in one line.
 g. Three planes intersect at one point.
 h. Three planes intersect in one line.
 i. Two points lie on the same line.
 j. Three points lie on the same line.
 k. Four points lie on the same line.
 l. Two points lie on the same plane.
 m. Three points lie on the same plane.
 n. Four points lie on the same plane.

Solve It

Solutions for 2-D Stuff Standing Up

1.

Statements	Reasons
1) $\overline{BD} \perp p$	1) Given.
2) $\overline{BD} \perp \overline{AC}$	2) If a line is perpendicular to a plane, then it's perpendicular to every line in the plane that passes through its foot.
3) $\angle BCA$ is a right angle $\angle DCA$ is a right angle	3) Definition of perpendicular.
4) $\angle BCA \cong \angle DCA$	4) All right angles are congruent.
5) $\overrightarrow{AC}$ bisects $\angle BAD$	5) Given.
6) $\angle BAC \cong \angle DAC$	6) Definition of bisect.
7) $\overline{AC} \cong \overline{AC}$	7) Reflexive.
8) $\triangle BAC \cong \triangle DAC$	8) ASA (4, 7, 6).
9) $\overline{BC} \cong \overline{DC}$	9) CPCTC.
10) C is the midpoint of $\overline{BD}$	10) Definition of midpoint.

2.

Statements	Reasons
1) $\overline{VX} \perp q$	1) Given.
2) $\overline{VX} \perp \overline{WZ}$ $\overline{VX} \perp \overline{WY}$	2) If a line is perpendicular to a plane, then it's perpendicular to every line in the plane that passes through its foot.
3) $\angle VWZ$ is a right angle $\angle VWY$ is a right angle	3) Definition of perpendicular.
4) $\triangle VYZ$ is isosceles with base $\overline{YZ}$	4) Given.
5) $\overline{VZ} \cong \overline{VY}$	5) Definition of isosceles triangle.
6) $\overline{VW} \cong \overline{VW}$	6) Reflexive.
7) $\triangle VWZ \cong \triangle VWY$	7) HLR (5, 6, 3).
8) $\angle ZVW \cong \angle YVW$	8) CPCTC.
9) $\overline{VX} \cong \overline{VX}$	9) Reflexive.
10) $\triangle ZVX \cong \triangle YVX$	10) SAS (5, 8, 9).
11) $\overline{ZX} \cong \overline{YX}$	11) CPCTC.
12) $\triangle XYZ$ is isosceles	12) Definition of isosceles triangle.

3

Statements	Reasons
1) $\overline{ST} \perp r$	1) Given.
2) $\overline{ST} \perp \overline{TV}$ $\overline{ST} \perp \overline{TU}$	2) If a line is perpendicular to a plane, then it's perpendicular to every line in the plane that passes through its foot.
3) ∠STV is a right angle ∠STU is a right angle	3) Definition of perpendicular.
4) ∠STV ≅ ∠STU	4) All right angles are congruent.
5) ∠TVU ≅ ∠TUV	5) Given.
6) $\overline{TV} \cong \overline{TU}$	6) If angles, then sides.
7) $\overline{ST} \cong \overline{ST}$	7) Reflexive.
8) △STV ≅ △STU	8) SAS (6, 4, 7).
9) $\overline{SV} \cong \overline{SU}$	9) CPCTC.
10) ∠SVU ≅ ∠SUV	10) If sides, then angles.

*4

Statements	Reasons
1) Circle O in plane p ∠AOZ and ∠BOZ are right angles	1) Given.
2) $\overline{OZ} \perp p$	2) If a line is perpendicular to two lines that lie in a plane and pass through its foot, then it is perpendicular to the plane.
3) Draw radius $\overline{OC}$	3) Two points determine a segment.
4) $\overline{OZ} \perp \overline{OC}$	4) If a line is perpendicular to a plane, then it's perpendicular to every line in the plane that passes through its foot.
5) ∠COZ is a right angle	5) Definition of perpendicular.
6) ∠AOZ ≅ ∠COZ	6) All right angles are congruent.
7) $\overline{OA} \cong \overline{OC}$	7) All radii are congruent.
8) $\overline{ZO} \cong \overline{ZO}$	8) Reflexive.
9) △AOZ ≅ △COZ	9) SAS (7, 6, 8).
10) $\overline{ZA} \cong \overline{ZC}$	10) CPCTC.
11) Draw chords $\overline{AB}$ and $\overline{BC}$	11) Two points determine a segment.
12) $\overset{\frown}{AB} \cong \overset{\frown}{BC}$	12) Given.
13) $\overline{AB} \cong \overline{BC}$	13) If arcs, then chords.
14) $\overline{ZB} \cong \overline{ZB}$	14) Reflexive.
15) △AZB ≅ △CZB	15) SSS (10, 13, 14).
16) ∠AZB ≅ ∠CZB	16) CPCTC.

To find out more about arcs and circles, read up on Chapter 8.

5

Statements	Reasons
1) $s \parallel y$	1) Given.
2) $\overleftrightarrow{IA}$ and $\overleftrightarrow{IU}$ determine a plane, AEIOU	2) Two intersecting lines determine a plane.
3) $\overleftrightarrow{EO} \parallel \overleftrightarrow{AU}$	3) If a plane intersects two parallel planes, the lines of intersection are parallel.
4) $\triangle EIO$ is isosceles with base $\overline{EO}$	4) Given.
5) $\overline{IE} \cong \overline{IO}$	5) Definition of isosceles triangle.
6) $\angle IEO \cong \angle IOE$	6) If sides, then angles.
7) $\angle IEO \cong \angle IAU$ $\angle IOE \cong \angle IUA$	7) If lines are parallel, corresponding angles are congruent.
8) $\angle IAU \cong \angle IUA$	8) Transitivity.
9) $\overline{IA} \cong \overline{IU}$	9) If angles, then sides.
10) $\triangle AIU$ is isosceles	10) Definition of isosceles triangle.

6 Here are the answers.

a.

Statements	Reasons
1) $x \parallel y$	1) Given.
2) $\overleftrightarrow{MR} \parallel \overleftrightarrow{ED}$	2) Given.
3) $\overleftrightarrow{MR}$ and $\overleftrightarrow{ED}$ determine plane MRED	3) Two parallel lines determine a plane.
4) $\overleftrightarrow{RE} \parallel \overleftrightarrow{MD}$	4) If a plane intersects two parallel planes, the lines of intersection are parallel.
5) MRED is a parallelogram	5) Definition of parallelogram.
6) $\overline{MR} \perp \overline{MD}$	6) Given.
7) $\angle RMD$ is a right angle	7) Definition of perpendicular.
8) MRED is a rectangle	8) A parallelogram with a right angle is a rectangle.

Answer: $\overline{MR}$ may or may not be perpendicular to plane y. (Remember, to know that a line is perpendicular to a plane, you must know that it is perpendicular to *two* lines in the plane that pass through its foot, not just one line. It's possible that MRED is slanting toward or away from us.) If $\overline{MR}$ is perpendicular to y, it's perpendicular to x as well. If it's not perpendicular to y, it's also not perpendicular to x.

b.

Statements	Reasons
1) $x \parallel y$ $\overline{MR} \parallel \overline{ED}$ $\overline{MR} \perp y$	1) Given.
2) $\overline{ED} \perp y$	2) If a plane is perpendicular to one of two parallel lines, it is perpendicular to the other.
3) $\overline{ED} \perp x$	3) If a line is perpendicular to one of two parallel planes, it is perpendicular to the other.

** 7. This proof begins just like the first three lines of the example proof: $\overline{GN}$ and $\overline{LE}$ are perpendicular to p (given), $\overline{GN}$ is parallel to $\overline{LE}$ (the two lines are perpendicular to the same plane), and $\overleftrightarrow{GN}$ and $\overleftrightarrow{LE}$ determine plane *NGLE*.

Then you state that $\overline{LE} \perp \overline{NE}$, because if a line is perpendicular to a plane, the line is perpendicular to any line in the plane that passes through its foot. And then you have that $\angle E$ is a right angle (you can show that $\angle N$ is a right angle the same way, but it doesn't help). Okay, so now you have a quadrilateral with two parallel sides ($\overline{GN}$ and $\overline{LE}$) — call them *bases* — where the other two sides are congruent and where one base angle is a right angle:

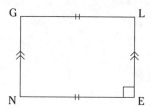

Now comes the tricky part.

Even though it's obvious that *NGLE* must be a rectangle, I couldn't find a way to prove it with ordinary techniques. I drew in diagonals $\overline{GE}$ and $\overline{NL}$, and I tried to use things like *alternate interior angles are congruent* to get congruent triangles (see Chapter 5). I wanted to show that $\overline{GL} \| \overline{NE}$ or that $\angle L$ is a right angle. Nothing worked. (If anyone out there finds a way to finish this proof with ordinary methods, please let me know about it.)

Here's how I finished the proof. The only quadrilaterals with parallel bases in which the other two sides are congruent are parallelograms and isosceles trapezoids (see Chapter 6). Assume *NGLE* is an isosceles trapezoid. Its base angles would therefore be congruent, and that would make $\angle L$ a right angle. But if $\angle E$ and $\angle L$ are right angles, then *NGLE* would be a rectangle, which contradicts the assumption (because a rectangle is not an isosceles trapezoid). Therefore, *NGLE* must be a parallelogram. This parallelogram has a right angle, and therefore it's a rectangle (one of the ways of proving that a parallelogram is a rectangle; see Chapter 5). Bingo. That's it.

8 Check it out.

a. Sometimes: You're working in three dimensions, so two lines that don't intersect could be parallel or skew. (*Skew* lines are lines that are neither parallel nor intersecting, like roads that don't meet because of a highway overpass; skew lines can also be defined as *non-coplanar* lines.)

b. Always

c. Sometimes (if they're coplanar); they could also be skew.

d. Always

e. Never: If two planes intersect, they intersect in a line or a plane.

f. Sometimes (if they intersect and the intersection is not a plane); they could also be parallel.

g. Sometimes: The intersection of three planes can be the null set (no intersection), a point, a line, or a plane (if the three planes are all identical).

h. Sometimes

i. Always

j. Sometimes

k. Sometimes

l. Always

m. Always

n. Sometimes

Chapter 11

Solid Geometry: Digging into Volume and Surface Area

In This Chapter
▶ Finding the surface area and volume of cylinders and prisms
▶ Calculating area and volume of cones and pyramids
▶ Having a ball with spheres

When working in flat, 2-D space earlier in this book, I introduce lines and angles and then move on to planar shapes like triangles and parallelograms. Now I delve into more than just flat things. In this chapter, you can take a look at all kinds of new and fun figures in the next dimension — cylinders, prisms, cones, pyramids, and spheres (got your 3-D glasses handy?).

Starting with Flat-Top Figures

Flat-top figure is my nontechnical name for a cylinder or a prism. Both figures have — guess what? — a flat top. This flat top is called a *base,* and it's congruent to and parallel to the other base at the bottom of the figure. I group prisms (whose bases are polygons) and cylinders together because computing their volume basically works the same way; ditto for computing surface area:

- **Volume of flat-top figures:** The volume of a prism or cylinder is given by the following formula:

    $\text{Vol}_{\text{Flat-Top}} = \text{area}_{\text{base}} \cdot \text{height}$

- **Surface area of flat-top figures:** To find the surface area of a prism or a cylinder, use the following formula:

    $\text{SA}_{\text{Flat-Top}} = 2 \cdot \text{area}_{\text{base}} + \text{lateral area}_{\text{rectangle(s)}}$

The *lateral area* (that's the area of the sides of the figure — namely, everything but the bases) of a right prism is made up of rectangles. The lateral area of a right cylinder is basically one rectangle rolled into a tube-shape — like one paper towel that rolls exactly once around a paper towel roll. The base of this rectangle (you know, its length) is thus the circumference of the cylinder.

Part V: 3-D Geometry and Coordinate Geometry

Q. A cylinder with a volume of 125π units^3 has a height equal to its radius. Find its surface area.

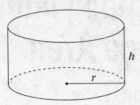

A. First use the volume formula:

$\text{Vol}_{\text{Flat-Top}} = \text{area}_{\text{base}} \cdot \text{height}$

$125\pi = \pi r^2 \cdot h$

$125\pi = \pi r^3$ (because $h = r$)

$r = 5$ (and therefore, $h = 5$)

Now you can compute the surface area:

$\text{SA}_{\text{Flat-Top}} = 2 \cdot \text{area}_{\text{base}} + \text{lateral area}_{\text{rectangle}}$

$= 2\pi r^2 + 2\pi r \cdot h$

$= 2\pi \cdot 5^2 + 2\pi \cdot 5 \cdot 5$

$= 50\pi + 50\pi$

$= 100\pi$ units^2

---

1. Find the volume and surface area of this prism.

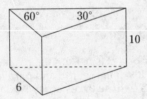

Solve It

2. Find the volume and surface area of the prism, whose bases are equilateral triangles.

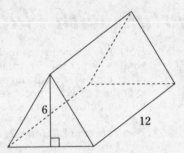

Solve It

*3. Find the volume and surface area of a box (technically a prism) with a height of 2, a width of $2\sqrt{3}$, and a diagonal of 8.

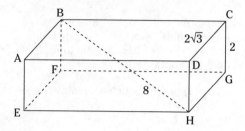

Solve It

4. Answer the following:

a. What's the volume and surface area of a cylinder with height and diameter both equal to 4?

*b. An ant crawls along the outside of the cylinder from A to C. If he goes straight across the top to B and then straight down to C, he goes a distance of 8. Is there a shorter route? If so, what's the shortest possible route, and how long is it?

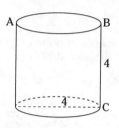

Solve It

5. A cylinder has a diameter of 6 and a lateral area of 60π. Find its volume and surface area.

Solve It

*6. A cylinder with a height of 6 has a surface area of 54π. Find its volume.

Solve It

Sharpening Your Skills with Pointy-Top Figures

Something tells me that you've already figured out that *pointy-top figures* are figures with pointy tops. This is my nontechnical name for pyramids and cones. And just like with prisms and cylinders, I group pyramids and cones together because computing volume basically works the same for both — as does computing surface area:

- **Volume of pointy-top figures:** The volume of a pyramid or cone is given by the following formula:

    $$\text{Vol}_{\text{Pointy-Top}} = \frac{1}{3} \, \text{area}_{\text{base}} \cdot \text{height}$$

- **Surface area of pointy-top figures:** The following formula gives you the surface area of a pyramid or cone:

    $$\text{SA}_{\text{Pointy-Top}} = \text{area}_{\text{base}} + \text{lateral area}_{\text{triangle(s)}}$$

The lateral area of a pyramid is made up of triangles whose areas work just like the area of any triangle: $\frac{1}{2} \, base \cdot height$. But note that the height of a triangle is perpendicular to its base, so you can't use the height of the pyramid for the height of one of its triangular faces. Instead, you use the *slant height*, which is just the ordinary height of the triangular face — if you look at the face like an ordinary flat, two-dimensional triangle. (The cursive letter ℓ is used to indicate slant height.)

Just like the lateral area of a cylinder is one rectangle rolled around into a tube-shape, the lateral area of a cone is one triangle (sort of — its bottom is curved) rolled around into a shape like a snow-cone cup. Its area works exactly like the area of one of the triangular faces of a pyramid — $\frac{1}{2} \, (base)(slant \, height)$ — where the base of this "triangle" (just like the base of the lateral rectangle in a cylinder) equals the circumference of the cone.

Q. Find the volume and surface area of these similar cones (that's *similar* in the technical sense — see Chapter 7 for more on similarity), one of which has dimensions that are double the other. What do you notice about the answers?

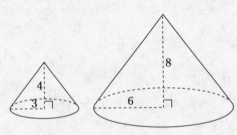

A. Doing volume first,

$$\text{Vol}_{\text{Small Cone}} = \frac{1}{3} \text{area}_{\text{base}} \cdot \text{height}$$
$$= \frac{1}{3} \pi r^2 h$$
$$= \frac{1}{3} \pi \cdot 3^2 \cdot 4$$
$$= 12\pi \text{ units}^3$$

$$\text{Vol}_{\text{Large Cone}} = \frac{1}{3} \pi r^2 h$$
$$= \frac{1}{3} \pi \cdot 6^2 \cdot 8$$
$$= 96\pi \text{ units}^3$$

The volume of the large cone is eight times the volume of the small one.

To find the cones' surface areas, you need their slant heights. If you use the height of a cone and one of its radii to form the legs of a right triangle, then the hypotenuse of the triangle is the cone's slant height. For the small cone in this problem, you have a 3-4-5 right triangle, so the slant height is 5 (see Chapter 3 for more on Pythagorean triples). In the large cone, the slant height is 10. Now you can compute their surface areas:

$$\text{SA}_{\text{Small Cone}} = \text{area}_{\text{base}} + \text{lateral triangle}$$

$$\left(\frac{1}{2} \cdot \text{base} \cdot \text{height}\right)$$

$$= \pi r^2 + \frac{1}{2}(\text{circumference})(\text{slant height})$$
$$= \pi r^2 + \frac{1}{2}(2\pi r)(5)$$
$$= 9\pi + \frac{1}{2}(2\pi \cdot 3)(5)$$
$$= 9\pi + 15\pi$$
$$= 24\pi \text{ units}^2$$

$$\text{SA}_{\text{Large Cone}} = \pi r^2 + \frac{1}{2}(2\pi r)(\text{slant height})$$
$$= \pi \cdot 6^2 + \frac{1}{2}(2\pi \cdot 6)(10)$$
$$= 36\pi + 60\pi$$
$$= 96\pi \text{ units}^2$$

The large cone has four times the surface area of the small cone. So the large cone, which is twice the size of the small cone, has eight (2^3) times as much volume and four (2^2) times as much surface area.

See the rule? Here it is:

If you enlarge a 3-D figure by a factor of k, its volume grows k^3 times and its surface area grows k^2 times.

A good way to remember this rule is to note the connection between the rule and the fact that volume is three-dimensional and is measured in units^3 and that surface area is two-dimensional and is measured in units^2.

248 Part V: 3-D Geometry and Coordinate Geometry

7. Find the volume and surface area of this rectangular, right pyramid.

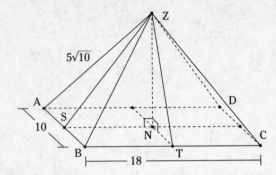

Solve It

***8.** Find the volume and surface area of a regular tetrahedron with edges of 6. (A regular *tetrahedron* is a pyramid with four equilateral triangle faces.)

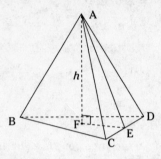

Solve It

9. The circumference of the base of a cone is 16π, and the cone's height is 6. Find the cone's volume and surface area.

Solve It

10. Try this one on for size:

*a. Find the volume of this double cone, which has a radius of 8 and a total height of 21.

b. If the surface area of the left-side cone is 80π, what's the surface area of the right-side cone?

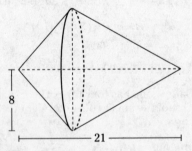

Solve It

Rounding Out Your Understanding with Spheres

I'm running short on space, so I better cut to the chase:

- **Volume of a sphere:** The volume of a sphere is given by the following formula:

$$\text{Vol}_{\text{Sphere}} = \frac{4}{3}\pi r^3$$

- **Surface area of a sphere:** Yada, yada, yada:

$$\text{SA}_{\text{Sphere}} = 4\pi r^2$$

Q. How do the volumes of a cube and an inscribed sphere compare? How do their surface areas compare?

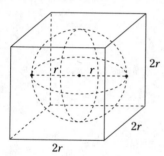

A. Check it out.

$$\text{Vol}_{\text{Sphere}} = \frac{4}{3}\pi r^3$$

$$\text{Vol}_{\text{Cube}} = (2r)^3$$
$$= 8r^3$$

$$\frac{\text{Vol}_{\text{Sphere}}}{\text{Vol}_{\text{Cube}}} = \frac{\frac{4}{3}\pi r^3}{8r^3}$$
$$= \frac{\pi}{6}$$
$$\approx 0.52$$

Thus, if you buy, say, a basketball that comes in a box, the basketball takes up about 52 percent of the volume of a box. (I'm sure you've been dying to know this.)

$$\text{SA}_{\text{Sphere}} = 4\pi r^2$$

$$\text{SA}_{\text{Cube}} = 6 \text{ sides} \cdot (2r)^2$$
$$= 24r^2$$

$$\frac{\text{SA}_{\text{Sphere}}}{\text{SA}_{\text{Cube}}} = \frac{4\pi r^2}{24r^2}$$
$$= \frac{\pi}{6} \text{ or about } 52\%$$

The very same percentage.

11. A cylinder with a radius of $\sqrt{5}$ and a height of 4 is inscribed in a sphere. Find the volume and surface area of the sphere.

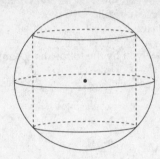

Solve It

12. The hemispherical (half-sphere) top of a 50-foot-tall grain silo has a surface area of 200π square feet. How many cubic feet of grain can the silo hold?

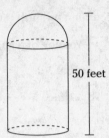

Solve It

Solutions for Solid Geometry

1. For both the volume and surface area, you need the sides of the 30°- 60°- 90° triangle. Its short leg is 6, so its long leg is $6\sqrt{3}$ and its hypotenuse is 12. You're all set to go.

$$\text{Vol} = \text{area}_{base} \cdot \text{height}$$
$$= \tfrac{1}{2}(6)(6\sqrt{3}) \cdot 10$$
$$= 18\sqrt{3} \cdot 10$$
$$= 180\sqrt{3} \text{ units}^3$$

$$\text{SA} = 2(\text{area}_{base}) + \text{three lateral rectangles}$$
$$= 2(18\sqrt{3}) + 6 \cdot 10 + 6\sqrt{3} \cdot 10 + 12 \cdot 10$$
$$= 36\sqrt{3} + 60 + 60\sqrt{3} + 120$$
$$= 180 + 96\sqrt{3} \text{ units}^2$$

2. All you need is the length of the base of the equilateral triangle. The triangle's altitude is 6, and that's the long leg of a 30°- 60°- 90° triangle. The short leg is therefore $\frac{6}{\sqrt{3}}$, or $2\sqrt{3}$, and the hypotenuse is twice that, or $4\sqrt{3}$ (see Chapter 3 for more on making this calculation). And that's the length, of course, of the sides of the equilateral triangle. Thus —

$$\text{Vol} = \text{area}_{base} \cdot \text{height}$$
$$= \tfrac{1}{2}(4\sqrt{3})(6) \cdot 12$$
$$= 144\sqrt{3} \text{ units}^3$$

$$\text{SA} = 2 \cdot \text{area}_{base} + \text{three lateral rectangles}$$
$$= 2(12\sqrt{3}) + 3(12 \cdot 4\sqrt{3})$$
$$= 24\sqrt{3} + 144\sqrt{3}$$
$$= 168\sqrt{3} \text{ units}^2$$

*3. Draw $\overline{CH}$; that's the hypotenuse of yet another 30°- 60°- 90° triangle ($\triangle CGH$). The length of the short leg, $\overline{CG}$, is 2, so CH is 4.

Now, note that $\triangle BCH$ is a right triangle with its right angle at C. One of its legs is 4 and its hypotenuse is 8, so — hold onto your hat — $\triangle BCH$ is *another* 30°- 60°- 90° triangle. The length of its long leg, $\overline{BC}$, is the length of the short leg, $\overline{CH}$, times $\sqrt{3}$, so BC is $4\sqrt{3}$. You have what you need to finish:

$$\text{Vol} = l \cdot w \cdot h \quad \text{(the same thing as } \text{area}_{base} \cdot \text{height)}$$
$$= 4\sqrt{3} \cdot 2\sqrt{3} \cdot 2$$
$$= 48 \text{ units}^3$$

$$\text{SA} = (2 \cdot \text{base}) + (2 \cdot \text{front}) + (2 \cdot \text{right side}) \quad \text{(the same thing as } 2 \cdot \text{area}_{base} + \text{lateral rectangles)}$$
$$= 2(4\sqrt{3} \cdot 2\sqrt{3}) + 2(4\sqrt{3} \cdot 2) + 2(2\sqrt{3} \cdot 2)$$
$$= 2 \cdot 24 + 2 \cdot 8\sqrt{3} + 2 \cdot 4\sqrt{3}$$
$$= 48 + 24\sqrt{3} \text{ units}^2$$

4. Here are the answers:

a. The diameter is 4, so the radius is 2; thus,

$$\text{Vol} = \text{area}_{base} \cdot \text{height}$$
$$= (\pi r^2)(h)$$
$$= (\pi \cdot 2^2)(4)$$
$$= 16\pi \text{ units}^3$$

$$\text{SA} = 2 \cdot \text{area}_{base} + \text{lateral area (which equals } circumference \cdot height)$$
$$= 2(\pi r^2) + (2\pi r)(h)$$
$$= 2(\pi \cdot 2^2) + (2\pi \cdot 2)(4)$$
$$= 8\pi + 16\pi$$
$$= 24\pi \text{ units}^2$$

***b.** This question is a great think-outside-the-box problem. Here's the trick: Imagine dividing the cylinder in half by cutting it along a plane that goes through *A*, *B*, and *C* and cuts the base along the dotted diameter. Now take the front half of the lateral area, uncurl it, and lay it flat. Here's what you get:

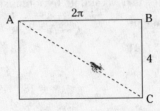

$\overline{AB}$ (in this rectangle, not across the top of the cylinder) has a length of half the circumference of the cylinder. That's 2π. The shortest path from *A* to *C* is straight, of course — that's $\overline{AC}$, which is the hypotenuse of right $\triangle ABC$. Its length is

$$c^2 = a^2 + b^2$$
$$= 4^2 + (2\pi)^2$$
$$= 16 + 4\pi^2$$
$$c = \sqrt{16 + 4\pi^2} = 2\sqrt{4 + \pi^2}$$
$$\approx 7.4$$

That's the shortest route; it curves along the outside of the cylinder, going diagonally down from *A* to *C*.

5 You need the cylinder's radius and height. The diameter is 6, so the radius is 3. To get the height, you use the fact that the lateral area is a rectangle with an area of *circumference · height*. So

$60\pi = \pi \cdot \text{diameter} \cdot \text{height}$

$60\pi = 6\pi h$

$10 = h$

Thus,

Vol = area$_{base}$ · height

$= \pi r^2 h$

$= \pi \cdot 3^3 \cdot 10$

$= 90\pi$ units^3

And

SA = 2 · area$_{base}$ + lateral area

$= 2\pi r^2 + 60\pi$ (this was given)

$= 2\pi \cdot 3^2 + 60\pi$

$= 78\pi$ units^2

***6** You're given the surface area, so you have to begin with that formula:

SA = 2 · area$_{base}$ + lateral rectangle

SA = $2\pi r^2 + 2\pi rh$

Now plug in the given information:

$$54\pi = 2\pi r^2 + 2\pi r \cdot 6$$
$$54\pi = 2\pi(r^2 + 6r)$$
$$27 = r^2 + 6r$$
$$r^2 + 6r - 27 = 0$$
$$(r+9)(r-3) = 0$$
$$r = -9 \text{ or } 3$$

You can reject –9, so r is 3. The rest is a walk in the park:

$$\text{Vol} = \text{area}_{base} \cdot \text{height}$$
$$= \pi r^2 h$$
$$= \pi \cdot 3^2 \cdot 6$$
$$= 54\pi \text{ units}^3$$

This cylinder is unusual and interesting because both the surface area and the volume are 54π. (For extra credit: Do you see why I didn't say that the surface area and the volume are *equal*?)

7. To get the surface area, you need the slant heights, the lengths of $\overline{ZS}$ and $\overline{ZT}$. (Note that because this is not a regular pyramid — a pyramid with a regular polygon as its base and congruent lateral edges — these slant heights are not equal.) Then you use one of the slant heights to get the pyramid's height.

Keep looking for right triangles — that's the key to problems like this. $\triangle ASZ$ is a right triangle with a leg of 5 (half of $\overline{AB}$) and hypotenuse of $5\sqrt{10}$, so

$$(ZS)^2 + (AS)^2 = (ZA)^2$$
$$(ZS)^2 + 5^2 = \left(5\sqrt{10}\right)^2$$
$$(ZS)^2 = 225$$
$$ZS = 15$$

$\triangle BTZ$ is another right triangle, with a leg of 9 (half the length of $\overline{BC}$):

$$(ZT)^2 + 9^2 = \left(5\sqrt{10}\right)^2$$
$$(ZT)^2 = 169$$
$$ZT = 13$$

You're all set to do the surface area:

$$\text{SA}_{\text{Pointy-Top}} = \text{area}_{base} + \text{four lateral triangles}$$
$$= 10 \cdot 18 + \underbrace{\tfrac{1}{2}(10)(15)}_{\text{left face}} + \underbrace{\tfrac{1}{2}(18)(13)}_{\text{front}} + \underbrace{\tfrac{1}{2}(10)(15)}_{\text{right (same as left)}} + \underbrace{\tfrac{1}{2}(18)(13)}_{\text{back (same as front)}}$$
$$= 180 + 75 + 117 + 75 + 117$$
$$= 564 \text{ units}^2$$

For the volume, you need the height, ZN. Well, $\triangle ZNS$ is a right triangle with a leg, $\overline{SN}$, that measures half of BC (9) and a hypotenuse, $\overline{ZS}$, that's 15 units long. You can finish with the Pythagorean Theorem, or if you're on your toes, you notice that this triangle is in the 3:4:5 family and that ZN is thus 12.

$$\text{Volume}_{\text{Pointy-Top}} = \tfrac{1}{3} \text{area}_{base} \cdot \text{height}$$
$$= \tfrac{1}{3}(10 \cdot 18) \cdot 12$$
$$= 720 \text{ units}^3$$

8. You can get the slant height, AE, fairly easily because $\triangle AEC$ is a 30°- 60°- 90° triangle. The short leg, $\overline{CE}$, is 3 units long, so AE is $3\sqrt{3}$.

To get the height, AF, imagine looking down on the base, $\triangle BCD$, like this:

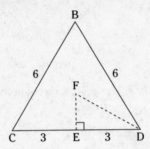

$\triangle EFD$ is, naturally, another 30°- 60°- 90° triangle. (You were expecting, maybe, a 29°- 57°- 94° triangle?) $\overline{ED}$ is half of edge $\overline{CD}$, so ED is 3, and that makes FE $\frac{3}{\sqrt{3}}$, or $\sqrt{3}$, and FD $2\sqrt{3}$.

Now you can use either $\overline{FE}$ with $\overline{AE}$ or $\overline{FD}$ with $\overline{AD}$ to get the height. I use $\overline{FE}$. Look back at the 3-D figure. $\triangle AFE$ is a right triangle, so you get AF with the Pythagorean Theorem:

$$(AF)^2 + (FE)^2 = (AE)^2$$
$$(AF)^2 + (\sqrt{3})^2 = (3\sqrt{3})^2$$
$$(AF)^2 + 3 = 27$$
$$AF = \sqrt{24} = 2\sqrt{6}$$

Time for the formulas, schmormulas:

$$\text{Vol}_{\text{Pointy-Top}} = \tfrac{1}{3} \, \text{area}_{\text{base}} \cdot \text{height}$$
$$= \tfrac{1}{3} \cdot \tfrac{s^2\sqrt{3}}{4} \cdot h$$
$$= \tfrac{1}{3} \cdot \tfrac{6^2\sqrt{3}}{4} \cdot 2\sqrt{6}$$
$$= 6\sqrt{18}$$
$$= 18\sqrt{2} \text{ units}^3$$

For surface area, I just realized that you don't need to use the formula. Instead, you can just use the fact that a regular tetrahedron is four equilateral triangles. Thus,

$$SA = 4 \cdot \tfrac{6^2\sqrt{3}}{4}$$
$$= 36\sqrt{3} \text{ units}^2$$

9. Circumference equals $2\pi r$, so

$$16\pi = 2\pi r$$
$$r = 8$$

Now, the height of 6 and radius of 8 form the legs of a right triangle with the slant height as its hypotenuse. You have a triangle in the 3:4:5 family, so the slant height is 10. Thus,

Chapter 11: Solid Geometry: Digging into Volume and Surface Area

$$\text{Vol}_{\text{Pointy-Top}} = \tfrac{1}{3} \text{area}_{\text{base}} \cdot \text{height}$$
$$= \tfrac{1}{3} \pi r^2 h$$
$$= \tfrac{1}{3} \pi r \cdot 8^2 \cdot 6$$
$$= 128\pi \text{ units}^3$$

$$\text{SA}_{\text{Pointy-Top}} = \text{area}_{\text{base}} + \quad \text{one lateral "triangle"}$$
$$\left(\tfrac{1}{2} \text{circumference} \cdot \text{slant height} \right)$$
$$= \pi r^2 + \tfrac{1}{2}(2\pi r)(\text{slant height})$$
$$= \pi \cdot 8^2 + \tfrac{1}{2}(2\pi \cdot 8)(10)$$
$$= 64\pi + 80\pi$$
$$= 144\pi \text{ units}^2$$

10 Here are the answers:

*a. Call the height of the left-side cone x; then the height of the right-side cone is $21 - x$.

$$\text{Total volume} = \text{vol}_{\text{left-side cone}} + \text{vol}_{\text{right-side cone}}$$
$$= \tfrac{1}{3} \cdot \text{area}_{\text{base}} \cdot \text{height} + \tfrac{1}{3} \cdot \text{area}_{\text{base}} \cdot \text{height}$$
$$= \tfrac{1}{3}(\pi \cdot 8^2)(x) + \tfrac{1}{3}(\pi \cdot 8^2)(21 - x)$$
$$= \left(\tfrac{64\pi}{3}\right)(x) + \left(\tfrac{64\pi}{3}\right)(21 - x)$$
$$= \tfrac{64\pi}{3}[x + (21 - x)]$$
$$= \tfrac{64\pi}{3} \cdot 21$$
$$= 448\pi \text{ units}^3$$

The way the x drops out tells you that the x is irrelevant and, therefore, that the volume of this shape will be the same regardless of how far to the left or right the circular "base" is. Pretty nifty, eh?

b. The surface area, on the other hand, does depend on where the "base" is. The lateral area of a cone equals $\tfrac{1}{2}$ (circumference)(slant height), so

$$80\pi = \tfrac{1}{2}(2\pi \cdot 8)(\text{slant height})$$
$$80\pi = 8\pi (\text{slant height})$$
$$\text{slant height} = 10$$

Will wonders never cease! You have another $3:4:5$ triangle here. So the height of the left-side cone is 6. Then, $21 - 6$ gives you 15, the height of the right-side cone. And then you notice, of course, that you have an 8-15-17 triangle on the right, so the right-side slant height is 17.

$$\text{SA} = \tfrac{1}{2}(\text{circumference})(\text{slant height})$$
$$= \tfrac{1}{2}(2\pi \cdot 8)(17)$$
$$= 136\pi \text{ units}^2$$

11 In many sphere problems (like with many circle problems), the key is finding the right radius or radii. Often, a radius becomes the hypotenuse of a right triangle. Find the right one? Here it is:

$r^2 = 2^2 + \sqrt{5}^2$
$r^2 = 9$
$r = 3$

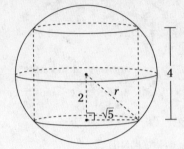

Thus,

$$\text{Vol}_{\text{Sphere}} = \frac{4}{3}\pi r^3$$
$$= \frac{4}{3}\pi \cdot 3^3$$
$$= 36\pi \text{ units}^3$$

$$\text{SA}_{\text{Sphere}} = 4\pi r^2$$
$$= 4\pi \cdot 3^2$$
$$= 36\pi \text{ units}^2$$

Only a sphere with a radius of 3 has a volume (in cubic units) equal to its surface area (in square units).

12 The surface area of a sphere equals $4\pi r^2$, so obviously, a hemisphere has a surface area of half that, or $2\pi r^2$:

$$\text{SA}_{\text{Hemisphere}} = 2\pi r^2$$
$$200\pi = 2\pi r^2$$
$$100 = r^2$$
$$r = 10$$

The radius of the cylinder is also 10, of course.

The "height" of the hemisphere (from its "peak" straight down to the center of its circular base) is just one of its radii, so that's 10. Because the total height is 50, the height of the cylinder is 50 – 10, or 40. Now you have what you need to finish:

$$\text{Total volume} = \text{vol}_{\text{cylinder}} + \text{vol}_{\text{hemisphere}}$$
$$= \text{area}_{\text{base}} \cdot \text{height} + \frac{1}{2}\text{vol}_{\text{sphere}}$$
$$= \pi \cdot 10^2 \cdot 40 + \frac{1}{2}\left(\frac{4}{3}\pi \cdot 10^3\right)$$
$$= 4{,}000\pi + \frac{2{,}000\pi}{3}$$
$$\approx 14{,}661 \text{ feet}^3$$

Chapter 12

Coordinate Geometry, Courtesy of Descartes

In This Chapter
- Line and segment formulas you may (or may not) fondly remember
- Completing coordinate proofs algebraically
- Working with handy equations for circles and lines

For someone who is said to have slept till 11 a.m. every day, Rene Descartes (1596–1650) — *not* pronounced "Dess-cart-eez" — sure achieved a lot: world famous philosopher, music theorist, physicist, and, of course, mathematician. Not too shabby, eh? Of interest here is the fact that he played a significant role in the evolution of geometry: He made the move from analyzing geometric shapes that exist independently of any location or orientation (the way the Greeks did geometry, and the way I've done problems up to this point in this book) to placing geometric shapes in the *x-y* coordinate system and using algebra to analyze them.

Formulas, Schmormulas: Slope, Distance, and Midpoint

Here are a few formulas you probably know (have a faint recollection of?) from Algebra I. You'll use these formulas to do the same sort of problems you have done in earlier chapters but in a completely different way:

- **Slope formula:** The slope of a line containing two points — (x_1, y_1) and (x_2, y_2) — is given by the following formula (don't ask me why, but the letter *m* is usually used for the slope):

  $$\text{Slope} = m = \frac{y_2 - y_1}{x_2 - x_1} = \frac{\text{rise}}{\text{run}}$$

- **Slope of horizontal lines:** The slope of a *horizontal* line is zero. Think about driving on a horizontal, flat road — the road has no steepness or slope.

- **Slope of vertical lines:** The slope of a *vertical* line is *undefined* (because the run is zero and you can't divide by zero). Think about driving up a vertical road — you can't do it; it's impossible. And it's impossible to compute the slope of a vertical line.

- **Slope of parallel lines:** The slopes of parallel lines are equal (unless both lines are vertical, in which case both of their slopes are undefined).

- **Slope of perpendicular lines:** The slopes of perpendicular lines are opposite reciprocals of each other, like 3 and $-\frac{1}{3}$ or $\frac{2}{5}$ and $-\frac{5}{2}$ (unless one line is horizontal [slope = 0] and the other line is vertical [slope is undefined]).

- **Midpoint formula:** The midpoint of the segment with endpoints at (x_1, y_1) and (x_2, y_2) is given by the formula

$$\text{Midpoint} = \left(\frac{x_1 + x_2}{2}, \frac{y_1 + y_2}{2}\right)$$

Just remember, the midpoint is the average of the *x*'s and the average of the *y*'s.

- **Distance formula:** The distance from (x_1, y_1) to (x_2, y_2) is given by the following formula:

$$\text{Distance} = \sqrt{(x_2 - x_1)^2 + (y_2 - y_1)^2}$$

The distance formula is simply the Pythagorean Theorem solved for *c*, the hypotenuse. The legs of the right triangle have lengths equal to the change in the *x*-coordinates and the change in the *y*-coordinates. If you just remember this connection, you can always solve a distance problem with the Pythagorean Theorem even if you forget the distance formula.

Q. Show that *ISOT* is an isosceles trapezoid.

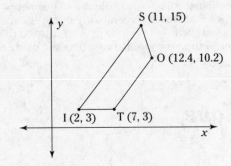

A. To prove that *ISOT* is an isosceles trapezoid, you must show that $\overline{IS} \parallel \overline{TO}$ (definition of a trapezoid; see Chapter 5) and that $\overline{IT} \cong \overline{SO}$ (the meaning of isosceles). (And for sticklers, there's one more thing to show: It's totally obvious from the diagram, but you have to show that $\overline{IT}$ is *not* parallel to $\overline{SO}$ — otherwise, *ISOT* would be a parallelogram by definition and thus not a trapezoid.)

A. (cont.)

First, check the slopes:

$$\text{Slope}_{\overline{IS}} = \frac{15 - 3}{11 - 2} = \frac{12}{9} = \frac{4}{3}$$

$$\text{Slope}_{\overline{TO}} = \frac{10.2 - 3}{12.4 - 7} = \frac{7.2}{5.4} = \frac{4}{3}$$

Check; $\overline{IS}$ is parallel to $\overline{TO}$.

Now check the lengths of $\overline{IT}$ and $\overline{SO}$ with the distance formula. Actually, although the distance formula works fine for *IT*, you don't need it. For vertical and horizontal segments, the distance is obvious. From *I* to *T*, you go straight across from 2 to 7, so the length is 5. For *SO*, you have

$$SO = \sqrt{(12.4 - 11)^2 + (10.2 - 15)^2}$$
$$= \sqrt{(1.4)^2 + (-4.8)^2}$$
$$= \sqrt{1.96 + 23.04}$$
$$= \sqrt{25}$$
$$= 5$$

Check; $\overline{IT} \cong \overline{SO}$.

You can easily check for yourself that $\overline{IT}$ is not parallel to $\overline{SO}$, so that does it: *ISOT* is an isosceles trapezoid.

Chapter 12: Coordinate Geometry, Courtesy of Descartes

1. Show that *ABCD* is a rhombus

   a. By showing the four sides congruent

   b. Without using the lengths of the sides

   Hint: Consider the other properties of a rhombus.

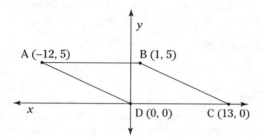

Solve It

2. Using the diagram,

   a. Show that *PLOG* is a parallelogram

   b. Find its area and perimeter

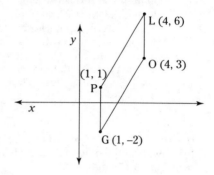

Solve It

3. Take a look at △*ABC*.

   a. What type of triangle is △*ABC*: acute, obtuse, or right? Equilateral, isosceles, or scalene?

   b. Find its area and perimeter

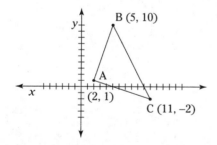

Solve It

4. Use the diagram and its labeled coordinates to

   a. Show that *KITE* is a kite

   b. Find its area

   c. Find the point where its diagonals intersect

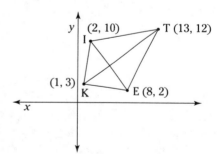

Solve It

Mastering Coordinate Proofs with Algebra

In this section, you get to see the power of doing geometry *analytically*, that is, with algebra. You prove the same types of things you proved in the chapter on quadrilaterals, but this time without any of the methods you used there (like congruent triangles, CPCTC, alternate interior angles, and so on). Sometimes proving something analytically is easier than with two-column proof methods. The second practice problem is a case in point. If you happen to see the trick, doing the proof the regular two-column way isn't that hard. But if you don't, you may not be able to do the proof the regular way. Doing it analytically, however, works like a charm.

Q. Use the isosceles trapezoid in the figure to prove that the diagonals in an isosceles trapezoid are congruent.

Note: I can't explain it fully here, so you have to take my word for it that this figure covers all conceivable isosceles trapezoids. You can place one vertex at the origin and another on the *x*-axis at $(a, 0)$ and put the whole trapezoid in the first quadrant "with no loss of generality," as mathematicians say. (*Caution:* It's probably not the best idea to use this phrase when you're out on a date.)

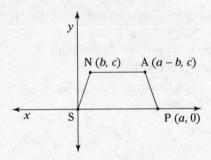

A. The proof is sort of one step long (or one idea long): You simply use the distance formula to show that the diagonals are congruent (for this property of isosceles trapezoids and more, check out Chapter 6):

$$SA \stackrel{?}{=} NP$$

$$\sqrt{(a-b-0)^2 + (c-0)^2} \stackrel{?}{=} \sqrt{(a-b)^2 + (0-c)^2}$$

$$\sqrt{(a-b)^2 + c^2} \stackrel{?}{=} \sqrt{(a-b)^2 + (-c)^2}$$

You know that c^2 is the same as $(-c)^2$, so these values are equal. That does it.

5. Given that the quadrilateral in the figure is a parallelogram, prove analytically that the diagonals of a parallelogram bisect each other.

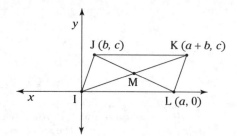

Solve It

6. Use the figure to prove that if you connect the midpoints of the sides of any quadrilateral, you create a parallelogram. (For an extra challenge, try to prove this with ordinary two-column proof methods.)

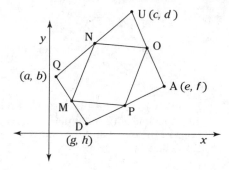

Solve It

Using the Equations of Lines and Circles

Lines and circles don't seem to have much in common at first glance; after all, lines are one-dimensional objects that go one forever in both directions, while circles are two-dimensional objects (if you count their interiors) that cover a definite amount of space. The major thing that lines and circles *do* have in common is that they become very important in further math classes, like trigonometry and calculus. These equations will keep popping up in your classes over and over again, so you might as well get used to 'em now and get ahead of the game.

Line equations: Here are the basic forms for equations of lines:

- **Slope-intercept form:**

    $y = mx + b$

    where m is the slope and b is the y-intercept $(0, b)$.

- **Point-slope form:**

    $y - y_1 = m(x - x_1)$

    where m is the slope and (x_1, y_1) is a point on the line.

- **Horizontal line:**

    $y = b$

    where b is the y-intercept.

- **Vertical line:**

  $x = a$

  where a is the x-intercept.

Circle equation. And here's the equation of a circle:

$(x - h)^2 + (y - k)^2 = r^2$

where (h, k) is the center of the circle and r is its radius.

Q. A circle whose center is at (6, 5) is tangent to a line at (2, 7). What are the equations of the circle and the line, and what is the line's y-intercept?

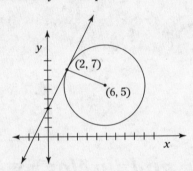

A. You have the circle's center, so all you need for the circle's equation is its radius. Use the distance formula:

$r = \sqrt{(6-2)^2 + (5-7)^2}$
$= \sqrt{4^2 + (-2)^2}$
$= \sqrt{20}$
$= 2\sqrt{5}$

Thus, the equation of the circle is

$(x-6)^2 + (y-5)^2 = (2\sqrt{5})^2$, or

$(x-6)^2 + (y-5)^2 = 20$

For the equation of the line, you have a point, so all you need is the slope. A line tangent to a circle is perpendicular to the radius drawn to the point of tangency, so first you need the slope of this particular radius:

$\text{Slope}_{\text{Radius}} = \frac{7-5}{2-6} = -\frac{1}{2}$

Because the line is perpendicular to the radius, their slopes are opposite reciprocals. The opposite reciprocal of $-\frac{1}{2}$ is 2, so that's the line's slope, and now you have everything you need to plug into the point-slope form:

$y - y_1 = m(x - x_1)$
$y - 7 = 2(x - 2)$

Finally, to get the y-intercept, just transform this equation into slope-intercept form:

$y - 7 = 2(x - 2)$
$y - 7 = 2x - 4$
$y = 2x + 3$

The y-intercept is (0, 3).

*7. A circle with equation $(x-7)^2 + y^2 = r^2$ is tangent to lines at $(4, 4)$ and $(11, -3)$. Find r and (a, b).

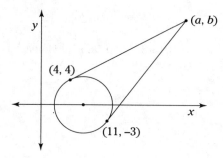

Solve It

**8. A line crosses the *x*- and *y*-axes at $(8, 0)$ and $(0, 6)$. A circle is inscribed in the triangle formed by the line and the two axes. Find

a. The equation of the circle. (***Hint:*** Draw in extra radii and "go for a walk.")

b. The point of tangency with the hypotenuse. (***Hint:*** Draw lines through the point of tangency so you can use similar triangles.)

c. Do part b. a second way. (***Hint:*** Use the equations of the hypotenuse and the radius to the point of tangency.) I think this is the easier of the two ways. It involves more math, but I suspect it's easier to see.

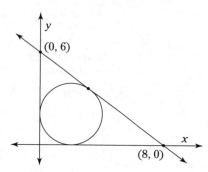

Solve It

Solutions for Coordinate Geometry, Courtesy of Descartes

1 Here's what you do:

a. From A to B, you go straight across from -12 to 1, so AB is 13. And DC is obviously also 13. Now use the distance formula for AD and BC (though you don't need to if you recognize the 5-12-13 triangles — see Chapter 3):

$$AD = \sqrt{(0-(-12))^2 + (0-5)^2}$$
$$= \sqrt{12^2 + (-5)^2}$$
$$= 13$$

$$BC = \sqrt{(13-1)^2 + (0-5)^2}$$
$$= \sqrt{12^2 + (-5)^2}$$
$$= 13$$

That's it. All four sides have a length of 13, so $ABCD$ is a rhombus.

b. You can show that $ABCD$ is a rhombus without using the lengths of the sides by first showing that $ABCD$ is a parallelogram and then that its diagonals are perpendicular (check out Chapter 6 for the properties of a rhombus).

$\overline{AB}$ and $\overline{DC}$ have slopes of 0, so they're parallel. Now check the slopes of $\overline{AD}$ and $\overline{BC}$:

$$\text{Slope}_{\overline{AD}} = \frac{0-5}{0-(-12)} = -\frac{5}{12}$$

$$\text{Slope}_{\overline{BC}} = \frac{0-5}{13-1} = -\frac{5}{12}$$

With two pairs of parallel sides, $ABCD$ must be a parallelogram.

Now check the slopes of the diagonals:

$$\text{Slope}_{\overline{AC}} = \frac{0-5}{13-(-12)} = -\frac{1}{5}$$

$$\text{Slope}_{\overline{DB}} = \frac{5-0}{1-0} = 5$$

Because 5 and $-\frac{1}{5}$ are opposite reciprocals, $\overline{AC} \perp \overline{DB}$. $ABCD$ is thus a rhombus, because a parallelogram with perpendicular diagonals is a rhombus.

2 Here's how this one unfolds:

a. $\overline{PG}$ and $\overline{LO}$ are both vertical, so they're parallel. Now check the other sides:

$$\text{Slope}_{\overline{PL}} = \frac{6-1}{4-1} = \frac{5}{3}$$

$$\text{Slope}_{\overline{GO}} = \frac{3-(-2)}{4-1} = \frac{5}{3}$$

That's all there is to it. $PLOG$ is a parallelogram.

b. You can use $\overline{PG}$ for the base of the parallelogram; its length is 3. The height of $PLOG$ is thus horizontal because it's perpendicular to the base, $\overline{PG}$; it goes straight to the right from $x = 1$ to $x = 4$. So the height is also 3, and the area of $PLOG$ is thus 3 times 3 (base times height), or 9 units^2 (see Chapter 6 for more on calculating the area of quadrilaterals).

The perimeter is a snap. $\overline{PG}$ and $\overline{LO}$ both have a length of 3. And

$$PL = \sqrt{(4-1)^2 + (6-1)^2}$$
$$= \sqrt{3^2 + 5^2}$$
$$= \sqrt{34}$$

Because you already know that *PLOG* is a parallelogram, $\overline{GO}$ has to be congruent to $\overline{PL}$, so it's also $\sqrt{34}$ units long. Thus, the perimeter of *PLOG* is $3 + 3 + \sqrt{34} + \sqrt{34}$, or $6 + 2\sqrt{34}$.

3. To solve these problems, you use the triangle basics I cover in Chapter 3.

   a. $\angle A$ looks like a right angle, so cross your fingers and check the slopes of $\overline{AB}$ and $\overline{AC}$ (if $\angle A$ is a right angle, this problem becomes much easier).

   $$\text{Slope}_{\overline{AB}} = \frac{10-1}{5-2} = 3$$

   $$\text{Slope}_{\overline{AC}} = \frac{-2-1}{11-2} = -\frac{1}{3}$$

   These answers are opposite reciprocals, so $\overline{AB} \perp \overline{AC}$, and thus $\angle A$ is a right angle; $\triangle ABC$ is a right triangle.

   Now compute the lengths of legs $\overline{AB}$ and $\overline{AC}$:

   $$AB = \sqrt{(5-2)^2 + (10-1)^2}$$
   $$= \sqrt{3^2 + 9^2}$$
   $$= \sqrt{90}$$
   $$= 3\sqrt{10}$$

   $$AC = \sqrt{(11-2)^2 + (-2-1)^2}$$
   $$= \sqrt{9^2 + (-3)^2}$$
   $$= 3\sqrt{10}$$

   $\overline{AB} \cong \overline{AC}$, so voilà, you have a 45°- 45°- 90° triangle, or, in other words, an isosceles right triangle.

   b. The area of a right triangle equals one half the product of its legs, so

   $$\text{Area}_{\triangle ABC} = \frac{1}{2}(3\sqrt{10})(3\sqrt{10})$$
   $$= 45 \text{ units}^2$$

   The legs are $3\sqrt{10}$, so the hypotenuse is $\sqrt{2} \cdot 3\sqrt{10}$, or $6\sqrt{5}$, and thus the perimeter of $\triangle ABC$ is $3\sqrt{10} + 3\sqrt{10} + 6\sqrt{5}$, or $6\sqrt{10} + 6\sqrt{5}$.

4. If you need a review, Chapter 5 explains the properties of kites.

   a. By definition, a kite must have two pairs of adjacent congruent sides. Use the distance formula:

   $$KI = \sqrt{(2-1)^2 + (10-3)^2}$$
   $$= \sqrt{1^2 + 7^2}$$
   $$= 5\sqrt{2}$$

   $$KE = \sqrt{(8-1)^2 + (2-3)^2}$$
   $$= 5\sqrt{2}$$

   So far, so good.

$$IT = \sqrt{(13-2)^2 + (12-10)^2}$$
$$= \sqrt{11^2 + 2^2}$$
$$= 5\sqrt{5}$$
$$ET = \sqrt{(13-8)^2 + (12-2)^2}$$
$$= \sqrt{5^2 + 10^2}$$
$$= 5\sqrt{5}$$

Bingo. *KITE* is a kite.

b. The area of a kite equals half the product of its diagonals (see Chapter 6), so you need their lengths:

$$KT = \sqrt{(13-1)^2 + (12-3)^2}$$
$$= \sqrt{12^2 + 9^2}$$
$$= 15$$

$$IE = \sqrt{(8-2)^2 + (2-10)^2}$$
$$= \sqrt{6^2 + (-8)^2}$$
$$= 10$$

$$\text{Area}_{KITE} = \frac{1}{2} d_1 d_2$$
$$= \frac{1}{2}(15)(10)$$
$$= 75 \text{ units}^2$$

c. $\overline{KT}$ is the perpendicular bisector of $\overline{IE}$ (property of a kite; see Chapter 5), so all you need is $\overline{IE}$'s midpoint:

$$\text{Midpoint}_{IE} = \left(\frac{x_1 + x_2}{2}, \frac{y_1 + y_2}{2}\right)$$
$$= \left(\frac{2+8}{2}, \frac{10+2}{2}\right)$$
$$= (5, 6)$$

5 The proof here is odd in a way, but it works. You might think that you have to first find where the diagonals cross and then show that this point bisects each diagonal. Instead, you simply show that the midpoints of the two diagonals are at the same point:

$$\text{Midpoint}_{JL} = \left(\frac{b+a}{2}, \frac{c+0}{2}\right)$$
$$= \left(\frac{b+a}{2}, \frac{c}{2}\right)$$
$$\text{Midpoint}_{IR} = \left(\frac{0+a+b}{2}, \frac{0+c}{2}\right)$$
$$= \left(\frac{a+b}{2}, \frac{c}{2}\right)$$

You're done. This simple procedure does, in fact, prove that the diagonals of any parallelogram bisect each other.

6 You need to get the coordinates of the midpoints using — hold onto your hat — the midpoint formula. Then use them to find the slopes of the sides of *MNOP*.

$$M = \left(\frac{a+g}{2}, \frac{b+h}{2}\right)$$

$$N = \left(\frac{a+c}{2}, \frac{b+d}{2}\right)$$

$$O = \left(\frac{c+e}{2}, \frac{d+f}{2}\right)$$

$$P = \left(\frac{g+e}{2}, \frac{h+f}{2}\right)$$

$$\text{Slope}_{MN} = \frac{\frac{b+d}{2} - \frac{b+h}{2}}{\frac{a+c}{2} - \frac{a+g}{2}}$$

$$= \frac{(b+d)-(b+h)}{(a+c)-(a+g)} \quad \text{(multiplying top and bottom by 2)}$$

$$= \frac{d-h}{c-g}$$

$$\text{Slope}_{PO} = \frac{\frac{d+f}{2} - \frac{h+f}{2}}{\frac{c+e}{2} - \frac{g+e}{2}}$$

$$= \frac{d-h}{c-g}$$

One pair of parallel sides down, one to go:

$$\text{Slope}_{MP} = \frac{\frac{h+f}{2} - \frac{b+h}{2}}{\frac{g+e}{2} - \frac{a+g}{2}}$$

$$= \frac{f-b}{e-a}$$

$$\text{Slope}_{NO} = \frac{\frac{d+f}{2} - \frac{b+d}{2}}{\frac{c+e}{2} - \frac{a+c}{2}}$$

$$= \frac{f-b}{e-a}$$

Bingo. It's a parallelogram. Pretty cool, eh? No matter what weird quadrilateral you begin with, you always get a parallelogram.

*7 The circle's equation gives you its center: (7, 0). Now use the distance formula to get the radius:

$$r = \sqrt{(4-7)^2 + (4-0)^2}$$
$$= \sqrt{(-3)^2 + 4^2}$$
$$= \sqrt{25} = 5$$

Well, bust my britches and bless my soul — another 3-4-5 triangle! What are the odds of that?

To find *(a, b)*, you need the equations of the tangent lines, and for that you need the slopes of the lines:

$$\text{Slope}_{\text{Radius to (4, 4)}} = \frac{4-0}{4-7} = -\frac{4}{3}$$

The tangent line is perpendicular to this radius, so its slope is the opposite reciprocal of $-\frac{4}{3}$, namely $\frac{3}{4}$. And now you have what you need for the point-slope form:

$$y - 4 = \frac{3}{4}(x - 4)$$

Use the same process for the other tangent line:

$$\text{Slope}_{\text{Radius to (11, -3)}} = \frac{-3 - 0}{11 - 7} = -\frac{3}{4}$$

The tangent line's slope is the opposite reciprocal of this, $\frac{4}{3}$, and thus its equation is

$$y - (-3) = \frac{4}{3}(x - 11)$$

Now find the point of intersection of the two lines by solving the system of equations with two unknowns. First solve each equation for y:

$$y - 4 = \frac{3}{4}(x - 4) \qquad y - (-3) = \frac{4}{3}(x - 11)$$
$$y = \frac{3}{4}(x - 4) + 4 \qquad y = \frac{4}{3}(x - 11) - 3$$

Now set the equations equal to each other and solve:

$$\frac{3}{4}(x - 4) + 4 = \frac{4}{3}(x - 11) - 3$$
$$9(x - 4) + 48 = 16(x - 11) - 36$$
$$9x - 36 + 48 = 16x - 176 - 36$$
$$9x + 12 = 16x - 212$$
$$224 = 7x$$
$$x = 32$$

Plugging this answer into either tangent line gives you a y-value of 25. Thus (a, b) is (32, 25).

** 8 And here's number 8.

   a. Jumpin' Jehosaphat! It's another 3 : 4 : 5 triangle! The leg along the y-axis is 6 and the leg along the x-axis is 8, so the hypotenuse is 10. Now, to find the radius, you need to "walk around" the triangle. Check out this figure.

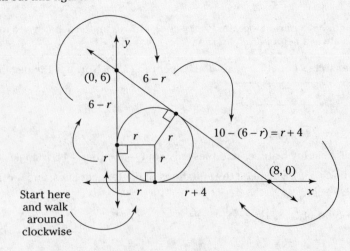

First, note that you have a square in the lower left-hand corner of the triangle and that all sides are equal to r, the radius. Next, because the vertical leg of the triangle is 6 and the little piece of that leg is r, the rest of it must be $6 - r$. Then, by the Dunce Cap Theorem, (see Chapter 8), the other side of the dunce cap (along the hypotenuse) is also $6 - r$. The whole hypotenuse is 10, so the rest of it is $10 - (6 - r)$, or $r + 4$. Finally, the other side of that dunce cap (along the *x*-axis) is also $r + 4$. You're all set to finish:

Along the bottom of the triangle,

$$r + (r + 4) = 8$$
$$2r = 4$$
$$r = 2$$

Knowing that r is 2, you can see that the circle's center is at (2, 2), so its equation is $(x - 2)^2 + (y - 2)^2 = 2^2$.

b. Draw a horizontal line through the point of tangency over to the *y*-axis (drawing a vertical line to the *x*-axis would work just as well), creating similar right triangles.

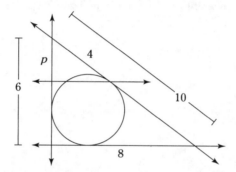

Look at the smaller right triangle in the figure above. In part *a.*, you learned that $r = 2$, so the hypotenuse of the small right triangle (which was set equal to $6 - r$ in part *a.*) has a length of 4. Now you can use similar triangles to find the coordinates of the point of tangency:

$$\frac{p}{6} = \frac{4}{10}$$
$$p = 2.4$$

Thus, the *y*-coordinate of the point of tangency is 2.4 below 6, or 3.6. You can find the *x*-coordinate with a related proportion, $\frac{q}{8} = \frac{4}{10}$ (or with the Pythagorean Theorem or by noticing that you have a 3-4-5 triangle shrunk down by a factor of 0.8). You should get an *x*-coordinate of 3.2, and thus, the point of tangency is (3.2, 3.6).

A simple way of thinking about this is that if you go, say, 40% of the way from one point to another in the *x-y* coordinate system (like you do in this problem, because the point of tangency is 40% of the way along the hypotenuse), you also go 40% in the *x*-direction (40% of 8 is 3.2) and 40% in the *y*-direction (40% of 6 is 2.4).

c. Another good way to find the point of tangency is to compute the intersection of the hypotenuse of the larger triangle and the radius drawn to the point of tangency. To do that, you need the equations of these two lines. The slope of the hypotenuse is $-\frac{3}{4}$ and its *y*-intercept is 6, so its equation is

$$y = -\frac{3}{4}x + 6$$

The radius is perpendicular to the hypotenuse, so its slope is $\frac{4}{3}$. This radius line contains the circle's center, (2, 2), so its equation is

$$y - 2 = \frac{4}{3}(x - 2), \text{ or}$$
$$y = \frac{4}{3}(x - 2) + 2$$

Set the right sides of the equations of the two lines equal to each other and solve:

$$-\frac{3}{4}x + 6 = \frac{4}{3}(x - 2) + 2 \quad \text{(Now multiply both sides by 12)}$$
$$-9x + 72 = 16(x - 2) + 24$$
$$-9x + 72 = 16x - 32 + 24$$
$$-25x = -80$$
$$x = 3.2$$

Plug this answer into $y = -\frac{3}{4}x + 6$, and you get a *y*-value of 3.6. Your point of tangency is thus (3.2, 3.6).

Chapter 13

Transforming the (Geometric) World: Reflections, Rotations, and Translations

In This Chapter
▶ A few reflections on reflections
▶ Shifting shapes with translations
▶ You spin me right round, Polly: Rotating polygons
▶ Reflecting thrice: Glide reflections

You can take any figure, say a triangle, and use a *transformation* to move it or change it in some way. You can slide it, flip it over, shrink it or blow it up, warp it into a different shape, and so on. In this chapter, you practice problems involving transformations that don't change the size or shape of a figure. Such transformations — called *isometries* — take a figure and move it, or *map* it, onto a congruent figure. The "before" figure is called the *pre-image,* and the "after" figure is called the *image*.

Reflections on Mirror Images

I begin this isometries journey with reflections, not because they're the simplest subject you need to tackle here but because they're the building blocks of all other isometries. In fact, you can use a series of reflections to perform all the other transformations I discuss later in this chapter. For example, in the next section, I show you that you can translate a figure in any direction (which you could do by just sliding it) by instead reflecting the figure over one line and then reflecting it again over another line. In fact, if you take, say, two congruent triangles and place them anywhere in the *x-y* coordinate system — one flipped over, if you like, and rotated to any angle — and you wanted to map one of the triangles onto the other by a series of transformations, you'd never have to rotate or slide the triangle. In one, two, or three reflections (you never need more than three), you can make the before triangle land exactly on the after triangle. I find this result interesting and somewhat surprising.

A couple more things before working through an example. (Egad! A sentence fragment!) First, check out Figure 13-1. △*ABC* has been reflected over line *l*. The result is congruent △*PQR*. △*ABC* has also been slid to the right (that move is a *translation,* if you were wondering), producing congruent △*XYZ*. △*PQR* and △*XYZ* are congruent, but there's a basic difference between them: their orientation. Figures like △*ABC* and △*XYZ* have the same *orientation* because you can make one stack perfectly on top of the other by sliding and/or rotating it onto the other. Figures like △*ABC* and △*PQR*, on the other hand, have opposite orientations because you can't possibly get △*ABC* to line up with △*PQR* without flipping △*ABC* over. Read on for some theorems.

272 Part V: 3-D Geometry and Coordinate Geometry

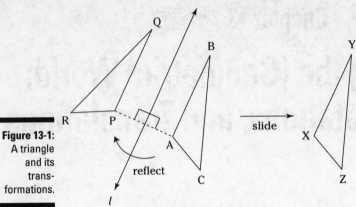

Figure 13-1: A triangle and its transformations.

Reflections and orientation:

- Reflecting a figure switches its orientation.
- If you reflect a figure and then reflect it again over the same line or a different line, the figure returns to its original orientation. More generally, if you reflect a figure an *even* number of times, the final result is a figure with the *same orientation*.
- Reflecting a figure an *odd* number of times produces a figure with the *opposite orientation*.

And here's one more thing about Figure 13-1. If you form $\overline{AP}$ by connecting pre-image A with its image point P (or B with Q or C with R), the reflecting line, l, is the perpendicular bisector of $\overline{AP}$. Pretty cool, huh?

Reflecting lines and connecting segments: When a figure is reflected, the *reflecting line* is the perpendicular bisector of all segments connecting points of the pre-image to corresponding image points.

After each transformation, you can label the image points with the *prime* symbol (′). If A is the pre-image point, the image point is A'.

Q. A specific transformation T maps (or sends) all points (x, y) to (y, x). Symbolically, $T(x, y) = (y, x)$. This transformation is a reflection. Given the coordinates of the vertices of $\triangle ABC$, find the coordinates of the reflection of $\triangle ABC$, which is $\triangle A'B'C'$, and find the equation of the reflecting line.

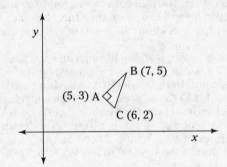

A. For vertex A, $T(5, 3) = (3, 5)$; that's A'

For B, $T(7, 5) = (5, 7)$; that's B'

For C, $T(6, 2) = (2, 6)$; that's C'

Chapter 13: Transforming the (Geometric) World: Reflections, Rotations, and Translations

Now sketch △ABC, △A'B'C', and the reflecting line.

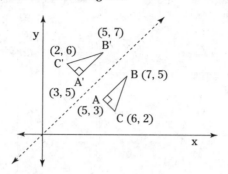

The reflecting line is the perpendicular bisector of $\overline{AA'}$ (and $\overline{BB'}$ and $\overline{CC'}$). To find this line, you first need the midpoint of $\overline{AA'}$; that's $\left(\frac{5+3}{2}, \frac{3+5}{2}\right)$, or (4, 4). Next, compute the slope of $\overline{AA'}$ (see Chapter 12 for more on slope and midpoints); that's $\frac{5-3}{3-5} = -1$. The perpendicular bisector is, of course, perpendicular to $\overline{AA'}$, so its slope is the opposite reciprocal of –1, which is 1. You have a point, (4, 4), and the slope, 1, of the perpendicular bisector, so you're all set to plug into the point-slope form (see Chapter 12 for more on line equations):

$y - y_1 = m(x - x_1)$

$y - 4 = 1(x - 4)$

$y - 4 = x - 4$

$y = x$

That's it.

---

1. Do the following pairs of figures have the same or opposite orientations?

a)

d)

b)

e)

c)

f)

Solve It

2. Reflect QRST over the line $y = x$.

a. Sketch Q'R'S'T', and give the coordinates of Q' and R'.

b. What shape is QQ'R'R?

c. What's the area and perimeter of QQ'R'R?

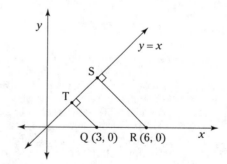

Solve It

274 Part V: 3-D Geometry and Coordinate Geometry

3. Sketch the reflected images and and give the coordinates of the following triangles.

  a. △ABC reflected over y = x to △A'B'C'
  b. △A'B'C' reflected over y = –x to △A"B"C"
  c. △A"B"C" reflected over the y-axis to △A'''B'''C'''

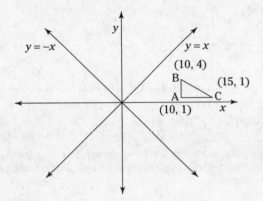

Solve It

4. Reflect △TUV over the line y = 3x + 2.

  a. Find the coordinates of △T'U'V'.

  Hint: You need the equations of $\overline{TT'}$, $\overline{UU'}$, and $\overline{VV'}$.

  b. Show that △T'U'V' ≅ △TUV.

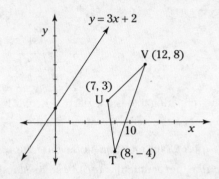

Solve It

Lost in Translation

Translating or sliding a figure is probably the simplest transformation to picture. It's so simple, in fact, that there wouldn't be much to say about it if it weren't for the fact that you can produce a translation with two reflections. You can picture how this works by imagining that you have a playing card — say, the ace of spades — face up in front of you on a table. Now, grab the bottom edge of the card and flip the card over (going up, away from you) leaving the top edge of the card where it is. You should now see a face down card whose bottom edge (the one close to you) is where the top edge was before you flipped it. Got it? If you repeat this flipping procedure, you should see the face-up ace again, pointing the same direction, and the card is now farther away from you a distance equal to twice the height of the card. Thus you see how two reflections (or flips) equals a slide.

Translations: A translation of a given distance along a given line is equivalent to two reflections over parallel lines that are perpendicular to the given line and separated by a distance equal to half the distance of the translation. As long as the parallel reflecting lines are separated by this distance, they can be located anywhere along the given line.

Chapter 13: Transforming the (Geometric) World: Reflections, Rotations, and Translations

Q. The translation $(x, y) \to (x - 12, y - 6)$ maps $\triangle TRI$ to $\triangle T'R'I'$.

  a. Find the distance the triangle has moved.

  b. Give the equations of two reflecting lines, l_1 and l_2, which — by reflecting $\triangle TRI$ first over l_1 and then over l_2 — will achieve the same result as the translation.

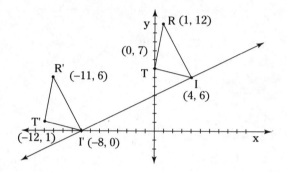

A. Here's how it all goes down:

  a. Piece o' cake. Just use the distance formula from Chapter 12 for II' (or TT' or RR'):

  $$II' = \sqrt{(-8-4)^2 + (0-6)^2}$$
  $$= \sqrt{144 + 36}$$
  $$= 6\sqrt{5}$$

  You can use a slight shortcut here if you realize that the translation instructions tell you that you're moving the figure 12 left and 6 down. If you see that, you just do $distance = \sqrt{12^2 + 6^2}$, and so on.

  b. You need two parallel lines perpendicular to $\overleftrightarrow{II'}$ and separated by half the length of $\overline{II'}$. There are, literally, an infinite number of correct answers. Here's an easy way to find a pair of lines that work:

  The pair of lines must be perpendicular to $\overleftrightarrow{II'}$, which has a slope of $\frac{6-0}{4-(-8)}$, or $\frac{1}{2}$, so the slope of the parallel lines is the opposite reciprocal of that, namely -2. The first line, l_1, can go through point I at $(4, 6)$. Its equation is thus

  $$y - 6 = -2(x - 4)$$
  $$y = -2x + 14$$

  Make the second line, l_2, parallel to l_1 (so its slope is also -2) and have it go through the midpoint of $\overline{II'}$. With this choice, you make the distance between l_1 and l_2 the required distance — half the length of $\overline{II'}$. The midpoint of $\overline{II'}$ is $\left(\frac{4+(-8)}{2}, \frac{6+0}{2}\right)$, or $(-2, 3)$. And thus, plugging those numbers into the point-slope form of a line (see Chapter 12), the equation of l_2 is

  $$y - 3 = -2[x - (-2)]$$
  $$y = -2x - 1$$

  Finito.

5. The translation $(x, y) \to (x, y + 5)$ maps *ISOC* onto *TRAP*. Find the equations of two reflecting lines that achieve the same result. Give three answers (in other words, three possible *pairs* of reflecting lines).

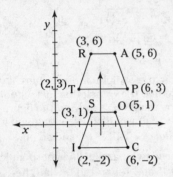

Solve It

6. The translation $(x, y) \to (x + 9, y + 2)$ maps $\triangle ABC$ onto $\triangle A'B'C'$. Find a pair of parallel reflecting lines that achieves the same result.

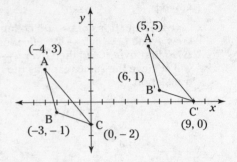

Solve It

So You Say You Want a... Rotation?

I can't give you a revolution, but I have a bunch of rotations waiting for you in this section. You know what *rotation* means, of course, but one thing you may not realize about rotation transformations is that they include not only spinning a figure where it is but also making it sort of move along an orbit centered at a point away from the figure (as in Figure 13-2). It might be more accurate to call this type of transformation a *revolution* instead of a rotation, but who am I to question the age-old terminology of geometry?

A rotation, just like a translation, can be achieved by a pair of reflections. Look at Figure 13-2.

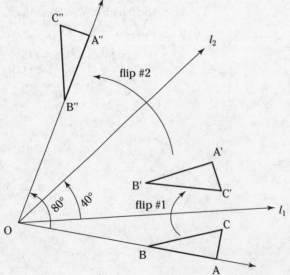

Figure 13-2: A rotation is equivalent to two reflections.

Chapter 13: Transforming the (Geometric) World: Reflections, Rotations, and Translations

You can see that △ABC has been rotated (revolved?) 80° to △A"B"C". Point O is called the *center of rotation*. It turns out that this same transformation can be achieved by reflecting △ABC over l_1 to △A'B'C' and then reflecting △A'B'C' over l_2 to △A"B"C". The reflecting lines must pass through the center of rotation, and the angle between them must be half the angle of rotation. Pretty nifty, eh?

Rotations: A rotation through a given angle around a center of rotation is equivalent to two reflections over lines passing through the center of rotation and forming an angle half the measure of the angle of rotation.

Q. △DEF has been rotated counterclockwise onto △D'E'F'. Find the center of rotation.

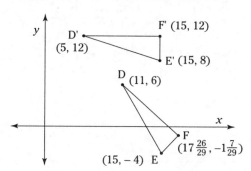

A. I haven't mentioned this process yet, so here I show you how to find a center of rotation. The trick is to use perpendicular bisectors. For this problem, the center of rotation lies at the intersection of the perpendicular bisectors of $\overline{DD'}$, $\overline{EE'}$, and $\overline{FF'}$. You need only two of these perpendicular bisectors, so use $\overline{DD'}$ and $\overline{EE'}$. (If you love working with fractions like $17\frac{26}{29}$, $\overline{FF'}$ would work as well.) The perpendicular bisector of $\overline{DD'}$ goes through its midpoint, which is $\left(\frac{11+5}{2}, \frac{6+12}{2}\right)$, or (8, 9). The slope of $\overline{DD'}$ is $\frac{12-6}{5-11}$, or –1, so the slope of the perpendicular bisector is the opposite reciprocal of that, which is 1. Write the equation of the line in point-slope form and convert to slope-intercept form (see Chapter 12). Thus, the equation of the perpendicular bisector of $\overline{DD'}$ is

$$y - 9 = 1(x - 8)$$
$$y = x + 1$$

Now do the same thing with $\overline{EE'}$. Its midpoint is $\left(\frac{15+15}{2}, \frac{-4+8}{2}\right)$, or (15, 2). $\overline{EE'}$ is vertical, so its perpendicular bisector is horizontal. The perpendicular bisector goes through (15, 2), so its equation is simply $y = 2$.

Finally, find the intersection of $y = 2$ and $y = x + 1$. That's (1, 2), the center of rotation. If you feel like it, locate (1, 2) on the figure, and then take a compass (no, not the kind a Boy Scout uses for orienteering, in case you were wondering) and place its point on (1, 2). You should be able to trace the circular arcs from D to D', E to E', and F to F'.

7. A clockwise rotation maps △ABC onto △A'B'C'. Find the center of rotation.
Tip: The math is a bit easier if you use $\overline{AA'}$ and $\overline{CC'}$.

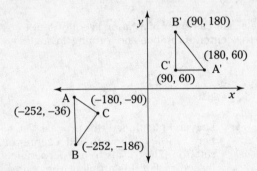

Solve It

8. △GHI has been rotated 90° counterclockwise onto △G'H'I'. The origin is the center of rotation. Give the equations of *three* pairs of reflecting lines that would achieve the same result.

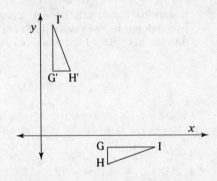

Solve It

Working with Glide Reflections

A *glide reflection* is, as its name suggests, a glide (that's a translation) followed by a reflection (or vice versa). It's also referred to as a *walk*. See Figure 13-3.

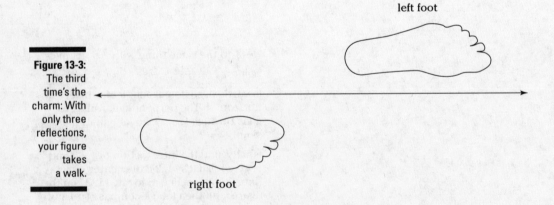

Figure 13-3: The third time's the charm: With only three reflections, your figure takes a walk.

Chapter 13: Transforming the (Geometric) World: Reflections, Rotations, and Translations

How can you map the right foot onto the left? Well, you can't do it with a translation or a rotation, because translations and rotations don't change orientation, and you can see that these feet have opposite orientations. Reflections do reverse orientation, but there's no reflecting line that you can use to map the right foot onto the left. As you may suspect from the title of this section, the answer is that only a glide reflection can accomplish the mapping. You can map the right foot onto the left foot by reflecting the right foot over the line and then sliding it to the right (or by sliding it first, then reflecting it).

As you can see in the previous section on translations, you can achieve a translation or slide with two reflections. Thus, the glide part of a glide reflection can be done with two reflections. And that means that you can do a glide reflection — like the right foot to left foot mapping in Figure 13-3 — with only three reflections. And three reflections is the most you ever need to map a figure to another congruent figure. To sum up, any two congruent figures are always one reflection, two reflections (a translation or a rotation), or three reflections (a glide reflection) away from each other.

After you find the reflecting line for a glide reflection, the transformation is a cinch, because it's just a reflection (which you should already know how to do) followed by a translation in the direction of the reflecting line (which you also already know how to do). The following theorem tells you the key to finding the reflecting line.

Glide reflections: In a glide reflection, the midpoints of all segments that connect pre-image points with their image points lie on the reflecting line.

Q. Find the reflecting line for the glide reflection.

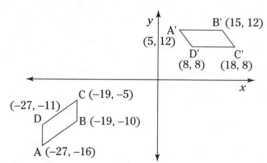

A. Pick any two point-image pairs, and make a segment out of each pair. Then, find the midpoints of these two segments. Next, find the slope of the line that goes through the two midpoints. (See Chapter 12 for info on slopes and midpoints.)

Using $\overline{AA'}$ and $\overline{BB'}$,

$$\text{Midpoint}_{\overline{AA'}} = \left(\frac{5-27}{2}, \frac{12-16}{2}\right) = (-11, -2)$$

$$\text{Midpoint}_{\overline{BB'}} = \left(\frac{15-19}{2}, \frac{12-10}{2}\right) = (-2, 1)$$

$$\text{Slope}_{\text{Reflecting Line}} = \frac{1-(-2)}{-2-(-11)} = \frac{1}{3}$$

Finally, just plug this slope and the point $(-2, 1)$ into the point-slope form for your equation:

$$y - 1 = \frac{1}{3}(x + 2)$$

That's all, folks.

*9. Use the transformation T(x, y) → (−x, y + 3).

  a. Transform △LEG using T(x, y). What are the new coordinates of L", E", and G"?

  b. Find the equation of the reflecting line.

  c. What are the coordinates of the image points (L', E', and G') obtained by reflecting L, E, and G over the line you found in part b., and what transformation, T$_{Reflect}$(x, y), achieves this reflection?

  d. After the reflection from part c. is completed, what transformation, T$_{Glide}$(x, y), completes the glide reflection?

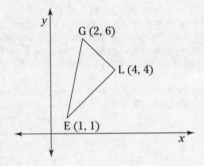

Solve It

*10. A glide reflection maps △PQR onto △P'''Q'''R'''.

  a. Find the equation of the reflecting line, l_1.

  b. The reflection of △PQR over the reflecting line, △P'Q'R', has the following coordinates: P' is at (−10, 8), Q' is at (−7, 14), and R' is at (−6, 6). Using the method from the section on translations, find the equations of two reflecting lines, l_2 and l_3, that will achieve the glide part of this glide reflection (the transformation that follows the reflection over the main reflecting line, l_1).

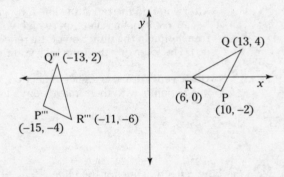

Solve It

Solutions for Transforming the World

1 Here are the answers concerning orientation:

  a. Opposite — you can't pair 'em up without a flip

  b. Same   **c.** Same

  d. These figures have neither the same nor opposite orientations because they're not congruent

  e. Opposite   **f.** Opposite

2 For $Q'R'S'T'$, here's what you get:

  a.

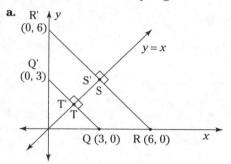

As you can see in the example problem, reflecting a figure over the line $y = x$ reverses the x- and y-coordinates of each point in the figure. Also note that S and S' are one and the same point. Ditto for T and T'. Any point that lies on the reflecting line stays put during a reflection.

  b. $QQ'R'R$ is an isosceles trapezoid.

  c. For the area of $QQ'R'R$, you could use the trapezoid area formula, but there's a much easier way. Call the origin point O. Now just subtract the area of right $\triangle OQ'Q$ from right $\triangle OR'R$:

$$\text{Area}_{QQ'R'R} = \text{area}_{OR'R} - \text{area}_{OQQ'}$$
$$= \tfrac{1}{2}bh - \tfrac{1}{2}bh$$
$$= \tfrac{1}{2}(6)(6) - \tfrac{1}{2}(3)(3)$$
$$= 18 - 4.5$$
$$= 13.5 \text{ units}^2$$

$\triangle OQ'Q$ and $\triangle OR'R$ are 45°- 45°- 90° right triangles, so that makes figuring the perimeter of $QQ'R'R$ a snap:

$$\text{Perimeter}_{QQ'R'R} = QQ' + Q'R' + R'R + RQ$$
$$= 3\sqrt{2} + 3 + 6\sqrt{2} + 3$$
$$= 6 + 9\sqrt{2}$$

3

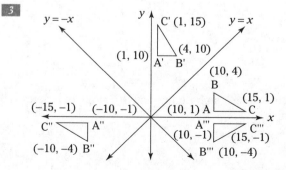

4 Here's what happens when you reflect $\triangle TUV$:

a. $\overline{TT'}$ must be perpendicular to the reflecting line $y = 3x + 2$, which has a slope of 3. Thus, $\overline{TT'}$ has a slope of $-\frac{1}{3}$, as do $\overline{UU'}$ and $\overline{VV'}$ (for more on finding slope, see Chapter 12). Plugging $-\frac{1}{3}$ and $(8, -4)$ into the point-slope form gives you the equation of $\overline{TT'}$:

$$y - (-4) = -\frac{1}{3}(x - 8)$$
$$y = -\frac{1}{3}x - \frac{4}{3}$$

Next, find where this line, $\overline{TT'}$, crosses $y = 3x + 2$:

$$-\frac{1}{3}x - \frac{4}{3} = 3x + 2$$
$$-x - 4 = 9x + 6$$
$$-10x = 10$$
$$x = -1$$

And plugging this answer into $y = 3x + 2$ gives you $y = -1$. So $\overline{TT'}$ crosses $y = 3x + 2$ at $(-1, -1)$. To get the coordinates of T', note that the reflecting line $y = 3x + 2$ must bisect $\overline{TT'}$, and thus $(-1, -1)$ must be the midpoint of $\overline{TT'}$. From T, $(8, -4)$, to $(-1, -1)$, you go left 9 and up 3. Do that again from $(-1, -1)$, and you get to T'. Left 9 from -1 brings you to -10, and up 3 from -1 brings you to 2. Thus, T' is at $(-10, 2)$.

In the interests of space, I'll skip the math for U' and V'. The procedure is identical to the one in the preceding paragraph. For the coordinates of U', you should get $(-5, 7)$, and for V', $(-6, 14)$.

b. You prove the triangles congruent with SSS, and to do that you just use the distance formula. Using the given coordinates of T, U, and V, you should get $5\sqrt{2}$, $5\sqrt{2}$, and $4\sqrt{10}$ for the lengths of the sides of $\triangle TUV$. And using the coordinates of T', U', and V' (which you calculated in part a.), you should get the same three lengths for $\triangle T'U'V'$. That does it.

5 Answers vary. The translation is vertical, so the reflecting lines must be horizontal. And the lines have to be separated by half the length of $\overline{IT'}$ (or $\overline{SR}$, $\overline{OA}$, or $\overline{CP}$), which is 5. Thus, any pair of horizontal lines separated by a distance of 2.5 will suffice. **Note:** The direction from l_1 to l_2 must be the same as the direction from the pre-image to the image.

Three possible answers are

✔ $l_1: y = -1$ and $l_2: y = 1.5$

✔ $l_1: y = 2$ and $l_2: y = 4.5$

✔ Or something crazy like $l_1: y = -1,002.5$ and $l_2: y = -1,000$

6 Find the slope and midpoint of $\overline{CC'}$ ($\overline{AA'}$ and $\overline{BB'}$ would work just as well):

$$\text{Slope}_{\overline{CC'}} = \frac{0 - (-2)}{9 - 0}$$
$$= \frac{2}{9}$$
$$\text{Midpoint}_{\overline{CC'}} = \left(\frac{0 + 9}{2}, \frac{-2 + 0}{2}\right)$$
$$= (4.5, -1)$$

You know l_1 and l_2 must be perpendicular to $\overline{CC'}$, so both lines have a slope of $-\frac{9}{2}$, or -4.5.

The first reflecting line, l_1, can go through C at $(0, -2)$:

$$y - (-2) = -4.5(x - 0)$$
$$y = -4.5x - 2$$

Chapter 13: Transforming the (Geometric) World: Reflections, Rotations, and Translations

Then, l_2 would go through the midpoint of $\overline{CC'}$:

$$y - (-1) = -4.5(x - 4.5)$$
$$y = -4.5x + 19.25$$

You're done.

7 You want to find the intersection of the perpendicular bisectors of $\overline{AA'}$ and $\overline{CC'}$. First, use the midpoint formula and the slope formula to compute the midpoint and slope of $\overline{AA'}$. You should get the following results:

$$\text{Midpoint}_{\overline{AA'}} = (-36, 12)$$
$$\text{Slope}_{\overline{AA'}} = \frac{2}{9}$$

The slope of the perpendicular bisector of $\overline{AA'}$ is the opposite reciprocal of the slope of $\overline{AA'}$, so its slope is $-\frac{9}{2}$, or -4.5. And thus, its equation is

$$y - 12 = -4.5[x - (-36)]$$
$$y = -4.5x - 150$$

Using the same method, you obtain the following for the equation of the perpendicular bisector of $\overline{CC'}$:

$$y = -1.8x - 96$$

The center of rotation lies at the intersection of these two perpendicular bisectors, so set the equation equal to each other:

$$-4.5x - 150 = -1.8x - 96$$
$$-45x - 1{,}500 = -18x - 960$$
$$-27x = 540$$
$$x = -20$$

Plugging $x = -20$ into either equation gives you a y value of -60, so the center of rotation is at $(-20, -60)$.

8 The rotation is 90° counterclockwise about the origin, so the reflecting lines must pass through the origin and form a 45° angle (half of 90°). Three possible answers are

- ✓ l_1: $y = 0$ (the x-axis) and l_2: $y = x$
- ✓ l_1: $y = x$ and l_2: $x = 0$ (the y-axis)
- ✓ l_1: $x = 0$ and l_2: $y = -x$

But any two lines work as long as they go through the origin and form a 45° angle. For example:

$$l_1: y = \frac{3}{4}x \quad \text{and} \quad l_2: y = 7x$$

*9 Here's what happens with $\triangle LEG$:

a. $T(x, y) = (-x, y + 3)$ sends points L, E, and G to the following image points:

$L'' = T(4, 4) = (-4, 7)$
$E'' = T(1, 1) = (-1, 4)$
$G'' = T(2, 6) = (-2, 9)$

b. You can use any two point-image pairs to find the reflecting line. How about $\overline{EE''}$ and $\overline{GG''}$?

$$\text{Midpoint}_{\overline{EE''}} = \left(\frac{1 + (-1)}{2}, \frac{1 + 4}{2} \right) = (0, 2.5)$$

$$\text{Midpoint}_{\overline{GG''}} = \left(\frac{2 + (-2)}{2}, \frac{6 + 9}{2} \right) = (0, 7.5)$$

Both midpoints are on the y-axis (if you realized that they would be before doing the math, you're a geometry natural), so the reflecting line must be the y-axis; its equation, of course, is $x = 0$.

c. Reflecting L, E, and G gives you

$L' = (-4, 4)$

$E' = (-1, 1)$

$G' = (-2, 6)$

The transformation that flips a figure over the y-axis is $T_{Reflect}(x, y) = (-x, y)$.

d. The transformation is just a slide straight up a distance of 3. That's achieved by $T_{Glide}(x, y) = (x, y + 3)$.

* 10 Here's how this glide reflection problem pans out:

a. Find two midpoints:

$$\text{Midpoint}_{\overline{QQ''}} = \left(\frac{13 + (-13)}{2}, \frac{4 + 2}{2}\right) = (0, 3)$$

$$\text{Midpoint}_{\overline{RR''}} = \left(\frac{6 + (-11)}{2}, \frac{0 + (-6)}{2}\right) = (-2.5, -3)$$

The reflecting line, l_1, connects these midpoints. Its slope is given by:

$$\text{Slope} = \frac{-3 - 3}{-2.5 - 0} = \frac{-6}{-2.5} = \frac{12}{5}$$

And you know its y-intercept is 3 from the midpoint of $\overline{QQ''}$, so its equation is

$$y = \frac{12}{5}x + 3$$

b. The translation following the reflection in a glide reflection is in the direction of the reflecting line. The theorem from the section on translations tells you that, therefore, any two reflecting lines, l_2 and l_3, that achieve this translation must be perpendicular to the main reflecting line, l_1. You know l_1 has a slope of $\frac{12}{5}$, so l_2 and l_3 must have slopes of $-\frac{5}{12}$.

An easy way to pick locations for l_2 and l_3 so they'll do the trick — using the method from the section on translations — is to run l_2 through one of the points you know, say R', and then run l_3 through the midpoint of $\overline{R'R'''}$. R' is at $(-6, 6)$, so that gives you the following equation for l_2:

$$y - 6 = -\frac{5}{12}(x - (-6))$$

$$y = -\frac{5}{12}x + 3.5$$

The midpoint of $\overline{R'R'''}$ is $\left(\frac{-6 - (-11)}{2}, \frac{6 + (-6)}{2}\right)$, or $(-8.5, 0)$, so the equation of l_3 is

$$y - 0 = -\frac{5}{12}(x - (-8.5))$$

$$y = -\frac{5}{12}x - \frac{85}{24}$$

Part VI
The Part of Tens

In this Part . . .

You get the top ten geometry formulas and the top ten strategies for doing proofs. The formulas come in handy for non-proof problems. The proof strategies can make many otherwise difficult proofs much easier to do. If you learn them well and have them at the ready when you're doing proofs, you won't get stuck very often, and if you do get stuck, you'll know what to do.

Chapter 14
Ten (Plus) Incredibly Fantastic Strategies for Doing Proofs

In This Chapter
- Looking over the diagrams before you start
- Thinking forwards and backwards
- Getting unstuck when you're stuck

For many students, doing two-column proofs is one of the most difficult exercises in the entire high school mathematics curriculum. So if proofs seem difficult to you, you're not alone. If you get stuck while doing a proof, don't panic. Just try some of the handy strategies in this chapter, which can often turn a difficult proof into an easy one.

The first four strategies involve features you should look for in the proof's diagram (or in the givens) before you try to figure out how to do the proof — unless, of course, the proof is very short or easy. The final six techniques involve planning and writing out the proof itself. And since you've been incredibly fantastic, I've thrown in an 11th bonus tip for no extra charge.

Look for Congruent Triangles

Glance at the diagram and look for congruent triangles. Try to find *all* pairs of congruent triangles. If you find any, proving one or more of these pairs of triangles congruent (with SSS, SAS, ASA, AAS, or HLR) will likely be an important part of the proof. After stating that the triangles are congruent, you'll almost certainly use CPCTC (corresponding parts of congruent triangles are congruent) on the very next line. (I cover all the fun you can have with triangles, congruent and otherwise, in Chapter 3.)

Try to Find Isosceles Triangles

Check out the diagram and look for all triangles that appear to be isosceles. If you find any, you'll very likely use one of the following isosceles theorems in the proof (as I mention in Chapter 4):

- If angles, then sides.
- If sides, then angles.

Look for Radii, and Draw More Radii

If the proof's diagram contains a circle, open your eyes and make sure you notice every radius. Mark all of them congruent, of course.

Penciling in another radius or two is often useful. Draw in new radii to labeled points on the circle or to points where a line intersects the circle or is tangent to it. Don't draw a new radius to "the middle of nowhere" on the circle — in other words, to a point on the circle where nothing else is going on. The new radii you draw may form one or more sides of a triangle that you need to solve the proof. (See Chapter 8 for more on this helpful radii tip.)

Look for Parallel Lines

Look for parallel lines in the proof's diagram or in the givens. If you find any, you'll probably use one or more of the parallel-lines-transversal theorems that I discuss in Chapter 5. And it's also quite likely that parallel lines will form the sides of congruent or similar triangles.

Make a Game Plan

After looking over the diagrams, you're ready to dive in and try to do the proof. Try to figure out how to get from the givens to the *prove* conclusion with your common sense before you worry about writing the formal, two-column proof and before you think about which theorems to use. If you can find a way to get from the givens to the prove (or at least most of the way), writing the formal proof will be easier.

Before looking for a game plan, make sure you've marked all congruent angles and segments on the proof diagram. Doing so makes the givens come to life and can help you see the logic of the proof. (If you want to read more about game plans, check out Chapter 2.)

Make Up Numbers for Segments and Angles

As part of making a game plan for a proof, it's often very helpful to make up arbitrary lengths for segments or degree measures for angles. This will help you see how the proof works.

Another great time to use this technique is when you're not sure which theorem, postulate, or definition to use in the reason column (for example, angle addition, segment subtraction, complements of congruent angles, Like Divisions, Like Multiples, and so on). Make up numbers for the segments and angles in the given and for unmentioned segments and angles, but don't make up numbers for the segments and angles in the *prove* conclusion. Then do some simple arithmetic. The way the segments or angles add up, subtract, and so on may give you some ideas about which reason to use. (Look at Chapter 2 for some examples.)

Use All the Givens

If you get stuck while doing a proof, don't throw in the towel until you've asked yourself why the problem includes each given. Geometry book authors don't put irrelevant givens in proofs. All the givens are there for a reason, as I tell you in Chapter 2.

Often, the use for a given is pretty obvious. Before you give up, put each given down in the statement column and write a statement that follows from the given, even if you're not sure where it'll lead. This step can trigger important insights.

Check Your If-Then Logic

Every reason (or just about every one) can appear in *if-then* form, as you can find out in Chapter 1. But after doing many proofs, most students start shortening their reasons, writing things like "All right angles are congruent" rather than "If two angles are right angles, then they're congruent." This short way of writing proofs is all fine and good; however, if you're not sure what reason to use or if a proof is stumping you, write out your reasons the long way in if-then form. And then, for each reason, check that

- All the ideas in the *if* clause appear in the statement column somewhere *above the line you're checking*.
- The single idea in the *then* clause appears in the statement column *on the same line*.

In addition to letting you check your reasons, this strategy can help you figure out which reason to use in the first place. If you feel stumped about what reason to use on a particular line, look at the lines above that one in the statement column. The *if* clause of the reason you're trying to come up with must contain an idea from above in the statement column. Often, you won't have too much to choose from and the reason will practically write itself.

Work Backwards

Thinking about how a proof will end can be very helpful. You know, of course, that the final statement has to be the *prove* conclusion. Then, using that knowledge, it's often pretty easy to see what the final reason should be (you should at least be able to make an educated guess about the final reason). And then, use your *if-then* logic to try to guess what the second-to-last statement should be. Sometimes you may be able to work backwards to the third-to-last statement or even further. This thought process can be extremely useful for getting all your ducks in a row. (Go to Chapter 2 to see more about how to use this idea.)

Think Like a Computer

You can't skip a step just because it's obvious to everyone. In a two-column proof, every single step in the chain of logic needs to be expressed. (I show you one example of this in Chapter 4.)

If you drop your coat in a puddle, everyone knows it'll get wet. You don't have to go through some logical proof to convince someone that your coat will get wet. You don't have to say, for example, "There's water in the puddle, and if the coat falls in the puddle, it will, therefore, get water on it; and if it gets water on it, it will, by definition, get 'wet.'" But this sort of proving the obvious is often what you have to do in a geometry proof.

One way to think about the step-by-step logic in a proof is what you'd have to go through to "convince" a computer. Computers don't have common sense. It wouldn't be obvious to a computer that a coat dropped in a puddle would get wet. You have to spell out every single step because computers know only what you tell them. That's the mindset you need while doing a proof.

Here's another way to think about the necessary logic: Imagine that you go to use your debit card at an ATM. Say your PIN is "house" (maybe I should call it a *PIW* for Personal Identification *Word*). Okay, so you put your card into the ATM, but you enter "home" instead of "house." Needless to say, even though people know that a house is the same as a home, the ATM won't accept your altered PIW. For machines, every little thing has to be precisely spelled out (no pun intended). That's how you have to think when writing geometry proofs.

Bonus! Number 11 (Like the Amp in This Is Spinal Tap That Goes Up to 11): Do Something!

This strategy may not sound legitimate, but in a sense, it's incredibly important. Doing proofs can sometimes seem overwhelming, and getting stuck now and then when trying to solve a proof is par for the course. But don't give up too easily, and don't let yourself leave a proof completely blank. If you patiently go through the previous strategies one by one, asking yourself whether each one can help you with a particular proof, you'll always be able to figure out at least part of the proof. When you do, put something down on paper even if you don't think you'll be able to finish the proof. Often, one idea leads to another, and before you know it, more and more of the proof gets done. And even if you can't finish the proof, your teacher may give you partial credit for what you do write.

Chapter 15

Ten Things You Better Know (for Geometry), or Your Name Is Mudd

In This Chapter
- Revisiting the best of triangles!
- Checking out area formulas
- Using volume and surface area formulas

1 actually don't have any problem with people named Mudd (for all you Mudds out there who are reading this book), but if you don't know these things, you really should go back and look through this book again! You need all the formulas and theorems in this chapter if you really want to be an expert in "the study of shapes."

The Pythagorean Theorem (the Queen of All Geometry Theorems)

The sum of the squares of the legs of a right triangle is equal to the square of the hypotenuse, or

$a^2 + b^2 = c^2$ (See Chapter 3.)

Special Right Triangles

The first four triangles in this section are so-called Pythagorean triple triangles. They're special because the lengths of all three sides are integers, which doesn't happen often with the Pythagorean Theorem (usually you get a square root of something for at least one of the sides):

- The 3-4-5 triangle
- The 5-12-13 triangle
- The 7-24-25 triangle
- The 8-15-17 triangle

The next two triangles are special because they're related to two of the most basic shapes in geometry: The first is half of a square, and the second is half of an equilateral triangle. They come up all the time in problems, so make sure you know them! (See Chapter 3 for details.)

- The 45°- 45°- 90° triangle, whose sides are in the ratio of $x:x:x\sqrt{2}$
- The 30°- 60°- 90° triangle, whose sides are in the ratio of $x:x\sqrt{3}:2x$

Area Formulas

The following formulas give you the area of triangles and special quadrilaterals (see Chapter 6):

- $Area_{Triangle} = \frac{1}{2} base \cdot height$
- $Area_{Parallelogram} = base \cdot height$

  (This formula also works for rectangles and squares because they're parallelograms.)
- $Area_{Kite} = \frac{1}{2} diagonal_1 \cdot diagonal_2$

  (This formula also works for rhombuses and squares because they're kites.)
- $Area_{Trapezoid} = \frac{base_1 + base_2}{2} \cdot height$

Sum of Angles

The sum of the *interior* angles of a polygon with n sides is $(n-2)180°$. The sum of the *exterior* angles of any polygon is 360°. (See Chapter 6 for more information.)

Circle Formulas

Try these equations (which you can find in Chapter 9) when you work with circumference and area:

- Circumference = $2\pi r = \pi d$
- $Area_{circle} = \pi r^2$

Angle-Arc Theorems

In some circle problems, you can have an angle whose vertex is *on* the circle or whose vertex is *outside* the circle or whose vertex is *inside* the circle. The following formulas give you the connection between the size of the angle and the arc it intercepts (see Chapter 9). Figure 15-1 gives examples of the types of angles these formulas apply to:

- Angle *on* a circle = $\frac{1}{2} arc_1$
- Angle *outside* a circle = $\frac{1}{2}(arc_2 - arc_3)$

- Angle *inside* a circle = $\frac{1}{2}(arc_4 + arc_5)$

  Note: You get an angle inside a circle when two chords cross each other, forming an *X*; for this formula, you use the arcs intercepted by the angle you want and its vertical angle.

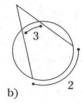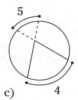

Figure 15-1: Angles (a) on, (b) outside, (c) inside a circle.

Power Theorems

Memorize the following theorems and become a geometry powerhouse (see Chapter 9):

- **Chord-chord:** part · part = part · part
- **Secant-secant:** whole · outside = whole · outside
- **Secant-tangent:** whole · outside = tangent^2

All three of these theorems follow the same simple rule:

(vertex to circle) · (vertex to circle) = (vertex to circle) · (vertex to circle)

Coordinate Geometry Formulas

Given two points in the coordinate plane, (x_1, y_1) and (x_2, y_2), you can compute the slope between the two points, the halfway point between the points, and the distance from one point to the other with the following formulas (see Chapter 12):

- Slope = $\frac{y_2 - y_1}{x_2 - x_1}$
- Midpoint = $\left(\frac{x_1 + x_2}{2}, \frac{y_1 + y_2}{2}\right)$
- Distance = $\sqrt{(x_2 - x_1)^2 + (y_2 - y_1)^2}$

Volume Formulas

Here's how to find the volume of spheres, flat-top solids like cylinders and prisms, and pointy-top solids like pyramids and cones (see Chapter 11).

- $\text{Vol}_{\text{Sphere}} = \frac{4}{3}\pi r^3$
- $\text{Vol}_{\text{Flat-Top solids}} = \text{area}_{\text{base}} \cdot \text{height}$
- $\text{Vol}_{\text{Pointy-Top solids}} = \frac{1}{3} \text{area}_{\text{base}} \cdot \text{height}$

Surface Area Formulas

And here's how to find the surface area of spheres, flat-top solids, and pointy-top solids (see Chapter 11):

- $\text{SA}_{\text{Sphere}} = 4\pi r^2$
- $\text{SA}_{\text{Flat-Top solids}} = 2 \cdot \text{area}_{\text{base}} +$ area of rectangle that wraps around

  recall that the area of this rectangle is equal to the height of the solid times its perimeter (if its a prism) or its circumference (if its a cylinder)

- $\text{SA}_{\text{Pointy-Top solids}} = \text{area}_{\text{base}} +$ area of "triangle" that wraps around

  recall that the area of this "triangle" is equal to one half of the slant height of the solid times its perimeter (if its a pyramid) or its circumference (if its a cone)

Index

• Numerics •

30°- 60°- 90° triangle, 70
45°- 45°- 90° triangle, 70

• A •

AAS (Angle-Angle-Side)
　proving triangles congruent, 96–97
　proving triangles similar, 168–169
acute angles, 52
adding angles and segments, 30
alternate exterior angles, 115
alternate interior angles, 115
altitude
　Altitude-on-Hypotenuse Theorem, 175–176
　defined, 55
　intersecting, 59
　and orthocenter, 60
Altitude-on-Hypotenuse Theorem, 175–176
Angle-Angle-Side (AAS)
　proving triangles congruent, 96–97
　proving triangles similar, 168–169
angle-arc theorems, 214–216, 292–293
Angle-Bisector Theorem, 178
Angle-Side-Angle (ASA) triangle theorem, 84–86
angles
　acute, 52
　adding, 30
　bisection, 22, 59, 60
　central, 199
　on circles, 214, 292
　complementary, 26–27
　congruent, 33–35, 37
　defined, 20
　dividing, 33–35
　inscribed, 214
　inside circles, 214–215, 292
　interior vs. exterior, 157–158
　multiplying, 33–35
　obtuse, 52
　outside circles, 215, 293
　right, 24, 52, 96–97
　subtracting, 30
　sum, 292
　supplementary, 26–27
　transversals and parallel lines, formed by, 115
　trisection, 22
　using theorems in proofs, 289
　vertical, 37
apothem, 155
Arc-Length Theorem, 211–212
arcs
　and central angles, 199
　and chords, 199
　defined, 199
　intercepted, 214–215
　length, 211–212
　theorems, 199
area formulas
　circles, 292
　equilateral triangles, 56
　kites, 292
　parallelograms, 151, 292
　polygons, 151
　rectangles, 292
　rhombuses, 292
　squares, 292
　trapezoids, 151, 292
　triangles, 55–56, 292
ASA (Angle-Side-Angle) triangle theorem, 84–86
assumptions about figures, 9–10

• B •

bisectors. *See* angles, bisection; perpendicular bisectors

• C •

center of rotation, 277
centers, triangle
　centroids as, 58, 60
　circumcenter point, 59–61
　incenter point, 59–61
　orthocenter point, 59–61

central angles
  and arcs, 199
  and chords, 199
  defined, 199
  theorems, 199
centroids, triangle, 58, 60
chord-chord angle, 214–215
Chord-Chord Power Theorem, 217, 293
chords
  and arcs, 199
  central angles, 199
  congruent, 195
  defined, 195
  theorems, 199
circles
  angle-arc theorems, 214–216, 292–293
  angles inside, 214–215, 292
  angles on, 214, 292
  angles outside, 215, 293
  arc length, 211–212
  area formula, 292
  circumference, 292
  completing proofs, 196
  congruent, 199
  defined, 261
  diameter, 195
  inscribed angles, 214
  measuring angle segment length, 217–218
  measuring angle size, 214–215
  properties, 195–196
  radius, 195
  sector area, 211–212
  and tangent line, 202–203
  theorems, 195, 199
circumcenter point, 59–61
circumference, 292
common-tangent problems, 203
complementary angles, 26–27
cones
  computing surface area, 246, 294
  computing volume, 246, 294
corresponding angles, 115
Corresponding Parts of Congruent Triangles are Congruent (CPCTC), 89–90
Corresponding Sides of Similar Triangles are Proportional (CSSTP), 172–173
cross-angles, 37

cylinders
  computing surface area, 243, 294
  computing volume, 243, 294

• D •

denominator, rationalizing, 72
diagonals, number of, 158
diameter, 195
distance formula, 258, 293
dividing angles and segments, 33–35
Dunce Cap Theorem, 203

• E •

equidistance theorems, 99–101
equilateral triangles
  altitude, 55
  area formula, 56
  defined, 51
exterior angles, 157–158, 292

• F •

45°- 45°- 90° triangle, 70

• G •

geometry overview, 9–11
givens, using, 31, 33, 89–90, 287, 288
glide reflections, 278–279

• H •

Hero's formula, 55
HLR (Hypotenuse-Leg-Right angle), 96–97
horizontal lines, 257, 261
hypotenuse
  Altitude-on-Hypotenuse Theorem, 175–176
  and distance formula, 258
  HLR (Hypotenuse-Leg-Right angle), 96–97
  in Pythagorean Theorem, 64, 70–72

• I •

if-then logic
  checking logic, 288–289
  overview, 12–13
  in two-column proof format, 14–16
incenter point, 59–61
interior angles, 157–158, 292
intersections, 20
isometries, 271–272
isosceles trapezoids
  area formula, 152
  congruent diagonals in, 260
  defined, 120
  illustrated, 120
  properties, 131
  as type of quadrilateral, 120
isosceles triangles
  altitude, 55
  defined, 51
  if angles, then sides theorem, 93–94
  if sides, then angles theorem, 93–94
  role in doing proofs, 287

• K •

kites
  area formula, 151, 292
  defined, 120
  illustrated, 120
  properties, 123–125
  proving quadrilaterals as, 133, 134
  as type of quadrilateral, 120

• L •

Like Divisions Theorem, 33, 35
Like Multiples Theorem, 33, 34
line segments. *See* segments
lines
  defined, 19
  foot of, 233
  forms for equations, 261
  horizontal, 257, 261
  midpoint, 22, 258, 293
  parallelism, 178, 233–235, 257, 288
  perpendicular, 24, 229–230, 258
  and planes, 233
  point-slope form, 261
  and slope, 257–258
  slope-intercept form, 261
  tangents to circles, 202–203
  transversals, 115–117, 178
  vertical, 257, 262

• M •

medians, triangle, 58, 60
midpoint formula, 22, 258, 293
multiplying angles and segments, 33–35

• O •

obtuse angles, 52
obtuse triangles
  altitude, 55
  defined, 52
orientation, 272
orthocenter point, 59–61

• P •

parallelism in lines
  intersecting, 178
  and planes, properties of, 233–235
  role in doing proofs, 288
  slope, 257
  transversals, 115–117, 178
parallelograms
  area formula, 151, 292
  defined, 120
  illustrated, 120
  properties, 123–125
  proving quadrilaterals as, 133–134
  as type of quadrilateral, 120
perpendicular, defined, 13
perpendicular bisectors
  and circumcenter, 60
  defined, 59
  and equidistance theorems, 99–101
  finding center of rotation, 277
perpendicular lines, 24, 229–230, 258

perpendicularity
  multiple, 233–235
  overview, 24
  properties, 233
  and tangent lines, 202–203
planes
  coordinate geometry formulas, 293
  defined, 229
  determining, 233
  intersecting, 233
  parallel, 233
points, defined, 19
points of tangency, 202–203
point-slope form, 261
polygons
  area formula, 151
  diagonals, 158
  exterior angles, 157–158, 292
  interior angles, 157–158, 292
  interior vs. exterior angles, 157–158
  minimum number of sides, 51
  perimeter ratio, 166
  similar, 165–166
  triangles as, 51
postulates, 12
prisms
  computing surface area, 243, 294
  computing volume, 243, 294
proofs
  how to approach writing, 27
  how to think about theorems, 30
  making game plans, 288
  postulates in, 12
  theorems in, 12–13
  tips for finishing, 290
  two-column format, 14–16, 287–290
  using all givens in, 31, 288
  working backwards in, 31, 289
proportions
  Angle-Bisector Theorem, 178
  parallel lines and transversals, 178
  Side-Splitter Theorem, 178
  similar polygons, 165–166, 172
pyramids
  computing surface area, 246
  computing volume, 246
Pythagorean Theorem, 64, 258, 291
Pythagorean triple, 67–68, 291–292

• Q •

quadratic formula, 52–53
quadrilaterals. *See also* isosceles trapezoids; kites; parallelograms; rectangles; rhombuses; squares; trapezoids
  area formulas, 151–152
  defined, 120
  illustrated, 120
  proving type, 132–134
  types, 120

• R •

radii
  bicycle spokes as, 202–203
  role in doing proofs, 288
radius, 195
rays, defined, 20
rectangles
  area formula, 152, 292
  defined, 120
  illustrated, 120
  properties, 127, 128
  proving quadrilaterals as, 136
  similar, 165–166
  as type of quadrilateral, 120
reflecting lines, 271, 272, 278, 279
reflections
  glide, 278–279
  and orientation, 272
  overview, 271–272
Reflexive property, 84
rhombuses
  area formula, 152, 292
  defined, 120
  illustrated, 120
  properties, 127, 128
  proving quadrilaterals as, 136, 137
  as type of quadrilateral, 120
right angles
  congruent, 24
  formed by perpendicular elements, 24
  in right triangles, 52
  theorems, 96–97
right triangles
  defined, 52
  as half a square, 70, 291

as half an equilateral triangle, 70, 291
and Pythagorean Theorem, 64
and Pythagorean triples, 67–68, 291–292
rotations, 276–277

• S •

same-side exterior angles, 115
same-side interior angles, 115
SAS (Side-Angle-Side) triangle theorem
  proving triangles congruent, 83–86
  proving triangles similar, 168–169
scalene triangles
  altitude, 55
  defined, 51
secant-secant angles, 215
Secant-Secant Power Theorem, 218, 293
secant-tangent angles, 215
Secant-Tangent Power Theorem, 218, 293
Sector-Area Theorem, 211–212
sectors, 211
segments
  adding, 30
  bisection, 22
  in circles, 212
  congruent, 33–35
  defined, 19
  dividing, 33–35
  midpoint, 22
  multiplying, 33–35
  perpendicular, 24
  subtracting, 30
  tip for using in proofs, 289
  trisection, 22
semiperimeter, 55
Side-Angle-Side (SAS) triangle theorem
  proving triangles congruent, 83–86
  proving triangles similar, 168–169
Side-Side-Angle (SSA) triangle theorem. *See*
  HLR (Hypotenuse-Leg-Right angle)
Side-Side-Side (SSS) triangle theorem
  proving triangles congruent, 83–86
  proving triangles similar, 168–169
Side-Splitter Theorem, 178, 179
similarity, 165–166
slant height, 246
slope
  formula, 257, 293
  horizontal lines, 257

parallel lines, 257
perpendicular lines, 258
vertical lines, 257
slope-intercept form, 261
spheres
  computing surface area, 249, 294
  computing volume, 249, 294
squares
  area formula, 151, 292
  defined, 120
  illustrated, 120
  properties, 128
  proving quadrilaterals as, 136
  as type of quadrilateral, 120
SSA (Side-Side-Angle) triangle theorem. *See*
  HLR (Hypotenuse-Leg-Right angle)
SSS (Side-Side-Side) triangle theorem
  proving triangles congruent, 83–86
  proving triangles similar, 168–169
Substitution property, 39–40
subtractintg angles and segments, 30
supplementary angles, 26–27
surface area
  cones and pyramids, 246
  cylinders and prisms, 243
  flat-top solids, 243–244, 294
  pointy-top solids, 246–247, 294
  spheres, 294

• T •

tangent lines, 202, 203
tangent-chord angle, 214
Tangent-Secant Power Theorem, 218, 293
tangent-tangent angle, 215
tangents, 202–203
30°- 60°- 90° triangle, 70
transformations
  overview, 271
  reflections, 271–272
  rotations, 276–277
  translations, 274–275
Transitive property, 39–40
transitivity, 39–40
translations, 274–275
transversals, 115–117, 178
trapezoids. *See also* isosceles trapezoids
  area formula, 151, 292
  defined, 120

trapezoids *(continued)*
  illustrated, 120
  median, 151
  properties, 130–131
  as type of quadrilateral, 120
triangles
  altitude, 55
  area formula, 55–56
  centers, 59–61
  centroids as true center, 58, 60
  circumcenter point, 59–61
  and CPCTC, 89–90
  and CSSTP, 172–173
  equilateral, 51, 55, 56
  height, 55
  incenter point, 59–61
  isosceles, 51, 55, 93–94, 287
  orthocenter point, 59–61
  overview, 51–53
  as polygons, 51
  proving congruent, 83–86
  proving similar, 168–169
  Pythagorean triples, 67–68, 291–292
  role in doing proofs, 287
  role of medians, 58, 60
  scalene, 51, 55
  special right, 70
  sum of angles, 52
trisectors, 22

• U •

unions, defined, 20

• V •

vertex, defined, 58
vertical angles, 37
vertical lines, 257, 262
volume
  cones and pyramids, 246
  cylinders and prisms, 243
  flat-top solids, 243–244, 294
  pointy-top solids, 246–247, 294
  spheres, 294

• X •

x-angles, 37

Notes

Notes